To my Omas and Opas

– Alain F. Zuur –

I would like to thank my family and especially my wife for patience and support

– Anatoly A. Saveliev –

To my family and friends for their constant love and support given throughout my career

– Elena N. Ieno –

Preface

This is our fourth book, following *Analysing Ecological Data* (Zuur *et al.* 2007), *Mixed Effects Models and Extensions in Ecology with R* (Zuur *et al.* 2009a), and *A Beginner's Guide to R* (Zuur *et al.* 2009b).

In our 2007 book, we describe a wide range of statistical techniques: data exploration, linear regression, generalised linear modelling (GLM) and generalised additive modelling (GAM), regression and classification trees, linear mixed effects modelling, multivariate techniques, time series analysis, and spatial analysis. The book comprises statistical theory along with 17 case studies for which complete statistical analysis of real data sets are presented.

Our second book focuses primarily on mixed effects modelling, but includes zero inflated models and GLM. Ten case studies are presented, applying techniques such as generalised linear mixed models (GLMM) and generalised additive mixed models (GAMM) on real data sets.

In our role as statistical consultants, we have frequently encountered data sets in which the response variable contains a high number of zeros; a 'high number' being from 25% to 95%. Most of these data sets involve an extra level of statistical complexity: temporal or spatial correlation, 1-way nested data, 2-way nested data, or a combination. Our original idea was to call the current book '*Shit Happens,*' as it expresses the sentiments that arise when analysing this type of data. Eventually we decided to opt for a more subtle, if less eye-catching, title.

So how do we analyse correlated and nested zero inflated data? What software is available and what level of statistical expertise is required to conduct and interpret the analyses? Zero inflated models are a combination of two GLMs or GAMs. We need a working knowledge of these models. Analysing correlated and nested zero inflated data means that we must extend the zero inflated models to GLMMs, so we also need to be familiar with mixed effect models.

There is no package available that can conveniently fit zero inflated GLMMs. In this book we use the software package R along with simulation techniques (Markov Chain Monte Carlo). This involves incorporating the concept of Bayesian statistics into proficiency in R. Bayesian statistics is not routinely taught at universities, and we include an introductory chapter on this topic.

The format of this book is different from our first two, as statistical theory is not presented as a separate entity, but is integrated into 9 case-study analyses. The common components of these data sets are zero inflation, 1-way and 2-way nested data, temporal correlation, and spatial correlation. As zero inflated models for such data consist of 2 GLMMs (or GAMMs), we discuss these types of models.

Outline and how to use this book

Chapter 1 provides a basic introduction to Bayesian statistics and Markov Chain Monte Carlo (MCMC), as we will need this for most analyses. If you are familiar with these techniques we suggest quickly skimming through it. In Chapter 2 we analyse nested zero inflated data of sibling negotiation of barn owl chicks. We explain application of a Poisson GLMM for 1-way nested data and discuss the observation-level random intercept to allow for overdispersion. We show that the data are zero inflated and introduce zero inflated GLMM. We recommend reading this chapter in detail, as we will refer often to it.

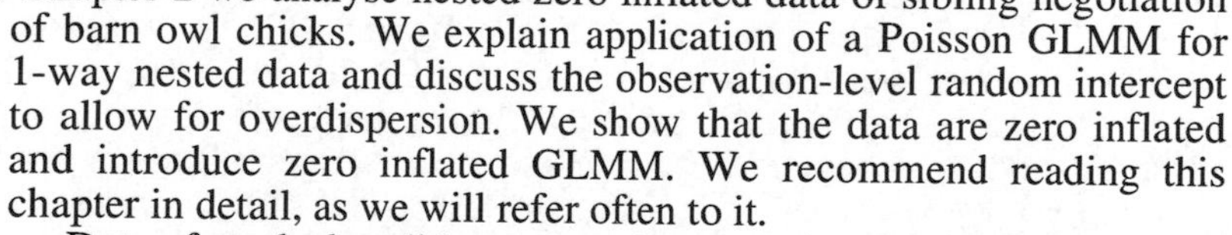

Data of sandeel otolith presence in seal scat is analysed in Chapter 3. We present a flowchart of steps in selecting the appropriate technique: Poisson GLM, negative binomial GLM, Poisson or negative binomial GAM, or GLMs with zero inflated distribution.

Chapter 4 is relevant for readers interested in the analysis of (zero inflated) 2-way nested data. The chapter takes us to marmot colonies: multiple colonies with multiple animals sampled repeatedly over time.

Chapters 5 – 7 address GLMs with spatial correlation. Chapter 5 presents an analysis of Common Murre density data and introduces hurdle models using GAM. Random effects are used to model spatial correlation. In Chapter 6 we analyse zero inflated skate abundance recorded at approximately 250 sites along the coastal and continental shelf waters of Argentina. Chapter 7 also involves spatial correlation (parrotfish abundance) with data collected around islands, which increases the complexity of the analysis. GLMs with residual conditional auto-regressive correlation structures are used.

In Chapter 8 we apply zero inflated models to click beetle data.

Chapter 9 is relevant for readers interested in GAM, zero inflation, and temporal auto-correlation. We analyse a time series of zero inflated whale strandings.

In Chapter 10 we demonstrate that an excessive number of zeros does not necessarily mean zero inflation. We also discuss whether the application of mixture models requires that the data include false zeros and whether the algorithm can indicate which zeros are false.

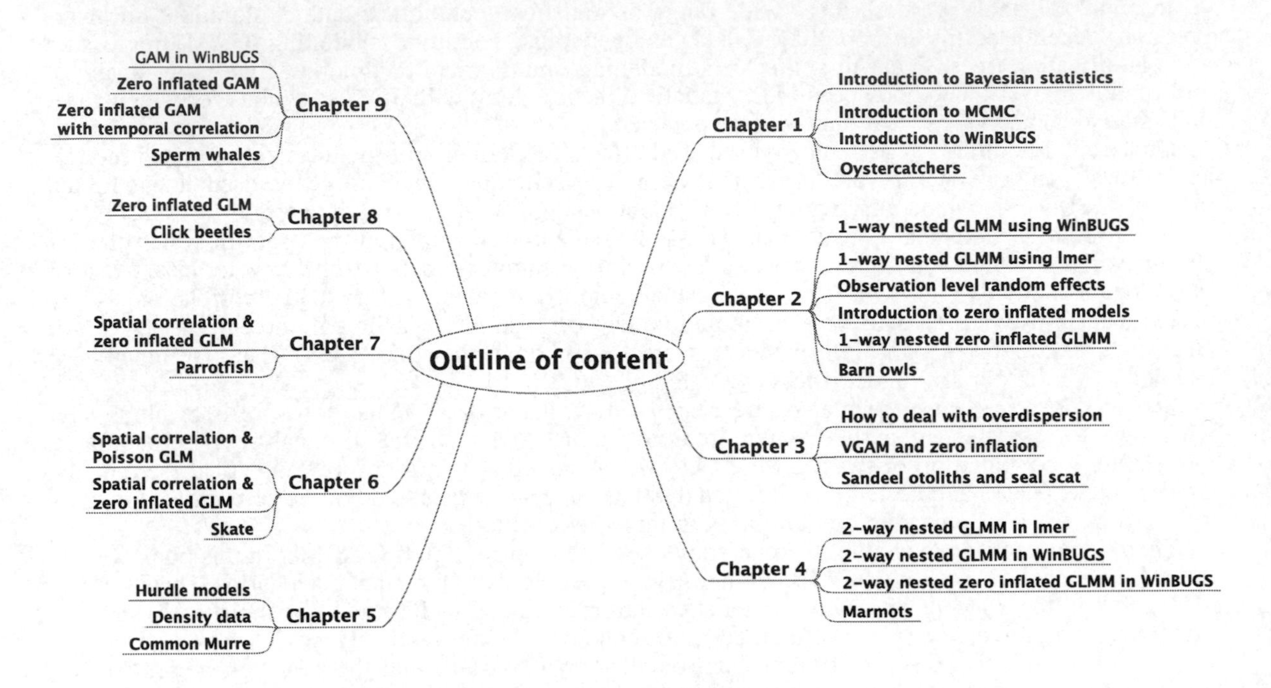

Outline of the material presented in this book.

Pre-requisite knowledge

Working knowledge of R, data exploration, linear regression, GLM, mixed effects modelling, and, for some material, GAM is required to make the best use of the material presented this book. Many excellent books deal with these techniques, among them Pineiro and Bates (2000), Dalgaard (2002), Wood (2006), Hardin and Hilbe (2007), Bolker (2008), Venables and Ripley (2002), and Hilbe (2011). Our own work (Zuur et al. 2007, 2009a, 2009b, 2010) can also be used. We strongly recommend the following chapters from our 2009a book: Chapters 8 – 10 on GLM, Chapter 11 on zero inflated models, Chapter 5 on random effects, and Chapters 6 and 7 on temporal and spatial correlation. If you are not familiar with R, consider our *Beginner's Guide to R*. Our 2010 paper on a protocol for data exploration is a recommended read, too.

Acknowledgements

The material contained in this book has been presented in courses in 2009, 2010, and 2011, and we are greatly indebted to all participants who helped improve the material. We would also like to thank those who read and commented on parts of earlier drafts: Aaron MacNeil, Jarrod Hatfield, and Joe Hilbe, as well as several anonymous referees. We thank them all for their positive, encouraging, and useful reviews. Their comments and criticisms greatly improved the book. Special thanks to the contributors to this book who kindly donated data sets for the case studies and helped to make it possible to explain these developing statistical techniques.

The cover drawing of oystercatchers was made by Jon Thompson (www.yellowbirdgallery.org). Mr Thompson was born in 1939 to Irish parents and has lived most of his life in Scotland. In the 1980s, he was instinctively drawn to the Orkney Islands and is continually inspired by the landscape and bird life of Orkney. He has been creating bird art for 30 years through drawing, painting, sculpture, and jewellery, never attempting to reproduce nature, but to draw parallels with it. A close up view of a bird feather is all the inspiration he needs.

We also thank Kathleen Hills of *The Lucidus Consultancy* for editing this book.

Newburgh, UK & Mount Irvine, Tobago
Kazan, Russia
Alicante, Spain

Alain F. Zuur
Anatoly A. Saveliev
Elena N. Ieno N

March 2012

Contents

Contributors

Pedro Afonso
Centre of IMAR
Dept. of Oceanography and Fisheries of the University of the Azores and LARSyS Associated Laboratory,
9901-862 Horta (Azores)
Portugal

Rod P Blackshaw
School of Biomedical and Biological Sciences
University of Plymouth
B426, Portland Square
Drake Circus
Plymouth
PL4 8AA
UK

Mary A Bishop
Prince William Sound Science Center
PO Box 705
Cordova, AK 99574
99574
USA

Federico Cortés
Programa Pesquerías de Condrictios
Instituto Nacional de Investigación y Desarrollo Pesquero (INIDEP)
Paseo Victoria Ocampo N^0 1,
(7600) Mar del Plata
Argentina

Alex Edridge
Fisheries Technology & Fish Behaviour
Marine Ecosystems
Marine Scotland - Science
Scottish Government, Marine Laboratory, PO Box 101
375 Victoria Road, Aberdeen, AB11 9DB
Scotland

Neil Dawson
Prince William Sound Science Center
PO Box 705
Cordova, AK 99574
99574
USA

Raúl A Guerrero
Proyecto Oceanografía Física
Instituto Nacional de Investigación y Desarrollo Pesquero (INIDEP)
Paseo Victoria Ocampo N^0 1
(7600) Mar del Plata
Argentina

Elena N Ieno
Highland Statistics LTD.,
Box No.82
Avda. Escandinavia 72
Local 6, BQ 4
Gran Alacant
03130 Santa Pola
Alicante
Spain

Andrés J Jaureguizar
Programa Pesquerías de Peces Demersales Costeros
Comisión de Investigaciones Científicas (CIC)
Instituto Nacional de Investigación y Desarrollo Pesquero (INIDEP)
Paseo Victoria Ocampo N^0 1
(7600) Mar del Plata
Argentina

Kathy J Kuletz
US Fish and Wildlife Service
Migratory Bird Management
1011 East Tudor Road
Anchorage, AK 99503
USA

Kathleen R Nuckolls
Environmental Studies Program
University of Kansas
252 Snow Hall
1460 Jayhawk Blvd
Lawrence, KS 66045-7523
USA

Graham J Pierce
University of Aberdeen
Oceanlab
Main Street
Newburgh
AB41 6AA
UK

Alexandre Roulin
Department of Ecology & Evolution
Building Biophore
University of Lausanne
CH-1015 Lausanne
Switzerland

M Begoña Santos
Instituto Español de Oceanografía
C.O. Vigo
P.O. Box. 1552
36280 Vigo
Spain

Anatoly A Saveliev
Ecology and Geography Institute
Kazan University
18 Kremlevskaja Street
Kazan, 420008
Russia

Mara Schmiing
Centre of IMAR
Dept. of Oceanography and Fisheries of the University of the Azores and LARSyS Associated Laboratory,
9901-862 Horta (Azores)
Portugal

Fernando Tempera
Centre of IMAR
Dept. of Oceanography and Fisheries of the University of the Azores and LARSyS Associated Laboratory,
9901-862 Horta (Azores)
Portugal

Florent Thiebaud
Agriculture and Agri-Food Canada
Pacific Agri-Food Research Centre
Agassiz Research Station
P.O. Box 1000
6947 Highway 7
Agassiz
British Columbia V0M 1AO
Canada

Robert S Vernon
Agriculture and Agri-Food Canada
Pacific Agri-Food Research Centre
Agassiz Research Station
P.O. Box 1000
6947 Highway 7
Agassiz
British Columbia V0M 1AO
Canada

Alain F Zuur
Highland Statistics LTD.
6 Laverock Road
Newburgh
AB41 6FN
UK

1 Introduction to Bayesian statistics, Markov Chain Monte Carlo techniques, and WinBUGS

1.1 Probabilities and Bayes' Theorem

This chapter provides a brief introduction to Bayesian statistics, Markov Chain Monte Carlo (MCMC) techniques, and WinBUGS (a Windows program for Bayesian inference Using Gibbs Sampling), for which we make extensive use of *Bayesian Statistics, Book 4* by The Open University (www.ouw.ac.uk). For the explanation of MCMC we use Ntzoufras (2009).

To demonstrate Bayes' Theorem and MCMC techniques we will use a data set on oystercatchers. These data are part of a 3-year study on the feeding ecology of the American oystercatcher *Haematopus palliatus* inhabiting coastal areas of Argentina (Ieno, unpublished data). For the present purpose, we present a subset of data consisting of observations of the shell length of clams eaten by oystercatchers at three sites during December and January. To break open a shell, an oystercatcher uses either a hammering technique or stabs it with the beak. Depending on the technique used, the bird is called a hammerer or a stabber. One of the underlying questions is whether clams eaten by hammerers are larger than those consumed by stabbers. Obviously, we need to take into account the location and time of year, as these may affect availability of clams, but we will begin simply and formulate the following proposition E:

E = Oystercatchers using hammering to crush shells consume larger clams.

We do not know for sure that this is true, but we can assign a probability to the statement. For example, $P(E) = 0.9$ means that most likely our statement is correct, and $P(E) = 0.1$ means that we are probably wrong. We will show how Bayesian statistics can be used to estimate the probability $P(E)$.

Without background knowledge it may be difficult to formulate a plausible statement at this point. However, if you are a field biologist and work with oystercatchers, you are likely to have formed an opinion on this. Perhaps you have seen oystercatchers using hammering techniques mainly to crush small shells. In that case, based on prior knowledge, you may say that the most likely value for $P(E)$ is approximately 0.4, perhaps between 0.2 and 0.6 if you want to be more conservative.

Conditional probability

Instead of directly addressing proposition E, we will develop a statistical model in which we use parameters that represent proposition E. Bayesian statistics will then be used to make probabilistic statements of the parameters. Before applying a statistical model to the data, we will try to derive probabilities for E, as this allows us to introduce Bayes' Theorem. One way of accomplishing this is to collect data and use relative frequencies to estimate probabilities. Table 1.1 shows the number of clams eaten by hammering oystercatchers observed at three feeding sites and records whether the shells were small or large.

Based on the data in Table 1.1, an estimate of $P(\text{Shell size} = \text{Small})$ for oystercatchers using hammering techniques is given by $138/172 = 0.80$. This is row total divided by the overall total. In the same way we can calculate $P(\text{Feeding site} = 1) = 60/172 = 0.35$. This is the first column total divided by the overall total. We want to know the probability that shell size is small, given the observations from site 1 for this group of birds. The notation for this is: $P(\text{Shell size} = \text{Small} \mid \text{Feeding site} = 1)$.

Table 1.1. Number of clams eaten by hammering oystercatchers per shell size (small versus large) and feeding site.

	Feeding site			
Shell size	1	2	3	**Total**
Small	46	35	57	**138**
Large	14	10	10	**34**
Total	**60**	**45**	**67**	**172**

The symbol | reads as 'given that.' The answer is 46 divided by 60, which is 0.77. This is the number of small shells divided by the total number of shells for feeding site 1. A basic rule in statistics states that two events A and B are independent if we have $P(A \mid B) = P(A)$. Let us see whether this is the case for shell size and feeding site.

$$0.77 = \text{P(Shell size = Small} \mid \text{Feeding site = 1)} \neq \text{P(Shell size = Small)} = 0.80$$

This means that shell size and feeding site are not independent. If the probability of A depends on that of B, we use *conditional* probability, which is defined as:

$$P(A \mid B) = \frac{P(A \text{ and } B)}{P(B)}$$

The $P(A \text{ and } B)$ is the joint probability of A and B, and $P(A \mid B)$ is the conditional probability for A, given B. This provides an alternative method of calculating the probability that small clams are selected by hammering birds, given that the feeding site is 1:

$$P(\text{Shell size = Small} \mid \text{Feeding site=1}) = \frac{P(\text{Shell size = Small and Feeding site = 1})}{P(\text{Feeding site = 1})}$$

We already calculated this probability directly as 0.77, but let us try to get the same answer using the definition of conditional probability. For this we need the nominator and denominator in the equation above. The denominator is simple; it is the 0.35 that we calculated above. As for the nominator, the probability that shell size is small *and* feeding site equals 1 is given by 46/172 = 0.27. Hence the conditional probability P(Shell size = Small | Feeding site = 1) is equal to 0.27/0.35 = 0.77. This is the same value as our original calculation, obtained using the definition of conditional probability.

Bayes' Theorem

The formula for conditional probability can be rearranged as:

$$P(A \text{ and } B) = P(A \mid B) \times P(B)$$

By changing the order of A and B we get:

$$P(B \text{ and } A) = P(B \mid A) \times P(A)$$

Because P(A and B) is the same as P(B and A) we get:

$$P(A \mid B) \times P(B) = P(B \mid A) \times P(A)$$

Rewriting this gives Bayes' Theorem:

$$P(A \mid B) = \frac{P(B \mid A) \times P(A)}{P(B)}$$

Provided that *P*(B) is not equal to 0, we can write the conditional probability of *A* given *B* as the conditional probability of *B* given *A* multiplied by the probability of *A* and divided by the probability of *B*. The advantage of this expression is that *P*(*A* | *B*) can be difficult to calculate for more complicated models. So if, for some reason, we cannot calculate *P*(Shell size = Small | site = 1), we can use Bayes' Theorem:

$$P(\text{Shell size = Small} \mid \text{Feeding site=1}) = \frac{P(\text{Feeding site} = 1 \mid \text{Shell size = Small}) \times P(\text{Shell size = Small})}{P(\text{Feeding size} = 1)}$$

$$= \frac{46/138 \times 138/172}{60/172} = 0.77$$

The result is the same as for our earlier calculation. We introduce this more complicated approach because Bayes' Theorem forms the basis for nearly everything we will present in this book. It allows us to find distributions for regression parameters in zero inflated models and in generalised linear mixed models, which, without this approach, would be a great deal more complicated. We present the equation once more with additional annotations. The expressions 'Prior' and 'Posterior' will be explained in the following section.

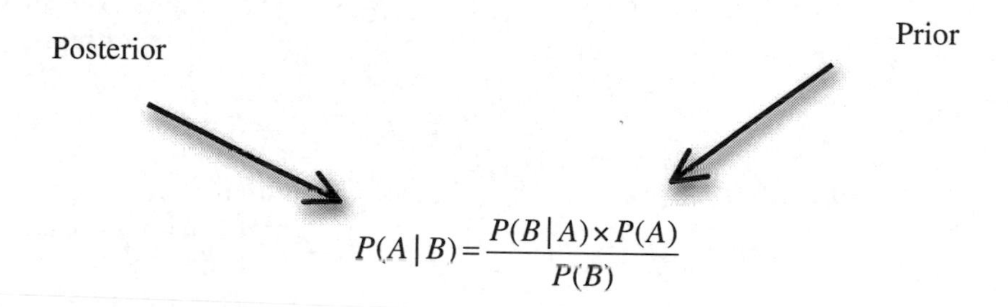

1.2 Likelihood functions

Instead of *P*(*A*) and *P*(*B*), we will use a statistical model that contains unknown parameters, θ, to make statements on $P(\theta)$ for all possible values of θ. We make a distinction as to whether θ takes discrete values or continuous values. If the former, it is called a probability distribution; if the latter, a density function.

Bayes' Theorem must be kept in mind. Essentially we begin with *A*. Subsequently, information on *B* becomes available, allowing calculation of the probability *P*(*A* | *B*). *P*(*A*) is called the prior information on *A*, and *P*(*A* | *B*) is the posterior information, or distribution. We begin with prior information on θ, denoted as $P(\theta)$. When appropriate data becomes available, we calculate the posterior distribution for θ given the data, $P(\theta \mid \text{data})$:

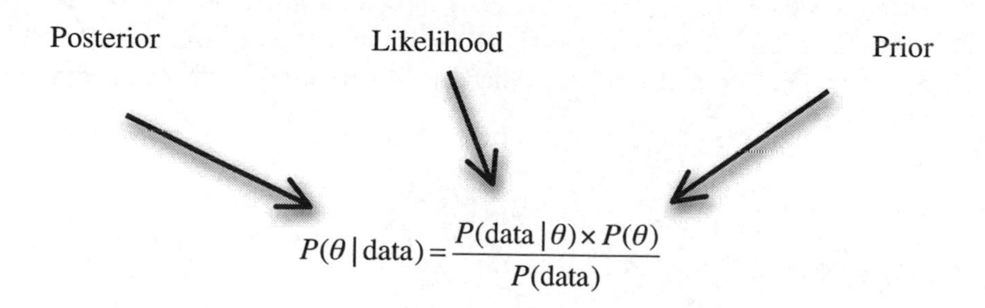

We will see that $P(\text{data} \mid \theta)$ is the likelihood of the data for given θ. We present an example to demonstrate this principle.

Example of prior, likelihood, and posterior for hammering oystercatchers

Returning to the oystercatcher data, ignore for the moment that there is a location effect. Suppose that we wish to know the proportion of larger clams preyed upon by hammering oystercatchers in the entire population; we call this proportion θ. Without knowing anything about these birds, we might assume that any value between 0 and 1 is equally probable. Other expressions for limited prior knowledge are a 'diffuse' or 'weak' prior.

On the other hand, we may estimate that θ is 0.5, with likely values between 0.2 and 0.8. This is not precise, and we call it a weak prior. Perhaps as ornithologists specialising in oystercatchers, we have observed in the past that birds using hammering techniques prey upon the larger clams. Based on our experience, we estimate θ at approximately 0.3, with likely values between 0.2 and 0.4. We are estimating a distribution for θ based on prior information. This is more precise, though it may be wrong, and we call it a 'strong prior.' We sketch three potential prior distributions $P(\theta)$ in Figure 1.1.

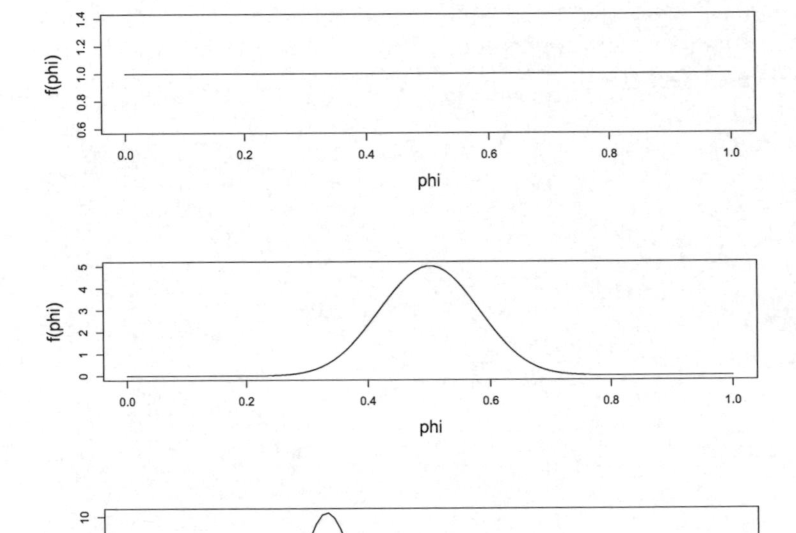

Figure 1.1. Three distributions for the prior $P(\theta)$. The top panel shows a uniform distribution; any value is equally likely for θ. The others represent stronger prior distributions.

So far we have only speculated on possible values of θ. Now suppose that we take a sample and find that, of the 172 eaten clams, 138 were small clams and 34 were large clams (see the row totals in Table 1.1). We have:

1. Prior estimations of θ, as represented by a prior distribution. Figure 1.1 shows three potential choices.
2. A sample with observed values.

Using the sample, we will specify the likelihood, $P(\text{data} \mid \theta)$. We have $N = 172$ observations and define X as the number of times that a large clam was preyed upon. The variable X is discrete and can take any value between 0 and 172. We want to specify the likelihood of X for a given θ; this is also called the distribution of X for a given θ. Let us denote this distribution as L, and, because it is a function of θ, write it $L(\theta)$. We need to specify $P(X = 0 \mid \theta)$, $P(X = 1 \mid \theta)$, and $P(X = 2 \mid \theta)$, …, $P(X = 172 \mid \theta)$. Given that we have only two sizes, a binomial distribution is a good candidate distribution to model these probabilities:

$$L(X = x \mid \theta) = \binom{172}{x} \times \theta^{x} \times (1-\theta)^{172-x}$$

For example, $L(X = 34 \mid \theta)$ can easily be calculated by substituting 34 (the observed counts for large shells) for x, resulting in:

$$L(X = 34 \mid \theta) = \binom{172}{34} \times \theta^{34} \times (1-\theta)^{172-34}$$

The problem is that we do not yet know the value of θ, so we cannot calculate $L(X = 34 \mid \theta)$. Hence the question is: What is a likely value of θ, given the observed data values? What we can do is to fill in every value of θ between 0 and 1 and calculate the likelihood (Figure 1.2). It seems that a value of θ around 0.2 is plausible.

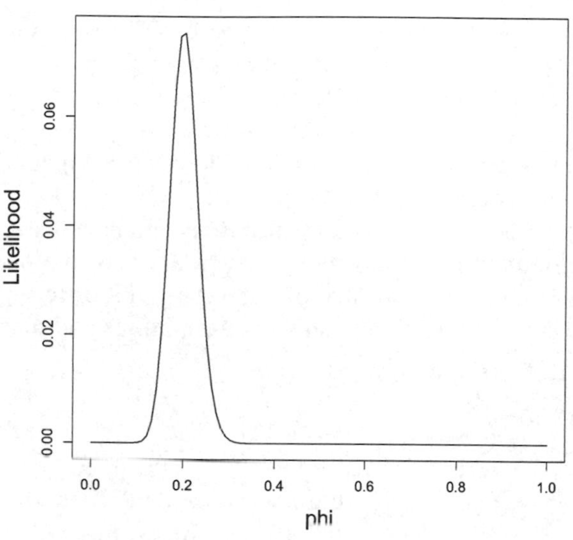

Figure 1.2. Likelihood $L(X = 34 \mid \theta)$ plotted versus θ.

Posterior

The curve in Figure 1.2 defines the term $P(\text{data} \mid \theta)$. In order to calculate the posterior distribution $P(\theta \mid \text{data})$, we need the denominator $P(\text{data})$. This is a constant value and does not depend on θ. Therefore it only affects the scale of the posterior distribution, not its shape, so we can eliminate it, and this gives:

$$P(\theta \mid \text{data}) \propto P(\text{data} \mid \theta) \times P(\theta)$$

The symbol $\propto$ means 'proportional to.' In words, we have: posterior $\propto$ likelihood $\times$ prior. So to arrive at the posterior distribution of θ, we need only multiply the curve in Figure 1.2 by one of the curves in Figure 1.1 (or, optionally, divide by $P(\text{data})$, if this term is easy to calculate). The resulting posterior distributions are presented in Figure 1.3.

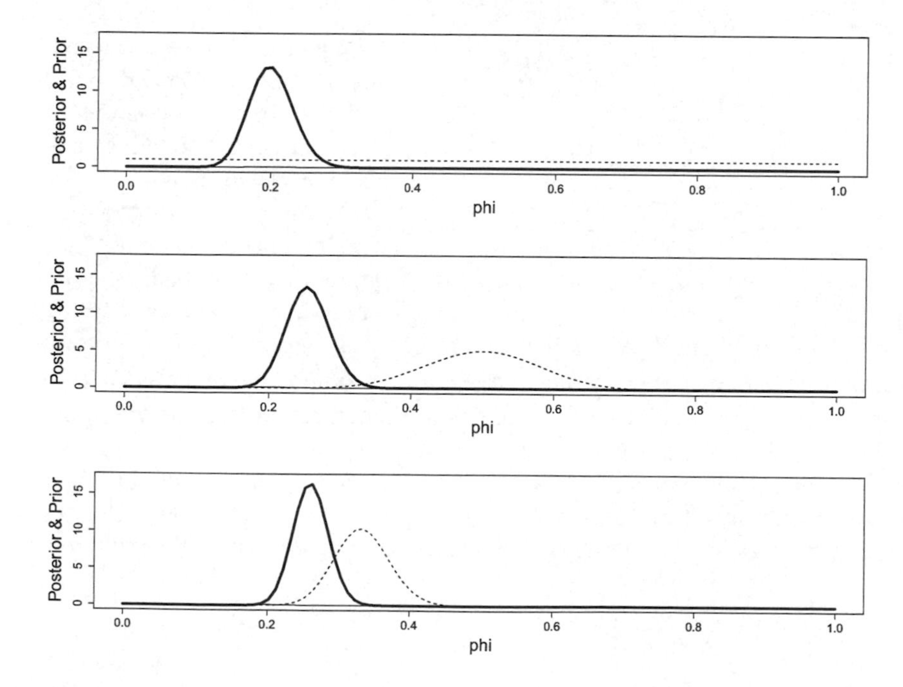

Figure 1.3. Posterior (solid thick line) and prior (dotted line). The top panel shows the posterior distribution obtained from a uniform prior distribution. For the other two posterior distributions we used stronger prior distributions. The priors in the centre and lower panel do not conform to the data, which pulls the posterior slightly to the right.

1.3 Conjugate prior distributions

In order to demonstrate Bayes's Theorem we simplified the oystercatcher data in the previous section. In fact, we know the length of clams that were preyed upon, so we can build a model for shell size of the form:

$$\text{Length}_i = \alpha + \text{Location}_i + \text{Feeding type}_i + \varepsilon_i$$

We could include an interaction effect and/or a month effect, but we will keep it simple for the time being. This model contains 4 regression parameters (1 intercept, 2 for location, and 1 for feeding type) and a variance. We could use Bayes' Theorem to obtain a prior distribution for each of these parameters. As a result we will obtain a posterior distribution for the intercept α, a posterior distribution for each of the slopes, and a posterior distribution for the variance. This implies that we need to specify a prior distribution for each of these parameters.

One option is to pretend that we have no information on the parameters and specify a diffuse prior. However, a series of special priors are available. If we choose one of these priors, the posterior distribution follows the same distribution. Such priors are called conjugate priors and are only useful in the simplest cases, in which we can derive expressions for the posterior distribution by calculus. For the more advanced models dealt with in this book, conjugate priors are not an option. However, conjugate priors provide a convenient bridge to explaining the more complex MCMC approach.

We will consider 4 distributions: the normal distribution, the uniform distribution, the beta distribution, and the gamma distribution. The normal distribution is the most well known and has two parameters; the mean, *a,* and the variance, *b*. Figure 1.4 shows three normal distributions with differing values of *a* and *b*. It is crucial to realise that the *x*-axis can contain any value between minus and plus infinity, but the majority of values will fall between $a-3\times\sqrt{b}$ and $a+3\times\sqrt{b}$.

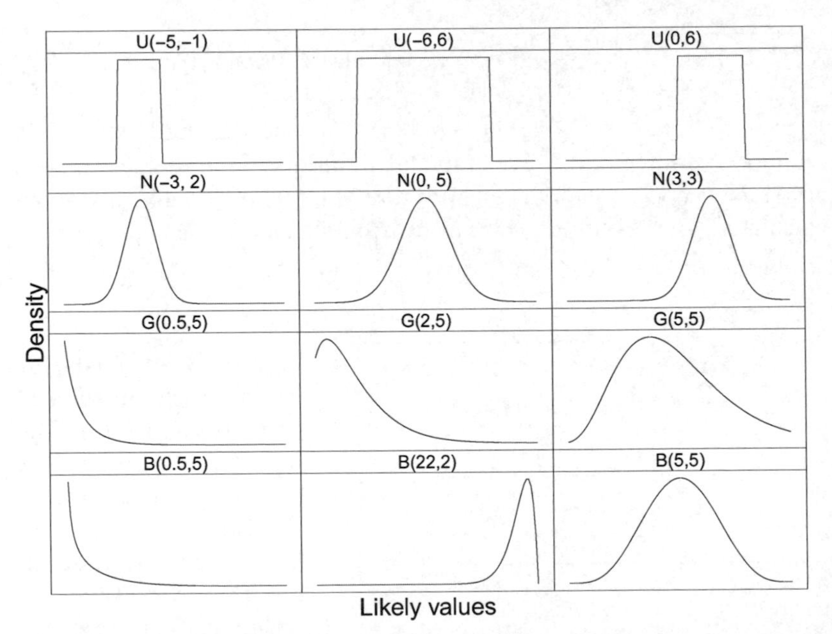

Figure 1.4. Uniform, normal, gamma, and beta distributions for values of *a* and *b*. The mean and variance for each distribution are simple functions of *a* and *b*.

We can use the normal distribution as a prior for a variable that is continuous and can take negative and positive values.

A uniform distribution with parameters *a* and *b* has probability 0 outside the interval [*a*, *b*] and probability 1 / (*b* – *a*) inside this interval. Three uniform distributions are presented in Figure 1.4. The uniform distribution is useful as a prior distribution if every value between *a* and *b* is equally likely for *θ*. If the interval [a, b] is large, it is a weak prior. If we know only that a parameter *θ* is somewhere between *a* and *b* with equal probability, this is a reasonable choice for a prior distribution.

A beta prior can only be used for a parameter *θ* that is between 0 and 1, so it is a good choice if we are working with proportion or probability. It contains two parameters and can take a wide variety of shapes (see Figure 1.4). If we want to use a beta prior, we can simulate some density curves with the function dbeta and see how choices for a and b determine the shape of the curve. The mean and variance of a beta distribution are simple functions of a and b (Table 1.2).

The gamma distribution can be used only for positive values. For example, it can be used for the prior of a variance term. Again, the mean and variance of a gamma (a, b) distribution are simple functions of a and b. Figure 1.4 shows three gamma distributions.

Table 1.2. Mean, variance, and density function of normal, uniform, beta, and gamma distributions.

Distribution	Mean	Variance	Density function	Domain θ
Normal(a, b)	a	B	$f(\theta)=\frac{1}{\sqrt{2\times\pi\times b}}\times e^{-\frac{1}{2\times b}\times(\theta-a)^2}$	$-\infty<\theta<\infty$
Uniform(a, b)	$\frac{1}{2}\times(a+b)$	$\frac{1}{12}\times(b-a)^2$	$f(\theta)=\frac{1}{b-a}$	$a\leq\theta\leq b$
Beta(a, b)	$\frac{a}{a+b}$	$\frac{a\times b}{(a+b)^2\times(a+b+}$	$f(\theta)=\text{Constant}\times\theta^{a-1}\times(1-\theta)^{b-1}$	$0\leq\theta\leq 1$
Gamma(a, b)	$\frac{a}{b}$	$\frac{a}{b^2}$	$f(\theta)=\text{Constant}\times\theta^{a-1}\times e^{-b\times\theta}$	$\theta>0$

How do we select one of these four distributions as the prior for θ, and what values of a and b do we use? The uniform distribution $U(a, b)$ is simple; we use it if we know only that θ falls somewhere between a and b. For the other three distributions we follow these steps:

1. Choose an appropriate distribution based on the values of θ. For a proportion, use the beta distribution; for a non-negative variable, use the gamma distribution; and the normal distribution can be used for variables between minus and plus infinity.
2. Determine (or guess) the centre of the distribution for θ.
3. Determine (or guess) the variation of the distribution for θ.
4. Based on 2 and 3, choose values for a and b.

The centre of a distribution is relatively easy to estimate using tools such as the mode (this is the value of the distribution with the highest value for θ), median, or mean. Assessing the variation is more challenging. For a normal distribution, designating b is relatively simple; we can choose a value for b such that 95% of the possible values of θ are between $a-2\times\sqrt{b}$ and $a+2\times\sqrt{b}$. For the beta and gamma distribution this is more difficult, and sketching these distributions for different a and b values may help.

Recall that we can model the posterior as: posterior $\propto$ likelihood $\times$ prior and that it can be demonstrated that, for certain likelihood functions, a 'conjugate prior' can be chosen such that the posterior has the same distribution as the prior. An example of this is a binomial distribution for the likelihood and a beta distribution for the prior. It is relatively easy to show that, in this case, the posterior also follows a beta distribution. It is a simple matter of multiplying two density functions and rewriting it as a new beta distribution, with slightly different a and b.

Example of a conjugate prior using oystercatcher data

We will calculate a conjugate prior for the proportion of large clams preyed upon by hammering oystercatchers, denoted θ. We will specify a prior for θ, the likelihood of the data given θ, and calculate a mathematical expression for the posterior distribution.

Suppose that, based on prior knowledge, we expect that θ is about 0.33, with possible values from 0.2 to 0.5. Because θ is a proportion, the beta distribution is a sensible choice for a prior distribution, but we need to select values for *a* and *b*. We sketch beta distributions for several values of *a* and *b*, and determine that beta(50, 100) is a reasonable choice for the prior. This is sketched in the bottom panel in Figure 1.1. A beta(50,100) distribution has a mean of 0.33 and a variance of 0.0015. (See equations for the mean and the variance of a beta distribution in Table 1.2.)

Now we will focus on the likelihood $P(\text{data} \mid \theta)$. We used the binomial distribution for this earlier. We have $N = 172$ observations, and 34 large clams were preyed upon by hammering oystercatchers. So the likelihood $P(\text{data} \mid \theta) = P(X = 34 \mid \theta)$ is given by:

$$L(X = 34 \mid \theta) = \begin{pmatrix} 172 \\ 34 \end{pmatrix} \times \theta^{34} \times (1-\theta)^{172-34}$$

The posterior $P(\theta \mid \text{data})$ is therefore equal to:

$$P(\theta \mid \text{data}) \propto P(\text{data} \mid \theta) \times f(\theta) = \begin{pmatrix} 172 \\ 34 \end{pmatrix} \times \theta^{34} \times (1-\theta)^{172-34} \times \text{Constant} \times \theta^{a-1} \times (1-\theta)^{b-1}$$
$$= C \times \theta^{34+a-1} \times (1-\theta)^{127-34+b-1}$$

where *C* is a constant. This expression is again a beta distribution with parameters $(a + 34)$ and $(b + 172 - 34)$, or, in a more general notation, beta$(a + x, b + N - x)$. Because our prior was beta(50, 100), the posterior distribution is beta(84, 238). The posterior distribution is of the same form as the prior, except for the parameters *a* and *b*; therefore this is a conjugate prior distribution.

Once we have the posterior distribution we need to present it. One option is to write 'the posterior distribution is beta(*a*, *b*),' but readers are unlikely to be able to visualise what this means. Sketching our beta(*a*, *b*) distribution may be a good way to depict the posterior distribution. We can use the `dbeta` function for this, or we can simulate a large number of observations from the beta distribution using the `rbeta` function, and construct a histogram (Figure 1.5). The graph was created with the R code:

```
> a <- 50
> b <- 100
> Z <- rbeta(1000, a + 34, b + 172 - 34 )
> hist(Z, main = "",
        xlab = "Simulated values from a beta(84, 238) distribution")
```

Alternatively, we could present the mean (or mode) and variance of the distribution. The equations to calculate these from *a* and *b* for a beta distribution are given in Table 1.2. An alternative method of summarising the posterior distribution is defining a credible interval. The interval (l, u) is a $100 \times (1 - \alpha)\%$ credible interval for a parameter θ, if the posterior probability that θ falls in this interval, given the data, is equal to $1 - \alpha$:

$$P(l \leq \theta \leq u \mid \text{data}) = 1 - \alpha$$

Note that in frequentist statistics, this probability is either 1 or 0. In Bayesian statistics we can say that the probability that θ is contained in a certain interval is $1 - \alpha$.

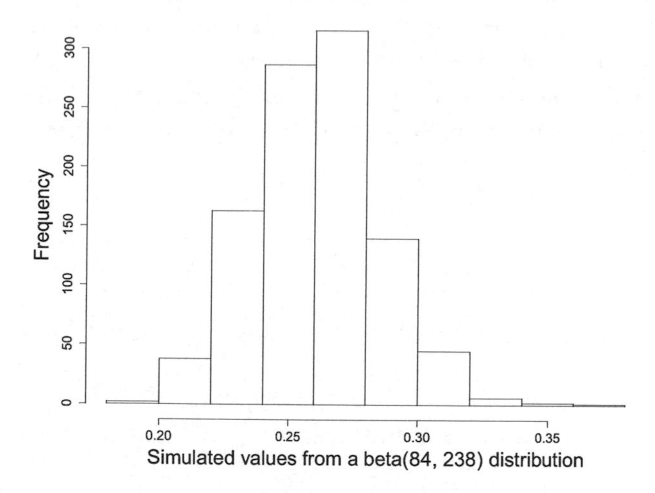

Figure 1.5. Simulated values from a beta(84, 238) distribution, representing the posterior distribution.

Other conjugate prior distributions

In the previous example we saw that under certain conditions the beta prior results in a beta posterior. This is also true of a prior that is gamma distributed with a likelihood that is a Poisson distribution; the posterior is also gamma distributed. A prior that is normally distributed, combined with a likelihood that is a normal distribution, gives a posterior that is normally distributed.

Non-conjugate prior distributions

So far we have discussed conjugate priors. The statistical models used in this book are too complicated to calculate a posterior; the posterior cannot always be expressed as a neat function. In the following section we will use simulation techniques to obtain samples from the posterior distribution. We applied simulation techniques when we illustrated the beta(84, 238) distribution in Figure 1.5. The R function `rbeta` did the hard work and simulated a large number of independent samples. This will not work for more complicated posteriors. Instead we will use the MCMC method to obtain samples from the posterior.

1.4 MCMC

Let us return to the linear regression example discussed earlier. We can simplify it to a bivariate linear regression model for shell length:

$$\text{Length}_i = \alpha + \beta \times \text{Feeding type}_i + \varepsilon_i$$

where ε_i is normally distributed with mean 0 and variance σ^2. The unknown parameters are α, β, and σ. To simplify notation we use: $\boldsymbol{\theta} = (\alpha, \beta, \sigma)$. As before, we can specify a prior $f(\boldsymbol{\theta})$ for $\boldsymbol{\theta}$, formulate the likelihood $f(\text{data} \mid \boldsymbol{\theta})$ for the data given $\boldsymbol{\theta}$, and try to find a mathematical expression for the posterior $f(\boldsymbol{\theta} \mid \text{data})$. The simplification is possible in this linear regression model but not for zero inflated models with temporal or spatial correlation. Instead we can use MCMC to generate a large number of samples for $f(\boldsymbol{\theta} \mid \text{data})$, and these can be summarized with numerical statistics such as the mean, mode, median, and standard deviation. Alternatively, we can create a histogram or boxplot to visualise samples of the posterior distribution.

1.4.1 Markov Chain

'Markov Chain' in MCMC refers to simulating values $\boldsymbol{\theta}^{(1)}, \boldsymbol{\theta}^{(2)}, \boldsymbol{\theta}^{(3)}, \ldots, \boldsymbol{\theta}^{(T)}$ in such a way that the density function for $\boldsymbol{\theta}^{(t+1)}$ depends only on $\boldsymbol{\theta}^{(t)}$ and not on any of the other samples (also called draws). In mathematical notation this is written as:

$$f(\boldsymbol{\theta}^{(t+1)} \mid \boldsymbol{\theta}^{(t)} \ldots, \boldsymbol{\theta}^{(1)}) = f(\boldsymbol{\theta}^{(t+1)} \mid \boldsymbol{\theta}^{(t)})$$

We need a mechanism that will provide many random samples for θ and will do this in such a way that each draw depends only on the previous draw. This is an iterative process. In order to initiate it we obviously need an initial value $\theta^{(1)}$, along with a tool, called the transition rule, that creates the draw $\theta^{(t+1)}$ for a given $\theta^{(t)}$. The transition rule creates the value of θ given the preceding value. Eventually this chain should converge with the target distribution, which is the posterior $f(\theta \mid \text{data})$. This is called equilibrium. The basic process is:

1. Choose an initial value, $\theta^{(1)}$.
2. Generate a large number (T) of iterations for θ, thousands, tens of thousands, or even millions of iterations. Designate these values $\theta^{(2)}$, $\theta^{(3)}$, etc.
3. Check for convergence and drop the first B iterations, as most likely these have not reached equilibrium.
4. Consider the iterations $\theta^{(B+1)}, \theta^{(B+2)}, \ldots, \theta^{(T)}$ as samples for the posterior distribution $f(\theta \mid \text{data})$.
5. Using the samples from step 4, present summary statistics and histograms for each parameter in θ, i.e. for α, β, and σ.

The 'burn-in period' refers to the first B iterations that were dropped in step 3. These iterations are discarded when creating histograms and summary statistics of the draws. Plotting the draws in sequential order (a chain) should give an indication of the value of B to choose. We provide an example in the following section.

Due to the way the samples are generated, there may be a correlation between $\theta^{(t)}$ and $\theta^{(t+1)}$ or between $\theta^{(t)}$ and $\theta^{(t+2)}$. Plotting the chains may indicate whether there is auto-correlation. Alternatively we can create an auto-correlation graph of the chain. If there is auto-correlation, we can select every 5th or every 10th iteration and store these for final presentation. This is called the thinning rate.

1.4.2 Transition rules*

Here is where things may potentially get tricky from a mathematical point of view. However, it is not necessary to know how MCMC creates new draws $\theta^{(t+1)}$ given $\theta^{(t)}$, as we can use a statistical software package called WinBUGS to do it for us. Books on MCMC and Bayesian statistics include such phrases as 'the Metropolis-Hasting algorithm' and the 'Gibbs sampler.' These are complex methods of creating draws $\theta^{(t+1)}$ given $\theta^{(t)}$. For the sake of satisfying curiosity, we will present an example here of MCMC, although it is not essential that you understand the material. Feel free to skip to the following subsection, in which we use WinBUGS to create draws $\theta^{(t+1)}$ given $\theta^{(t)}$. It applies more sophisticated implementations of these techniques.

Suppose we have the function $g(x) = 100 + 2 \times x - 3 \times x^2$ and want to know the value of x at which $g(x)$ has its maximum. Yes, that is high school mathematics. How was it calculated?

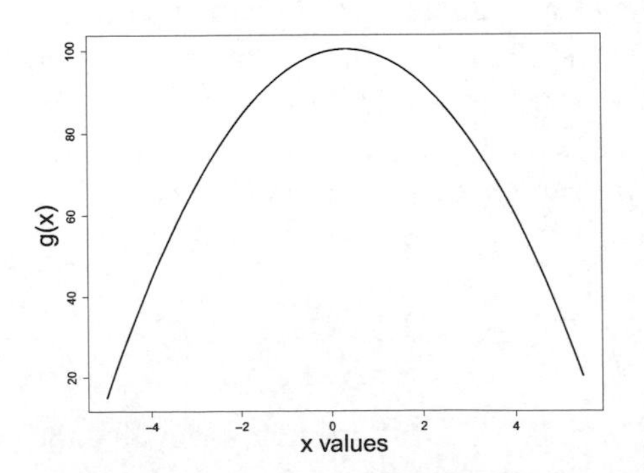

One option is to take the derivate of $g(x)$ with respect to x, set it to zero, and solve it. That gives: $2 - 6 \times x = 0$, which means that $x = 1/3$ is the solution. In linear regression we use a similar approach to obtain estimates for the intercept and slope, and even in generalised linear modelling we work with derivatives that are set to zero. A different approach would be to use one of the many available numerical optimisation routines to find the maximum value of x. As an example, we will use a basic strategy. Suppose our initial guess for the value of x for which $g(x)$ has its maximum is 5. For this value of x we have $g(5) = 35$. Now suppose that we have a mechanism that gives us a new value (a transition rule) based on the old value. The current value is 5, so let us try reducing by one, x = 4. This gives: $g(4) = 60$. The value of $g(4)$ is larger than that of $g(5)$, so $x = 4$ is probably closer to the optimum than $x = 5$, and our next guess should be $x = 3$, and so on. Instead of looking at the difference, we could look at the ratio of $g(4)/g(5)$. If it is less than one, we are moving in the wrong direction.

To recap, to find the x value for which $g(x)$ is optimal (be it a minimum or a maximum) we can try an x value, try another that is close to the original, compare the ratio of the corresponding g values, and eventually we will end up with $x = 1/3$. In this case $g(x)$ is a simple function. The mathematical models discussed later may contain complicated posterior likelihood functions with a great deal of local optima; hence we will need something more advanced, but we have sketched the basic principles of MCMC: try two x values, compare the optimisation criteria, and choose a new x value; essentially simple number crunching.

The Metropolis-Hasting algorithm

The Metropolis-Hasting algorithm can be used to generate draws $\theta^{(t+1)}$ given $\theta^{(t)}$. We will present the underlying algorithm, provide R code, and apply it to the oystercatcher data. Variations and extensions of this algorithm can be found in Ntzoufras (2009). The algorithm consists of the following steps:

1. Start with initial values $\theta^{(1)}$.
2. Using the current value of θ, which is $\theta^{(t)}$, sample a new candidate, θ^*. This is done with a so-called 'proposal distribution,' $q(\theta^* \mid \theta^{(t)})$. Other terms are 'jumping distribution' and 'candidate-generating distribution.' Most often it is the normal distribution, and we will draw random values from it, using the current value of θ. The resulting draw is denoted by θ^*. This value is potentially the subsequent value of θ in the chain; hence the term 'candidate generating distribution.'
3. We now have a candidate sample θ^*, and we need to decide whether it will be the next value in the chain. In order to do this we determine the ratio of the posterior density calculated at θ^* and at $\theta^{(t)}$. The common practice is to use the symbol α for this ratio, but we have already used α for the intercept in the linear regression model specified above, so we will designate this ratio R. R is given by:

$$R = \frac{f(\theta^* \mid \text{data})}{f(\theta^{(t)} \mid \text{data})}$$

4. If R is greater than 1, the posterior distribution for θ^* is greater than that for $\theta^{(t)}$, and we accept θ^* as the next draw in the chain $\theta^{(t+1)} = \theta^*$ and continue with step 2 for the subsequent draw. If the ratio R is less than 1, let chance decide whether θ^* should be the next draw in the chain. We will draw a 'yes' or 'no' from a random number generator. If 'yes,' set $\theta^{(t+1)} = \theta^*$ and continue with step 2 for the following draw. If 'no,' discard θ^* and go to step 2.

We will implement this algorithm for the linear regression model:

$$\text{Length}_i = \alpha + \beta \times \text{Feeding type}_i + \varepsilon_i$$

Before doing this, we estimate the parameters of this model using the `glm` function in R. The R code below loads the data from the tab delimited ascii file `OystercatcherData.txt`. Our code contains an extra command, not show here, that we ran prior to the `read.table` function; namely the `setwd` command. It sets the working directory. Its syntax is:

```
> setwd("C:/AnyName")
```

where `AnyName` is the complete path to the directory in which the text file is saved. We will not present the `setwd` command in the R code in this book, as its argument will be different for every reader, since it depends on the directory in which the data is stored. If you encounter problems importing the data, see our *Beginner's Guide to R* (Zuur et al., 2009b). The `read.table` command assumes that the decimal separator is a point and not a comma. We use the `glm` function with the `family = gaussian` argument. We could also have used the `lm` function.

```
> OC <- read.table(file = "OystercatcherData.txt", header = TRUE)
> M1 <- glm(ShellLength ~ FeedingType, family = gaussian, data = OC)
> summary(M1)

                    Estimate Std. Error t value Pr(>|t|)
(Intercept)          2.09788    0.02569  81.670   <2e-16
FeedingTypeStabbers -0.03100    0.06374  -0.486    0.627
(Dispersion parameter for gaussian family taken to be 0.1088741)
```

The intercept and slope are estimated at 2.098 and -0.031, respectively, and the variance (indicated as dispersion) is 0.109. The `logLik(M1)` command gives the log likelihood value of -60.095. Let us see whether we can reproduce these values using a homemade MCMC algorithm. To simplify our code below, we will (wrongly) assume that we know the variance and set it equal to 0.1088741, the dispersion parameter in the numerical output. When we use WinBUGS later in this chapter, we will estimate the variance.

Priors

We need to specify priors for the regression parameters α and β (recall that we are proceeding on the basis that we know the variance), and will use a normal distribution of the form:

$$\alpha \sim N(\mu_\alpha, \sigma_\alpha^2)$$
$$\beta \sim N(\mu_\beta, \sigma_\beta^2)$$

operating on the principle that this is the first time applying this model and we do not know what to expect. The slope may be positive or negative, so we use weak normal priors for α and β. This means using relatively large values for the variances and setting the mean to 0:

$$\alpha \sim N(0, 10^2)$$
$$\beta \sim N(0, 10^2)$$

The algorithm dictates that we calculate R. Perhaps it is easier to work with the log of R:

$$\begin{aligned}\log(R) &= \log\left(\frac{f(\theta^* \mid \text{data})}{f(\theta^{(t)} \mid \text{data})}\right) \\ &= \log\left(f(\theta^* \mid \text{data})\right) - \log\left(f(\theta^{(t)} \mid \text{data})\right) \\ &= \log\left(f(\text{data} \mid \theta^*) \times f(\theta^*)\right) - \log\left(f(\text{data} \mid \theta^{(t)}) \times f(\theta^{(t)})\right) \\ &= \log\left(f(\text{data} \mid \theta^*)\right) + \log\left(f(\theta^*)\right) - \log\left(f(\text{data} \mid \theta^{(t)})\right) - \log\left(f(\theta^{(t)})\right)\end{aligned}$$

Note that the f(data) disappears. We have two components here, the log likelihood function and the log of the prior, and we need to calculate each of these for the candidate value $\boldsymbol{\theta}^*$ and the $\boldsymbol{\theta}$ from the current iteration, $\boldsymbol{\theta}^{(t)}$. Instead of comparing the ratio R to 1 we compare log(R) to 0.

Calculating the log likelihood for the data, given the parameters $\boldsymbol{\theta}^*$ (and $\boldsymbol{\theta}^{(t)}$), is simple. Length is a continuous variable, and its values are not close to 0; hence the normal distribution is a reasonable choice. The likelihood $f(\text{data} \mid \boldsymbol{\theta}^*)$ is given by:

$$f(\text{data} \mid \theta^*) = \prod_{i=1}^{N}\left(\frac{1}{\sqrt{2\pi\sigma^2}} \times e^{-\frac{(Length_i - \alpha^* - \beta^* \times FeedingType_i)^2}{2\sigma^2}}\right)$$

This expression can be found in many undergraduate statistics textbooks explaining maximum likelihood. Using the log changes the product to a sum, and some basic rewriting gives:

$$\log(f(\text{data} \mid \theta^*)) = \log\left(\prod_{i=1}^{N} \frac{1}{\sqrt{2\pi\sigma^2}} \times e^{-\frac{(Length_i - \alpha^* - \beta^* \times FeedingType_i)^2}{2\sigma^2}}\right)$$

$$= \sum_{i=1}^{N} \log\left(\frac{1}{\sqrt{2\pi\sigma^2}} \times e^{-\frac{(Length_i - \alpha^* - \beta^* \times FeedingType_i)^2}{2\sigma^2}}\right)$$

$$= -\sum_{i=1}^{N}\left(\frac{(Length_i - \alpha^* - \beta^* \times FeedingType_i)^2}{2\sigma^2}\right) - \frac{N}{2}\log(2\pi\sigma^2)$$

So for given $\boldsymbol{\theta}^*$ (i.e. for given α^* and β^*) we can easily calculate the log likelihood and can also calculate it for $\boldsymbol{\theta}^{(t)}$. We will write a function in R that for given parameters calculates the log likelihood.

```
MyLogLik <- function(theta, Length, FeedingType){
      alpha <- theta[1]
      beta  <- theta[2]
      sigma <- 0.3299608  #Taken from the glm function
      N <- length(Length)
      L <- -sum(1 / (2*sigma * sigma) *
            (Length - alpha-beta * FeedingType) ^ 2)  -
            (N / 2) * log(2 * pi * sigma^2)
      L }
```

As mentioned, we took the variance (sigma^2) from the `summary` output of `M1`. To check whether this function is correct, we run it for the optimal parameters, hoping to find the same likelihood as the `glm` function. First we convert the categorical variable `FeedingType` to a variable with zeros and ones (`glm` did this for us, but we must do it manually for our homemade MCMC code), and then we call our function:

```
> OC$FT01 <- as.numeric(OC$FeedingType) - 1
> MyLogLik(c(2.09788,-0.03100), OC$ShellLength, OC$FT01)

-60.1009
```

The -60.10 is close enough to the log-likelihood given by the `glm` function; our code seems to be correct. This means we can now calculate two of the four components of log(R). The remaining two components are the log of the priors for $\boldsymbol{\theta}^*$ and $\boldsymbol{\theta}^{(t)}$. The prior itself is given by:

$$f(\theta^*) = f(\alpha^*, \beta^*) = f(\alpha^*) \times f(\beta^*)$$

We need to obtain the prior for each parameter. For the intercept and slope this is simple:

$$f(\alpha^*) = \frac{1}{\sqrt{2\pi \times 10^2}} \times \exp\left(-\frac{(\alpha^* - 0)^2}{2 \times 10^2}\right)$$

$$f(\beta^*) = \frac{1}{\sqrt{2\pi \times 10^2}} \times \exp\left(-\frac{(\beta^* - 0)^2}{2 \times 10^2}\right)$$

This is based on the diffuse priors *N*(0, 100) that we specified earlier. Multiply the two individual prior distributions and we get the prior $f(\boldsymbol{\theta}^*)$. We can do the same for $f(\boldsymbol{\theta}^{(t)})$. We wrote a simple function to calculate the prior:

```
MyPrior <- function(theta){
   alpha <- theta[1]
   beta  <- theta[2]
   fprior.a <- (1/sqrt(2 * pi * 100)) * exp(-alpha^2/ (2*100))
   fprior.b <- (1/sqrt(2 * pi * 100)) * exp(-beta^2 / (2*100))
   fprior <-  fprior.a * fprior.b
   log(fprior) }
```

If we want to consider σ as an unknown parameter (which we didn't, but should have done, and will do later), we need to choose an appropriate prior for it and add it into the multiplication above. We now have the essential tools for building up the Metropolis-Hasting algorithm. We start by specifying the number of iterations (`nT`) and create a matrix `Theta.t` that should contain the draws for the posterior. This matrix contains the information that we will need later for making histograms, etc.

```
> nT <- 10000
> Theta.t <- matrix(nrow = nT+1, ncol = 2)
```

Based on some initial runs, we decide to use 10,000 iterations. In step 1 of the algorithm we need to specify starting values and will set the starting value for the intercept and the slope to 0.

```
> Theta.t[1,] <- c(0, 0)
```

`Theta.t` contains all $\boldsymbol{\theta}^{(t)}$; its first row is $\boldsymbol{\theta}^{(1)}$, its second row is $\boldsymbol{\theta}^{(2)}$, etc. Next we prepare a vector of length 2 representing $\boldsymbol{\theta}^*$, the new proposed value. We only create the vector; we do not give it a value at this time.

```
> Theta.star <- vector(length = 2)
```

The first two lines inside the loop draw a new value for $\boldsymbol{\theta}^*$ given the current value, $\boldsymbol{\theta}^{(1)}$. We use the `rnorm` function to draw new values for $\boldsymbol{\theta}^*$. The crucial aspect is that $\boldsymbol{\theta}^{(1)}$ is being used in `rnorm`. This is the 'given the current value of $\boldsymbol{\theta}^{(t)}$, obtain a new sample' part of the algorithm. There are some aspects of this step that we will skip, e.g. the value of the variance to generate $\boldsymbol{\theta}^*$. We will address this later.

In step 3 we calculate `logR`, and in step 4 we need to use its value to assess whether our new sample $\boldsymbol{\theta}^*$ represents a greater likelihood. If `logR` is larger than 0, we have a set of parameters that have a higher log likelihood, which means that the new set of parameters are better than in $\boldsymbol{\theta}^{(t)}$. In this case we keep them and store them in the matrix `Theta.t`. Otherwise we let chance decide whether we keep them or draw another set of parameters. In `u <- runif(1)` and `logu <- log(u)` the value of *u* is always between 0 and 1. So if $u < R$, either *R* is larger than 1 (keep the new sample) or *R* is slightly smaller than 1 (in which case we may keep the new sample, depending on the value of u). This can be implemented in the R code with `u < R` or with `logu < logR`. The index *j* is used to ensure that any new draws are stored in the correct row of `Theta.t`.

```
> j <- 1
> for (i in 1:nT) {
     #Step 2: Draw samples from a proposal distribution
     Theta.star[1] <- rnorm(1, Theta.t[j,1], 0.1) #Intercept
     Theta.star[2] <- rnorm(1, Theta.t[j,2], 0.1) #Slope
```

```
    #Step 3: Calculate log(R)
    logR <- MyLogLik(Theta.star, OC$ShellLength, OC$FT01)  +
            MyPrior(Theta.star) -
            MyLogLik(Theta.t[j,], OC$ShellLength, OC$FT01) -
            MyPrior(Theta.t[j,])

    #Step 4: Keep the draw or not?
    u <- runif(1)
    logu <- log(u)
    if (logu < logR) {
           j <- j + 1
           Theta.t[j,] <- Theta.star
  } }
```

The code creates many draws, which are stored in the columns of `Theta.t`. The first column contains the draws for α and the second column for β. Let us look at the first 10 rows:

```
> head(Theta.t, 10)

          [,1]        [,2]
 [1,] 0.00000000  0.00000000
 [2,] 0.03888729 -0.05009651
 [3,] 0.16272722  0.04581325
 [4,] 0.21043045  0.08175466
 [5,] 0.25959099  0.03535716
 [6,] 0.38466520  0.17141646
 [7,] 0.53560719  0.04204862
 [8,] 0.56090129  0.19663108
 [9,] 0.74994716  0.16412715
[10,] 0.82914454  0.15980657
```

The results of the linear regression indicated that the estimated intercept and slope should be approximately 2.097 and -0.031, respectively. Clearly we haven't reached those values in the first 10 iterations, but this will most likely never happen so quickly. The top 2 panels in Figure 1.6 show the first 500 iterations (accepted draws) for the intercept and slope. After approximately 50 iterations, the draws for the intercept and slope approach those obtained by the `glm` function. We can see some auto-correlation in the chains. We dropped the first 500 iterations for each parameter and plotted all remaining iterations (see the bottom two panels in Figure 1.6). Detecting possible patterns is a challenging task. If clear cyclic patterns emerge, we have a problem. Figure 1.7 shows an auto-correlation function (ACF) of the chains for the intercept and slope. The ACF of the slope shows auto-correlation, and it may be better to select every 5th or 10th iteration and use these for the sample of the posterior.

The mean of all iterations (except for the first 500) for the intercept is 2.099, which is close to the value obtained by the `glm` function. The mean of all iterations for the slope is -0.034, which is also close to the value obtained by `glm`. The standard deviations of these chains are also close to the standard errors obtained by the `glm` function, 0.028 for the chain of α and 0.069 for the chain of β.

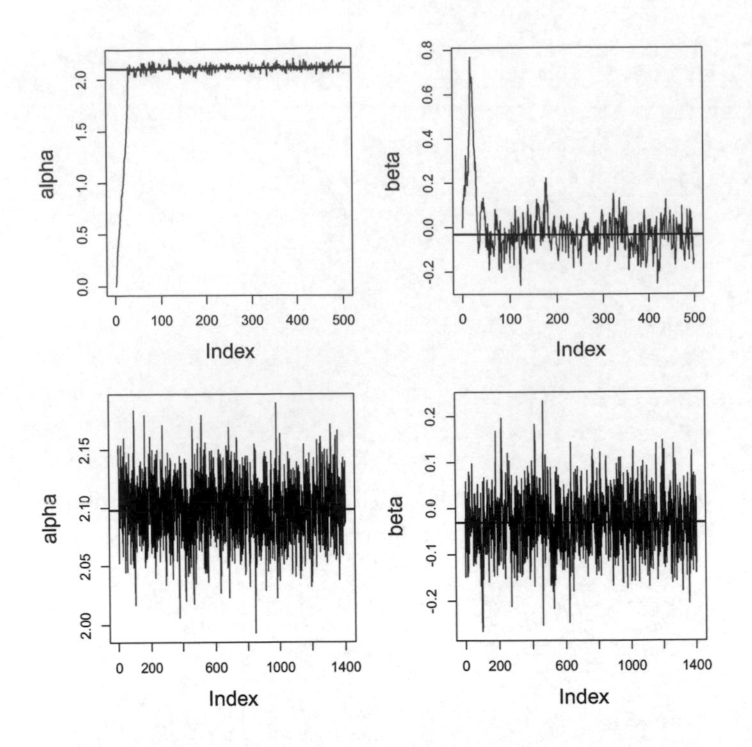

Figure 1.6. Top panels: The first 500 iterations from each chain. The horizontal line represents the estimated value obtained by the `glm` function. Bottom panels: Iterations 500 to 1,808 for each parameter.

Our homemade algorithm for MCMC has several flaws and should be improved. For example, the standard deviation used in generating new draws in step 2 needs modification. If we use a small value, new draws are close to the current draw, and a large value means that new values differ more widely. A discussion and solution can be found in Ntzoufras (2009). Another improvement would be to obtain `nT` stored draws instead of drawing `nT` draws and discarding some.

The Metropolis-Hastings algorithm can be extended in various ways, resulting in the Random-walk Metropolis, the Independence Sampler, the Component-wise Metropolis-Hasting, and the Gibbs Sampler, among other approaches. All these methods provide improvements of the proposal distribution and other steps in the algorithm. The Gibbs Sampler is the technique used by WinBUGS.

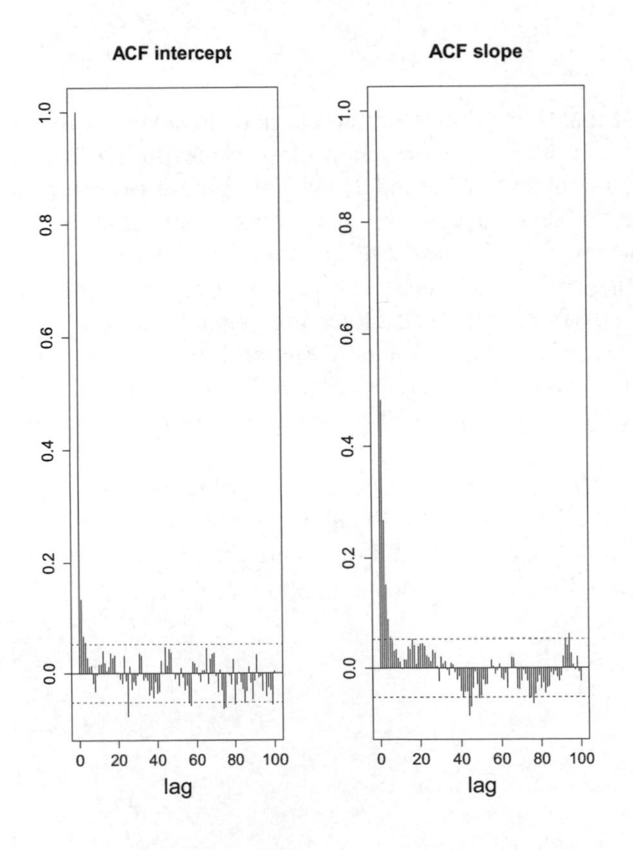

Figure 1.7. Auto-correlation of the iterations for the intercept and slope. The first 500 iterations were dropped as burn-in. The ACF for the slope shows strong auto-correlation.

1.5 Using WinBUGS

The BUGS in WinBUGS stands for Bayesian inference Using Gibbs Sampling. It is cited as Lunn et al. (2000). WinBUGS can be run as a standalone under a Windows operating system, or it can be accessed from other software packages such as `R2WinBUGS` (Stutrz et al., 2005) in R, which is what we will do in this book. It is also possible, at least theoretically, to run WinBUGS and `R2WinBUGS` under Wine in Mac and Linux operating systems. WinBUGS can be downloaded from: http://www.mrc-bsu.cam.ac.uk/bugs/winbugs/contents.shtml. You will need to install the patch and a free key, see the URL above for instructions. It may be best to install it under C:/WinBUGS14 and not under C:/Program Files/WinBUGS14. The R package `R2WinBUGS` is an add-on package and can be installed from CRAN (see Zuur et al., 2009b for installing packages).

The following demonstrates the use of WinBUGS and `R2WinBUGS` to fit a linear regression model for the oystercatcher data. In the previous section we used:

$$\text{Length}_i = \alpha + \beta \times \text{Feeding type}_i + \varepsilon_i$$

This time we will also estimate the variance of the residuals. To fit this model in WinBUGS via R we need the following code. First we load the package `R2WinBUGS` and specify the directory in which we installed WinBUGS. On our computer this is C:/WinBUGS14.

```
> library(R2WinBUGS)
> MyWinBugsDir <- "C:/WinBUGS14/"
```

Next we need to put all the necessary variables into an object, which we will call `win.data`. You can use any name you like.

```
> win.data <- list(ShellLength  = OC$ShellLength,
                   FeedingType  = OC$FT01,
                   N            = nrow(OC))
```

These are the variables that we need in our MCMC code. Note that the `FT01` variable was defined earlier and is not part of the original data. The following block of code is essential; it tells WinBUGS which variable is the response variable and specifies the statistical distribution it follows so WinBUGS can calculate the likelihood of the data given the parameters. The prior distributions are specified for the regression parameters α, β, and σ. The `sink` and `cat` functions put everything from `model {` to the closing bracket } in a text file called `LM.txt`. We could have typed it directly into this file using a text editor. So in principle you can ignore the first two and last two rows below.

```
> sink("LM.txt")
  cat("
  model {
    for (i in 1:N) {
          ShellLength[i]   ~  dnorm(mu[i], tau1)
          mu[i] <- beta1[1] + beta1[2] * FeedingType[i]
       }

    #Priors
    for (i in 1:2) { beta1[i] ~ dnorm(0.0, 0.01) }
    tau1 <- 1/sigma2
    sigma2 <- pow(sigma,2)
    sigma ~ dunif(0.001,10)
        }
  ",fill = TRUE)
  sink()
```

The `for` loop specifies that each length observation is normally distributed, with mean μ_i and precision τ. We will discuss the precision τ in a moment. It is not a loop in R itself; it is a shorthand way of writing:

```
ShellLength[1]   ~   dnorm(mu[1], tau1)
mu[1] <- beta1[1] + beta1[2] * FeedingType[1]

ShellLength[2]   ~   dnorm(mu[2], tau1)
mu[2] <- beta1[1] + beta1[2] * FeedingType[2]
```

etc. The order in which the commands are entered is irrelevant. In R, the code above will crash, as `mu[i]` is being used before it is defined.

The mean μ_i is modelled as `beta1[1] + beta1[2] * FeedingType[i]`, which represents $\alpha + \beta \times$ Feeding type$_i$. So the intercept is `beta1[1]` and the slope is `beta1[2]`. We did not want to use the name `beta`, because it is an existing R function. Earlier in this chapter we specified the following priors for the intercept and the slope:

$$\alpha \sim N(0,10^2)$$
$$\beta \sim N(0,10^2)$$

This corresponds to `beta1[i] ~ dnorm(0.0, 0.01)` in the second loop, which is slightly confusing. We wanted a large variance, representing uncertainty, yet the code uses a small value. WinBUGS specifies precision and not variance. The precision value is defined as 1/variance. So a variance of 10^2 should be entered as a precision of 1/100 = 0.01. This explains why we use the argument 0.01 in the `dnorm` function. For the same reason we use: `ShellLength[i] ~ dnorm(mu[i], tau1)`. The `tau1` (called τ in the text above) in the code is the precision parameter linked to the variance via: $\tau = 1/\sigma^2$. We can either specify the prior of τ, which is typically a gamma distribution, or continue working with σ and let WinBUGS do the conversion $\tau = 1/\sigma^2$ each time it draws a new proposal sample. The code above does this. We used a uniform prior with values between 0.001 and 10 for σ.

We also need to specify initial values of the chain for each regression parameter. For the intercept and slope (both stored in `beta1`) we draw two samples from a normal distribution with mean 0 and standard deviation 0.01. For more complicated models it is better to use realistic starting values (see also Chapters 2 and 4).

```
> inits <- function () {
     list(beta1 = rnorm(2, 0, 0.01),
          sigma = runif(1,0.001, 10))}
```

We need to tell WinBUGS the parameters from which to retain the chains. These are `beta1` and `sigma`. There is no need to extract information from `tau1`.

```
> params <- c("beta1", "sigma")
```

Next we specify the number of draws, the burn in, and the thinning rate. Getting the appropriate values requires some artistic skill. With MCMC it is critical to obtain chains that are well mixed. Outliers, weak patterns, complicating models, correlated parameters, and collinearity are all factors that may result in chains that are not well mixed, and we must take appropriate action (e.g. remove collinear variables). If we cannot find the cause of poor mixing, increasing the number of draws, the burn-in, or even using multiple chains may be an option. The goal is to reach several thousands of iterations so that we can make a histogram to visualise the posterior distribution. Our choices are:

```
> nc <- 3               #Number of chains
> ni <- 100000          #Number of draws from posterior (for each chain)
> nb <-  10000          #Number of draws to discard as burn-in
> nt <-    100          #Thinning rate
```

This model takes a few minutes to run. For such a simple model we could have reduced `ni`, `nb`, and `nt` each by a factor of 10, but for the GLMM and zero inflated models discussed later we should multiply them by at least 10. The number of draws for the posterior of each parameter is:

```
> 3 *(ni - nb) / nt

2700
```

This is an ample number of draws for a posterior distribution. Finally, the *Supreme Moment* has arrived. We can start WinBUGS. In order to do so we need:

```
> out1 <- bugs(data = win.data,
               inits = inits,
               parameters = params,
               model = "LM.txt",
               n.thin = nt,
               n.chains = nc,
               n.burnin = nb,
               n.iter = ni,
               debug = TRUE,
               bugs.directory = MyWinBugsDir)
```

The `debug = TRUE` option means that WinBUGS stays open when it is finished. Provided R can locate the directory to which WinBUGS was installed, the WinBUGS window will pop up. If you are lucky there will be a message in the lower left of the window saying 'Updating model.' If this happens the first time you run the code, cherish the moment of victory. More often you will get vague error messages, and it will take some trial and error to fix the problem. If you have a complicated model, begin by programming a simplified model, copy and paste code from existing working projects that work, and slowly build up the complexity of the model. Once WinBUGS runs, the bad news is that it may need to run for a long time. For more complicated models (e.g. zero inflated GLMM with spatial correlation), expect it to run for a full night. Windows must be set to not automatically update and restart during this process to prevent loss of the data. Users of Mac laptops using Parallels (a Windows emulator) should be careful; the first author of this book overheated his laptop and burned it out while running WinBUGS on Parallels.

When WinBUGS is finished it shows multiple graphs and displays of output. Close it and return to R, as that is where we will present the output. The numerical output is obtained by:

```
> print(out1, digits = 3)

Inference for Bugs model at "LM.txt", fit using WinBUGS,
 3 chains, each with 1e+05 iterations (first 10000 discarded), n.thin = 100
 n.sims = 2700 iterations saved
            mean    sd    2.5%     25%     50%     75%   97.5%  Rhat n.eff
beta1[1]   2.098 0.026   2.046   2.080   2.097   2.115   2.147 1.001  2700
beta1[2]  -0.031 0.064  -0.161  -0.071  -0.031   0.011   0.094 1.000  2700
sigma      0.332 0.017   0.300   0.320   0.331   0.343   0.367 1.002  1800
deviance 123.238 2.557 120.400 121.400 122.600 124.400 130.000 1.000  2700

For each parameter, n.eff is a crude measure of effective sample size,
and Rhat is the potential scale reduction factor (at convergence, Rhat=1).
```

```
DIC info (using the rule, pD = Dbar-Dhat)
pD = 3.0 and DIC = 126.2
DIC is an estimate of expected predictive error (lower deviance is better).
```

The object `out1` contains the draws for the posterior of each parameter along with other information. The `print` function presents the mean and standard deviation as well as various quartiles for the draws of each posterior. The posterior means for α, β, and σ are similar to the estimated values obtained by the `glm` function. The standard errors for the intercept and slope are also similar. The 2.5% and 97.5% quartiles can be used to derive a 95% credible interval. For example, the 95% credible interval for β is from -0.16 to 0.09. This means that there is a 95% chance that β falls in this interval. Since this interval contains 0, β is not important, so there is no strong effect of feeding behaviour of oystercatchers on the length of clams.

The output contains information on the Deviance Information Criteria (DIC), which is a similar tool to the Akaike Information Criterion (AIC). We will discuss DIC in Chapters 2 and 4. There are 2,700 draws for each posterior. Before presenting them we need to assess whether our choices for the number of iterations, the burn-in, and the thinning rate are correct. If the chains (three per posterior) show non-mixing, we have a problem. We can either plot the chains or use the summary statistics `Rhat` and `n.eff`. The `n.eff` is the number of iterations divided by a measure of auto-correlation. If a chain shows auto-correlation, `n.eff` will be smaller than 2,700. This is the case for σ; so we should investigate the chains of this parameter. Graphics tools are also useful. Figure 1.8 shows a plot of each of the three chains for the intercept. Each chain is of length 900. We should not see patterns in these chains.

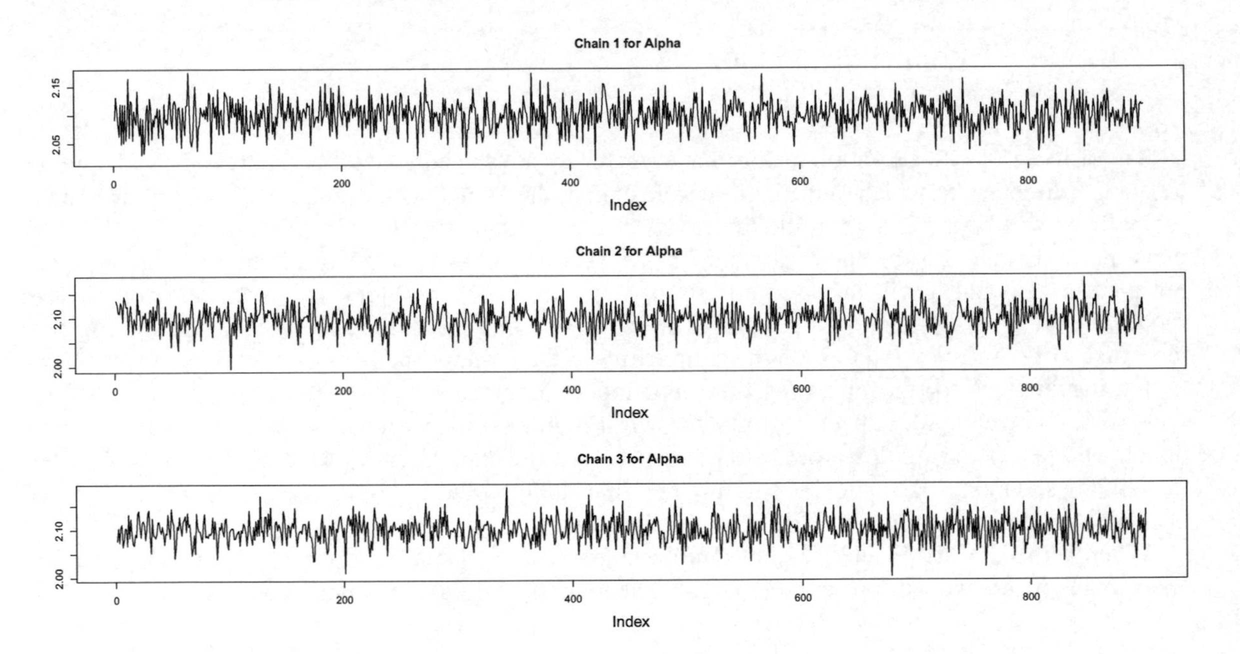

Figure 1.8. Three chains for the posterior of the intercept.

The R code to generate this graph follows. The crucial part is accessing the three chains from the intercept in the `out1` object. They are in `$sims.array`. The rest of the code is elementary plotting. Instead of plotting the chains in three panels, we can plot them in a single graph with different colours. We need to repeat this for the slope (replace `beta1[1]` with `beta1[2]` on the second line) and the variance (replace `beta1[1]` with `sigma`).

```
> par(mfrow = c(3,1))
> Alpha.Chains <- out1$sims.array[,, "beta1[1]"]
```

```
> plot(Alpha.Chains[,1], type = "l", col = 1,
        main = "Chain 1 for Alpha", ylab = "", cex.lab = 1.5)
> plot(Alpha.Chains[,2], type = "l", col = 1,
        main = "Chain 2 for Alpha", ylab = "", cex.lab = 1.5)
> plot(Alpha.Chains[,3], type = "l", col = 1,
        main = "Chain 3 for Alpha", ylab = "", cex.lab = 1.5)
```

If we enter `acf(Alpha.Chains[,1])` we get the auto-correlation function for the first chain of the intercept. The graph (not shown) does not show auto-correlation. We can repeat it for the other two chains for the intercept and all other parameters. Results can be presented in a multi-panel scatterplot made with the `xyplot` of the `lattice` package.

The mean posterior values can be accessed by typing:

```
> out1$mean$beta1

2.09750481 -0.03066264

> out1$mean$sigma

0.3322197
```

Finally we plot the histograms of the 2,700 draws of each posterior (Figure 1.9). The crucial point is to know that the draws are in `$sims.list`. We plot the three histograms adjacent to one another, but later we will use more advanced code to present them in a multi-panel histogram. R code to make this graph is:

```
> par(mfrow = c(1,3))
> hist(out1$sims.list$beta1[,1], xlab = "Draws of posterior",
      main = "Intercept")
> hist(out1$sims.list$beta1[,2], xlab = "Draws of posterior",
      main = "Slope")
> hist(out1$sims.list$sigma, xlab = "Draws of posterior",
      main = "Sigma")
```

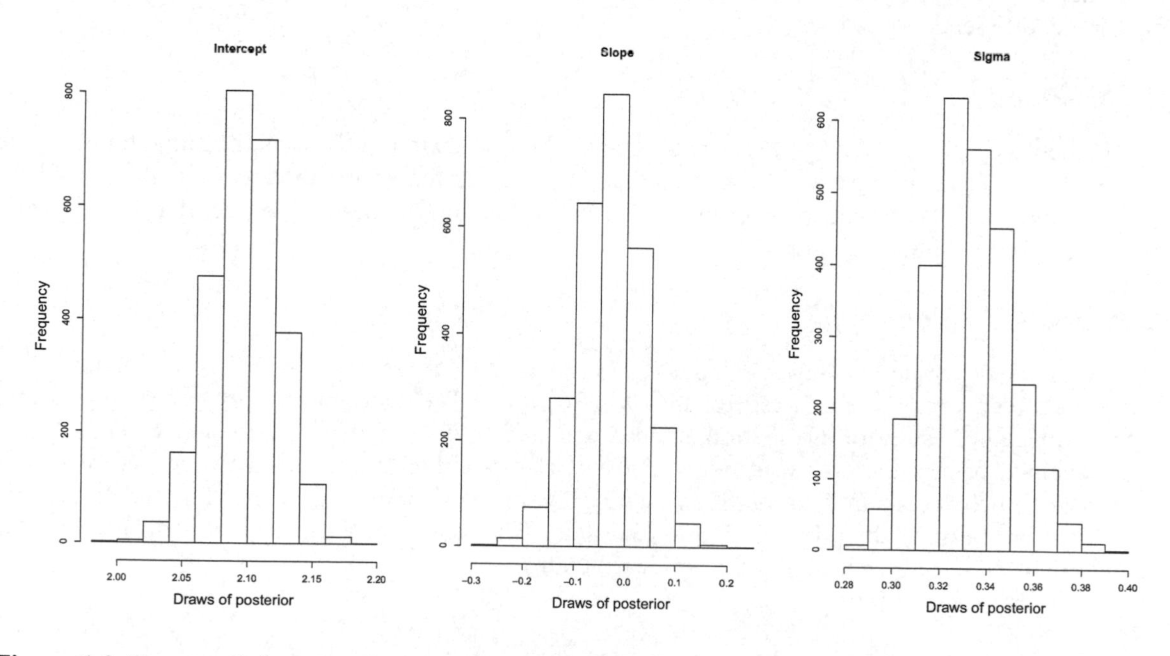

Figure 1.9. Draws of the posterior for each parameter.

There are more things that can and should be done, for example plotting the draws of each parameter versus the others to determine any correlation among them. We will do this in Chapter 4. In Chapter 2 we discuss how to obtain residuals.

1.6 Summary

Before continuing with the case studies, we briefly summarise the preceding sections on Bayesian statistics and MCMC. In our 2007 and 2009a books we used frequentist statistics. By this we mean that we adopted a philosophy in which we formulate a hypothesis for the regression parameters; apply a regression type model; and estimate parameters, standard errors, 95% confidence intervals, and *p*-values. We can then say that if we were to repeat this experiment a large number of times, in 95% of cases the real population regression parameters would fall within the estimated confidence intervals. The *p*-values also tell us how *often* (frequency) we find an identical or larger test statistic. Key elements of frequentist statistics are:

- The parameters, such as mean, variance, and regression parameters, that determine the behaviour of the population are fixed, but unknown.
- Based on observed data, these unknown parameters are estimated in such a way that the observed data agree well with our statistical model. In other words, the parameter estimates are chosen such that the likelihood of the data is optimised. This is maximum likelihood estimation.
- Frequentist approaches are objective in that only the information contained in the current data set is used to estimate the parameters.

Bayesian statistics is based on a different philosophy. The main difference is the assumption that the parameters driving the behaviour of the population are not fixed. Instead, it is presumed the parameters themselves follow a statistical distribution.

The four main components of Bayesian statistics are:

1. Data

Suppose we have a stochastic variable Y, with density function $f(Y \mid \boldsymbol{\theta})$, and let $\mathbf{y} = (y_1,\ldots,y_n)$ denote n observations of Y. Now $\boldsymbol{\theta}$ is a vector containing unknown parameters which are to be estimated from the observed data $\mathbf{y}$. In the case of the oystercatcher data, $\mathbf{y}$ is the length (of shells) and $\boldsymbol{\theta}$ the regression coefficients.

2. Likelihood

The density $f(\mathbf{y} \mid \boldsymbol{\theta}) = \prod_i f(y_i \mid \boldsymbol{\theta})$ is the likelihood. When maximum likelihood estimation is carried out, $\boldsymbol{\theta}$ is chosen to maximise $f(\mathbf{y} \mid \boldsymbol{\theta})$. For example, for a linear regression model, the maximum likelihood estimates of the parameters are obtained from calculating the derivatives, setting them to zero, and solving the resulting equations.

3. Prior distribution

The major difference in Bayesian statistics is that instead of assuming that $\boldsymbol{\theta}$ is an unknown, but fixed, parameter vector, we now assume that $\boldsymbol{\theta}$ is stochastic. The distribution of $\boldsymbol{\theta}$, before the data are obtained, is called the prior distribution, and we denote it by $f(\boldsymbol{\theta})$. It reflects knowledge about $\boldsymbol{\theta}$, perhaps obtained from previous experiments, but it is also possible to choose $f(\boldsymbol{\theta})$ such that it reflects very little knowledge about $\boldsymbol{\theta}$, so that the posterior distribution is mostly influenced by the data. In the latter case we say that the prior distribution is vague. The terms non-informative or diffuse are also commonly used in reference to vague prior distributions.

4. Posterior distribution

This forms the final component of a Bayesian analysis setting. Using a simple statistical theory (namely, Bayes' Theorem), the prior information is combined with information from the data to give us the posterior distribution $f(\boldsymbol{\theta} \mid \mathbf{y})$. This represents the information about $\boldsymbol{\theta}$ after observing the data $\mathbf{y}$:

$$f(\boldsymbol{\theta} \mid \mathbf{y}) = f(\mathbf{y} \mid \boldsymbol{\theta}) \times f(\boldsymbol{\theta}) \,/\, f(\mathbf{y})$$
$$\propto f(\mathbf{y} \mid \boldsymbol{\theta}) f(\boldsymbol{\theta})$$

The second equation follows because $f(\mathbf{y})$, which is the marginal density of the data, is constant. In contrast to maximum likelihood, where a point estimate for $\boldsymbol{\theta}$ is obtained, with Bayesian statistics a density of $\boldsymbol{\theta}$ is the final result. This density averages the prior information with information from the data. Gelman et al. (2003) provide an accessible book covering the basics of Bayesian analysis.

We have discussed the main components of Bayesian statistics: The prior distribution of $\boldsymbol{\theta}$, our observed data $\mathbf{y}$, and finally, how these two pieces of information are combined to obtain the posterior distribution of $\boldsymbol{\theta}$. In only a limited number of cases is the posterior distribution in a known form. More often than not, this distribution is complex, making it difficult to obtain summaries. For many years this was the chief reason that Bayesian statistics was not widely used. However, with the advent of computers, simulation tools such as Markov Chain Monte Carlo have become widely available, and Bayesian analysis is more accessible. The development of freeware software to implement such simulation tools has also helped greatly to popularise Bayesian approaches.

Markov Chain Monte Carlo techniques

The aim is to generate a sample from the posterior distribution. This is the Monte Carlo part. Often the precise form of the posterior distribution is unknown, but fortunately another stochastic device, the Markov chain, can be used to deal with this. Assume we start with an initial value of $\boldsymbol{\theta}$, denoted by $\boldsymbol{\theta}^1$. The next state of the chain, $\boldsymbol{\theta}^2$, is generated from $P(\boldsymbol{\theta}^2 \mid \boldsymbol{\theta}^1)$, where $P(\,.\, | \,.\,)$ is the so-called transition kernel of the chain. $\boldsymbol{\theta}^3$ is generated from $P(\boldsymbol{\theta}^3 \mid \boldsymbol{\theta}^2)$,... and $\boldsymbol{\theta}^t$ is generated from $P(\boldsymbol{\theta}^t \mid \boldsymbol{\theta}^{t-1})$. Under certain regularity conditions, the distribution of $P(\boldsymbol{\theta}^t \mid \boldsymbol{\theta}^0)$ will converge into a unique stationary distribution $\mathbf{f}(.)$. An important property of a Markov chain is that once it has reached its stationary distribution it will have 'forgotten' its initial starting value, so it does not matter how inaccurate the initial value $\boldsymbol{\theta}^1$ was.

Assuming we can define an appropriate Markov chain (i.e. that an appropriate distribution $P(\boldsymbol{\theta}^t \mid \boldsymbol{\theta}^{t-1})$ can be constructed), we can generate *dependent* draws, or realisations, from the posterior distribution. The samples are not independent, as the distribution of $\boldsymbol{\theta}^t$ depends on the value of $\boldsymbol{\theta}^{t-1}$. In turn, the distribution of $\boldsymbol{\theta}^{t-1}$ depends on the value of $\boldsymbol{\theta}^{t-2}$, and so on. This has the following consequences:

- The initial part of the chain should be discarded (commonly referred to as 'burn-in') so that the influence of an arbitrary initial value $\boldsymbol{\theta}^1$ is eliminated.
- The MCMC samples are less variable compared to independent samples; therefore the variance of estimated summary statistics, such as the sample mean, is larger than would be the case if the samples had been independent.
- When a stationary state has been reached (that is, the realisations no longer depend on the initial value $\boldsymbol{\theta}^1$) a large number of samples is needed to cover the entire region of the posterior distribution, as small portions of consecutive samples tend to be concentrated in small regions of the posterior distribution.

When we generate many samples after the burn-in, they will be distributed appropriately from the entire posterior distribution. This distribution can be summarised by summary statistics such as the sample mean and sample quantiles. A useful property of MCMC is that statistics calculated from the MCMC sample will converge to the corresponding posterior distribution quantities. For example, the sample mean converges with the posterior mean and the sample quantiles converge with the posterior quantiles.

It may seem complicated to generate samples from the posterior distribution, but fortunately there are algorithms available to simplify the task. The Metropolis-Hastings algorithm (Metropolis et al. 1953; Hastings 1970) and the Gibbs Sampler (Geman and Geman 1984; Gelfand and Smith 1990), which is a special case of the former, are two commonly used algorithms for creating appropriate Markov chains. Gilks et al. (1996) provide an accessible introduction to various MCMC techniques illustrated with many examples. Although these algorithms are easily programmed in R, there are technical complexities, and it is better to use specialised software such as the freeware package WinBUGS. The R package `R2WinBUGS` is an interface to WinBUGS, and is what we use in this book.

2 Zero inflated GLMM applied to barn owl data

Alain F Zuur, Anatoly A Saveliev, Elena N Ieno, and Alexandre Roulin

2.1 Introduction

2.1.1 Vocal begging behaviour of nestling barn owls

Using microphones and a video recorder, Roulin and Bersier (2007) gathered data from 27 owl nests to investigate vocal behaviour of nestlings. Response variables were defined as the amount of time a parent spends on the perch, the amount of time a parent is in the nestbox, sibling negotiation, and begging. Zuur et al. (2009a) used sibling negotiation to explain linear mixed effects models and additive mixed effects models with a Gaussian distribution.

Sibling negotiation was defined as the number of calls made by all offspring, in the absence of the parents, during 30 second time periods recorded at 15 minute intervals. The number of calls from the recorded period in the 15 minutes preceding their arrival was allocated to a visit from a parent. This number was divided by the number of nestlings, which was between 2 and 7. Thus sibling negotiation was the mean number of calls per nestling in the 30-second sampling period prior to arrival of a parent (Figure 2.1). These calls are a form of sibling to sibling communication.

By vocalizing in the absence of parents, siblings inform one another of their relative hunger level. Vocal individuals are hungrier than their siblings. By vocalising loudly, a hungry owlet informs its siblings that it is willing to compete to monopolize the next delivered food item. Being informed, its siblings temporarily retreat from the contest, thereby reducing the level of competition. This sibling/sibling communication process, termed *sibling negotiation,* is adaptive, since food items delivered by parents are indivisible and hence consumed by a single nestling. For this reason, the effort invested in sibling competition will be paid back to a single nestling. Sibling competition therefore allows siblings to optimally adjust effort in competition for parent-provided resources.

23.00	23.15	23.30	23.45	0.00	0.15	0.30	...

Figure 2.1. Sampling process: A parent arrives at 23.45 at a nest. The number of calls for this observation is taken from the 30 second recording made between 23.30 and 23.45.

Half of the nests were given extra prey in the morning preceding recording of behaviour, and prey remains were removed from the other half of the nests. This was called 'food-satiated' and 'food-deprived,' respectively, since in half of the nests nestlings were food-satiated before the first parental feeding visit of the night, and in the other half the nestlings were hungry. Sampling was conducted on two nights, with the food treatment reversed on the second night, so that each nest was subjected to both protocols. Recordings were made between 21.30 h and 05.30 h. Arrival time reflects the time when a parent arrived at the perch with prey. Further information can be found in Roulin and Bersier (2007).

The underlying biological question is whether sibling negotiation is related to the sex of the parent, food treatment, or arrival time of the parent. Of particular interests are the interaction between food treatment and sex of the parent and the interaction between sex of the parent and arrival time.

2.1.2 Previous analyses of the owl data

We have multiple observations from each of the 27 nests. It is likely that observations from the same nest are more similar than observations from different nests. We are not particularly interested in the nest effect *per se*, but to allow for correlation among observations from an individual nest, we need to apply mixed effects models (Pinheiro and Bates, 2000). Roulin and Bersier (2007) and Zuur et al. (2009a) used linear mixed effects models of the form:

$$\begin{aligned}\text{Negotiation} = \ &\text{Sex effect} + \text{Food treatment effect} + \text{Arrival time effect} + \\ &\text{Sex} \times \text{Food treatment interaction} + \text{Sex} \times \text{Arrival time interaction} + \text{noise}\end{aligned} \quad (2.1)$$

The two interaction terms were included because of the underlying biological questions raised by Roulin and Bersier. The residuals of this model were heterogeneous, and we attempted to model the heterogeneity using generalised least squares. See, for example, Pinheiro and Bates (2000) or Zuur et al. (2009a). However, no suitable variance covariate could be found to model the heterogeneity; therefore a log-transformation was applied to the response variable. The expression in Equation (2.1) is formulated in words, and the statistical notation is given by:

$$\begin{aligned}LogNeg_{ij} = \ &\alpha + \beta_1 \times SexParent_{ij} + \beta_2 \times FoodTreatment_{ij} + \beta_3 \times ArrivalTime_{ij} + \\ &\beta_4 \times SexParent_{ij} \times FoodTreatment_{ij} + \beta_5 \times SexParent_{ij} \times ArrivalTime_{ij} + \\ &a_i + \varepsilon_{ij}\end{aligned} \quad (2.2)$$

$LogNeg_{ij}$ is the log-10 transformed sibling negotiation for observation *j* at nest *i*. $SexParent_{ij}$ and $FoodTreatment_{ij}$ are categorical variables of two levels, and $ArrivalTime_{ij}$ is a continuous variable. Note that minutes and hours must be converted to a 1 to 100 scale; an observation at 01.15 am is coded as 25.25. The term a_i is a random intercept and is assumed to be normally distributed with mean 0 and variance σ^2_{Nest}. The residuals ε_{ij} are assumed to be normally distributed with mean 0 and variance σ^2_{ε}. The random terms are assumed to be independent of each other.

Using a 10-step protocol, a model selection was applied in Zuur et al. (2009a), and the optimal model was given as:

$$LogNeg_{ij} = \alpha + \beta_2 \times FoodTreatment_{ij} + \beta_3 \times ArrivalTime_{ij} + a_i + \varepsilon_{ij} \quad (2.3)$$

There are serious criticisms of model selection. See Johnson and Omland (2004) or Whittingham et al. (2006). A discussion of model selection can be found in Chapter 4 (marmot case study).

The residuals of the model in Equation (2.3) showed a non-linear pattern when plotted versus arrival time; therefore Zuur et al. (2009a) continued to apply a Gaussian additive mixed effects model of the form:

$$LogNeg_{ij} = \alpha + \beta_2 \times FoodTreatment_{ij} + f(ArrivalTime_{ij}) + a_i + \varepsilon_{ij} \quad (2.4)$$

The resulting smoother for arrival time showed a clear non-linear pattern (Figure 2.2). Investigation showed that multiple smoothers (e.g. a smoother for arrival time of the males and one for arrival time of the females) were unnecessary.

If we plotted residuals versus fitted values, the result was a graph as in Figure 2.3. Note the clear band of points at the lower left area. These points correspond to the zeros in the response variable. For discrete data we will always have such bands of points. The chief concern here is the large number of zeros.

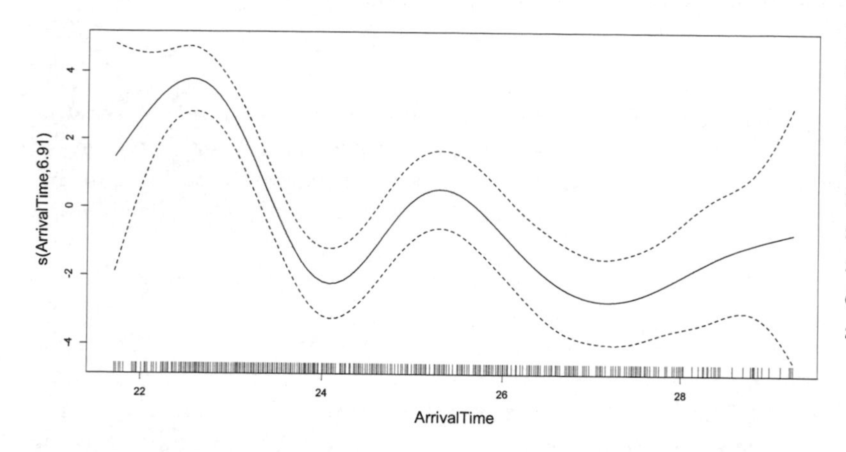

Figure 2.2. Shape of the smoother for arrival time in an additive mixed effects model. The amount of smoothing was 6.9 degrees of freedom. The shape of the smoother suggests that there are two periods in which the chicks make the most noise: around 22.30 and 01.30.

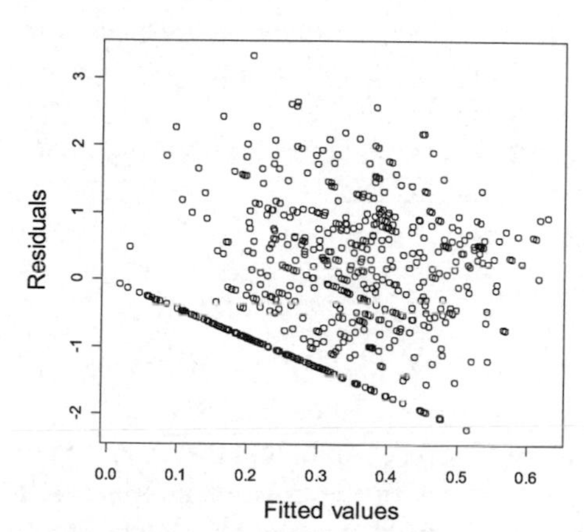

Figure 2.3. Graph of residuals versus fitted values for the linear mixed effects model in Equation (2.3). The band of points on the lower left corresponds to zeros in the response variable, i.e. no vocal negotiation recorded during the 30 sec recorded period.

The disadvantage of analysing the sibling negotiation data with Gaussian models was that we could predict negative fitted values, and a log transformation was needed to deal with the heterogeneity. In Chapter 13 of Zuur et al. (2009a) generalized linear mixed models (GLMM) and generalized additive mixed effects models (GAMM) were used to analyse the number of calls using a Poisson distribution. The advantage of these approaches is that we always have positive fitted values, and heterogeneity is automatically modelled via the Poisson distribution.

The GLMM used in Zuur et al. (2009a) was of the form:

$$\begin{aligned}
&NCalls_{ij} \sim Poisson(\mu_{ij}) \\
&\log(\mu_{ij}) = \eta_{ij} \\
&\eta_{ij} = \alpha + offset(LBroodSize_{ij}) + \beta_1 \times SexParent_{ij} + \beta_2 \times FoodTreatment_{ij} + \\
&\quad \beta_3 \times ArrivalTime_{ij} + \\
&\quad \beta_4 \times SexParent_{ij} \times FoodTreatment_{ij} + \beta_5 \times SexParent_{ij} \times ArrivalTime_{ij} + a_i \\
&a_i \sim N(0, \sigma^2_{Nest})
\end{aligned} \tag{2.5}$$

The first line states that the number of calls for observation j at nest i, $NCalls_{ij}$, is Poisson distributed with mean μ_{ij}. The linear predictor function η_{ij} looks similar to that of an ordinary Poisson GLM, except that we use the log transformed brood size as an offset, and a_i is added at the end, allowing for a different intercept for each nest. We assume that the term a_i is normally distributed with a mean 0 and variance σ^2_{Nest}. Zuur et al. (2009a) used the function `lmer` from the `lme4` package to fit this model. They also used a GAMM with a Poisson distribution. The optimal GAMM is given in Equation (2.6). The shape of the estimated smoother for arrival time was similar to that in Figure 2.2.

$$\begin{aligned} &NCalls_{ij} \sim Poisson(\mu_{ij}) \\ &\log(\mu_{ij}) = \eta_{ij} \\ &\eta_{ij} = \alpha + offset(LBroodSize_{ij}) + \beta_1 \times FoodTreatment_{ij} + f(ArrivalTime_{ij}) + a_i \\ &a_i \sim N(0, \sigma^2_{Nest}) \end{aligned} \tag{2.6}$$

For all these models we plotted residuals versus fitted values. The patterns in these graphs are similar to those in Figure 2.3. The GLMM and GAMM models were overdispersed. A possible reason for overdispersion is zero inflation (Hilbe 2011), and zeros in the response variable are at 25%. In this chapter we will investigate whether zero inflated GLMM and zero inflated GAMM models can be used to improve the models.

It is difficult to say whether 25%, 50%, or 80% zeros makes a data set zero inflated. It also depends on the mean value of the response variable. If the mean, without the zeros, is relatively high (> 10), it is unlikely that a Poisson distribution can be used to model a response variable with 25% zeros, and a zero inflated Poisson distribution (this is a different distribution) may be more suitable. On the other hand, if most observations are < 10, perhaps 25% zeros complies with the distribution.

There is also the problem of dependence. Some nests may be located close to others and the parents may be competing for food and interacting with other owls. Owls breeding close to one another may experience similar environmental conditions. This results in spatial correlation. Spatial correlation was not considered in the models presented in Zuur et al. (2009a).

The aim of this chapter is to investigate whether zero inflated models provide model improvements compared to the GLMMs. We will also investigate potential spatial correlation.

2.1.3 Prerequisite knowledge for this chapter

The required knowledge for this chapter is included in Chapter 5, 8–11, and 13 in Zuur et al. (2009a). Chapter 5 explains mixed effects modelling, Chapters 8–10 introduce generalised linear modelling, Chapter 11 provides a detailed discussion of zero inflated models, and Chapter 13 explains GLMM and GAMM techniques. We also recommend reading Chapter 1 in the current volume as it explains Bayesian statistics. For the data exploration we use techniques described in Zuur et al. (2010). Excellent alternative references are Hilbe (2011) for GLM, Wood (2006) for GAM, Pinheiro and Bates (2000) for mixed modelling, and Kéry (2010) for an introduction to Bayesian statistics. We also recommend Bolker (2008) and Hardin and Hilbe (2007).

2.2 Importing and coding the data

We first import the data and inspect the data frame using standard functions such as `read.table`, `names`, and `str`.

```
> Owls <- read.table("Owls.txt", header = TRUE)
> names(Owls)

[1] "Nest"                "Xcoord"        "Ycoord"
[4] "FoodTreatment"       "SexParent"     "ArrivalTime"
[7] "SiblingNegotiation"  "BroodSize"     "NegPerChick"
```

```
> str(Owls)

'data.frame':    599 obs. of 9 variables:
 $ Nest              : Factor w/ 27 levels  ...
 $ Xcoord            : int  556216 556216 556216 ...
 $ Ycoord            : int  188756 188756 188756 ...
 $ FoodTreatment     : Factor w/ 2 levels  ...
 $ SexParent         : Factor w/ 2 levels  ...
 $ ArrivalTime       : num  22.2 22.4 22.5 22.6 ...
 $ SiblingNegotiation: int  4 0 2 2 2 2 18 4 18 0 ...
 $ BroodSize         : int  5 5 5 5 5 5 5 5 5 5 ...
 $ NegPerChick       : num  0.8 0 0.4 0.4 0.4 0.4 ...
```

The categorical variables `FoodTreatment` and `SexParent` were coded as character strings in the spreadsheet; hence they are imported automatically as factors (categorical variables). As explained in Section 2.1, the variable `NegPerChick` was log transformed and analysed using Gaussian linear mixed effects models in Roulin and Bersier (2007) and Zuur et al. (2009a), and the variable `SiblingNegotiation` was analysed with a Poisson distribution in Chapter 13 in Zuur et al. (2009a). Because the latter variable name is long we rename it:

```
> Owls$NCalls <- Owls$SiblingNegotiation
```

This saves typing and synchronises the R code with the statistical equations.

2.3 Data exploration

Data exploration graphs not included in Chapter 5 of Zuur et al. (2009a) are discussed here. First we present the number of observations per nest:

```
> table(Owls$Nest)

AutavauxTV            Bochet      Champmartin         ChEsard
        28                23               30              20
  Chevroux  CorcellesFavres         Etrabloz           Forel
        10                12               34               4
    Franex              GDLV       Gletterens         Henniez
        26                10               15              13
     Jeuss       LesPlanches           Lucens           Lully
        19                17               29              17
   Marnand            Moutet           Murist          Oleyes
        27                41               24              52
   Payerne            Rueyes            Seiry           SEvaz
        25                17               26               4
   StAubin              Trey         Yvonnand
        23                19               34
```

Note that we have two nests with considerably fewer observations, namely Forel and SEvaz. Before we code zero inflated GLMMs, we first write R code for a Poisson GLMM and estimate the regression parameters in a Bayesian context using WinBUGS (Lunn et al. 2000) as well as using a frequentist approach with the function `lmer` from the `lme4` package (Bates et al. 2011) in R. The estimated random effects obtained by these two techniques are different, and the mixing of the chains in WinBUGS is not good, as there is auto-correlation in the chains. When we omit the two nests with only 4 observations, along with the nests Chevroux and GDLV, the estimated parameters and random effects obtained by the two estimation techniques are similar, and we also obtain a good mixing (no auto-correlation) of the chains. Hence it is practical to drop the 4 nests with 10 or fewer observations (zeros and non-zeros). Note that it is not possible to provide a general cut-off level for the number of observations per level for a random effect term. It depends on the complexity of the model and the data.

The fact that Bayesian and frequentist approaches produce differing results does not, in itself, justify removing nests with limited observations. Discrepancies can arise due to influences of the prior, the way the posterior is characterised, or algorithmic issues. Our approach is pragmatic, but it is better to determine the source of the differences.

The following R code removes the observations of the nests Cheroux, Forel, GDLV, and SEvaz:

```
> Owls2 <- Owls[Owls$Nest != "Chevroux" & Owls$Nest != "Forel" &
                Owls$Nest != "GDLV"     & Owls$Nest != "SEvaz", ]
> Owls2$Nest <- factor(Owls2$Nest)
```

The final line ensures that the four names are also removed from the levels of the factor `Nest`. A design plot is given in Figure 2.4 and an interaction plot in Figure 2.5. There is substantial variation among nests. The interaction plot suggests that the interaction of food treatment and sex of the parent is weak. R code to make these two graphs is:

```
> plot.design(NCalls ~ FoodTreatment + SexParent + Nest,
              data = Owls2, cex.lab = 1.5)
> interaction.plot(Owls2$FoodTreatment, Owls2$SexParent,
             Owls2$NCalls, cex.lab = 1.5,
             ylab = "Sibling negotiation", xlab = "FoodTreatment",)
```

The multi-panel scatterplot in Figure 2.6 shows the relationship between sibling negotiation and arrival time for the food treatment and parent sex combinations. A smoother (locally weighted regression smoother; LOESS) was added to aid visual interpretation. There seems to be a non-linear relationship between sibling negotiation and arrival time. R code for the multi-panel scatterplot is: (detailed explanation in Zuur et al. (2009b)).

```
> library(lattice)
> xyplot(NCalls ~ ArrivalTime | SexParent * FoodTreatment,
    data  = Owls2, ylab = "Sibling negotiation",
    xlab  = "Arrival time (hours)",
    panel = function(x, y){
       panel.grid(h = -1, v = 2)
       panel.points(x, y, col = 1)
       panel.loess(x, y, span = 0.3, col = 1, lwd =2)})
```

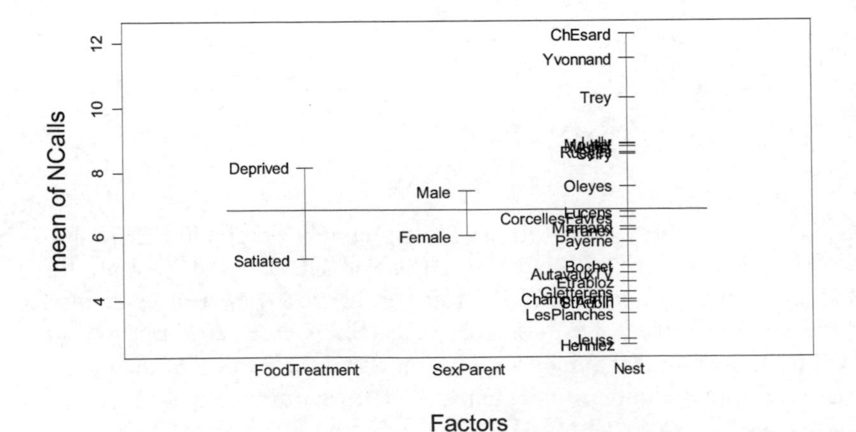

Figure 2.4. Design plot. Each vertical line represents a categorical variable, and the short horizontal lines are mean values per level. The solid horizontal line is the overall average. The variation in mean number of calls per nest is greater than for food treatment or sex of the parent.

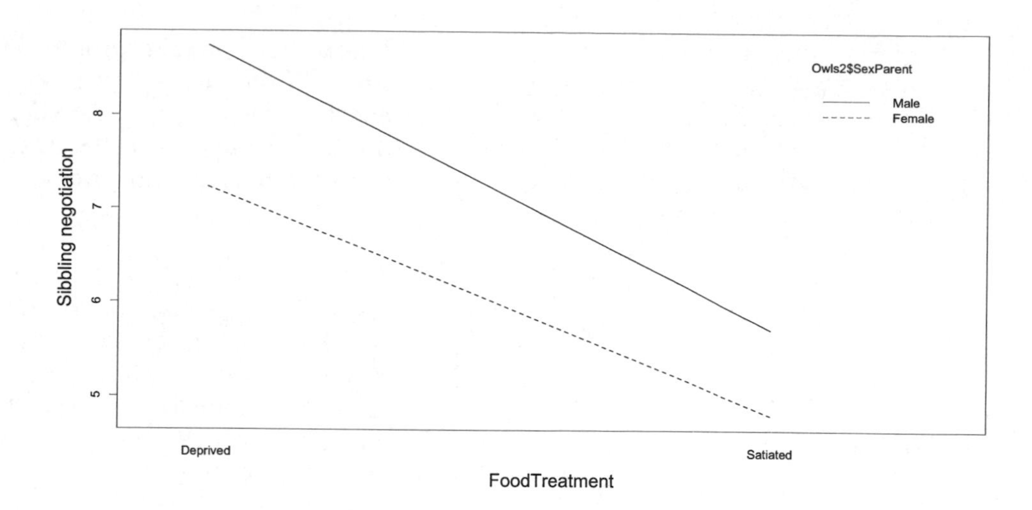

Figure 2.5. Interaction of food treatment and sex of the parent. Mean values per level indicate weak interaction.

Finally we present the spatial position of the sites (Figure 2.7). All sites are within 30 kilometres of one another; hence we may expect some spatial correlation. Spatial correlation may be due to owls interacting with one another (e.g. competing for food), or due to a small-scale covariate (e.g. habitat) effect.

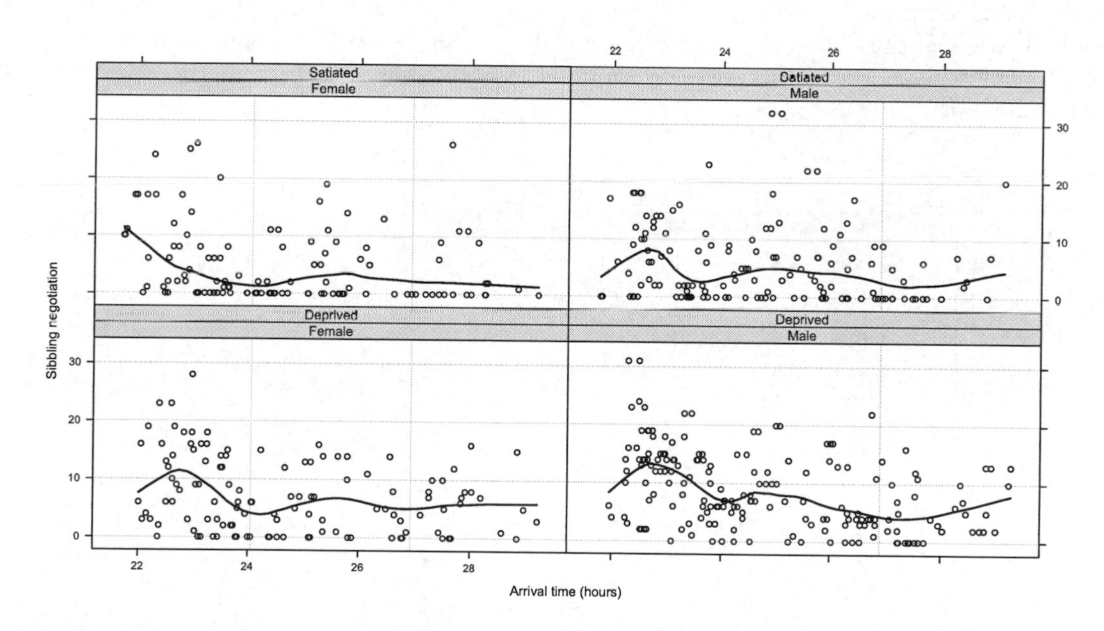

Figure 2.6. Multi-panel scatterplot of sibling negotiation (counts) versus arrival time for varying combinations of food treatment and sex of the parent. A LOESS smoother was added to aid visual interpretation. For instance, the upper left panel (Satiated/Female) indicates the number of negotiation calls per food-satiated nestling in the 30 second period recorded in the 15 minute interval prior to the arrival of their mother at the nest.

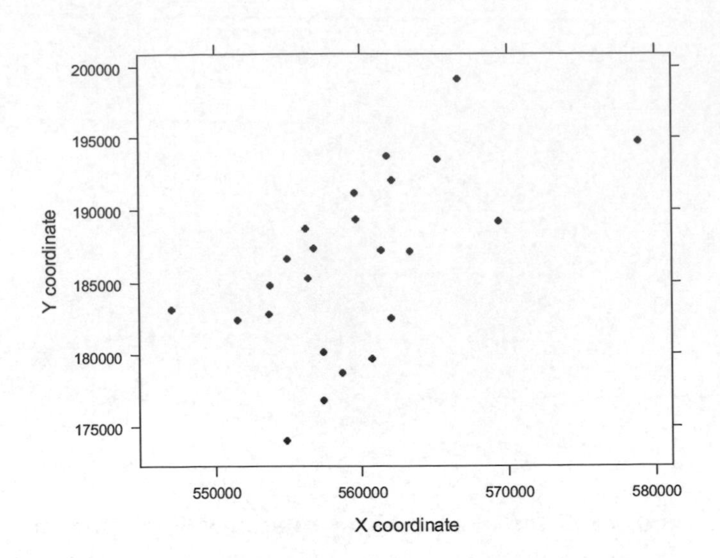

Figure 2.7. Spatial position of the sites. All sites are within 30 km of one another. The `xyplot` function from the `lattice` package (Sarkar 2008) was used to create the graph.

The following three graphs are time series plots. The data are time series data, where arrival time is the time axis. We first plot the number of calls versus arrival time for each nest using the `xyplot` function (Figure 2.8). The number of calls is plotted using vertical lines. At first glance it appears that each panel shows the temporal pattern at a particular nest, but this is incorrect, as each panel represents the data from two nights. Hence two adjoining observations in a panel might be separated by 24 hours. To see the data in the real time, we can either add 24 hours to observations from the second night, or we can make two multi-panel scatterplots, one for the first night and one for the second night. Instead of using night we could use food treatment, as this is different on the two nights. If we redraft Figure 2.8 using only data from a particular food treatment, we can see results at a nest during a single night, although not all nest data is from the same night. These graphs are presented in Figure 2.9 and Figure 2.10. At some nests the chicks are noisy during the night, whereas at other nests the number of calls is recorded only during condensed time periods. A biological explanation for this might be the biomass of the prey. A big mouse may keep the siblings quiet for a while.

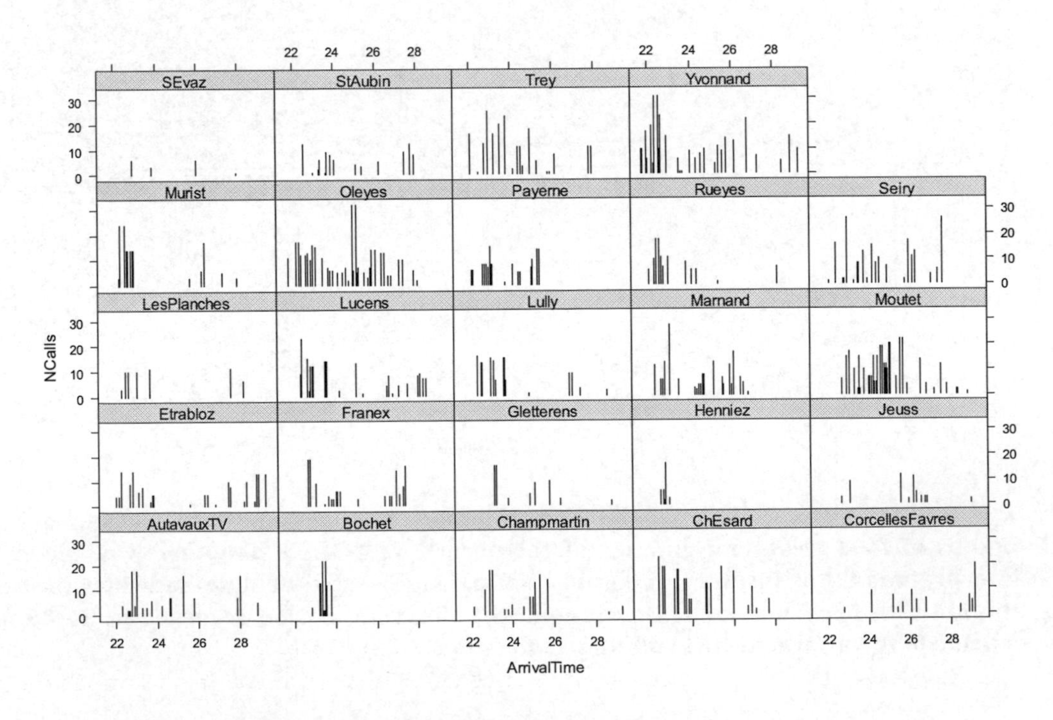

Figure 2.8. Number of calls versus arrival time for each nest.

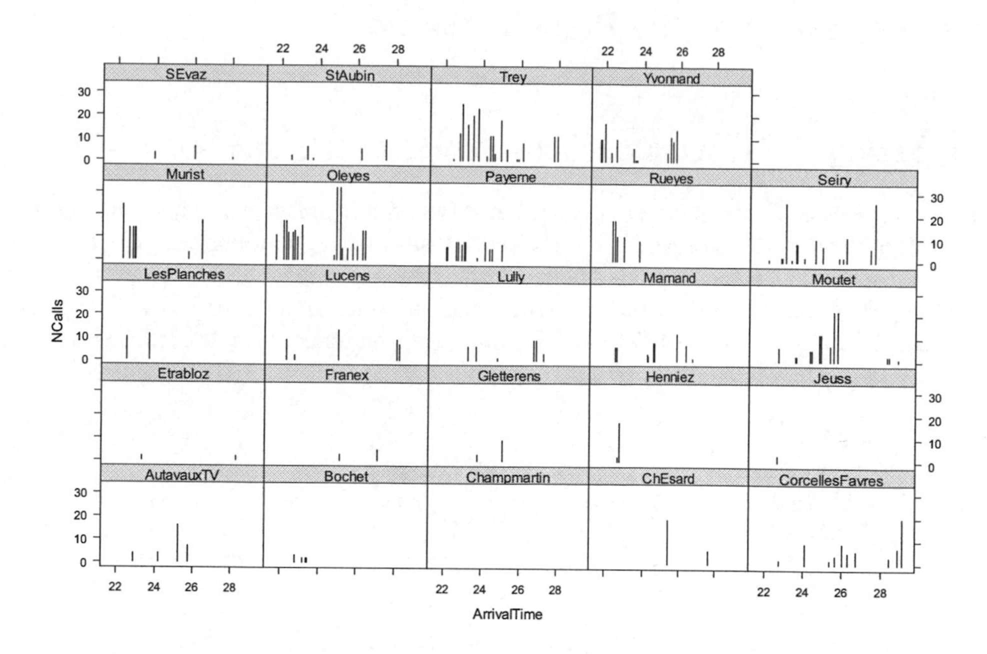

Figure 2.9. Number of calls versus arrival time for each satiated nest.

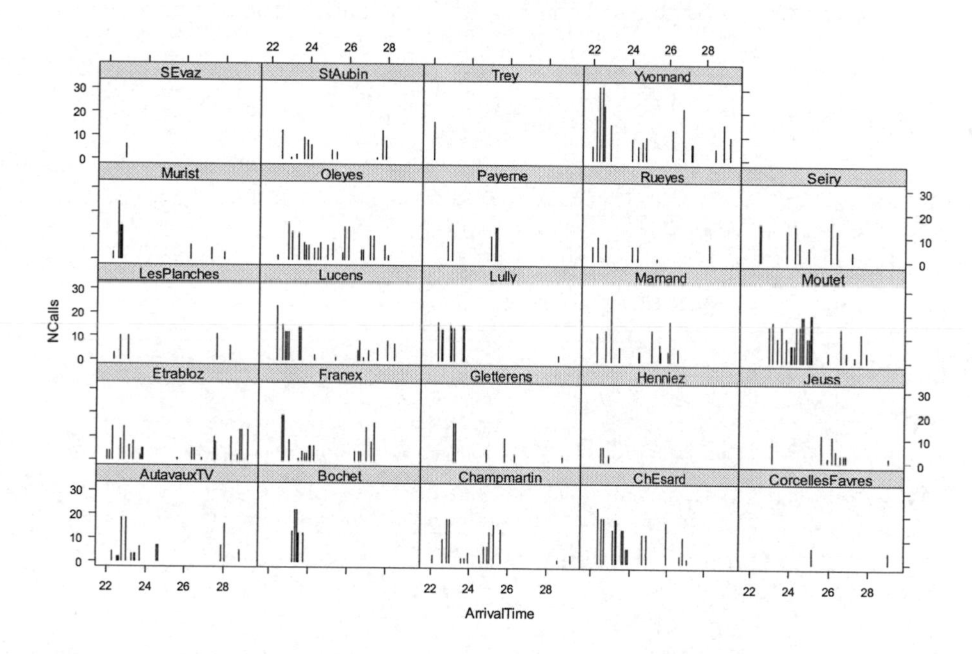

Figure 2.10. Number of calls versus arrival time for each deprived nest.

R code to produce Figure 2.8 is:

```
> library(lattice)
> xyplot(NCalls ~ ArrivalTime | Nest, type = "h", data = Owls2)
```

The `subset` argument in the `xyplot` function was used to generate Figure 2.9 and Figure 2.10. Add `subset = (FoodTreatment == "Satiated")` or `subset = (FoodTreatment == "Deprived")` to the `xyplot` code.

2.4 Overdispersion in the Poisson GLMM

2.4.1 Assessing overdispersion using Pearson residuals

We have multiple observations per nest, and observations from the same nest may be more similar than observations from different nests. To analyse the sibling negotiation data it is therefore obvious that we must use mixed effect models. Because the response variable is a count, we should use a Poisson or negative binomial distribution. It is better to start simple, so we will first consider the Poisson GLMM in Equation (2.5). We will estimate the parameters in this model using the `lmer` function and inspect the residuals for overdispersion.

```
> library(lme4)
> Owls2$LogBroodSize <- log(Owls2$BroodSize)
> M1 <- glmer(NCalls ~ SexParent * FoodTreatment +
        SexParent * ArrivalTime + offset(LogBroodSize) + (1| Nest),
        family = poisson, data = Owls2)
> summary(M1)

Generalized linear mixed model fit by the Laplace approximation
Formula: NCalls ~ SexParent * FoodTreatment +
        SexParent * ArrivalTime + offset(LogBroodSize) + (1 | Nest)

 Data: Owls2
  AIC  BIC logLik deviance
 3209 3239  -1597     3195

Random effects:
 Groups Name        Variance Std.Dev.
 Nest   (Intercept) 0.12802  0.3578
Number of obs: 571, groups: Nest, 23

Fixed effects:
                              Estimate Std. Error z value Pr(>|z|)
(Intercept)                     3.6480     0.3619   10.08   <2e-16
SexParentMale                   0.4907     0.4549    1.08    0.281
FoodTreatmentSatiated          -0.6553     0.0563  -11.64   <2e-16
ArrivalTime                    -0.1218     0.0145   -8.39   <2e-16
SexPMale:FoodTreatmentSatiated  0.1776     0.0710    2.50    0.012
SexPMale:ArrivalTime           -0.0213     0.0186   -1.14    0.254
```

We have shortened the variable names so that the output for the fixed effects fits on a single line. We assess the overdispersion by calculating the residuals, taking its sum of squares, and dividing by $N - p$, where N is the sample size and p the number of parameters (i.e. regression parameters and parameters in the random part of the model). This can be done in R as follows:

```
> E1 <- residuals(M1)
> p1 <- length(fixef(M1)) + 1   #+1 due to random intercept variance
> Overdisp1 <- sum(E1^2) / (nrow(Owls2) - p1)
> Overdisp1

5.538433
```

There is an ongoing discussion of the number of parameters that should be used for the degrees of freedom for statistical inference in mixed effect models and also for the denominator for the overdispersion parameter. The basic source of conflict is that the number of random intercepts is ignored in $N - p$. If we include the number of levels of the random effect term (23) in p, the overdispersion value is 5.77.

2.4.2 Assessing overdispersion using an observation level random effect term

Yet another option for assessing and modelling the overdispersion is to add an extra residual term, ε_{ij}, to the predictor function (Elston et al., 2001), leading to:

$$
\begin{aligned}
&NCalls_{ij} \sim Poisson(\mu_{ij}) \\
&\log(\mu_{ij}) = \eta_{ij} \\
&\eta_{ij} = \alpha + offset(LBroodSize_{ij}) + \beta_1 \times SexParent_{ij} + \\
&\quad \beta_2 \times FoodTreatment_{ij} + \beta_3 \times ArrivalTime_{ij} + \\
&\quad \beta_4 \times SexParent_{ij} \times FoodTreatment_{ij} + \beta_5 \times SexParent_{ij} \times ArrivalTime_{ij} + \\
&\quad a_i + \varepsilon_{ij} \\
&a_i \sim N(0, \sigma^2_{Nest}) \quad \text{and} \quad \varepsilon_{ij} \sim N(0, \sigma^2_{\varepsilon})
\end{aligned}
\tag{2.7}
$$

Perhaps it is easier to reformulate this in words:

$$\log(\mu) = \eta = \text{intercept} + \text{offset} + \text{fixed covariates} + \text{random intercept} + \text{extra noise}$$

The extra noise term, ε_{ij}, acts as a latent variable. It will pick up any variation that cannot be explained by the covariates and random intercepts and would otherwise result in overdispersion. We will demonstrate this with a small simulation study.

2.4.3 Simulation study demonstrating observation level random effect

To explain ε_{ij}, we will simulate some data and show the results of `lmer`. We first set the random seed to a fixed value, enabling replication of results. We then select a sample size of $N = 300$, choose covariate values X from a uniform distribution, select values for the intercept and slope, and simulate Poisson distributed counts Y. The resulting data are plotted in Figure 2.11.

```
> set.seed(12345)
> N     <- 300
> X     <- runif(N, 0, 5)
> alpha <- 3
> beta  <- -1
> eta   <- alpha + beta * X
> mu    <- exp(eta)
> Y     <- rpois(N, lambda = mu)
> plot(x = X, y = Y)
```

If we fit a Poisson GLM on these data we get estimated parameters that match the original variables with only small differences due to drawing from a Poisson distribution (see the `summary` output below).

```
> M1 <- glm(Y ~ X, family = poisson)
> summary(M1)
```

```
Coefficients:
            Estimate Std. Error z value Pr(>|z|)
(Intercept)  3.01454    0.04232   71.23   <2e-16
X           -1.00681    0.03149  -31.98   <2e-16

(Dispersion parameter for poisson family taken to be 1)
    Null deviance: 1922.30  on 299  degrees of freedom
Residual deviance:  274.44  on 298  degrees of freedom
```

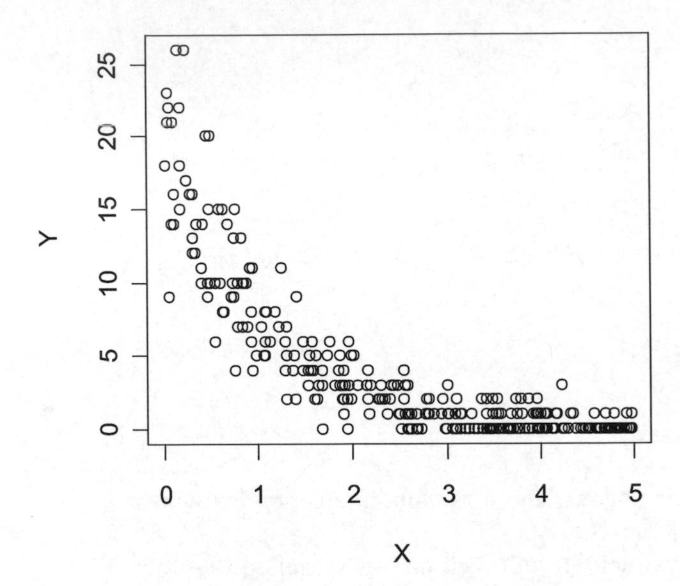

Figure 2.11. Simulated data from a Poisson distribution.

Let us add random intercepts to the data. We take 20 values from a random distribution with mean 0 and variance 0.1^2. Each value is repeated 15 times. This means that the predictor function is of the form:

$$\log(\mu_{ij}) = \alpha + \beta \times X_{ij} + a_i \quad \text{where} \quad a_i \sim N(0, \sigma_a^2)$$

The R code to simulate data for this is:

```
> a1 <- rnorm(20, mean = 0, sd = 0.1)
> a  <- rep(a1, each = 15)
> RE <- rep(1:20, each = 15)
> cbind(a, RE) #Results not printed here
> eta1 <- alpha + beta * X + a
> mu1  <- exp(eta1)
> Y1   <- rpois(N, lambda = mu1)
```

The variable *a* contains the simulated random intercepts, and RE is a variable with the values 1 1 … 1 2 2 2 … 20 20 20. To fit the GLMM in `lmer` we use:

```
> library(lme4)
> M2 <- glmer(Y1 ~ X + (1|RE) , family = poisson)
> summary(M2)

...
Random effects:
 Groups Name        Variance Std.Dev.
 RE     (Intercept) 0.016961 0.13023
```

```
Number of obs: 300, groups: RE, 20

Fixed effects:
             Estimate Std. Error z value Pr(>|z|)
(Intercept)   3.06963    0.05305   57.86   <2e-16
X            -1.03469    0.03259  -31.75   <2e-16
```

The estimated regression parameters are close to the original variables, and the same holds for the estimated variance of the random intercepts. There is no overdispersion, as expected.

```
> E2 <- resid(M2, type = "pearson")
> N  <- length(Y1)
> p  <- length(fixef(M2)) + 1
> sum(E2^2) / (N - p)

0.9334311
```

In the previous subsection we explained that we can add an extra residual term, ε_{ij}, to the predictor function, leading to:

$$
\begin{aligned}
&Y_{ij} \sim Poisson(\mu_{ij}) \\
&\log(\mu_{ij}) = \alpha + \beta \times X_{ij} + a_i + \varepsilon_{ij} \\
&a_i \sim N(0, \sigma_a^2) \quad \text{and} \quad \varepsilon_{ij} \sim N(0, \sigma_\varepsilon^2)
\end{aligned}
$$

If we fit this model on our `Y1` and `X` data, we would expect the variance σ^2_ε to be small, as the generated data has only Poisson variation and variation due to the random intercept a_i. We verify this:

```
> Eps <- 1:length(Y1)
> M3 <- glmer(Y1 ~ X + (1|RE) + (1|Eps) , family = poisson)

Number of levels of a grouping factor for the random effects
is *equal* to n, the number of observations
```

This is a warning message that we can ignore. The output is obtained with:

```
> summary(M3)

Generalized linear mixed model fit by the Laplace approximation
Formula: Y1 ~ X + (1 | RE) + (1 | Eps)
   AIC   BIC logLik deviance
 286.6 301.4 -139.3    278.6

Random effects:
 Groups Name        Variance Std.Dev.
 Eps    (Intercept) 0.000000 0.00000
 RE     (Intercept) 0.016961 0.13023
Number of obs: 300, groups: Eps, 300; RE, 20

Fixed effects:
             Estimate Std. Error z value Pr(>|z|)
(Intercept)   3.06963    0.05305   57.86   <2e-16
X            -1.03470    0.03259  -31.75   <2e-16
```

The estimated σ_ε is small. The likelihood ratio also indicates that we do not need the observation level random intercept, although keep in mind that we are testing on the boundary (Zuur et al., 2009a).

```
> anova(M2, M3)

Models:
M2: Y1 ~ X + (1 | RE)
M3: Y1 ~ X + (1 | RE) + (1 | Eps)

   Df    AIC    BIC  logLik Chisq Chi Df Pr(>Chisq)
M2  3 284.63 295.74 -139.31
M3  4 286.63 301.44 -139.31     0      1          1
```

Now let us add extra noise to the predictor function and create data for which we do need the observation level random intercept.

```
> EpsNoise <- rnorm(N, mean = 0, sd = 0.4)
> eta <- alpha + beta * X + a + EpsNoise
> mu <- exp(eta)
> Y2 <- rpois(N, lambda = mu)
```

Fitting the model with the log link function $\log(\mu_{ij}) = \alpha + \beta \times X_{ij} + a_i$ gives an overdispersion parameter of 1.28 and an estimated value of σ_a of 0.148. This indicates a small amount of overdispersion. The intercept and slope are slightly different from the original values.

```
> M4 <- glmer(Y2 ~ X + (1|RE), family = poisson)
> summary(M4) # Results not printed here
```

Finally, we fit the model with the random intercept and the extra residual term:

```
> M5 <- glmer(Y2 ~ X + (1|RE) + (1|Eps) , family = poisson)
```

The AIC of model `M5` is lower than that of `M4`, and the likelihood ratio test indicated that σ_ε is significantly different from 0 at the 5% level. Hence model `M5` is better than `M4`. The estimated parameters are:

```
> summary(M5)

Generalized linear mixed model fit by the Laplace approximation
Formula: Y2 ~ X + (1 | RE) + (1 | Eps)
   AIC BIC logLik deviance
 397.1 412 -194.6    389.1

Random effects:
 Groups Name        Variance  Std.Dev.
 Eps    (Intercept) 0.1455394 0.381496
 RE     (Intercept) 0.0023355 0.048327
Number of obs: 300, groups: Eps, 300; RE, 20

Fixed effects:
            Estimate Std. Error z value Pr(>|z|)
(Intercept)  3.10704    0.06971   44.57   <2e-16
X           -1.05141    0.04020  -26.15   <2e-16
```

The estimated variance of the extra noise term is close to the original value, but the variance of the random intercept is half of the original value. Repeat running of the code (including the `rnorm` and `rpois` commands, but not the `set.seed` command) will give an indication of how much the estimated variances change due to the random sampling from a Poisson distribution and the two normal distributions. It is a useful exercise to vary the values of σ_a and σ_ε to see how the overdispersion and estimated parameters change. Also try changing the sample size. If you run the code again, the estimated variance terms will change slightly due to the variation caused by drawing random numbers from the Poisson and the two normal distributions. We therefore program a loop around our code and repeat the process 100 times, each time extracting the estimated σ_a and σ_ε. The average of the 100 estimated values for σ_a is 0.10, and the average of the 100 estimated σ_ε is 0.40. So they are close to the original values of 0.10 and 0.40, although not identical if more decimals places are used.

We also investigate what happens if we change the ratios of these two variances. We put a double loop around our 100 simulations and use different values for σ_a and σ_ε to generate the counts. For both σ_a and σ_ε we use the values 0.1, 0.3, 0.5, 0.7, 0.9, 1.5, 2, and 3. Hence we generate data under 64 variance combinations. The sketch of the simulation study is:

For any value of σ_a
  For any value of σ_ε
    Simulate 100 data sets (as described above).
    Apply a GLMM with random intercept and individual level random intercept on each data set.
    Estimate σ_a and σ_ε for each data set and store these values.
  }
}

The average of the 100 estimated values of σ_ε are presented in Table 2.1. If $\sigma_a = 0.1$ and $\sigma_\varepsilon = 0.1$, the average of the 100 estimated σ_ε values is 0.11. To interpret this table, look at the values along the columns and compare them to the original σ_ε in the column heading. Based on the values in this table, it seems that the estimated parameters of the GLMM with the observation level random intercept are unbiased, at least for the scenario that we use here: a single covariate with 300 observations and no strong residual patterns due to non-linear covariate effects. Table 2.2 shows the average of the 100 estimated σ_a values. If $\sigma_a = 0.1$ and $\sigma_\varepsilon = 3$, the average of the 100 estimated σ_a values is 0.15. To interpret this table, look along the rows and compare the values to the row headings. Note that the average of the 100 estimated σ_a can be as much as 30% different from the original value. Most of the values are approximately 10% different from the original values. The upper right corner of Table 2.2 shows that for a small σ_a and alarge σ_ε the mismatch between the estimated and original σ_a is larger.

We suggest repeating this simulation study with different values for σ_a and σ_ε using 1000 simulated data sets.

Table 2.1. Average of 100 estimated values of σ_ε.

	Values for σ_ε							
Values for σ_a	0.1	0.3	0.5	0.7	0.9	1.5	2	3
0.1	0.11	0.30	0.49	0.70	0.88	1.5	2.0	3.0
0.3	0.01	0.29	0.50	0.72	0.91	1.5	2.0	3.1
0.5	0.10	0.29	0.50	0.69	0.89	1.5	2.0	3.1
0.7	0.11	0.30	0.49	0.71	0.89	1.5	2.0	2.9
0.9	0.11	0.30	0.50	0.73	0.90	1.5	2.0	3.0
1.5	0.11	0.33	0.54	0.70	0.91	1.5	2.0	3.3
2	0.10	0.230	0.49	0.69	0.88	1.5	2.0	3.0
3	0.11	0.29	0.51	0.70	0.89	1.5	2.0	2.9

Table 2.2. Average of 100 estimated values of σ_a.

	Values for σ_ε							
Values for σ_a	0.1	0.3	0.5	0.7	0.9	1.5	2	3
0.1	0.09	0.09	0.10	0.11	0.11	0.11	0.09	0.13
0.3	0.26	0.26	0.27	0.25	0.22	0.20	0.18	0.08
0.5	0.48	0.48	0.47	0.47	0.47	0.41	0.32	0.27
0.7	0.75	0.74	0.74	0.75	0.76	0.70	0.67	0.48
0.9	0.71	0.70	0.71	0.71	0.70	0.67	0.64	0.54
1.5	1.13	1.13	1.12	0.11	1.11	1.07	1.05	0.90
2	2.17	2.18	2.20	2.15	2.15	2.15	2.12	2.13
3	2.75	2.74	2.75	2.74	2.79	2.75	2.75	2.70

2.4.4 A GLMM with observation level random effect

The following R code applies the GLMM with a random effect nest and an observation level random intercept in Equation (2.7) to the owl data. The likelihood ratio test indicates that the σ_ε is significantly different from 0 at the 5% level.

```
> Eps <- 1:nrow(Owls2)
> M2 <- glmer(NCalls ~ SexParent * FoodTreatment +
              SexParent * ArrivalTime + offset(LogBroodSize) +
              (1| Nest) + (1|Eps), family = poisson, data = Owls2)
> anova(M1,M2)

Data: Owls2
Models:
M1: NCalls ~ SexParent * FoodTreatment + SexParent * ArrivalTime +
             offset(LogBroodSize) + (1 | Nest)
M2: NCalls ~ SexParent * FoodTreatment + SexParent * ArrivalTime +
             offset(LogBroodSize) + (1 | Nest) + (1 | Eps)

   Df    AIC    BIC   logLik  Chisq Chi Df Pr(>Chisq)
M1  7 3208.8 3239.3 -1597.42
M2  8 1776.3 1811.0  -880.13 1434.6      1  < 2.2e-16
```

We cannot use the Pearson residuals of the model M2 to assess whether we still have overdispersion. In Equation (2.7) we model the overdispersion with a latent process ε_{ij}, but it does not allow us to determine why we have the overdispersion. Instead of adding an extra residual term to capture the overdispersion, we prefer to investigate its cause. Is it because of outliers or non-linear patterns in the residuals? Are we using the wrong link function? Should we use a different distribution? Is there correlation in the data, missing covariates, or do we have zero inflation? Hilbe (2011) provides an extensive list of possible causes of overdispersion.

In this case there are no obvious outliers, as can be seen in the multi-panel scatterplots in Figure 2.6. We applied a GAMM to these data (see Zuur et al., 2009a), but it also showed overdispersion. A GAMM allows for non-linear patterns, so that is not the source of the overdispersion. The random intercept introduces a correlation between observations from the same nest, so, in theory, correlation has been dealt with and cannot be the cause of the overdispersion. However, an option may be to implement a more advanced correlation structure, e.g. an auto-regressive correlation of order 1. This is difficult, as the data are irregularly spaced. There may be spatial correlation among the nests, something we will investigate later, but will see that it is of no concern. Given the fact that 25% of the data equals zero, it is possible that zero inflation may be the source of the overdispersion, and we will investigate this in the following sections.

2.5 Why zero inflated models?

In the previous section we saw that the Poisson GLMM with a random intercept for nest showed overdispersion. Instead of focussing on the GLMM with an observation level random intercept, we ask ourselves what is driving the overdispersion. Could the overdispersion in the GLMM be due to zero inflation?

Kuhnert et al. (2005) and Martin et al. (2005) make a distinction between true zeros and false zeros, quantities that may be of interest or may be nuisance parameters. False zeros are recorded zeros that are not really zeros. This may occur due to a poor experimental design (e.g. the sampling period is too short) or errors in observation (counting zero calls whereas, in reality, there were calls). In the owl study the observers are excellent scientists and the experimental design is as good as it can be. However, the sentence describing the response variable states, 'Sibling negotiation was defined as the number of calls made by all offspring, in the absence of the parents, during 30 second time periods recorded at 15 minute intervals. The number of calls from the recorded period in the 15 minutes preceding the arrival was allocated to a visit from a parent.' It could be that the siblings were noisy in the 4 minutes before arrival of the parent, but that the 30-second sampling period was at 5 minutes before, and chicks were quiet for that 30 seconds (Figure 2.12). In this case the measured number of calls is 0, but we argue that this is a false zero because the sampling period was too short. To avoid this, longer time periods should be sampled, but this considerably increases the amount of time required for the experiment and may be impractical.

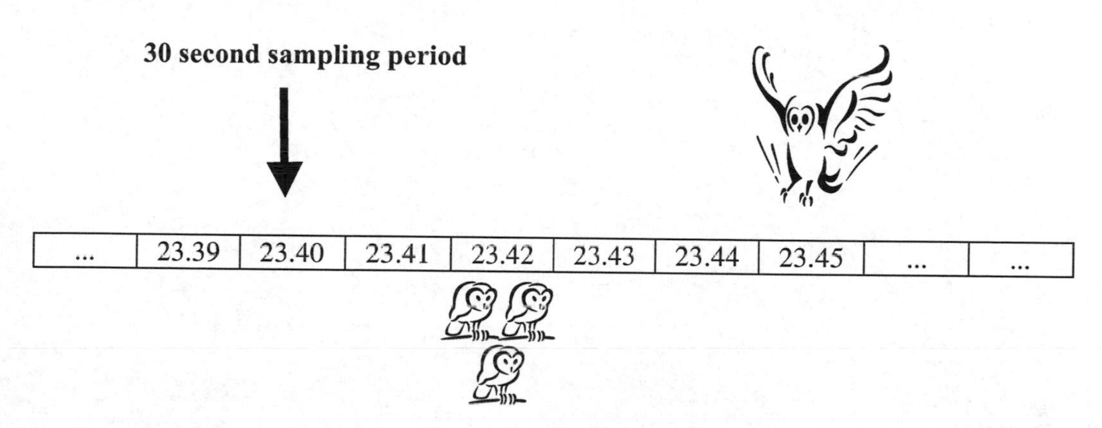

Figure 2.12. A parent arrives at 23.45 at a nest. The number of calls allocated to this observation is taken from a single 30 second period made between 23.30 and 23.45. Suppose this was at 23.40, the chicks were quiet during this period, but made noise at 23.42. It would be recorded as 0 calls.

Figure 2.13 gives a schematic overview of the models for zero inflated data. There are two processes taking place. One process is false zeros versus all other data (true zeros and the non-zero data). We will call the 'other data' the *count process*. The count process consists of zeros and positive counts. In the current context, the true zeros are situations where the chicks were quiet, perhaps because they just had food, or because they were satiated.

Instead of applying GLMM with the observation level random intercept, perhaps we should start with a zero inflated model. The applicable zero inflated Poisson (ZIP) model is given in Equation (2.8). To model the number of calls and the positive zeros (lower branch in Figure 2.13), Poisson GLM is used. The log link is used with the same predictor function as in Equation (2.5).

$$
\begin{aligned}
& NCalls_{ij} \sim ZIP(\mu_{ij}, \pi) \\
& \log(\mu_{ij}) = \eta_{ij} \\
& \eta_{ij} = \alpha + offset(LBroodSize_{ij}) + \beta_1 \times SexParent_{ij} + \beta_2 \times FoodTreatment_{ij} + \\
& \quad \beta_3 \times ArrivalTime_{ij} + \beta_4 \times SexParent_{ij} \times FoodTreatment_{ij} + \\
& \quad \beta_5 \times SexParent_{ij} \times ArrivalTime_{ij} + a_i \\
& a_i \sim N(0, \sigma^2_{Nest}) \\
& \text{logit}(\pi) = \gamma_1
\end{aligned} \tag{2.8}
$$

Figure 2.13. Visualisation of the zero inflation process in the owl data. The sampling period of 30 seconds may be too short to record sibling calls. The presence of false zeros is not a *requirement* for the application of ZIP models, see Chapter 10 for a discussion.

So what are potential sources of the false zeros? We have discussed the possibility that the sampling period was too short. Another scenario may be that the chicks change their negotiation patterns during the night. Perhaps the calls become more clustered later in the night or due to the food treatment. In this case one of the covariates can be used to model the probability that a zero is a false zero. This means that we can also use covariates in the predictor function for the logistic link function. However, this increases the complexity of the model and may cause estimation problems. For the moment, we will assume that the probability of a false zero, π, does not change. This means that we use an intercept γ_1 in the predictor function of the logistic link function. Note that if its estimated value is highly negative, the ZIP GLMM converges to a Poisson GLMM.

We leave it as an exercise for the reader to extend the logistic link function with a random effect and covariates. However, biological knowledge is critical in this process. Suppose that food deprivation causes chicks to be noisier in clustered time periods. This would increase the probability of false zeros (measuring a 0 whereas in reality they did make calls). Such a hypothesis would mean that we could include food treatment as a covariate in the logistic link function and test its significance.

Instead of a parametric model for the counts we could also use a predictor function with the smoothing function for arrival time, leading to:

$$\begin{aligned}
&NCalls_{ij} \sim ZIP(\mu_{ij},\pi) \\
&\log(\mu_{ij}) = \eta_{ij} \\
&\eta_{ij} = \alpha + offset(LBroodSize_{ij}) + \beta_1 \times SexParent_{ij} + \beta_2 \times FoodTreatment_{ij} + \\
&\qquad \beta_3 \times SexParent_{ij} \times FoodTreatment_{ij} + f(ArrivalTime_{ij}) + a_i \\
&a_i \sim N(0,\sigma^2_{Nest}) \\
&\text{logit}(\pi) = \gamma_1
\end{aligned} \tag{2.9}$$

The models in Equations (2.8) and (2.9) contain a random intercept for `nest`, which introduces a correlation among observations from an individual nest (see Chapter 4 for the precise form of this correlation for a ZIP model). However the models in Equations (2.8) and (2.9) do not allow for spatial correlation. We will extract the residuals of the ZIP GLMM and inspect these for spatial correlation, hoping that there is none. If spatial correlation is present, we will need to further extend the models to allow for it.

The parameters in the Poisson GLMM can be estimated with `lmer`, as we showed earlier, and the `gamm` function in `mgcv` can be used for the Poisson GAMM. However, ZIP GLMMs and ZIP GAMMs cannot be fitted with these packages. Ordinary ZIP GLMs can be fitted with the function `zeroinfl` from the `pscl` package (Zeileis et al., 2008), but this function cannot deal with random effects. We will use the R package `R2WinBUGS` (Sturtz et al., 2005) to fit ZIP GLMMs and ZIP GAMMs, which will also allow us to add more complicated spatial correlation structures, should this be needed. Please ensure that you are familiar with the material presented in Chapter 1 before going on to the following section.

2.6 Implementing a Poisson GLM in WinBUGS

Before attempting the ZIP GLMM in Equation (2.7) in WinBUGS, we will start simple and implement the following Poisson GLM:

$$\begin{aligned}
&NCalls_{ij} \sim Poisson(\mu_{ij}) \\
&\log(\mu_{ij}) = \eta_{ij} \\
&\eta_{ij} = \alpha + offset(LBroodSize_{ij}) + \beta_1 \times SexParent_{ij} + \beta_2 \times FoodTreatment_{ij} + \\
&\qquad \beta_3 \times ArrivalTime_{ij} + \beta_4 \times SexParent_{ij} \times FoodTreatment_{ij} + \\
&\qquad \beta_5 \times SexParent_{ij} \times ArrivalTime_{ij}
\end{aligned} \tag{2.10}$$

Why would we start with a Poisson GLM if we want to create a zero inflated Poisson GLMM? The issue is that programming code in WinBUGS is challenging. Once the GLM is running, we will add a random intercept so that we have a Poisson GLMM, and once the Poisson GLMM is running, we will add the zero inflation information. For the enthusiast we will discuss the ZIP GAMM at the end of this chapter.

2.6.1 Converting vectors to matrices

In the statistical equations presented earlier we used the notation NCalls_{ij}, where j is the jth observation of nest i. A complication arises in that the `NCall` variable in R contains all the data in a single long vector. For a GLM model this is not a problem, but the intent is to construct a GLMM (in the next section), and the WinBUGS code would be more straightforward if we can use a double loop, one for the *i* index and one for the *j* index, and create a two-dimensional matrix. For balanced data we can convert a vector to a matrix with one command, for example, with `as.matrix`. The problem here is that the number of observations per nest differs, so we have written a simple R function to convert a vector to a matrix. As input it needs a vector with data that we want to convert to a matrix as well as a variable telling which observations belong to a given nest:

```
> Vec2Mat <- function(FNest, FY){
   AllNests <- levels(FNest)
   NNest    <- length(unique(FNest))
   MaxObs   <- max(table(FNest))
   Y.ij     <- matrix(NA, nrow = MaxObs, ncol = NNest)
   for (i in 1:NNest) {
     Selection     <- FNest == AllNests[i]
     ni            <- sum(Selection)
     Y.ij[1:ni,i] <- FY[Selection]
   }
   Y.ij}
```

First, we determine the names of the nests (`AllNests`), the number of nests (23), and the maximum number of observations for a nest (`MaxObs`, with a value of 52). We then create a matrix of dimension 52 by 23 (`Y.ij`) and fill it out with NAs. Finally we initiate a loop, extract the data from nest *i*, and store it in the matrix `Y.ij` in rows 1 to `ni`, where `ni` is the number of observations for nest *i*. So, `Y.ij` will be a matrix where each column contains the data of a single nest. Because each nest has a different number of observations, `Y.ij` is filled with NAs at the bottom of columns showing data for nests with lower numbers of observations. The output of the function is the matrix. We can execute our function with:

```
> NCalls.ij <- Vec2Mat(Owls2$Nest, Owls2$NCalls)
```

We can inspect the first few rows and columns of `NCalls.ij` to ensure that it works properly:

```
> NCalls.ij[1 : 15, 1 : 5]

      [,1] [,2] [,3] [,4] [,5]
 [1,]    4    3    3   23    2
 [2,]    0    2    0   19    9
 [3,]    2   13    0   19    7
 [4,]    2   13   10   14    2
 [5,]    2   22   18   18    4
 [6,]    2    2   18   18    9
 [7,]   18    2    0   14    5
 [8,]    4    2    0   14    6
 [9,]   18   22    2    6    3
[10,]    0   12    2    6    7
[11,]    0    0    4   12    5
[12,]    3   12    0   12   20
[13,]    0   12    0   12   NA
[14,]    3    0    3   19   NA
[15,]    3    0    7    0   NA
```

For nest 5 we have only 12 observations, which are stored in the first 12 rows. The other rows for this column are NAs. We need to do this conversion for each variable in the model:

```
> LBroodSize.ij     <- Vec2Mat(Owls2$Nest, Owls2$LogBroodSize)
```

```
> SexParent.ij      <- Vec2Mat(Owls2$Nest, Owls2$SexParent)
> FoodTreatmwnt.ij <- Vec2Mat(Owls2$Nest, Owls2$FoodTreatment)
```

WinBUGS will carry out simulations and draw samples. To ensure that these samples are unrelated, it is advisable to centre continuous covariates (i.e. subtract its mean) and even better to standardise them (subtract its mean and divide by the standard deviation):

```
> Owls2$Cen.AT    <- Owls2$ArrivalTime - mean(Owls2$ArrivalTime)
> ArrivalTime.ij <- Vec2Mat(Owls2$Nest, Owls2$Cen.AT)
```

Sex of the parent and food treatment are categorical variables, and our function converts their levels to ones and twos. Using the WinBUGS code is less complicated if these values are zeros and ones; therefore we subtract the value of 1 from both matrices. The reason that we need so-called 0-1 dummy variables is explained in Chapter 5 in Zuur et al. (2007). The R coding is not elegant, but it meets our requirements:

```
> iSexParent.ij      <- SexParent.ij -1
> iFoodTreatment.ij <- FoodTreatment.ij - 1
```

The value of 0 for food treatment corresponds to deprived, and the value of 0 for sex of the parent represents a female. The final information needed is the number of observations per nest:

```
> NObservationsInNest <- as.numeric(tapply(Owls2$NCalls,
                              FUN = length, INDEX = Owls2$Nest))
```

2.6.2 Data for WinBUGS

We will now build up code for WinBUGS. First, we need the input variables in a list.

```
> win.data <- list(
      NCalls              = NCalls.ij,
      ArrivalTime         = ArrivalTime.ij,
      LBroodSize          = LBroodSize.ij,
      NNest               = length(levels(Owls2$Nest)),
      NObservationsInNest = NObservationsInNest,
      iSexParent          = iSexParent.ij,
      iFoodTreatment      = iFoodTreatment.ij)
```

The object `win.data` contains the response variable, all covariates, and additional information such as the number of observations per nest.

2.6.3 Modelling code for WinBUGS

The essential code for the model is:

```
sink("modelglm.txt")
cat("
model{

#Priors
for (i in 1:5) { beta[i]  ~ dnorm(0, 0.001) }
alpha ~ dnorm(0, 0.001)
```

```
#Likelihood
for (i in 1:NNest) {
 for (j in 1:NObservationsInNest[i]) {
   NCalls[j,i] ~ dpois(mu[j,i])
   log(mu[j,i]) <- max(-20, min(20, eta[j,i]))
   eta[j,i] <- alpha + beta[1] * iSexParent[j,i] +
                     beta[2] * iFoodTreatment[j,i] +
                     beta[3] * ArrivalTime[j,i] +
                     beta[4] * iFoodTreatment[j,i] * iSexParent[j,i] +
                     beta[5] * iSexParent[j,i] * ArrivalTime[j,i] +
                     1 * LBroodSize[j,i]
 }}}
",fill = TRUE)
sink()
```

This is a great deal of information to absorb. The `sink` and `cat` functions write the entire block of WinBUGS code in an ascii file called `modelglm.txt`. The alternative is to copy and paste the text from "`model{`" to the final curly bracket into an ascii file. Within the "model{}" block of code there are essentially two blocks of code. Later we will add more blocks, but this code is the critical part.

The `Likelihood` part looks reasonably familiar. We can recognise the predictor function η_{ij} (note that we indexed it the other way around) with the offset and all the covariate terms exactly as defined in Equation (2.10). There are also two loops: i for the nests and j for the observations within a nest. According to Kéry (2010, pg 280) the manner in which the loops are built can make a considerable difference in computing speed. He recommends placing the loop for the longest index first and that for the shortest index last. In this case the i loop ranges from 1 to 23 and the j loop from 1 to the number of observations per nest, which can be anything from 13 to 52. Hence it will possibly be faster to exchange the order of the i and j loops.

The log link function contains a `max` and a `min` function. The MCMC algorithm will create random samples for the regression parameters `alpha` and `beta[1]` to `beta[5]`, corresponding to β_1 to β_5. The `max` and `min` functions ensure that, for a particular draw, we will not get a value for `eta[j,i]` (η_{ij} in the statistical equation) sufficiently large to cause numerical problems due to the exponential function.

The `Priors` part specifies the prior distribution for the regression parameters. We use uninformative prior distributions for all parameters. Note that the `dnorm` notation used in WinBUGS is different from that used in R. The most important difference is that the variance in the `dnorm` function is specified as 1/variance. This is a Bayesian convention with 1/variance being the so-called precision of the distribution, usually denoted by τ. So a variance of 10^3 is entered into the model framework as a precision of 10^{-3}. The reason for working with precision is that the posterior distribution of our parameters of interest will be a weighted combination of the prior distribution and the distribution of the data, with weights given by their respective precisions (i.e. 1/prior variance and 1/data variance). Thus, a large prior variance means that its precision is close to zero, and, as a consequence, the prior distribution will receive almost no weight in determining the posterior distribution of the parameters. This again reflects our non-informative prior distribution, and its contribution to the posterior outcome is negligible. As a consequence, our posterior distribution should be similar to that obtained from maximum likelihood estimation.

2.6.4 Initialising the chains

Having formulated the model, we can now generate a sample from the posterior distribution using MCMC techniques. As described in Chapter 1, from a given starting value $\boldsymbol{\theta}^1$, the MCMC routine will generate values $\boldsymbol{\theta}^2$, $\boldsymbol{\theta}^3$, etc. When the chain is run for a sufficient period of time, it will have 'forgotten' the initial value, and, from that point on, the values drawn will represent samples from our posterior distribution of interest. We use three chains, each of which is initialised with:

```
> inits <- function () {
    list(alpha  = rnorm(1),
         beta   = rnorm(5)) }
```

For alpha we draw one sample from a normal distribution with a mean 0 and variance 1. Beta is a vector of length 5, also with values from a normal distribution. Instead of initial values from a normal distribution, we could use estimated regression parameters obtained by a linear regression model, and add some random noise to the estimated parameters. We could also use any other type of random number generator, provided the values are realistic.

2.6.5 Parameters, thinning rate and length of the chains

We need to tell WinBUGS from which parameters to save the simulated values (chains). For the moment we want only output for alpha and beta:

```
> params <- c("alpha", "beta")
```

We also need to specify the number of chains, the length of the chains, the burn-in, and the thinning rate:

```
> nc <- 3
> ni <- 100000
> nb <- 10000
> nt <- 100
```

This means that we will be using 3 chains each of length 100,000. The burn-in is 10,000; hence we will discard the first 10,000 samples. Because sequential draws may be auto-correlated, we will store every 100th realisation. As a result we will have 3 × (100,000 – 10,000) / 100 = 2,700 observations for the posterior distribution.

2.6.6 Starting WinBUGS from within R

We are now ready to run WinBUGS from within R.

```
> library(R2WinBUGS)
> out1 <- bugs(data    = win.data,
               inits      = inits,
               parameters = params,
               model      = "modelglm.txt",
               n.thin     = nt,
               n.chains   = nc,
               n.burnin   = nb,
               n.iter     = ni,
               debug      = TRUE)
```

When this code is copied and pasted into R, WinBUGS appears as a pop-up window, and in the lower left corner it states 'Model is updating.' From this point it will run fine, and you can make some coffee. Depending on the speed of your computer, you may want to go to the beach as well. However, you may encounter numerical problems, non-convergence of the chains, etc. The debug = TRUE option ensures that WinBUGS stays open when it is finished (allowing a first impression of the results). If it is set to FALSE, WinBUGS will automatically close when it is finished and the R console will appear.

2.6.7 Assessing convergence of the chains

We will close WinBUGS when it is finished and assess the convergence of the chains in R. The object `out1$sims.array` contains all the saved draws from the posterior distribution, and we can use it to see whether the chains have converged. For example, the following code extracts the three chains for `beta[1]`:

```
> Chains <- out1$sims.array[,, beta[1]]
```

It is tedious to type this for each parameter, and we prefer to write code that extracts each of the three chains for every parameter and plots them against the index number. We have 5 betas, hence our loop goes from 1 to 5. This would need to be changed for models with a different number of regression parameters. The `paste` command is used to access parameter `beta[,i]` from the object `out1$sims.array`.

```
for (i in 1:5) {
  par(mfrow = c(1,1))
  Chains <- out1$sims.array[,, paste("beta", "[", i, "]", sep = "")]
  plot(Chains[,1], type = "l", col = 1, ylab = "",
         main = paste("beta", i, sep = ""))
  lines(Chains[,2], col = 2)
  lines(Chains[,3], col = 3)
  win.graph() }
```

This produces 5 graphic panels, one for each regression parameter. One of these graphs is shown in Figure 2.14. It contains three lines, one for each chain. In R these are in colour, but we converted them to greyscale in Word. The key question is whether we can see a pattern in these realisations. If a pattern is apparent, rerunning the model with a higher thinning rate may help, although there may be other factors causing the patterns. Assessing whether there is a pattern in the graphs is, to some extent, subjective. In this case they appear satisfactory, although frequently these graphs look acceptable, but the next tool, the auto-correlation function, indicates problems.

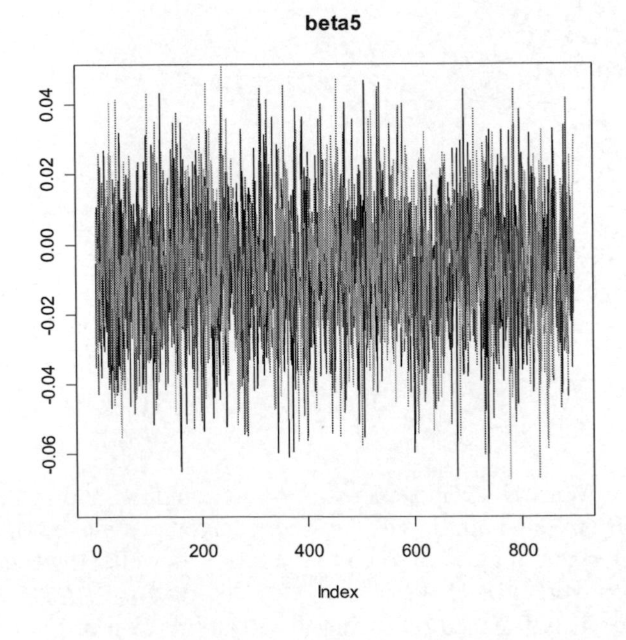

Figure 2.14. Samples for each of the three chains for β_5 overlaid.

A more useful tool is the auto-correlation function applied to each chain (Figure 2.15). There is no significant auto-correlation, so that is good news. We must use the `sims.array` data to assess auto-correlation, as the order of the rows in this object match the order of the iterations.

R code for Figure 2.15 follows. Note that the samples in the object `out1$sims.array` are in the original order, so we can apply the `acf` function to it.

```
for (i in 1:5) {
   par(mfrow = c(2,2))
   Chains <- out1$sims.array[,,paste("beta", "[", i, "]", sep = "")]
```

```
acf(Chains[,1], cex.lab = 1.5,
    main = paste("ACF ", "beta", i, " chain 1", sep = ""))
acf(Chains[,2], cex.lab = 1.5,
    main = paste("ACF ", "beta", i, " chain 2", sep = ""))
acf(Chains[,3], cex.lab = 1.5,
    main = paste("ACF ", "beta", i, " chain 3", sep = ""))
win.graph()}
```

More formal tests to assess the chain convergence are available, and are provided in the R package CODA (Plummer et al., 2006). One such test is the Gelman-Rubin statistic, which compares the variation between chains to the variation within a chain. Initially, the value of the Gelman-Rubin statistic will be large, but when convergence has been reached, it will have decreased to a value close to 1. Gelman (1996) suggests a value less than 1.1 or 1.2 is acceptable. This test should be applied to each of the model parameters. The value of the Gelman-Rubin test is presented as `Rhat` in the output presented in the following subsection. All values are less than 1.002.

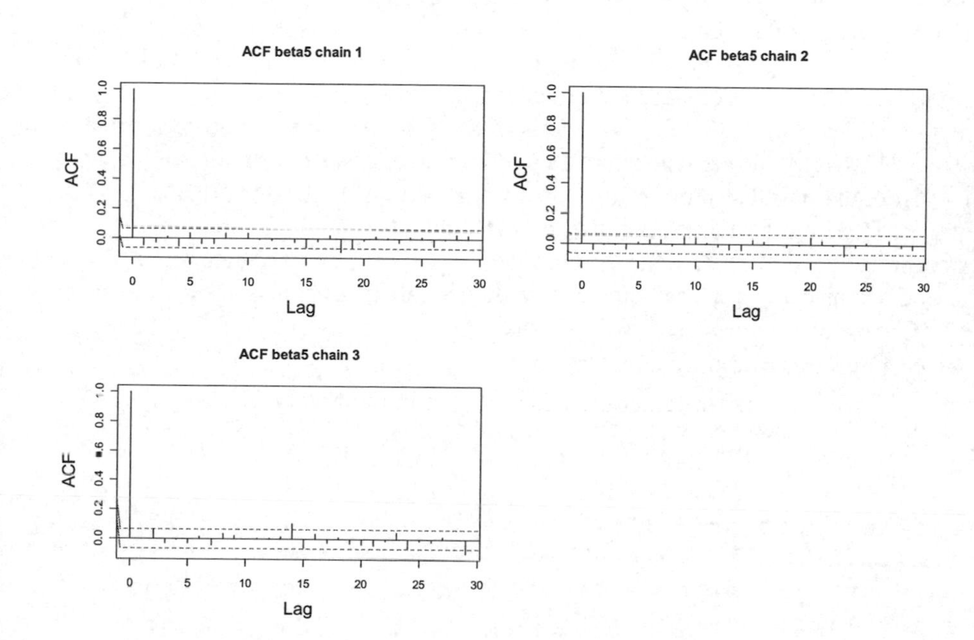

Figure 2.15. Autocorrelation functions for each of the three chains for β_5. Dotted lines are 95% confidence bands and all autocorrelations (except at lag 0) should fall within these bands.

2.6.8 Summarising the posterior distributions

The samples from the posterior distribution are summarised by the mean, the median, and the 2.5 and 97.5 percentiles. The standard deviation of the posterior distribution, given under the header `sd`, is the Bayesian equivalent of the standard error of the mean (recall that the standard error of the mean is defined as the standard deviation of the mean values if the study were repeated many times). Results are presented below. In the next subsection we will see that the results are nearly identical to those obtained by the `glm` function.

```
> print(out1, digits = 2)

Inference for Bugs model at "modelglm.txt", fit using WinBUGS, 3
chains, each with 1e+05 iterations (first 10000 discarded), n.thin =
100. n.sims = 2700 iterations saved
```

```
          mean   sd    2.5%     25%     50%     75%   97.5% Rhat n.eff
alpha     0.56 0.03    0.49    0.54    0.56    0.59    0.63    1  2700
beta[1]   0.06 0.04   -0.02    0.03    0.06    0.09    0.15    1  2700
beta[2]  -0.57 0.05   -0.68   -0.61   -0.57   -0.53   -0.47    1  2700
beta[3]  -0.13 0.01   -0.16   -0.14   -0.13   -0.12   -0.10    1  2700
beta[4]   0.14 0.07    0.01    0.09    0.14    0.18    0.27    1  2700
beta[5]  -0.01 0.02   -0.04   -0.02   -0.01    0.01    0.03    1  2700
devia  5142.28 3.48 5138.00 5140.00 5142.00 5144.00 5151.00    1  1900

For each parameter, n.eff is a crude measure of effective sample size, and
Rhat is the potential scale reduction factor (at convergence, Rhat=1.

DIC info (using the rule, pD = Dbar-Dhat)
pD = 5.9 and DIC = 5148.2

DIC is an estimate of expected predictive error (lower deviance is better).
```

A 95% credible interval (the Bayesian version of a confidence interval) for β_5 is given by -0.04 – 0.03. It allows us to say that there is a 95% chance that β_5 falls in this interval. Note that this is a considerably more useful statement than that obtained by a frequentist approach. In this case 0 falls in the 95% credible interval, indicating that β_5 is not significantly different from 0.

Instead of numerical summaries, we could plot histograms of the samples from the posterior distribution for each parameter. This can be done using the `hist` function, or we can make multi-panel histograms (Figure 2.16). Note that the distribution for β_2 does not contain 0, nor does that of β_3. For β_4 a small proportion of the posterior distribution is equal to or less than 0.

R code to make this graph follows. First we need the simulated values for all parameters in a single long vector (`Beta15`), and we also need a character vector that identifies which samples are from which parameter (`ID15`). Once we have these two vectors, the `histogram` function from the `lattice` package can be used to make the multi-panel histogram.

```
> Beta15 <- as.vector(out1$sims.list$beta[,1:5])
> ID15   <- rep(c("beta1", "beta2", "beta3", "beta4", "beta5"),
                each = nrow(out1$sims.list$beta))
> histogram( ~ Beta15 | factor(ID15), type = "count", nint = 100,
          layout = c(1, 5), col = gray(0.5),
          xlab   = "Posterior distribution", ylab   = "Frequencies",
          scales = list(alternating = T,
                        x = list(relation = "same"),
                        y = list(relation = "free")),
          panel = function(x, ...) {
                panel.histogram(x, ...)
                panel.abline(v = 0, lwd = 3, col =2)})
```

We will discuss the DIC later. The MCMC output contains thousands of realisations of the model parameters, which can be used to calculate various quantities of interest. For example, the correlation among the parameters can be obtained. High correlation is commonly expected for the constant and factor effects only. If there is a high correlation among regression coefficients associated with continuous variables, standardising these variables will reduce the correlation and improve mixing of the MCMC chains so that consecutive realisations will be less dependent, shortening the burn-in period and the total number of iterations to be run.

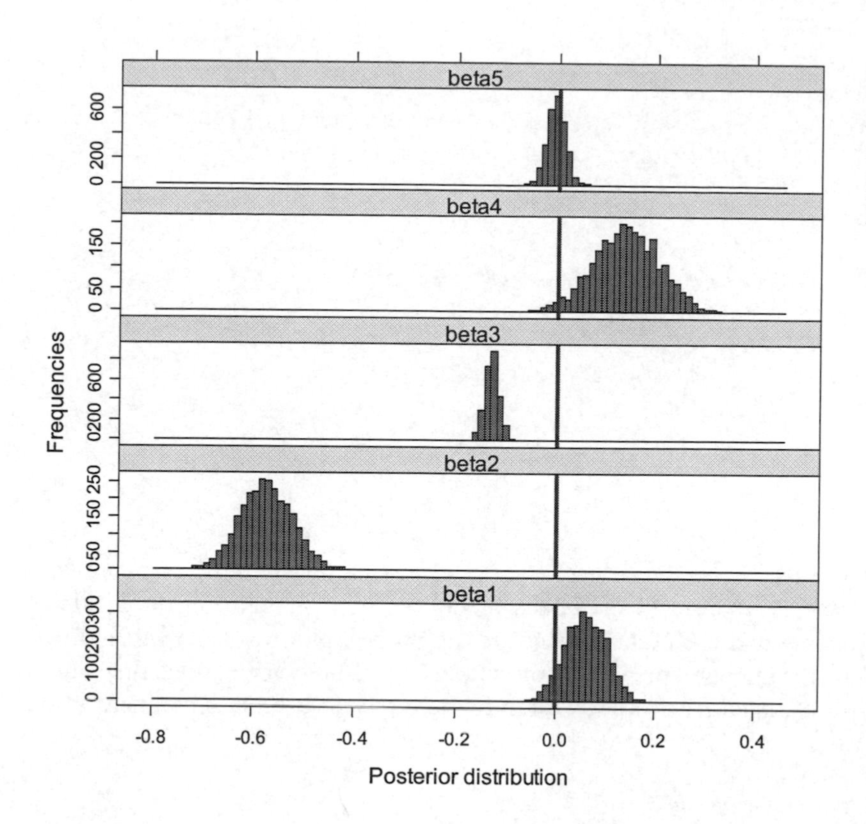

Figure 2.16. Histogram of the samples from the posterior for each regression parameter β_1 to β_5. A vertical line is drawn at $x = 0$.

2.6.9 Pearson residuals

We can also obtain Pearson residuals. The simplest method is to take the mean values of the regression parameters from the MCMC samples and calculate the expected value μ and then the Pearson residuals, but it is more informative to calculate the Pearson residuals individually for each MCMC realisation. In addition, we can generate 'predicted' residuals in each MCMC realisation, obtained by simulating the number of calls from a Poisson distribution (Congdon 2005). These simulated counts will be properly Poisson distributed so will not display overdispersion. The WinBUGS model code is:

```
sink("modelglm.txt")
cat("
model{

#Priors
for (i in 1:5) { beta[i]  ~ dnorm(0, 0.001) }
alpha ~ dnorm(0, 0.001)

#Likelihood
for (i in 1:NNest) {
 for (j in 1:NObservationsInNest[i]) {
   NCalls[j,i] ~ dpois(mu[j,i])
   log(mu[j,i]) <- max(-20, min(20, eta[j,i]))
   eta[j,i] <- alpha    + LBroodSize[j,i]      +
               beta[1]  * iSexParent[j,i]      +
               beta[2]  * iFoodTreatment[j,i] +
               beta[3]  * ArrivalTime[j,i]     +
               beta[4]  * iFoodTreatment[j,i] * iSexParent[j,i] +
```

```
               beta[5] * iSexParent[j,i] * ArrivalTime[j,i]

   #Discrepancy measures
   YNew[j,i] ~ dpois(mu[j,i])
   Pres[j,i]      <- (NCalls[j,i] - mu[j,i]) / sqrt(mu[j,i])
   PResNew[j,i] <- (YNew[j,i] - mu[j,i]) / sqrt(mu[j,i])
   D[j,i]         <- pow(PRes[j,i], 2)
   DNew[j,i]      <- pow(PResNew[j,i], 2)
 }
 Fiti[i]     <- sum(D[1:NObservationsInNest[i], i])
 FitiNew[i] <- sum(DNew[1:NObservationsInNest[i], i])
}
Fit <- sum(Fiti[1:NNest])
FitNew <- sum(FitiNew[1:NNest])
}
",fill = TRUE)
sink()
```

In order to obtain the realisations, we need to add `PRes`, `Fit`, and `FitNew` to the `params` vector. The variable `PRes` gives the Pearson residuals of the model, and we should plot the residuals against each covariate in the model and each covariate not in the model, plot residuals versus fitted values, and inspect the residuals for temporal or spatial correlation. A problem arises since they are in a matrix format and a vector format is easier to work with in R. Below is an R function to convert the matrix to a vector:

```
Mat2Vec <- function(NumInNest, FY){
   nc <- ncol(FY)
   N  <- sum(NumInNest)
   Y  <- vector(length = N)
   a1 <- 1
   for (i in 1:nc) {
     a2        <- a1 + NumInNest[i] -1
     Y[a1:a2] <- FY[1:NumInNest[i],i]
     a1        <- a1 + NumInNest[i]
   }
   Y}
```

This does the opposite of the function presented earlier for converting vectors to matrices. Columns in the matrix `FY` (internal function name) are extracted and concatenated in a vector `Y` (internal function name). The following code extracts the mean of the posterior distribution of the Pearson residuals, calls our function `Mat2Vec` to convert the output from WinBUGS to a vector, and makes a boxplot of the Pearson residuals conditional on nest. This boxplot is not presented here, but it shows that residuals for some nests are all less than, and from other nests all more than, 0. This means we should include the covariate `Nest` in the model, preferably as a random intercept.

```
> E1.mat <- out1$mean$PRes
> E1     <- Mat2Vec(NObservationsInNest, E1.mat)
> boxplot(E1 ~ Owls2$Nest, xlab = "Nest", ylab="Pearson residuals")
> abline(0, 0, lty = 1, lwd = 1) #Boxplot not shown here
```

The additional code also allows us to perform a posterior predictive check. It compares the lack of fit of the model fitted to the actual data (`Ncalls`) with the lack of fit of the model fitted to ideal data (`YNew`). The ideal data is simulated from a Poisson distribution with mean `mu[j,i]`. The MCMC code produces a large number of replicates for the summary statistics of the actual and ideal data; these are `Fit` and `FitNew`. If the model fits the actual data reasonably well, we would expect the two summary statistics to be approximately the same. The number of times that the summary statistic for the 'ideal' data is larger than that of the actual data provides a Bayesian p-value for the model. Values close to 0.5 indicate a good model fit, whereas values close to 0 or 1 indicate problems (Gelman et al., 2006). The Bayesian p-value is calculated by:

```
> mean(out1$sims.list$FitNew > out1$sims.list$Fit)

0
```

In all cases, the sum of squared residuals of the model fitted to the actual data is larger than that fitted to ideal data; hence we have serious model miss-specification. Most likely we have overdispersion, and we need to add the random intercept for nests.

Our discrepancy measure is the sum of squared Pearson residuals. In the absence of overdispersion, this sum of squares will follow a χ^2 distribution with N - k degrees of freedom. N is the sample size and k is the number of regression parameters in the model. The MCMC summary statistics for `FitNew` and `Fit` are:

```
            mean     sd    2.5%      50%     97.5% Rhat n.eff
Fit      3556.50  69.14 3429.00 3555.00  3699.00    1  2700
FitNew    576.66  35.21  510.39  575.40   647.55    1  2700
```

For comparison, the true 2.5% and 97.5% percentiles of the χ^2 distribution with N = 575 degrees of freedom are:

```
> qchisq(c(0.025, 0.975), 569)

504.7963 636.9914
```

Note that the percentiles of the simulated residual sum of squares values, `FitNew`, correspond to the theoretical percentiles. There are two means of assessing overdispersion. The first is to compare the distribution of `Fit` to the distribution of the predicted `FitNew` in the absence of overdispersion. This is what we have just described and labelled as a Bayesian p-value. Clearly, the two distributions do not match, indicating substantial overdispersion. The second approach is to compare the `Fit` distribution to the χ^2 distribution with $N - k$ degrees of freedom. This gives exactly the same information. The average sum of squared Pearson residuals is 3556.50 and is similar to what is observed with the `glm` fit (see the next subsection).

2.6.10 WinBUGS versus GLM results

The GLM can be fitted using a frequentist approach with the `glm` function:

```
> M2 <- glm(NCalls ~ SexParent * FoodTreatment +
                SexParent * Cen.AT +offset(LogBroodSize),
                family = poisson, data = Owls2)
> summary(M2)

Coefficients:
                        Estimate Std. Error z value Pr(>|z|)
(Intercept)             0.565528   0.034389  16.445   <0.001
SexParentMale           0.061976   0.042611   1.454   0.1458
FoodTSatiated          -0.570892   0.053626 -10.646   <0.001
Cen.AT                 -0.131268   0.014422  -9.102   <2e-16
SexPMale:FoodTrSatiated 0.137736   0.068518   2.010   0.0444
SexPMale:Cen.AT        -0.007354   0.018414  -0.399   0.6896
```

```
(Dispersion parameter for poisson family taken to be 1)

    Null deviance: 3973.8  on 574  degrees of freedom
Residual deviance: 3506.5  on 569  degrees of freedom
AIC: 5148.3
```

This is less code than needed for WinBUGS and is also much faster. The advantage of WinBUGS is that we can easily extend it to more complicated models, as we will see in the following sections. If we compare the estimated parameters obtained by WinBUGS and those obtained by the `glm` function we get:

```
                     beta WinBUGS sd WinBUGS beta glm sd glm
Intercept                   0.563      0.034    0.566  0.034
SexP.Fem                    0.064      0.043    0.062  0.043
FoodT.Satiated             -0.570      0.053   -0.571  0.054
ArrivalTime                -0.131      0.014   -0.131  0.014
SexP.Fem x FoodT.Sat        0.136      0.068    0.138  0.069
SexP.Fem x ArrTime         -0.008      0.018   -0.007  0.018
```

The results of the two estimation techniques are nearly identical; however, the interpretation of the parameters is different (see Chapter 1). The Bayesian results allow us to say that there is a 95% chance that the regression parameter for the interaction of sex of the parent and arrival time is within the interval $-0.008 \pm 2 \times 0.018$; whereas, using the frequentist approach, we can say that if we were to repeat the experiment a large number of times, in 95% of the cases the real β would be within the interval specified by the estimated value plus/minus twice the standard error (which also changes each time).

2.7 Implementing a Poisson GLMM in WinBUGS

Now that the model in Equation (2.10) is running properly in WinBUGS we can program the Poisson GLMM with `nest` as random effect. This is the model in Equation (2.5), and we saw the `lmer` output in the previous section. The WinBUGS modelling code is:

```
sink("modelglmm.txt")
cat("
model{

#Priors
for (i in 1:5) { beta[i]   ~ dnorm(0, 0.001) }
for (i in 1:NNest) { a[i] ~ dnorm(0, tau.Nest) }
alpha ~ dnorm(0, 0.001)

tau.Nest  <- 1 / (sigma.Nest * sigma.Nest)
sigma.Nest ~ dunif(0,10)

#Likelihood
for (i in 1:NNest) {
 for (j in 1:NObservationsInNest[i]) {
   NCalls[j,i] ~ dpois(mu[j,i])
   log(mu[j,i]) <- max(-20, min(20, eta[j,i]))
   eta[j,i] <- alpha + LBroodSize[j,i]       +
               beta[1] * iSexParent[j,i]     +
               beta[2] * iFoodTreatment[j,i] +
               beta[3] * ArrivalTime[j,i]    +
```

```
                beta[4] * iFoodTreatment[j,i] * iSexParent[j,i] +
                beta[5] * iSexParent[j,i] * ArrivalTime[j,i] + a[i]

    #Discrepancy measures
    YNew[j,i] ~ dpois(mu[j,i])
    PRes[j,i]     <- (NCalls[j,i]- mu[j,i]) / sqrt(mu[j,i])
    PResNew[j,i] <- (YNew[j,i]   - mu[j,i]) / sqrt(mu[j,i])
    D[j,i]        <- pow(PRes[j,i], 2)
    DNew[j, i]    <- pow(PResNew[j,i], 2)
    ExpY[j,i] <- exp(eta[j,i]) * exp(sigma2.Nest / 2)

  VarY[j,i] <- exp(eta[j,i]) * (exp(eta[j,i]) *
      (exp(sigma2.Nest) - 1) * exp(sigma2.Nest) + exp(sigma2.Nest/2))
    PResEQ[j,i] <- (NCalls[j,i] -ExpY[j,i]) / sqrt(VarY[j,i])
    Disp1[j,i]   <- pow(PResEQ[j,i], 2)
    }
  Disp2[i]    <- sum(Disp1[1:NObservationsInNest[i],i])
  Fiti[i]     <- sum(D[1:NObservationsInNest[i], i])
  FitiNew[i] <- sum(DNew[1:NObservationsInNest[i], i])
 }
 Dispersion <- sum(Disp2[1:NNest])
 Fit         <- sum(Fiti[1:NNest])
 FitNew      <- sum(FitiNew[1:NNest])
 }
 ",fill = TRUE)
 sink()
```

The only change is the addition of a random intercept, `a[i]`, to the predictor function. This random intercept has a diffuse prior, as can be seen from the N(0, τ_{Nest}) command. We have set τ_{Nest} (which is called `tau.Nest` in the code) to:

$$\tau_{Nest} = \frac{1}{\sigma_{Nest} \times \sigma_{Nest}}$$

Some authors (e.g. Ntzoufras, 2009) use a gamma distribution for the prior of σ_{Nest} (called `sigma.Nest` in the code) whereas other authors (Kéry 2010) use a uniform distribution (See also Gelman et al., 2006). For the data sets and models used in this chapter the uniform distribution gives slightly better mixed chains, indicating less auto-correlation, so that is what we use here.

Mixing of the chains is acceptable, but would benefit from a slightly higher thinning rate. A few regression parameters show minor auto-correlation. Comparing the results of WinBUGS and `lmer` gives:

	beta WinBUGS	sd WinBUGS	beta glm	sd glm
Intercept	0.664	0.093	0.664	0.088
SexParent.Fem	-0.034	0.048	-0.035	0.047
FoodT.Satiated	-0.654	0.058	-0.654	0.056
ArrivalTime	-0.122	0.015	-0.122	0.015
SexPar.Fem x FoodT.Sat	0.180	0.073	0.180	0.071
SexPar.Fem x ArrTime	-0.021	0.018	-0.021	0.019

The estimated value for `sigma.Nest` (σ_{Nest} in the equations) is:

```
> out2$mean$sigma.Nest
```

```
0.4208897
```

The corresponding value from `lmer` is 0.38701. From the `lmer` results we know that the Poisson GLMM in Equation (2.5) is overdispersed, but suppose we had not fitted the model with `lmer`. How can we know whether our GLMM is overdispersed when using WinBUGS? As explained in detail in Chapter 4, if we use a Poisson GLMM with a random intercept that is N(0, σ^2_{Nest}), we can derive expressions for the expected value and variance of $NCalls_{ij}$. These are given by:

$$E(NCalls_{ij}) = e^{\eta_{ij}} \times E(e^{a_i}) = e^{\eta_{ij}} \times e^{\frac{\sigma^2_{Nest}}{2}}$$

$$\text{var}(NCalls_{ij}) = e^{\eta_{ij}} \times \left(e^{\eta_{ij}} \times (e^{\sigma^2_{Nest}} - 1) \times e^{\sigma^2_{Nest}} + e^{\frac{\sigma^2_{Nest}}{2}} \right)$$

We can easily calculate these two expressions in each MCMC iteration. They are the expressions for `ExpY[j,i]` and `VarY[j,i]` in the WinBUGS code above. Once we have made these calculations, in order to assess the level of overdispersion, we can calculate Pearson residuals, calculate the sum of their squares, and divide by $N - k$, where N is the sample size and k the number of regression parameters. However it is not clear what value we should use for k. If we use the number of fixed parameters, we will be ignoring the random effects. If we use the number of observations (N), the overdispersion is estimated at 2.37. This is approximately the same value of overdispersion obtained with `lmer`.

The problem with this approach for assessing overdispersion is that these equations cannot be derived for a binomial GLMM (see Chapter 4). We need something more general. Alternative options are to:

1. Inspect the difference between the lack of fit for the observed data and for the ideal data again.
2. Compare the DIC of our current Poisson GLMM with that of a Poisson GLMM that also contains the observation level random intercept ε_{ij}. This error term is normally distributed, with mean 0 and variance σ^2_ε (see Equation (2.7)).
3. Calculate and extract Pearson residuals from the GLMM with nest as random intercept and assess the overdispersion by the usual method, i.e. the sum of squared Pearson residuals divided by $N - k$. Because we have a complete distribution for each Pearson residual, we need to work with the mean for each residual.

With option 2, the DIC of the GLMM with a random intercept `Nest` is 4794.9, and that of the GLMM with random intercept `nest` along with the observation level term is 2749.8, which is a substantial reduction. The mean of the posterior samples is 1.09, and the 2.5% and 97.5% quartiles are 0.987 and 1.198, respectively. The Bayesian p-value of the GLMM model with the observation level random intercept and random effect `nest` is 0.30, indicating that this is a reasonable model. Apparently adding the residual term substantially improved the model. However, the term ε_{ij} is a latent variable, and it does not explain why the model is better. Our next step is to try a zero inflated model and see whether it explains the overdispersion.

2.8 Implementing a zero inflated Poisson GLM in WinBUGS using artificial data

In the previous section we implemented a Poisson GLMM for 1-way nested data in WinBUGS, applied it to the owl data, and noted model miss-specification, potentially due to an excessive number of zeros. We now need to extend the Poisson GLMM to a zero inflated Poisson GLMM, that in Equation (2.8).

Before presenting the WinBUGS code for the ZIP GLMM and applying it to the owl data, we will simulate artificial data and go through the essential steps of a ZIP GLM. Let us assume we have the following model:

$$\begin{aligned} &Y_i \sim ZIP(\mu_i, \pi) \\ &\text{logit}(\pi) = \gamma_1 \\ &\log(u_i) = \beta_1 + \beta_2 \times X_i \end{aligned} \quad (2.11)$$

This is a simple ZIP GLM. Compare it to the sketch of zero inflated data in Figure 2.13. The probability of a false zero (the upper part of the branch in Figure 2.13) is denoted by π, and the probability of a true zero and the positive counts (the lower branch in Figure 2.13) is given by $1 - \pi$. We use the logit link function to model π:

$$\text{logit}(\pi) = \gamma_1 \quad \Rightarrow \quad \pi = \frac{e^{\gamma_1}}{1+e^{\gamma_1}}$$

To keep things simple we use only an intercept in the predictor function for the logistic link function, but we can easily add covariates to this part of the model. Kéry (2010) uses an identical model and estimates π directly. We will do it via the intercept. Once we know γ_1 we can calculate π. The advantage of working with the intercept is that we can assess its significance more easily. For the count process (see lower right Figure 2.13) we use a Poisson GLM with mean μ_i and log link function. In this link function we use a covariate X_i.

$$\log(\mu_i) = \beta_1 + \beta_2 \times X_i \quad \Rightarrow \quad \mu_i = e^{\beta_1 + \beta_2 \times X_i}$$

The unknown regression parameters are β_1, β_2, and γ_1. These can be estimated using the `zeroinfl` function from the `pscl` package (Zeileis et al, 2008). There are no random effects in this model, and there is only one covariate, X_i. To demonstrate estimating these parameters with WinBUGS we will carry out the following steps:

- First we will choose values for the regression parameters and simulate Y_i and X_i values. Let's say that we want a sample size of 250 observations.
- Using these 250 Y_i and X_i values, we will estimate the parameters using the function `zeroinfl`.
- Using the same Y_i and X_i values, we present and run the WinBUGS code to estimate the parameters.
- For comparison we will apply ordinary Poisson GLM and ignore the zero inflation.

So let us simulate some data. We arbitrarily choose $\beta_1 = 2$, $\beta_2 = 2.5$, and $\gamma_1 = 2$. For the covariate values we choose 250 observations between 0 and 1. Setting the seed for the random number generator to a fixed value ensures that you will get the same results as we do.

```
> set.seed(12345)
> beta1  <- 2
> beta2  <- 2.5
> gamma1 <- 2
> N      <- 250
> X      <- runif(N, min = 0, max = 1)
```

Next we simulate the process that generates the false and true zeros. Using $\gamma_1 = 2$ means that:

$$\pi = \frac{e^2}{1+e^2} = 0.88$$

Hence we assume that there is an ecological process working in the background that creates false zeros at a probability of 0.88. As Kéry (2010) states, 'Nature flips a coin with probability 0.88.' Let's do this:

```
> psi <- exp(gamma1) / (1 + exp(gamma1))
> W   <- rbinom(N, size = 1, prob = 1 - psi)
```

The vector `W` contains 250 values consisting of zeros and ones. They are obtained by drawing from a Bernoulli distribution with probability $1 - \pi = 0.12$. Note that Kéry (2010) draws the `W` from a Bernoulli distribution with probability π. That is not wrong, but makes the interpretation slightly somewhat more complicated. Most values of `W` are equal to 0, as expected.

```
> table(W)

  0   1
210  40
```

The random number generator gave us 210 0s and 40 1s. For those 40 observations, we will generate count data following a Poisson distribution:

```
> mu     <- exp(beta1 + beta2 * X)
> mu.eff <- W * mu
> Y      <- rpois(N, lambda = mu.eff)
```

The first line is the log link, and on the second line we multiply the expected values of the Poisson process by the vector of zeros and ones. A draw from a Poisson distribution with mean 0 is always 0; hence the Y_i values follow a ZIP distribution; 84% of the observations in Y are equal to 0. The Poisson process may also have generated zeros, but these are considered to be true zeros. The zeros generated by the count process are likely to be at the lower values of X_i. A graph of Y_i versus X_i is shown in Figure 2.17.

What will happen if we apply an ordinary Poisson GLM to these data and ignore the fact that the data are ZIP distributed?

```
> T1 <- glm(Y ~ X, family = poisson)
> summary(T1)

Coefficients:
              Estimate Std. Error z value Pr(>|z|)
(Intercept)    0.87317    0.07539  11.582   <2e-16
X              1.04552    0.11009   9.497   <2e-16

(Dispersion parameter for poisson family taken to be 1)
    Null deviance: 4595.0  on 249  degrees of freedom
Residual deviance: 4500.7  on 248  degrees of freedom

> deviance(T1)/ T1$df.res

18.14809
```

We have an overdispersion of 18.15, and the estimated parameters are biased. We could use a negative binomial distribution for these data, but the parameters would still be biased. Furthermore, we get model validation graphs with residual patterns (Figure 2.18). Note that the zeros appear as a clear band in the graph of residuals versus fitted values. For the owl data we saw a similar situation in Figure 2.3, though not as strong.

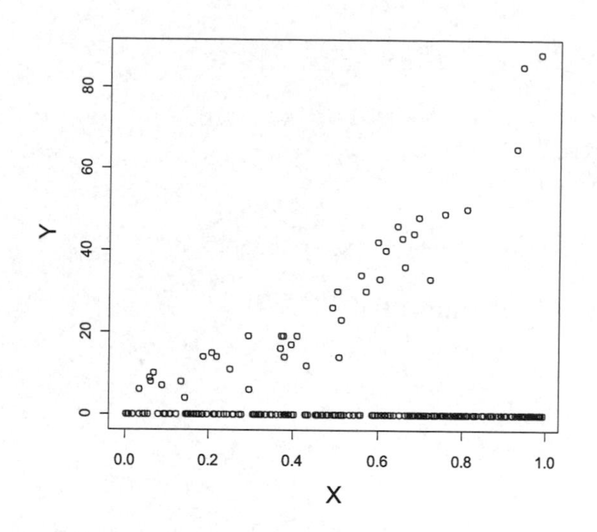

Figure 2.17. Scatterplot of Y_i versus X_i. The Y_i data are simulated from a ZIP distribution.

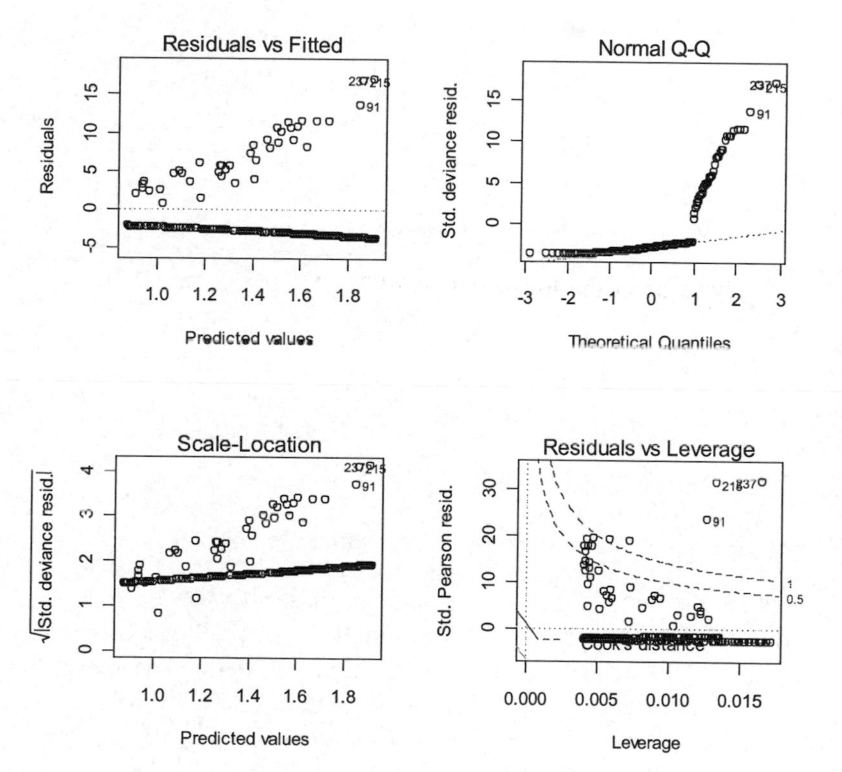

Figure 2.18. Model validation graphs for a GLM applied to the Y_i and X_i data. Note the clear patterns in the residuals.

This simple simulation exercise shows that we should not analyse data that are truly zero inflated Poisson with a quasi-Poisson or a negative binomial distribution.

Let us estimate the parameters with the `zeroinfl` function from the `pscl` package.

```
> library(pscl)
> Z1 <- zeroinfl(Y ~ X | 1)
```

The formula code in the `zeroinfl` function ensures that the log link function uses an intercept and a slope for X_i, and the logistic link function uses only an intercept. Estimated parameters are given below. Note that the estimated values for β_1, β_2, and γ_1 are similar to the original values. Any differences are due to the process of generating random numbers.

```
> summary(Z1)
```

```
Count model coefficients (poisson with log link):
            Estimate Std. Error z value Pr(>|z|)
(Intercept)  1.87709    0.08278   22.68   <2e-16
X            2.65495    0.12280   21.62   <2e-16

Zero-inflation model coefficients (binomial with logit link):
            Estimate Std. Error z value Pr(>|z|)
(Intercept)   1.6582     0.1725   9.611   <2e-16
```

We will now fit the model in WinBUGS. First we need to bundle the required data.

```
> win.data <- list(Y = Y, X = X, N = length(Y))
```

The essential WinBUGS modelling code is:

```
sink("ziptest.txt")
cat("
model{
  #Priors
  beta[1] ~ dnorm(0, 0.01)
  beta[2] ~ dnorm(0, 0.01)
  gamma1  ~ dnorm(0, 0.01)

 #Likelihood
 for (i in 1:N) {
   #Binary part
   W[i] ~ dbern(psi.min1)
   #Count process
   Y[i] ~ dpois(mu.eff[i])
   mu.eff[i]  <- W[i]  * mu[i]
   log(mu[i]) <- beta[1] + beta[2] * X[i]
 }
 psi.min1   <- 1 - psi
 logit(psi) <- gamma1
}
",fill = TRUE)
sink()
```

First we specify the diffuse priors for the three regression parameters. In the `Likelihood` part, `W[i]` follows a Bernoulli distribution with probability `1 - psi`. Finally, `Y[i]` follows a Poisson distribution with mean `W[i] * mu[i]`. The log and logistic link functions are defined as above. In order to run the code and estimate the parameters we need to provide initial values. As part of the initial values we need a vector `W` consisting of zeros and ones that is of the same length as the Y_i data:

```
> W         <- Y
> W[Y > 0] <- 1
> inits <- function () {
     list(beta   = rnorm(2),
          gamma1 = rnorm(1),
          W      = W)}
```

We need to tell WinBUGS which parameters to save:

```
> params <- c("beta", "gamma1")
```

We include the previously used values for number of chains, chain length, burn-in, and thinning rate and are ready to run the code:

```
> nc <- 3         #Number of chains
> ni <- 100000 #Number of draws from posterior
> nb <- 10000   #Number of draws to discard as burn-in
> nt <- 100      #Thinning rate
> ZipSim1 <- bugs(data = win.data, inits = inits,
                   parameters = params, model = "ziptest.txt",
                   n.thin = nt, n.chains = nc, n.burnin = nb,
                   n.iter = ni, debug = TRUE)
```

The numerical output is:

```
> print(ZipSim1, digits = 2)

...
           mean   sd  2.5%   50%   75% 97.5% Rhat n.eff
beta[1]    1.88 0.08  1.72  1.88  1.93  2.03    1  2700
beta[2]    2.66 0.12  2.43  2.65  2.74  2.89    1  2700
gamma1     1.67 0.18  1.33  1.66  1.78  2.04    1  2700
...
```

The estimated parameters are similar to the original parameters. The value of the precision of the prior distributions can cause numerical problems in the log link and logistic link. If the precision parameter chosen is too small, WinBUGS crashes. It may be wise to use:

```
log(mu[i]) <- max(-20, min(20, eta[i]))
psi.min1   <- min(0.99999, max(0.00001,(1 - psi)))
logit(psi) <- max(-20, min(20, gamma1))
```

in the WinBUGS code. This avoids any WinBUGS crashes due to infinity values after calculating an exponential.

In the process of writing this section, we noticed that the difference between the parameters estimated by an ordinary Poisson GLM and the original parameters was small when the mean value of Y_i was small. In other words, if the Y_i values are near zero, there is little difference between the Poisson GLM and the ZIP results. At the beginning of this section we stated that we would arbitrarily choose values for β_1, β_2, and γ_1. This is not true. Since small values for β_1 and β_2 produce Poisson GLM results nearly the same as the estimated parameters, we chose larger values. From an instructive point of view we considered it better to show an example that produced different results.

With real data, we do not know which zeros are false and which are true. From Figure 2.17 it can be seen that the true zeros are most likely those that can be fitted by the model, and all other zeros are deemed to be false zeros.

Although this book mainly uses real data sets that can be demanding to analyse, using a simulated data set helped us to comprehend how ZIP functions. We suggest experimenting with the code for simulating ZIP data and comparing results from the `glm` function, `zeroinfl`, and WinBUGS code. Change the values of β_1, β_2, and γ_1 and even add a covariate to the logistic link function and compare resulting scatterplots. Before continuing with the analysis of the owl data, it may be wise to initialize the random seed.

2.9 Application of ZIP GLMM in WinBUGS using the owl data

Now that we know how to do a ZIP GLM in WinBUGS, we can easily extend the code so that we can fit the model in Equation (2.8). We need only change the predictor function for the log link function, use two loops, and change the number of betas. The WinBUGS modelling code is:

```
sink("modelzipglmm.txt")
cat("
model{

  #Priors
  for (i in 1:5) { beta[i]  ~ dnorm(0, 0.001) }
  for (i in 1:NNest) { a[i] ~ dnorm(0, tau.Nest) }
  alpha  ~ dnorm(0, 0.001)
  gamma1 ~ dnorm(0, 0.001)
  tau.Nest <- 1 / (sigma.Nest * sigma.Nest)
  sigma.Nest ~ dunif(0, 10)

  #Likelihood
  for (i in 1:NNest) {
   for (j in 1:NObservationsInNest[i]) {
    #Logit part
    W[j,i] ~ dbern(psi.min1)

    #Poisson part
    NCalls[j,i] ~ dpois(eff.mu[j,i])
    eff.mu[j,i]  <- W[j,i] * mu[j,i]
    log(mu[j,i]) <- max(-20, min(20, eta[j,i]))
    eta[j,i] <- alpha + beta[1] * iSexParent[j,i] +
                beta[2] * iFoodTreatment[j,i] +
                beta[3] * ArrivalTime[j,i] +
                beta[4] * iFoodTreatment[j,i] * iSexParent[j,i] +
                beta[5] * iSexParent[j,i] * ArrivalTime[j,i] +
                1 * LBroodSize[j,i] + a[i]
    #Extract Pearson residuals
    EZip[j,i]   <- mu[j,i] * (1 - psi)
    VarZip[j,i] <- (1 - psi) * (mu[j,i] + psi * pow(mu[j,i], 2))
    PRES[j,i]   <- (NCalls[j,i] - EZip[j,i]) / sqrt(VarZip[j,i])
  }}
  psi.min1   <- min(0.99999, max(0.00001,(1 - psi)))
  eta.psi    <- gamma1
  logit(psi) <- max(-20, min(20, eta.psi))
}
",fill = TRUE)
sink()
```

The initial values of the matrix **W** are a copy of NCalls, except that all values larger than 0 have been set to 1. The `min` and `max` functions in the logistic link function prevent the algorithm from crashing. The code above has an extra set of commands to calculate Pearson residuals. Chapter 11 in Zuur et al. (2009a) and Chapters 3 and 4 in this book present equations for the mean and variance of a ZIP GLM. The code above calculates the mean and variance in each MCMC iteration.

The code for assembling the data and set initial parameters is:

```
> MyW <- NCalls.ij
> MyW[NCalls.ij > 0] <- 1
```

```
> inits <- function () {
   list(
    alpha      = rnorm(1),
    beta       = rnorm(5),
    a          = rnorm(length(levels(Owls2$Nest)),0, 1),
    sigma.Nest = rlnorm(1),
    W          = MyW,
    gamma1     = rnorm(1) )  }
> params <- c("alpha", "beta", "a", "sigma.Nest", "gamma1", "PRES")
```

We used the same number of chains, thinning rate, and chain length as previously. The chains were well mixed with no auto-correlation.

Our main interest is in the regression parameter `gamma1` (γ_1 in Equation (2.8)). If it is large and negative, the probability of a false zero, π, converges to 0, and we should adopt the Poisson GLMM. A histogram for the samples of the posterior distribution of `gamma1` is presented in Figure 2.19. The numerical output below shows that the mean value of `gamma1` is -1.04 and the 2.5% and 97.5% quartiles are -1.28 and -0.90, respectively. Hence, the probability that γ_1 is within -1.28 and -0.90 is 95%. The estimated probability of a false zero is given by π = exp(-1.04) / (1 + exp(-1.04)) = 0.26. We should put a hat, or a short line, above π to indicate that it is an estimator.

The 2.5% quartile, median, and 97.5% quartile for the regression parameters and variance of the random intercepts are:

```
              2.5% median  97.5%
beta1       -0.178 -0.086  0.011
beta2       -0.403 -0.290 -0.172
beta3       -0.099 -0.069 -0.041
beta4       -0.020  0.127  0.271
beta5       -0.056 -0.022  0.015
sigma.Nest   0.235  0.326  0.471
```

These results show that `beta4` (interaction of food treatment and sex of the parent) and `beta5` (interaction of sex of the parent and arrival time) are not significantly different from 0, but `beta2` (food treatment) and `beta3` (arrival time) are significantly different from 0. If we wanted to engage in a model selection, we could take the DIC of this model, drop each of the two interactions in turn, obtain the two DICs, and do a backward selection.

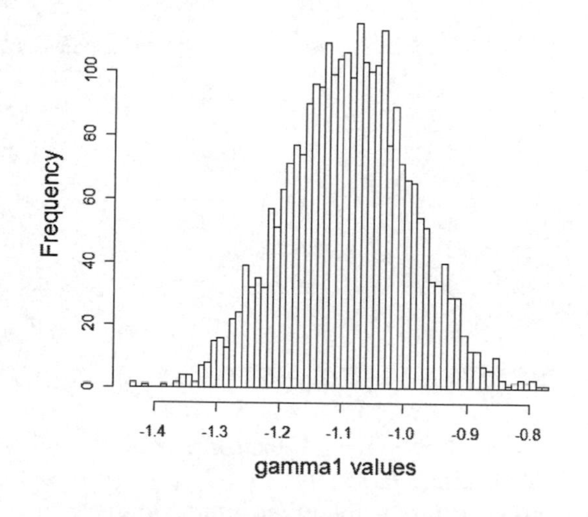

Figure 2.19. The posterior distribution of γ_1. The mean value is -1.04, and the 95% credible interval does not contain zero.

The final step is to extract residuals, plot them against arrival time, and inspect the residuals for spatial correlation. The mean value of the posterior distribution of the Pearson residuals is used for the model validation graphs. We plot the Pearson residuals versus arrival time (Figure 2.20A) and add a smoother to aid visual interpretation, but the smoother is not significantly different from 0. Hence, we are not sure whether the little bump at around 11.00 h is important. Without the zero inflation we end up with a significant pattern in the residuals. So it may be that, in the GAMM without zero inflation, the shape of the smoother is driven by false zeros.

We will also examine the Pearson residuals for spatial correlation using a variogram. First we investigate the distribution of the distances between the sites (Figure 2.20B). Most sites are separated by approximately 10 kilometres, so it would make sense to truncate the variogram at about 8 – 10 kilometres. The function `variogram` from the `gstat` package was used to make the variogram in Figure 2.21. There is no obvious spatial correlation. This means there is no need to apply ZIP GLMMs with spatial correlation to the owl data.

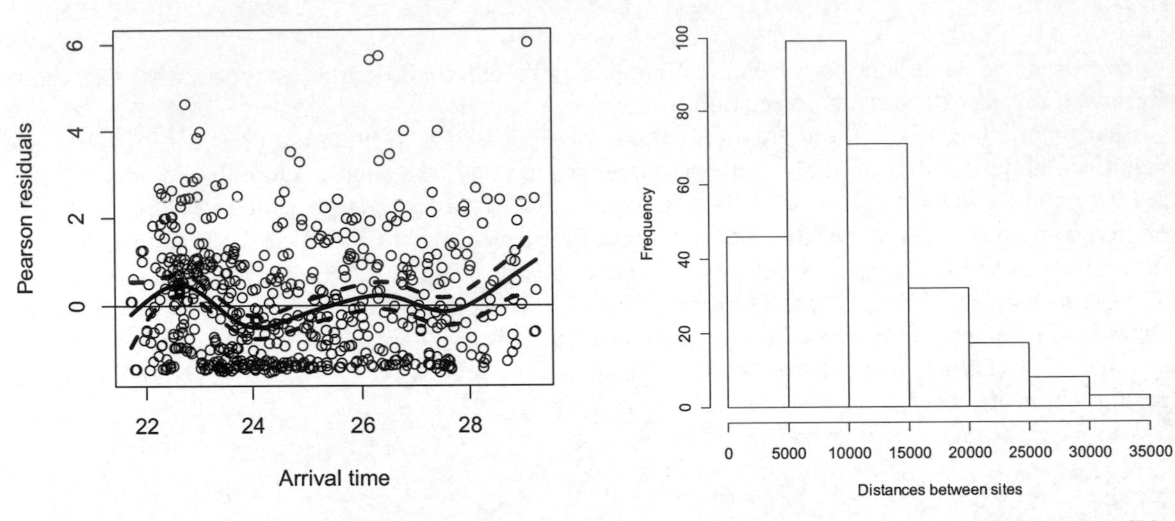

Figure 2.20. Left: Pearson residuals plotted versus arrival time for the ZIP GLMM. Right: Distance between the sites in metres.

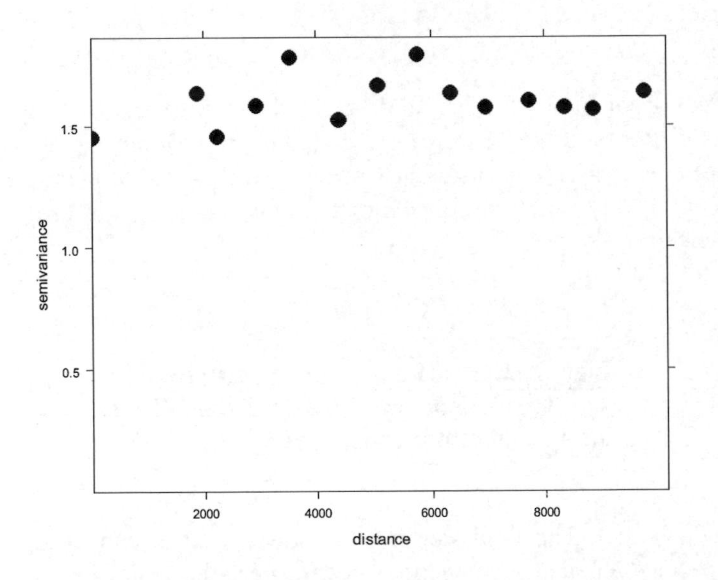

Figure 2.21. Sample variogram of the Pearson residuals of the ZIP GLMM.

2.10 Using DIC to find the optimal ZIP GLMM for the owl data

Results of the ZIP GLMM presented in the previous section show no significant interactions, and we might consider the analysis finished. However, we could continue the analysis by dropping one of the 2-way interactions and observe its effect on the remaining terms. If the other interaction term is not significant, we could also eliminate it. This can involve the main terms. This process is called model selection and is condemned by some scientists and routinely applied by others. Applying model selection may be a reason for referees to reject a paper, and alternative methods such as the Information Theory approach (IT, Burnham and Anderson, 2002) can be used. In the IT approach 10 – 15 *a priori* chosen models are compared using the Akaike Information Criterion (AIC), and it is possible to calculate an average model. However, the IT method is also criticised by some scientists.

In any statistical model, interactions are difficult to understand, and perhaps doing a model selection on only the interactions while keeping the main terms in the model, whether they are significant or not, is a compromise. In frequentist statistics, we have tools such as comparison of deviances (for nested models) and AIC at our disposal for comparing models. With Bayesian statistics, one such tool is the Deviance Information Criterion (DIC, Spiegelhalter et al., 2002). It is an extension of AIC and works as follows: Let $D(\boldsymbol{\theta})$ be defined as $-2 \times \log(f(\mathbf{y}|\boldsymbol{\theta}))$, where $\boldsymbol{\theta}$ contains the model parameters and $f(\mathbf{y}|\boldsymbol{\theta})$ is the likelihood. $D(\boldsymbol{\theta})$ is calculated for each realisation of the MCMC chain. Let $\bar{D}$ be the average of $D(\boldsymbol{\theta})$. The effective number of parameters, p_D, is calculated as the posterior mean deviance minus the deviance evaluated at the posterior mean

$$p_D = \bar{D} - D(\bar{\theta})$$

where $\bar{\theta}$ is the average of all realisations of $\boldsymbol{\theta}$. Then, analogous to AIC, the DIC is defined as

$$DIC = D(\bar{\theta}) + 2 \times p_D = \bar{D} + p_D$$

As with AIC, models with a small DIC are preferred. This works well for Poisson or Bernoulli GLMs fitted in WinBUGS, but our ZIP distribution is not a standard built-in distribution. As a result, the DIC reported by WinBUGS is not adequate for our needs, and we must calculate it ourselves. For this we need an expression for the likelihood function. Once we have an expression for the likelihood, we can calculate the deviance $D(\boldsymbol{\theta})$ within the MCMC iterations and, when WinBUGS is finished, we can use the posterior mean of these deviances to obtain $\bar{D}$. We also need to obtain the mean of the estimated regression parameters and use these to calculate $D(\bar{\theta})$. We give the underlying equations and R code in Chapter 4.

The DIC of various ZIP GLMM models are given in Table 2.3. Using all three main terms and the two 2-way interactions produces a DIC of 3859.75. Dropping each of the two 2-way interactions gives similar DIC values, and the model with only the three main terms has a similar DIC. This means that we can drop both interaction terms. If you rerun the model with only the three main terms, the 95% credible intervals give the same message: sex of the parent is not significant, while the other two terms are significant.

The overdispersion value of the ZIP GLMM for 1-way nested data with the three main terms is 1.6, which is slightly high. Three potential solutions are to fit a smoother for arrival time, to extend the binary part with covariates, or to improve the correlation structure. Using a smoother means that we need to implement a ZIP GAMM for 1-way nested data, and this will be discussed in the next section. Adding covariates to the binary model involves careful consideration of which covariates can cause false zeros. This should be driven by biology. As to the correlation, the random intercept in the count model imposes a correlation structure on observations from the same nest. The form of the correlation is such that correlation between two observations of the same nest is always the same, regardless of the time difference between the observations. Observations close in time (e.g. within an hour) may be more similar than observations separated by hours, and adopting a more complicated correlation may reduce the overdispersion, but it would also increase the model complexity.

The use of the DIC for models with random effects is not without its critics (Spiegelhalter et al., 2003). Model selection for mixed models using DIC is the subject of intense debate in recent years, and it may be advisable to use the DIC with great care.

Table 2.3. DIC values rounded to the nearest integer. The symbol X means that a term is included. The lower the DIC the better the model.

Terms							
Sex parent	X	X	X	X		X	X
Food treatment	X	X	X	X	X		X
Arrival time	X	X	X	X	X	X	
Sex parent × Food treatment	X		X				
Sex parent × Arrival time	X	X					
DIC	**3860**	**3860**	**3860**	**3860**	**3858**	**3884**	**3930**

2.11 ZIP GAMM for 1-way nested data

In order to fit a GAM or GAMM in WinBUGS we need material that is described in Chapter 9, in which a sperm whale stranding time series is analysed using a ZIP GAM model with temporal correlation. It demonstrates how a smoother can be rewritten into a linear mixed effects modelling notation so that we can use the WinBUGS code explained in this chapter to estimate the parameters and smoothers. Instead of presenting the zero inflated Poisson GAM with `nest` as random intercept, we leave it as an exercise for the reader. The overdispersion of the ZIP GAMM for 1-way nested data is also 1.6. The shape of the smoother is similar to that in Figure 2.2.

2.12 What to present in a paper?

What would we present in a paper? In our opinion the data are zero inflated, and the zero inflation is most likely due to false zeros (short sampling interval). Hence, we would present a paper based on the ZIP GLMM or ZIP GAMM, whichever model makes more sense from a biological standpoint. It may be interesting to add covariates to the logistic link function, which means that further analyses must be carried out. We may want to drop some of the non-significant terms from the predictor function of the count model.

In more detail, present some time series plots of the data and the spatial position of the nests. Give the statistical equation of the model that you will be using, most likely (Equation 2.8), and state that estimation of parameters was in a Bayesian context. The Result section will need to include the number of chains used, the thinning rate, and the length of the chains. Make statements of convergence of the chains and present the estimated parameters and some of their summary statistics. It would be useful to sketch the fitted values of the model in a graph. We will demonstrate this in other chapters in this book.

3 A roadmap for analysis of overdispersed sandeel counts in seal scat

Elena N Ieno, Graham J Pierce, M Begoña Santos, Alain F Zuur, Alex Edridge, and Anatoly A Saveliev

3.1 Sandeel otoliths and seal scat

In this chapter we analyse data on the summer diet of harbour seals (*Phoca vitulina*) on Orkney in northeast Scotland, based on faecal sampling at haul-out sites on the island of Eynhallow, 1986 – 2006. Time series of dietary data can potentially provide information on how predator diets respond to changing prey abundance, e.g. due to overfishing and/or climate change. The study originated as part of a survey of seal diet in Scotland in 1986 – 1989 and a follow-up study in 1993/1994. Subsequent sampling has been mainly opportunistic, so there are gaps in the series. An additional complication is that the island is also used by grey seals (*Halichoerus grypus*). However, in summer, there are few grey seals present, and the haul-out sites are used by as many as several hundred harbour seals. We focus on data collected from May through August in 1986 – 1988, 1993 – 1996, 1998, 2002 – 2003, and 2005 – 2006.

Seal scat was collected from the haul-out sites. Each scat was individually bagged, labelled, transferred to the laboratory, and stored at -20°C. Thawed scat was sieved (minimum mesh size 0.355 mm) and all identifiable prey remains extracted: fish otoliths, jaw bones, vertebrae, and other skeletal elements; crustacean exoskeletons; polychaete jaws; mollusc shells; and cephalopod mandibles. Remains were identified to the lowest possible taxon. Otoliths and cephalopod mandibles were counted and measured to estimate original prey size.

In general, diet selection is likely to reflect the relative availability of prey types. This depends on prey abundance and the levels of interspecific and intraspecific competition for food. The most important prey of harbour seals on Eynhallow is the sandeel (Ammodytidae). There are several species, which are difficult to distinguish based on hard parts – they have similar otoliths – but it is reasonable, for this study, to treat them as a group, and it is likely that the majority are the most abundant species, *Ammodytes marinus.* Results of stock assessments of the entire North Sea suggest that the abundance of *A. marinus* has fluctuated widely, with peaks in 1987 and 1998 and a sharp decline since 1998. Stock in 2006 was less than 10% of its 1998 abundance (ICES, 2009). It is therefore of interest to ask whether the importance of sandeels in the harbour seal diet changed over the study period.

The significance of sandeels in the diet can be expressed in various ways, based, for example, on presence as indicated by numbers or biomass, expressed in absolute terms or relative to other prey species at the population or individual level. Unlike stomach contents, faecal samples or scat cannot usually be assigned to a particular individual seal (unless defecation is observed or, hypothetically, DNA identification is used). In addition, scat may represent a meal, part of a meal, or parts of several meals. Nevertheless, it is a convenient unit of sampling. There are issues with all quantification measures. For example, prey biomass involves errors, since it must be estimated from measures of hard parts such as otoliths, which will have undergone some degree of digestive erosion. Use of counts avoids this error and provides more information than presence/absence data, but counts may be misleading if there is wide variation in prey size. Comparisons among prey species are problematic in general, since prey species display differing degrees of digestive erosion and loss when passing through a seal's digestive tract. However, by focusing on the absolute importance of single prey groups, we can argue that any inherent biases are likely to be consistent.

This chapter will address identifying patterns in the number of sandeels in the samples. We will start simple and apply a generalised linear model (GLM) to the data and show which tools indicate that it is not an appropriate model, as it is overdispersed. The aim of this chapter is to provide a roadmap to deal with overdispersion. Do we use a negative binomial distribution? Do we apply a generalised additive model (GAM), or is the overdispersion caused by zero inflation?

We assume the reader is familiar with GLM and GAM (Wood 2006, Zuur et al. 2009a, Hilbe 2011). We also suggest that you familiarise yourself with the theory underlying the zero inflated models (Chapter 11 in Zuur et al., 2009a), although a brief review is given in Section 3.3.

An overview of this chapter is sketched in Figure 3.1. You may find it useful to review the figure as you read sections of this chapter. In the Discussion we present improved a revised roadmap.

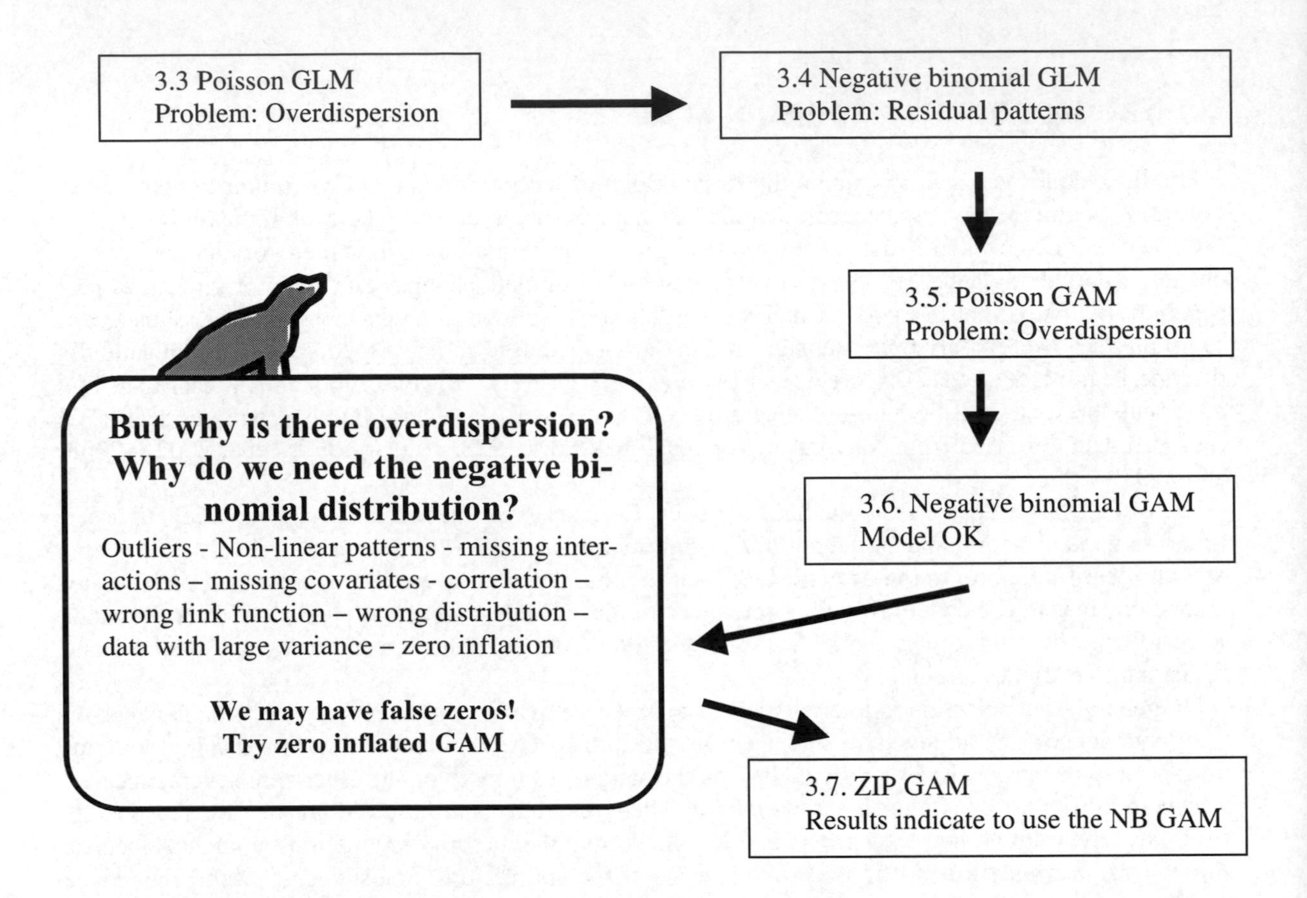

Figure 3.1. Steps in analysis of the seal scat data. We first apply a Poisson GLM, which shows overdispersion. A potential solution is a negative binomial GLM, however this model shows residual patterns, and we apply a Poisson GAM. Again, the model is overdispersed, and we continue with a negative binomial GAM. This model shows no residual patterns and no overdispersion, but we wonder why the negative binomial distribution was needed. We apply a zero inflated Poisson GAM, because we may have false zeros. It makes more sense to apply the Poisson GLM followed by a detailed model validation and, at that point, decide what to do. An improved roadmap is presented in the Discussion section.

3.2 Data exploration

We initiate analysis with data exploration following Zuur et al. (2010). The data are in the text file `SandeelsSeals.txt` and are imported into R with the `read.table` function. The `names` and `str` functions show the variables and how they are coded. The variable `Otoliths` contains the total number of otoliths per scat. Because fish have two otoliths, the variable Sandeels is the number of otoliths divided by two (rounded, if not an integer), and is an estimation of the number of sandeels in the scat. We presume that 10 otoliths in a scat means that the seal ate 5 fish.

```
> Fish <- read.table(file = "SandeelsSeals.txt", header = TRUE)
> names(Fish)

[1] "ID"        "Month"     "Otoliths" "Year"      "Day"      "Sandeels"

> str(Fish)

 'data.frame':   845 obs. of  6 variables:
 $ ID       : int  6 20 34 48 62 76 90 104 118 132 ...
 $ Month    : int  8 8 8 8 8 8 8 8 8 8 ...
 $ Otoliths: int  1 100 117 355 0 0 236 534 0 313 ...
 $ Year     : int  2006 2006 2006 2006 2006 2006 ...
 $ Day      : int  7533 7533 7533 7533 7533 7533 ...
 $ Sandeels: int  1 50 59 178 0 0 118 267 0 157 ...
```

The variable Day represents the numbers of days since 1 January 1986, the year in which the first samples were collected. This variable allows assessment of trends over time. Figure 3.2 shows a simple scatterplot of the number of sandeels versus time (Day). There is no clear trend in the data. Note that three observations have a relatively large number of sandeels.

It is also interesting to investigate change in the number of sandeels consumed within a year and when counts were taken within a year. Figure 3.3 shows scatterplots of number of sandeels versus sampling day, by year. The graph indicates that sampling in different years did not always take place in the same month. A change in sampling month can also be seen from the following table.

```
> table(Fish$Year, Fish$Month)

      2  3  4  5  6  7  8  9 11 12
1986  0  0  0  0  0  1 53  0  0  0
1987  0 16  0 18 58  0 24  0  0  0
1988  3  4  0  0 15 16  0  0 19  0
1993  0  0  0 11  0 90  0  0  0  0
1994  0 55  0  0  0  0  0 98  0  0
1995  0 16  1 17  9  0 65  0  0 23
1996  0  0  0 22  0  0  0  0  0  0
1998  0  0  0  0 21  0 25  0  0  0
2002  0  0  0 19  0  0 62  0  0  0
2003  0  0  0  0  0  0 21  0  0  0
2005  0  0  0  0  0  0 39  0  0  0
2006  0  0  0  0  0  0 24  0  0  0
```

In 1986, 53 scats were collected in August and 1 in July. If the objective is to assess trends over time, we need to compare like with like. In 2003 – 2006 sampling took place only in August. Therefore, perceived alterations in long term trends may be due to differences in the sampling period over time. Because these data are basically data collected by opportunity, we cannot alter this situation. We could focus on the summer and use only the May – August data, but that would still entail comparing May – August data from 1986 and 1987 with August-only data from 2003 to 2006.

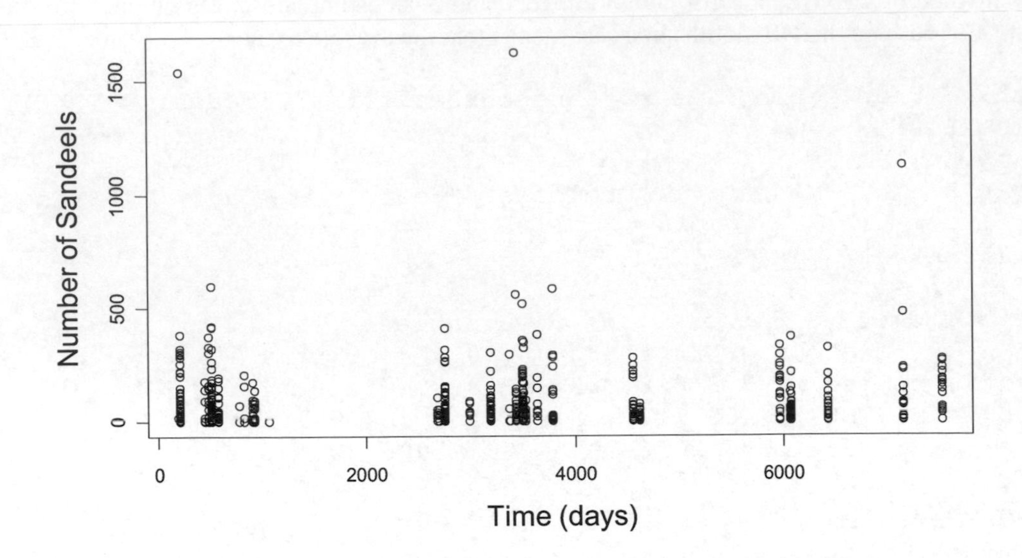

Figure 3.2. Number of sandeels versus time in days. Day 1 corresponds to 1 January 1986.

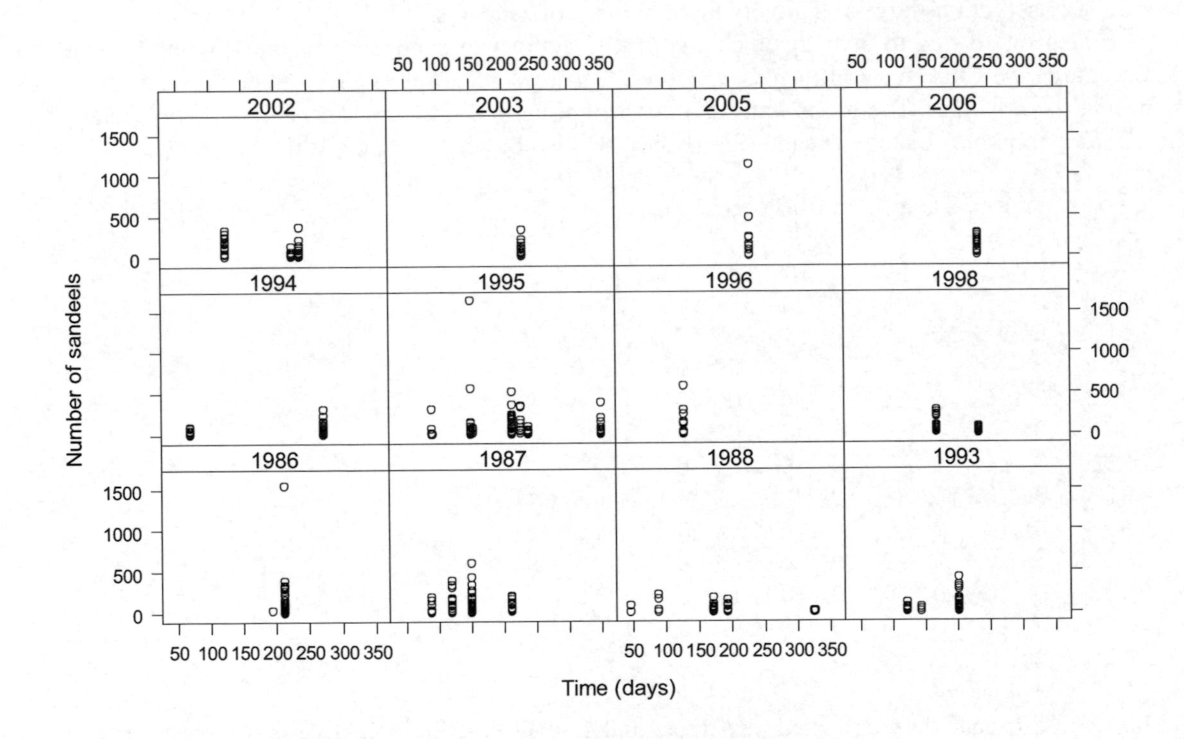

Figure 3.3. Sandeel consumption over time. The horizontal axes show the sampling day in a year and the vertical axes the number of sandeels. Each panel shows data for a particular year.

The data exhibit seasonality, as can be seen from Figure 3.4. If we are to analyse the full data set, the models need to include a seasonal component. The software we use for the zero inflated models later in this chapter may have numerical estimation problems for complicated models. Therefore, we will keep the models as simple as possible and only use the May – August data. An additional biological reason for focussing on May – August data is that in summer only harbour seals are present, as opposed to winter, when grey seals are included. We will use the May – August data in the following analysis. The data is obtained with the code:

```
> Fish2 <- Fish[Fish$Month > 4 & Fish$Month < 9,]
```

The following elementary R code was used to create Figure 3.2.

```
> plot(y = Fish$Sandeels , x = Fish$Day, xlab = "Time (days)",
                    ylab = "Number of Sandeels")
```

The `xyplot` function from the `lattice` package (Sakar, 2008) was used to make Figure 3.3, and the `boxplot` function was used for Figure 3.4.

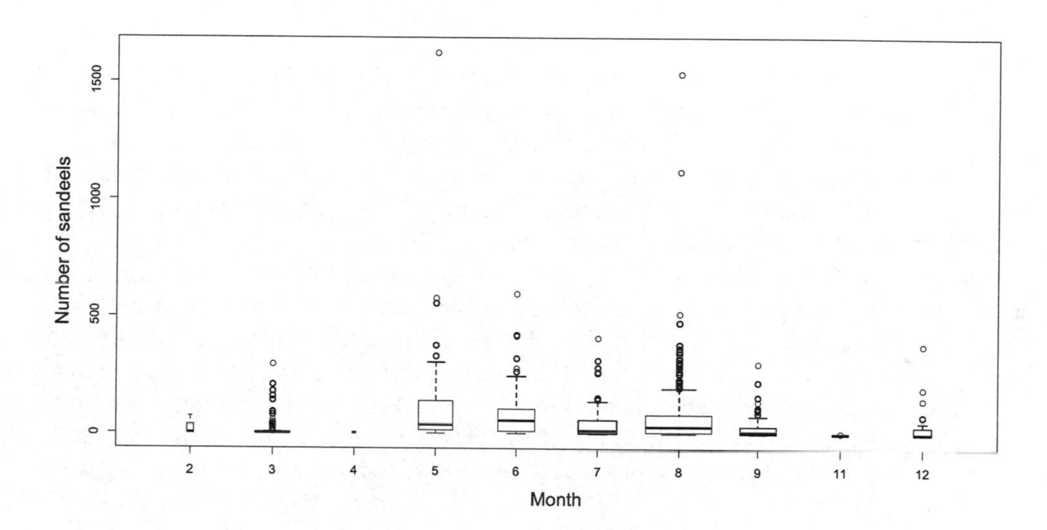

Figure 3.4. Boxplot of the number of sandeels conditional on month.

3.3 GLM with a Poisson distribution

The primary biological question is whether there is a change in sandeel numbers in the seal diet over time. This means that we should consider models of the form *Sandeels* = function(*Time*). The question is which parameters to include in the function of time. In this case there are two options. We can use the form *Sandeels* = function(*Year*, *Month*) or *Sandeels* = function(*Day*). If a function of year and month is used, we ignore the within-month information. Hence a function of day gives more detail. By concentrating only on the May – August data, there should be only limited seasonality in the data, so we may be able to use only function(*Day*) and not function(*Day, Month*). The response variable is the number of sandeels; therefore we begin with a GLM with a Poisson distribution:

$$Sandeels_{is} \sim Poisson(\mu_{is})$$
$$E(Sandeels_{is}) = \mu_{is} \quad \text{and} \quad \text{var}(Sandeels_{is}) = \mu_{is}$$
$$\log(\mu_{is}) = \alpha + \beta \times Day_s$$

$Sandeels_{is}$ is the number of sandeels in scat i on day s. The model is implemented in R with the following code:

```
> M1 <- glm(Sandeels ~ Day, family = poisson, data = Fish2)
> summary(M1)

Coefficients:
              Estimate Std. Error z value Pr(>|z|)
(Intercept)  4.439e+00  7.912e-03  561.08   <2e-16
Day         -3.995e-05  2.011e-06  -19.87   <2e-16

(Dispersion parameter for poisson family taken to be 1)
    Null deviance: 79357  on 609  degrees of freedom
Residual deviance: 78959  on 608  degrees of freedom
AIC: 81871

> sum(resid(M1, type = "pearson")^2) / M1$df.res

243.0135
```

Note that the effect of `Day` is negative and significant. However, the ratio of the sum of squared Pearson residuals to the residual degrees of freedom (Hilbe, 2011) is 243.01, and indicates serious overdispersion. This overdispersion is too large for a quasi-Poisson GLM. Hilbe (2011) mentioned various reasons for overdispersion, namely outliers, missing covariates, non-linear patterns, correlation, data with a large variance, or zero inflation. Figure 3.5 presents a flowchart of the options and solutions. We call outliers and missing covariates trivial solutions.

To check whether there are outliers with a high impact on the estimated regression parameters, we create a graph of the Cook's distances. It shows that there are observations with a Cook's distance greater than 1 (Figure 3.6). Some of the observations with high Cook values are those with large numbers of sandeels. Instead of classifying all these observations as outliers and dropping them, we suspect that the data are more variable than allowed for in a Poisson distribution. This means that we should consider applying a negative binomial GLM. This is the first option in the 'relatively easy solution' branch in Figure 3.5. We will implement it in the next section.

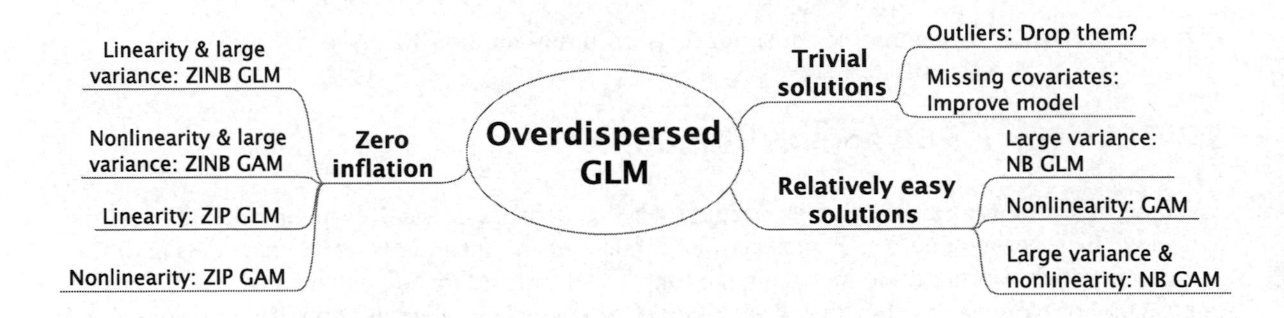

Figure 3.5. Procedures to deal with an overdispersed GLM model when a quasi-Poisson model is not an option. NB denotes negative binomial distribution. Hurdle models are an option and are discussed in Chapter 5.

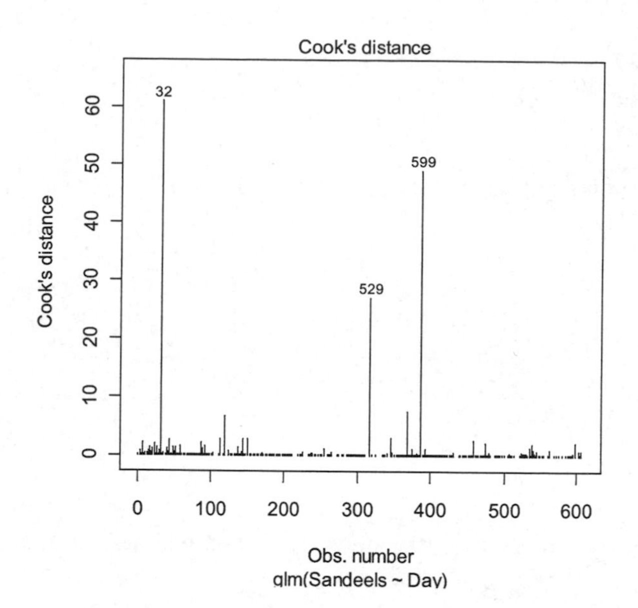

Figure 3.6. Cook's distance values of the Poisson GLM. The graph was created with the command: `plot(M1, which = 4)`

3.4 GLM with a negative binomial distribution

Since we have overdispersion but do not want to drop a large number of observations, we apply a negative binomial GLM with the model specification:

$$Sandeels_{is} \sim NB(\mu_{is}, k)$$

$$E(Sandeels_{is}) = \mu_{is} \quad \text{and} \quad \text{var}(Sandeels_{is}) = \mu_{is} + \frac{\mu_{is}^2}{k} = \mu_{is} + \alpha \times \mu_{is}^2$$

$$\log(\mu_{is}) = \alpha + \beta \times Day_s$$

The negative binomial distribution is similar to that of a Poisson distribution except for the extra quadratic term in the variance. The parameter k is unknown and needs to be estimated by the software. If it is large, the negative binomial distribution converges to a Poisson distribution. Most major statistical software packages (e.g. Stata, SAS, SPSS, Genstat, Limdep) and textbooks (e.g. Hilbe 2011; Cameron and Trivedi 1998) use the notation $\mu + \alpha \times \mu^2$ for the variance of the negative binomial distribution (for simplicity we omitted the indices) where $\alpha = 1/k$. This is more intuitive as $\alpha = 0$ implies that we obtain the Poisson distribution and $\alpha > 0$ means that the variance is larger than that of the Poisson distribution. The function `glm.nb` in R uses the notation with k (in fact it is called theta in the numerical output), and in this book we will do the same.

The model is implemented in R with the code:

```
> library(MASS)
> M2 <- glm.nb(Sandeels ~ Day, link = "log", data = Fish2)
> summary(M2)
```

```
              Estimate Std. Error z value Pr(>|z|)
(Intercept)  4.428e+00  1.060e-01  41.785   <2e-16
Day         -3.658e-05  2.567e-05  -1.425    0.154

(Dispersion parameter for Negative Binomial(0.4487) family taken to
be 1)
```

```
    Null deviance: 751.74  on 609  degrees of freedom
Residual deviance: 749.54  on 608  degrees of freedom
AIC: 6189.9
Number of Fisher Scoring iterations: 1
Theta:  0.4487    Std. Err.:  0.0230
2 x log-likelihood:  -6183.8790
```

The effect of Day is still negative but is not now significant. The estimated value for k is 0.4487. Further numerical output is obtained with:

```
> drop1(M2, test = "Chi")

Single term deletions
Model: Sandeels ~ Day
       Df Deviance    AIC    LRT Pr(Chi)
<none>      749.54 6187.9
Day     1   751.74 6188.1 2.2027  0.1378
```

We can check for overdispersion using the ratio of the sum of the squared Pearson residuals to the residual degrees of freedom.

```
> E2 <- resid(M2, type = "pearson")
> sum(E2^2) / M2$df.res

1.462926
```

This shows marginal overdispersion. The extent of the overdispersion is such that one may decide to ignore it. As part of the model validation, we can also plot Pearson residuals versus day and Pearson residuals versus month. Any pattern means that the model needs further improvement. Instead of viewing the graphs and subjectively judging whether there is a pattern, it is possible to apply a smoothing model with the Pearson residuals used as response variables and day as smoother. A significant smoother would indicate model misspecification for the NB GLM. Such a smoother is given by:

$$E_{is} = \alpha + f(Day_{is}) + noise_{is}$$

The term E_{is} contains the Pearson residuals from the NB GLM, and noise_{is} means normally distributed noise. This model validation step is carried out with the R code:

```
> library(mgcv)
> tmp <- gam(E2 ~ s(Day), data = Fish2)
> anova(tmp)

Approximate significance of smooth terms:
         edf Ref.df     F p-value
s(Day) 5.347  6.313 2.124  0.0457
```

There is a weak temporal pattern in the residuals. This is an indication that the effect of day is slightly non-linear and both the Poisson GLM and negative binomial GLM are possibly inappropriate. However, the pattern is not strong, and the summary output (not presented here) shows that the smoother explains only about 3% of the variation in the residuals. Moreover, Wood (2006) argues that a p-value between 0.01 and 0.05 for a smoother should be doubled. We apply a linear regression model in which the Pearson residuals are used as response variable and month is a categorical explanatory variable. In this case, month is highly significant, but the explained variance is again low (2%).

A pragmatic and sensible approach might be to present the NB GLM, state that there are marginal patterns in the residuals, and stop at this point. However, we are curious and would like to know what will happen if we continue with a GAM to model a non-linear trend. We expect that an NB GAM will result in a model with no overdispersion, a slightly non-linear day effect, and no residual patterns. Note that we are still in the 'relatively easy solutions' branch of Figure 3.5. Instead of a GAM, we could also apply a GLM with polynomial functions of Day.

3.5 GAM with a Poisson distribution

In this section we apply a GAM with a Poisson distribution in which Day is used as smoother. The basic model is of the form:

$$
\begin{aligned}
&Sandeels_{is} \sim Poisson(\mu_{is}) \\
&E(Sandeels_{is}) = \mu_{is} \quad \text{and} \quad \text{var}(Sandeels_{is}) = \mu_{is} \\
&\log(\mu_{is}) = \alpha + f(Day_s)
\end{aligned}
$$

The term $f(Day_s)$ is a smoother, and the optimal amount of smoothing is estimated with cross-validation. We use the package mgcv (Wood 2006) to estimate the smoother. The R code is:

```
> M3 <- gam(Sandeels ~ s(Day), family = poisson, data = Fish2)
> summary(M3)

Family: poisson
Link function: log
Formula: Sandeels ~ s(Day)

Parametric coefficients:
            Estimate Std. Error z value Pr(>|z|)
(Intercept) 4.244442   0.005023     845   <2e-16

Approximate significance of smooth terms:
         edf Ref.df Chi.sq p-value
s(Day) 8.971      9   4898  <2e-16

R-sq.(adj) =   0.0205   Deviance explained = 6.78%
UBRE score = 120.31     Scale est. = 1         n = 610
```

The smoother has 8.9 degrees of freedom and is highly significant. Note that the smoother explains about 7% of the variation in the sandeel data. That is not a great deal, but is more than with the NB GLM, as that model explained almost nothing. The deviance of the Poisson GLM in M1 and the deviance of the Poisson GAM in M3 are 78959.17 and 73976.88, respectively. The deviance for M3 is slightly lower, so the GAM was a good choice. The plot(M3) command shows the smoother (Figure 3.7). There is an interesting upward trend; however the effect is spoiled, because we still have high overdispersion:

```
> sum(resid(M3, type = "pearson")^2) / M3$df.res

209.3184
```

This means that the model is not suitable. Another source of uncertainty is the large gap in the data. Note that there were no observations around day 2000, and this is just where the smoother goes down. This is disturbing and is probably caused by a downward trend in the first sampling period (up to day 1000) and an upward trend in the second period (between days 2500 and 3500). The smoother is interpolating these two blocks of sampling periods. It is valid to ask whether the large drop in the smoother around day 2000 makes sense. It may be reasonable to manually reduce the degrees of freedom.

The next logical step is to apply a GAM with a negative binomial distribution (see Figure 3.5), which is what we will do in the next section.

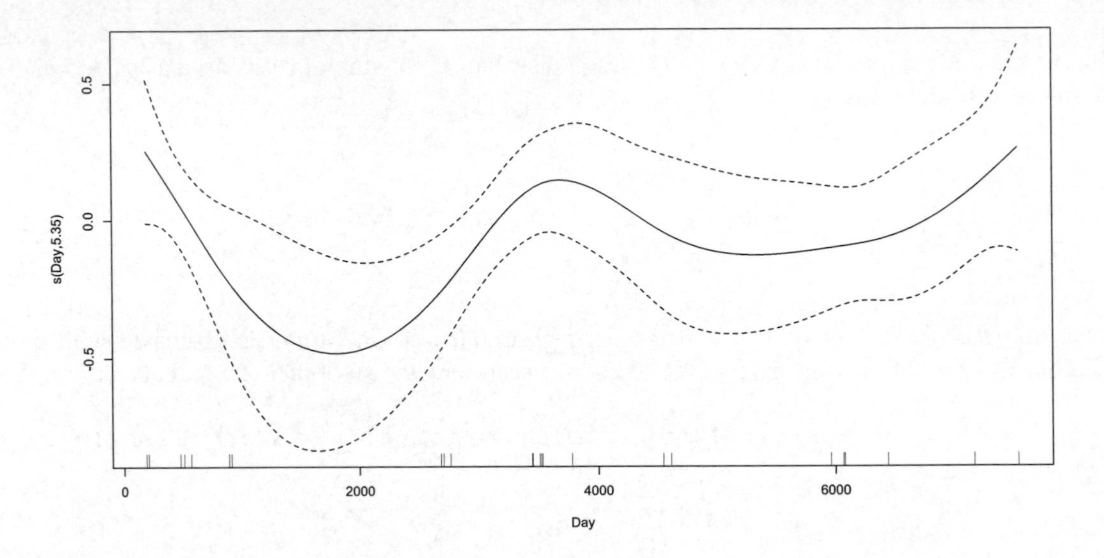

Figure 3.7. Smoother obtained by the Poisson GAM. Note the large confidence bands at time periods with no data.

3.6 GAM with a negative binomial distribution

3.6.1 Model and R code

The following negative binomial GAM is applied:

$$Sandeels_{is} \sim NB(\mu_{is}, k)$$

$$E(Sandeels_{is}) = \mu_{is} \quad \text{and} \quad \text{var}(Sandeels_{is}) = \mu_{is} + \frac{\mu_{is}^2}{k} = \mu_{is} + \alpha \times \mu_{is}^2$$

$$\log(\mu_{is}) = \alpha + f(Day_s)$$

To run this model in R we use the `negbin` option for the `family` option. We need to specify a range of possible values for k (called theta in the numerical output), and we choose values between 0.1 and 10.

```
> M4 <- gam(Sandeels ~ s(Day) , data = Fish2,
            family = negbin(c(0.1, 10), link = log))
> summary(M4)
```

```
Family: Negative Binomial(0.467)
Link function: log
Formula: Sandeels ~ s(Day)

Parametric coefficients:
            Estimate Std. Error z value Pr(>|z|)
(Intercept)  4.24948    0.05945   71.48   <2e-16

Approximate significance of smooth terms:
        edf Ref.df Chi.sq p-value
s(Day) 6.67  7.595   31.3   9e-05
R-sq.(adj) =  0.0209   Deviance explained = 4.29%
UBRE score = 0.24926   Scale est. = 1        n = 610

> sum(resid(M4, type = "pearson")^2) / M4$df.res

1.257864
```

The overdispersion is marginal, 1.26. Hence, in principle, this is a candidate model to present. The estimated value of k is 0.467 (which means that $\alpha = 1/0.467 = 2.141$). At this point we can compare all estimated models. The following code presents the deviances, AIC values, log likelihood values, and dispersion parameters for the models `M1` to `M4` in a single table. The AIC values indicate that the NB GAM is indeed better than the NB GLM.

```
> Z <- cbind(
    c(deviance(M1), deviance(M2), deviance(M3), deviance(M4)),
    c(deviance(M1)/M1$df.res, deviance(M2)/M2$df.res,
      deviance(M3)/M3$df.res, deviance(M4)/M4$df.res),
    c(AIC(M1), AIC(M2), AIC(M3), AIC(M4)),
    c(logLik(M1), logLik(M2), logLik(M3), logLik(M4)))
> colnames(Z) <- c("Deviance", "Dispersion", "AIC", "LogLik")
> rownames(Z) <- c("Poisson GLM", "NB GLM", "Poisson GAM", "NB GAM")
> Z
```

	Deviance	Dispersion	AIC	LogLik
Poisson GLM	78959.1733	129.867061	81871.389	-40933.694
NB GLM	749.5422	1.232800	6189.879	-3091.940
Poisson GAM	73976.8826	123.288892	76905.041	-38442.549
NB GAM	746.7102	1.239704	6168.698	-3076.678

3.6.2 Model validation of the negative binomial GAM

Before judging whether the NB GAM is indeed a good candidate, we need to apply a model validation and check the residuals for patterns. We extract Pearson residuals and apply a linear regression model in which the residuals are the response variable, and `month` is a categorical variable. Results show a weak, but significant, `month` effect ($F_{3,606} = 3.59$, $p = 0.01$), although the variance in the Pearson residuals explained by `month` is small (1.75%). Hence, it may be an option to add it as a categorical covariate to the model. We do not present such a model here, but results would be similar to that of `M4`.

Next we discuss independence. It is possible that multiple scats from the same seal were collected. This causes dependence, as an individual seal may have specific predation skills, and otolith counts of scat from an individual seal may be more similar than those of scat from different seals. However, no information is available on samples from a single seal, nor on the spatial positioning of scat. Hence we cannot take this type of dependence into account. We can only hope that sampling was done in such a way that not more than one scat from an individual seal was collected.

Another source of dependence is temporal dependence. Scat was collected repeatedly from the same location. Theoretically, it is possible that scat from the same seal was collected in sequential months, although it is unlikely that this happened. To double-check that the residuals of model M4 do not exhibit temporal correlation, we create a sample variogram of its Pearson residuals (Figure 3.8) using the following code. There is no obvious correlation structure in the residuals. Because time is one-dimensional, we use a column with ones in the variogram function.

```
> library(gstat)
> Fish2$Ones <- rep(1, length = length(Fish2$Day))
> mydata      <- data.frame(E4, Fish2$Day, Fish2$Ones)
> coordinates(mydata) <- c("Fish2.Day", "Fish2.Ones")
> Vario1 <- variogram(E4 ~ 1, mydata)
> plot(Vario1, col = 1 , pch = 16)
```

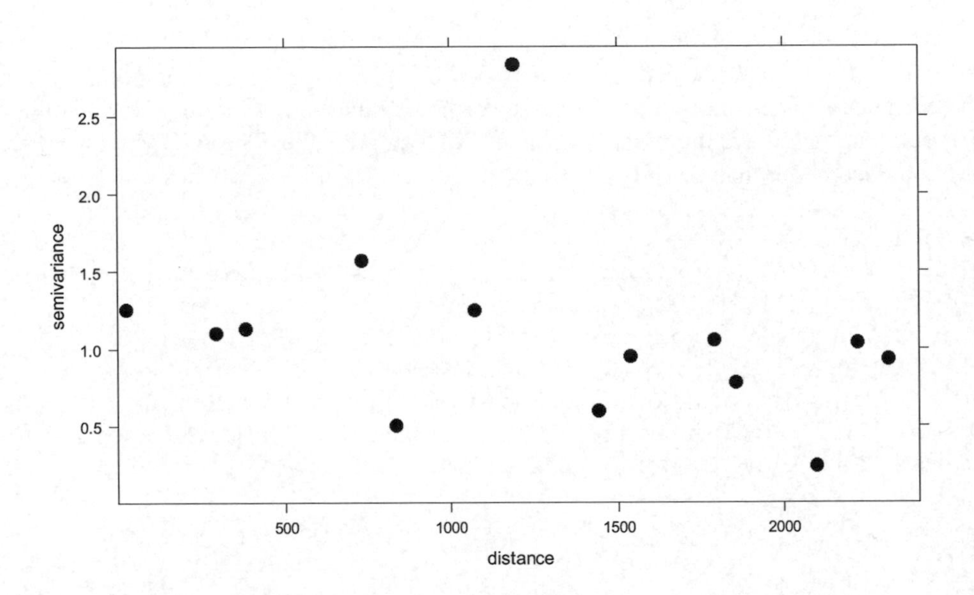

Figure 3.8. Sample variogram of the residuals plotted versus distance (days).

3.6.3 Model interpretation

The graph with the smoother for the NB GAM is obtained with the plot(M4) command (Figure 3.9). The smoother is significant at the 5% level, though we need to be cautious in interpreting those parts of the smoother where there is no data.

So this could be the end point of the statistical analysis. We applied a Poisson GLM and Poisson GAM on the data and discovered overdispersion; therefore we applied NB GLM and NB GAM models. These were better, and the NB GAM seems to be satisfactory, but the estimated value of k in the NB variance term is relatively small, and this means that the model is allowing for extra variance. In other words, the large overdispersion in the Poisson GLM (and GAM) is being captured by the quadratic variance term in the NB distribution. That is like shooting at a mosquito with a cannon. It does the job if it kills the mosquito, but do we really need it? Perhaps there is a reason for the overdispersion, and hiding it behind a quadratic variance component may mean that we miss some important information. In the simulation study in Chapter 2 we saw that, if the data are zero inflated, which can cause overdispersion, an NB GLM will give biased parameters. We need to ask why we have overdispersion. Why do we need the NB distribution? Do we have data with wide variation, which would justify the NB distribution, or is there something else causing overdispersion?

Overdispersion can be caused by zero inflation. Figure 3.10 shows a frequency plot, and the spike at 0 may indicate that we have zero inflation. However, zooming in on the smaller values (between 0 and 30) gives Figure 3.11, and we are not sure whether there is zero inflation. It seems that both zeros and ones were observed more frequently than other values. If the mean of the data is close to 1, all the zeros would comply with a Poisson distribution. However, if we sketch the fitted value (not shown here) of the Poisson or NB GAM, the resulting mean is not close to 1. Summarizing, the two spikes in Figure 3.11 are suspicious. We may have zero inflation, but we are not sure. For the owl data in Chapter 2 we created a similar frequency plot, and the results were more convincing.

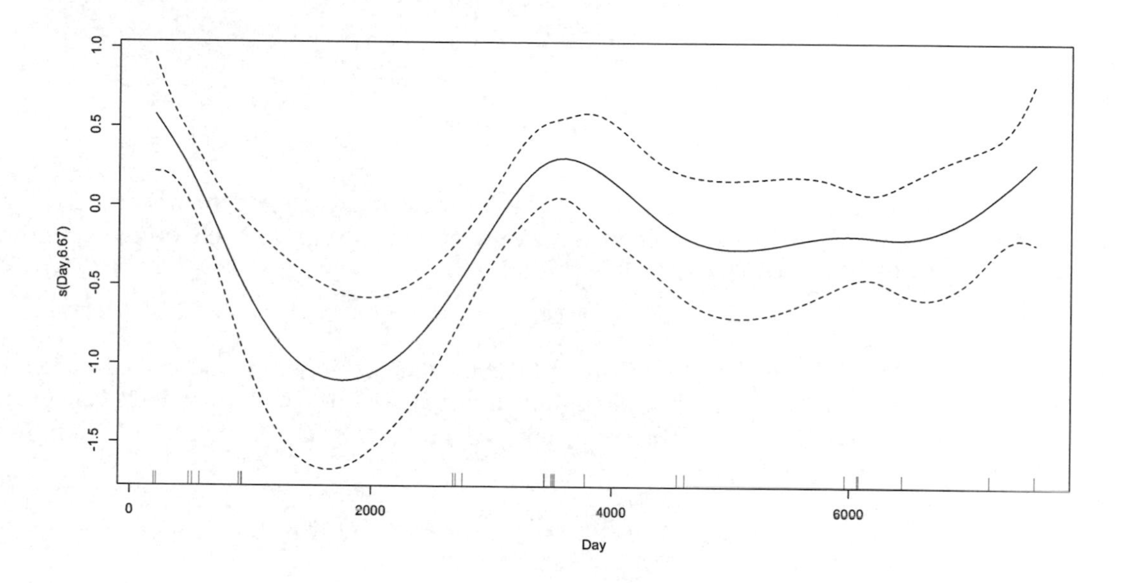

Figure 3.9. Smoother obtained by the negative binomial GAM.

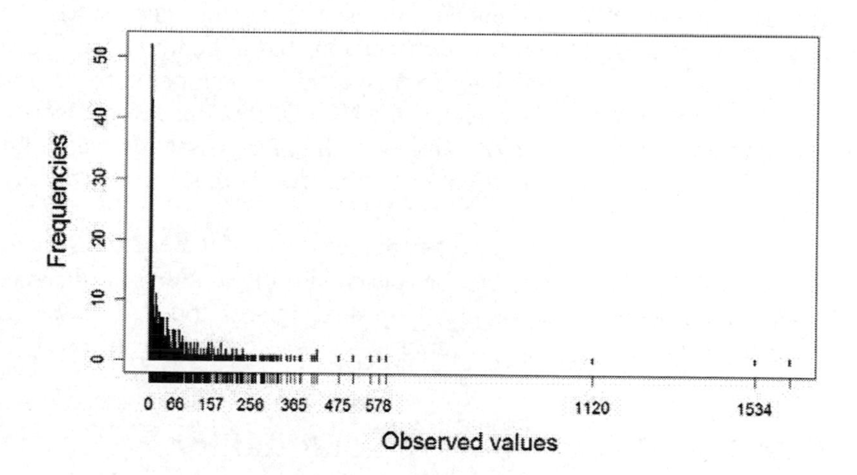

Figure 3.10. Frequency plot for the sandeel data. The horizontal axis shows the observed values, and the vertical axis shows frequency of observation of each value.

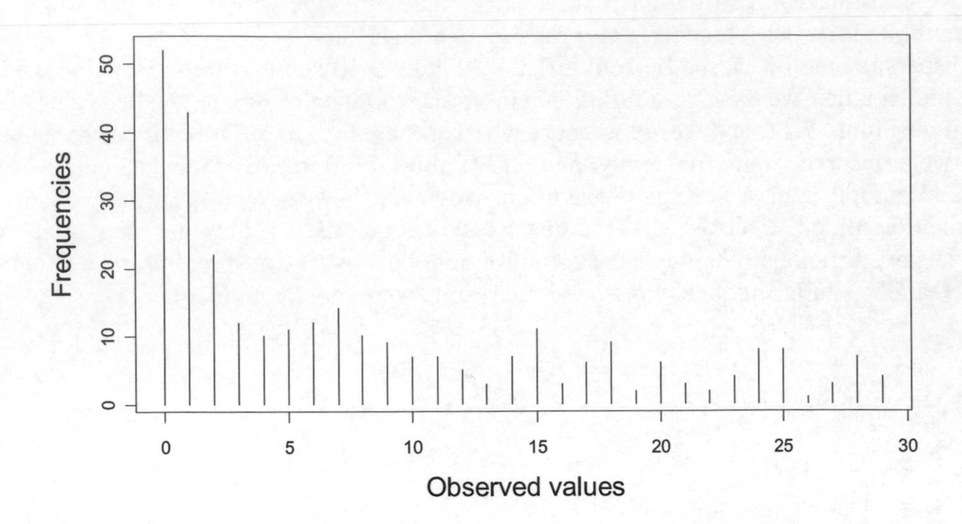

Figure 3.11. Frequency plot for the sandeel data. The horizontal axis shows the observed values between 0 and 30, and the vertical axis shows frequency of observation of each value. Note that 0 is the most frequently observed value.

We could have included Figure 3.10 and Figure 3.11 in the data exploration section, but wanted to present a learning experience by withholding the information. Figure 3.10 and Figure 3.11 were created with simple R code. The second `plot` function applies the correct labels starting from 0 (not from 1).

```
> plot(table(Fish2$Sandeels), xlab = "Observed values",
       ylab = "Frequencies", type = "h")
> plot(as.numeric(names(table(Fish2$Sandeels)[1:30])),
       table(Fish2$Sandeels)[1:30], xlab = "Observed values",
       ylab = "Frequencies", type = "h")
```

So, where does this leave us? We first applied a Poisson GLM and then an NB GLM. We noted slightly non-linear residual patterns, marginal overdispersion, and a month effect. We continued with an NB GAM, finding a non-linear trend and no overdispersion, which could be the end point of the analysis. In attempting to understand why we needed the NB distribution, we consider 'zero inflation,' but do we really want to enter the world of zero inflation? In the next section we will discuss the nature and origin of zeros. Based on that discussion, one could decide to start immediately with zero inflated models for these data.

Another consideration is that the `month` effect in the residuals of the NB GAM is reduced. It may be that the trend in Figure 3.9 merely reflects that the month of sampling is different. Plotting the sampling months superimposed on the figure may help shed light on this. If it is the case we may as well stop the analysis here and truncate the data further.

3.7 Zero inflated GAM with a Poisson distribution

We begin this section with a short summary of Chapter 11 in Zuur et al. (2009a), explain the principle of true and false zeros, and give the equations for a zero inflated Poisson and negative binomial GLM. There will also be a small overlap with Chapter 2 on owls.

3.7.1 True and false zeros

As we saw in the previous section, the value of 0 sandeel otoliths was more often observed than the value of 1 otolith, 2 otoliths, etc., and the fundamental question is why we have these zeros. Kuhnert et al. (2005) and Martin et al. (2005) classified types of errors that may lead to zeros: structural errors, design errors, observer errors, and 'animal' errors. Austin and Meyers (1996) added a fifth class, the naughty zeros. We will discuss these types of errors in more detail in Chapter 4, but will address them here in the context of the seals.

1. The naughty zeros are probably the easiest to identify in a data set. These correspond to samples with a probability of 0 of having any non-zero number of sandeel otoliths in a scat. Sampling for sandeel otoliths in seal scat in an area where there are no sandeels is an example. Such observations should be removed from the data before analysis. The key point with naughty zeros is that, when the probability of positive counts is 0, we will always and only measure zeros. That is not the case with the other types of zeros.
2. A design error means that poor experimental design or sampling practices are thought to be the reason for zero values. As an example, imagine collecting seal scat after a storm; all scat may have washed away. Sampling for an insufficient period of time, or over too small an area, reduces the chance of detecting a species even if it is present.
3. Another cause for zeros is observer error. In studies where otoliths of different species look similar, an inexperienced observer may end up with zero counts for relatively rare species, for those with otoliths that are difficult to identify, or simply because they don't recognize them among all the other individuals.
4. An animal error occurs when there are sandeels, but the seals don't eat them. Note that this is not a naughty zero, as we may potentially have sandeel otoliths in the scat. The probability of measuring positive counts is not 0.
5. A structural error means that the otolith is not present because the habitat is not suitable for sandeels.

The zeros due to design, survey, animal, and observer errors are also called false zeros, or false negatives. In a perfect world, we should not have them. The structural zeros are called positive zeros, true zeros, or true negatives.

For sandeel otoliths in seal scat data set, it can be argued that the zeros consist of true zeros and false zeros. A false zero may be due to a digested sandeel otoliths (theoretically a design error but there is not much you can do about this), or the otoliths may be in another scat from the same seal (design error). If a count data set consists of true and false zeros, zero inflated Poisson (or negative binomial) models should be applied. This argument can be used to initiate a statistical analysis with a zero inflated model, ignoring ordinary GLMs and GAMs. On the other hand, one could decide to start with ordinary GLMs and GAMs. This is what we did in the previous sections.

The presence of false zeros is not a *requirement* for the application of ZIP models; see Chapter 10 for a discussion.

3.7.2 The ZIP model

In this subsection we present the equations for a ZIP model. They are similar to those used in Chapter 2 for the owl data. We begin with the question, 'What is the probability of obtaining zero sandeel counts?' Let $Pr(Y_i)$ be the probability that, in scat i, we measure a certain number, Y_i, sandeels. The answer to the question is:

$$\begin{aligned} Pr(Y_i = 0) = & \; Pr(\text{False zeros}) + \\ & (1 - Pr(\text{False zeros})) \times Pr(\text{Count process gives a zero}) \end{aligned} \tag{3.1}$$

The component Pr(False zeros) is the probability that a false zero is obtained. The second component is derived from the probability that it is not a false zero, multiplied by the probability that it is a true zero. Basically, we divide the data into two imaginary groups. The first group contains only zeros (the false zeros). The second group is made up of the count data, which may produce zeros (true zeros) as well as values larger than zero.

We assume that the probability that Y_i is a false zero is binomially distributed with probability π_i; therefore, the probability that Y_i is not a false zero is equal to $1 - \pi_i$. Using this assumption, we can rewrite Equation (3.1):

$$\Pr(Y_i = 0) = \pi_i + (1 - \pi_i) \times \Pr(\text{Count process at site } i \text{ gives a zero}) \quad (3.2)$$

So, what do we do with the term Pr(Count process gives a zero)? We assume that the counts follow a Poisson or negative binomial distribution. Let us assume for simplicity that the count Y_i follows a Poisson distribution with expectation μ_i. Its probability function is:

$$f(y_i;\mu_i \mid y_i \geq 0) = \frac{\mu^{y_i} \times e^{-\mu_i}}{y_i!} \quad (3.3)$$

For a Poisson distribution, the term Pr(Count process gives a zero) is given by:

$$f(y_i = 0;\mu_i \mid y_i \geq 0) = \frac{\mu^0 \times e^{-\mu_i}}{0!} = e^{-\mu_i} \quad (3.4)$$

Hence Equation (3.2) can now be written as:

$$\Pr(y_i = 0) = \pi_i + (1 - \pi_i) \times e^{-\mu_i} \quad (3.5)$$

The probability that we measure a 0 is equal to the probability of a false zero, plus the probability that it is not a false zero, multiplied by the probability that we measure a true zero.

This was the probability that $Y_i = 0$. Let us now discuss the probability that Y_i is a non-zero count. This is given by:

$$\Pr(Y_i = y_i) = (1 - \Pr(\text{False zero})) \times \Pr(\text{Count process}) \quad (3.6)$$

Because we assumed a binomial distribution for the binary part of the data (false zeros versus all other types of data) and a Poisson distribution for the count data, we can write Equation (3.6) as:

$$\Pr(Y_i = y_i \mid y_i > 0) = (1 - \pi_i) \times \frac{\mu^{y_i} \times e^{-\mu_i}}{y_i!} \quad (3.7)$$

Hence, we have the following probability distribution for a ZIP model. The notation Pr() stands for probability.

$$\begin{aligned} \Pr(y_i = 0) &= \pi_i + (1 - \pi_i) \times e^{-\mu_i} \\ \Pr(Y_i = y_i \mid y_i > 0) &= (1 - \pi_i) \times \frac{\mu^{y_i} \times e^{-\mu_i}}{y_i!} \end{aligned} \quad (3.8)$$

The final step is to introduce covariates. Just as in Poisson GLM, we model the mean μ_i of the positive count data as:

$$\mu_i = e^{\alpha + \beta_1 \times X_{i1} + \text{L} + \beta_q \times X_{iq}} \quad (3.9)$$

Hence covariates are used to model the positive counts. What about the probability of a false zero, π_i? The easiest approach is to use a logistic regression model with only an intercept:

$$\pi_i = \frac{e^{\nu}}{1+e^{\nu}} \tag{3.10}$$

where ν is an intercept. If the process generating false zeros depends on covariates, we can include covariates in the logistic model:

$$\pi_i = \frac{e^{\nu+\gamma_1 \times Z_{i1} + \text{L} \ \gamma_q \times Z_{iq}}}{1+e^{\nu+\gamma_1 \times Z_{i1} + \text{L} \ \gamma_q \times Z_{iq}}} \tag{3.11}$$

We used the symbol Z for the covariates, as these may be different from the covariates that influence the positive counts. The γ represents regression coefficients.

We have a probability function in Equation (3.8) and unknown regression parameters α, β_1, ..., β_q, ν, γ_1, ..., γ_q. It is now necessary to formulate the likelihood based on the probability functions in Equation (3.8), take the logarithm and use a good optimisation routine to obtain parameter estimates and standard errors. We do not present all the mathematics here. See pg. 126 in Cameron and Trivedi (1998), or pg.174 in Hilbe (2007).

The only difference between a ZIP and ZINB is that the Poisson distribution for the count data is replaced with the negative binomial distribution. This allows for overdispersion from the non-zero counts. In Chapter 11 in Zuur et al. (2009[a]) the mean and variance of a ZIP and ZINB models are given. For the ZIP, we have:

$$\begin{aligned} E(Y_i) &= \mu_i \times (1-\pi_i) \\ \text{var}(Y_i) &= (1-\pi_i) \times (\mu_i + \pi_i \times \mu_i^2) \end{aligned} \tag{3.12}$$

The equations for the ZINB are: (the equation for the variance of a ZINB in Zuur et al. (2009a) has a small mistake).

$$\begin{aligned} E(Y_i) &= \mu_i \times (1-\pi_i) \\ \text{var}(Y_i) &= (1-\pi_i) \times \mu_i \times (1+\pi_i \times \mu_i + \frac{\mu_i}{k}) \end{aligned} \tag{3.13}$$

Now that we have expressions for the mean and variances of ZIP and ZINB models, we can calculate Pearson residuals:

$$\text{Pearson residual}_i = \frac{Y_i - (1-\pi_i) \times \mu_i}{\sqrt{\text{var}(Y_i)}}$$

Depending on whether a ZIP or ZINB is used, we substitute the appropriate variance.

3.7.3 Result for the ZIP GAM

In the previous section, we argued that the overdispersion may potentially be due to zero inflation of the number of sandeel otoliths in seal scat. We also argued that these zeros consist of true zeros and false zeros, and that therefore we should apply a ZIP GLM or ZIP GAM. Earlier analyses indicated a non-linear effect of `Day`; hence we will apply a ZIP GAM. If this model is still overdispersed, we will consider a ZINB GAM. We apply the ZIP GAM model:

$$Sandeels_{is} \sim ZIP(\pi_{is}, \mu_{is})$$
$$E(Sandeels_{is}) = \mu_{is} \times (1-\pi_{is}) \quad \text{and} \quad \text{var}(Sandeels_{is}) = (1-\pi_{is}) \times (\mu_{is} + \pi_{is} \times \mu_{is}^2)$$
$$\log(\mu_{is}) = \alpha + f_1(Day_s)$$
$$\text{logit}(\pi_{is}) = \nu + f_2(Day_s)$$

We use the VGAM package (Yee and Wild, 1996) to fit the zero inflated GAM in R. To avoid conflict between the mgcv and VGAM packages it is best to close R, restart it, and reload the data.

```
> Fish <- read.table(file = "SandeelsSeals.txt",  header = TRUE)
> Fish$Day <- Fish$dDay2
> Fish2 <- Fish[Fish$Month > 4 & Fish$Month < 9,]
> library(VGAM)
> M5 <- vgam(Sandeels ~ s(Day), family = zipoisson, data = Fish2)
```

To check whether the ZIP GAM is overdispersed, we need the expected value and variance in Equation (3.12). The following code extracts the expected values for the logistic link function and log link functions. The predict function produces fitted values (M5.prd) for the binary part of the model (first column of M5.prd) and the count part of the model (second column of M5.prd). These are on the scale of the linear predictor function. Therefore, we need to convert them to the original scale. For the binary part we write a one-line function, my.logit.inv, that converts the predictor function in the logistic link function to the scale of the original data; the results are in M5.prd.pi. For the log link function, we simply take the exponential; the results are in M5.prd.mu. Using these we can calculate the expected values and variance, and therefore the Pearson residuals. What remains is a matter of calculating the overdispersion parameter as the ratio of the squared Pearson residuals to the residual degrees of freedom. Note that there is still overdispersion.

```
> my.logit.inv <- function(x) {(1 / (1 + exp(-x)))}
> M5.prd        <- predict(M5)
> M5.prd.pi     <- my.logit.inv(M5.prd[,1])
> M5.prd.mu     <- exp(M5.prd[,2])
> M5.E          <- M5.prd.mu * (1 - M5.prd.pi)
> M5.Var   <- (1 - M5.prd.pi) * (M5.prd.mu + M5.prd.pi * M5.prd.mu^2)
> M5.res        <- (Fish2$Sandeels - M5.E)/sqrt(M5.Var)
> Dispersion    <- sum(M5.res^2) / (nrow(Fish2)-3-3 -2)
> Dispersion

32.70879
```

In case of non-convergence of the function vgam, it may be an option to use an intercept only model for the logistic part of the ZIP.

3.7.4 Result for the ZINB GAM

Because the ZIP GAM is overdispersed, we fitted a ZINB GAM. The code for this is:

```
> M6 <- vgam(Sandeels ~ s(Day), family = zinegbinomial,data = Fish2,
             control = vgam.control(maxit = 100, epsilon = 1e-4))
```

We encounter numerical optimisation problems and, therefore, lower the convergence criteria using the vgam.control argument (see above). We increase the number of iterations and lower the precision of the optimisation criteria (see help file of vgam). We again assess the overdispersion:

```
> M6.prd     <- predict(M6)
> M6.prd.pi <- my.logit.inv(M6.prd[,1])
> M6.prd.mu <- exp(M6.prd[, 2])
> M6.prd.k  <- exp(M6.prd[, 3])
> M6.E      <- M6.prd.mu * (1 - M6.prd.pi)
> M6.Var    <- (1-M6.prd.pi) * M6.prd.mu *
               (1 + M6.prd.pi * M6.prd.mu + M6.prd.mu / M6.prd.k)
> M6.res     <- (Fish2$Sandeels - M6.E) / sqrt(M6.Var)
> Dispersion <- sum(M6.res^2) / (nrow(Fish2) - 3 - 3 - 3)
> Dispersion

1.433458
```

The dispersion is sufficiently small to consider the ZINB GAM a suitable candidate model. However, plotting the smoothers for the binary and count parts (Figure 3.12) results in large standard errors around the binary smoother.

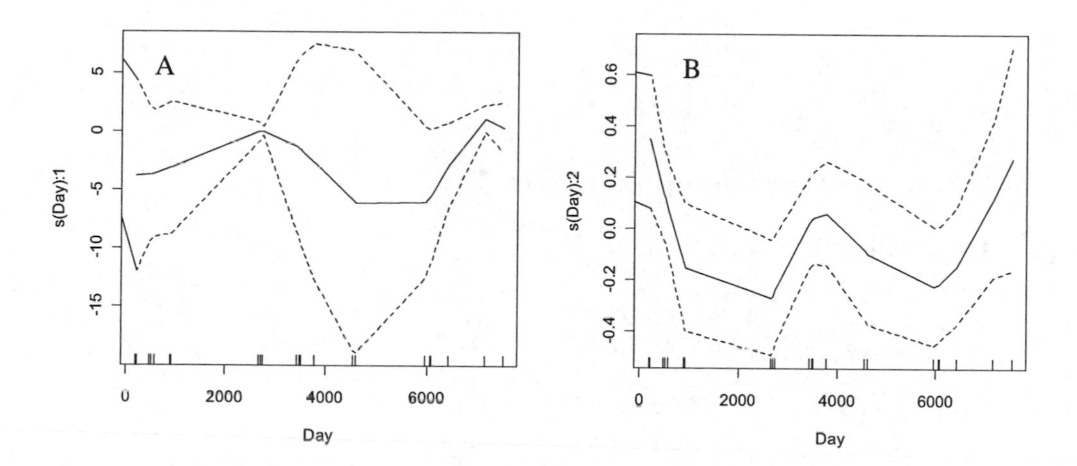

Figure 3.12. A: Binary smoother for the ZINB GAM. B: Smoother for count process of the ZINB GAM. Dotted lines indicate 95% confidence bands.

Based on these results, we fit a ZINB GAM model without a smoother in the predictor function of the binary model, resulting in the model:

$$
\begin{aligned}
&Sandeels_{is} \sim NB(\pi_{is}, \mu_{is}, k) \\
&E(Sandeels_{is}) = \mu_{is} \times (1 - \pi_{is}) \\
&\text{var}(Sandeels_{is}) = (1 - \pi_i) \times \mu_i \times (1 + \pi_i \times \mu_i + \frac{\mu_i}{k}) \\
&\log(\mu_{is}) = \alpha + f_1(Day_s) \\
&\text{logit}(\pi_{is}) = \nu
\end{aligned}
$$

Note that the logistic link function has only an intercept. The log link function has not changed. To fit this model in R we need a bit of VGAM magic using the `constraints` function:

```
> constraints(M6)
```

```
$`(Intercept)`
     [,1] [,2] [,3]
[1,]    1    0    0
[2,]    0    1    0
[3,]    0    0    1

$`s(Day)`
     [,1] [,2]
[1,]    1    0
[2,]    0    1
[3,]    0    0
```

The `` $`(Intercept)` `` tells us that we have an intercept in the logistic link function (first column), an intercept in the log link function (second column), and a parameter k (third column). The `` $`s(Day)` `` is used to tell R that there is a smoother in the logistic link function and a smoother in the log link function. If we could change the second part to:

```
$`s(Day)`
     [,1] [,2]
[1,]    0    0
[2,]    0    1
[3,]    0    0
```

there would be no smoother in the logistic link function. To do this, we use:

```
> m.SecondSmoother <- rbind(c(0,0),
                            c(0,1),
                            c(0,0))
> M7.clist <- list("(Intercept)" = diag(3),
                   "s(Day)"      = m.SecondSmoother)
> M7 <- vgam(Sandeels ~ s(Day), family = zinegbinomial,
             constraints = M7.clist, data = Fish2,
             control = vgam.control(maxit = 100,
                          trace = TRUE, epsilon = 1e-4))
```

The `constraints` option ensures that there is an intercept in the logistic link function, an intercept and smoother in the log link function, and also a parameter k. The numerical output from model `M7` is:

```
> coef(M7, matrix = TRUE)

             logit(phi)      log(munb)     log(k)
(Intercept)   -101.8284  4.331404e+00 -0.9101475
s(Day)           0.0000 -3.169173e-05  0.0000000
```

The first column contains the estimated intercept for the logistic link function. Note that its estimated value is -101.83. This means that the probability of a false zero, π_i, is 0. Substituting $\pi_i = 0$ in the equations for the ZINB GAM gives an NB GAM. Hence there is no need for a ZINB distribution, and we can present the results of model `M4`.

3.8 Final remarks

In the previous section we applied a ZINB GAM and noted that it is better to stay with the NB GAM. But suppose model M6 had been better. Although the VGAM package contains many useful tools, it is not as easy to work with as packages such as mgcv. In this section we present R code that is useful for working with VGAM. We will use model M5.

The ZIP GAM gives two sets of smoothers, one for the binomial part of the model and one for the Poisson. The binomial part gives information on the probability of measuring false zeros versus other types of data. The Poisson GAM provides information on the effects of the covariates (a smoother over time, in this case) while filtering out the false zeros, although we are not really filtering them out but merely dividing the data into two imaginary groups. It is the Poisson GAM that analyses the interesting part of the data; hence we should write our biological discussion around it.

To use the ZIP GAM to predict values, we need the expressions for the mean and the variance of a ZIP model (see also Equation (3.12):

$$E(Sandeel_{is}) = \mu_{is} \times (1 - \pi_{is})$$
$$\text{var}(Sandeels_{is}) = (1 - \pi_{is}) \times (\mu_{is} + \pi_{is} \times \mu_{is}^2)$$

The μ_{is} and π_{is} contain the smoothers. Because, at the time of writing, VGAM is still under development, it is difficult to obtain these smoothers, but we use the code:

```
> M5.prd.response   <- predict(M5, type = "response")
> M5.prd.terms      <- predict(M5, type = "terms")
> M5.prd.terms.raw <- predict(M5, type = "terms",
                              raw = TRUE, se.fit = TRUE)
> M5.const <- attr(M5.prd.terms, "constant")
> bin.eta.raw <- M5.const[1] + M5.prd.terms.raw$fitted.values[,1]
> poi.eta.raw <- M5.const[2] + M5.prd.terms.raw$fitted.values[,2]
> Mu <- exp(poi.eta.raw)
> Pi <- exp(bin.eta.raw)/ (1 + exp(bin.eta.raw))
> Fit <- Mu * (1 - Pi)
> cbind(fitted(M5), Mu * (1 - Pi))
```

This is intimidating R code. The object M5.const contains the estimated intercepts, and bin.eta.raw and poi.eta.raw are the two predictor functions. These are used to calculate μ_{is} (Mu) and π_{is} (Pi). To verify results, we compare the fitted values from M5 with Mu * (1 - Pi), and they are identical. The variance can be calculated in a similar way.

The problem is that if we want to present confidence intervals for the predictions, we can't simply take the square root of the variance, present a confidence interval as +/- 1.96 × SE, and pretend that the predictions are normally distributed. We may even get a confidence interval containing negative values.

One option to estimate a confidence interval around the predicted values is bootstrapping. The ZIP GAM is basically an advanced combination of a logistic GAM and a Poisson GAM. Bootstrapping GLMs and GAMs is discussed in Davison and Hinkley (1997), and their approach can be applied to each of the two GLM or GAM components of the zero inflated model. An alternative is to adopt a Bayesian approach. In Chapter 9 we show how to use MCMC to estimate the smoothers in a zero inflated GAM model. Within the MCMC code it is relatively easy to simulate expected values, and these may be used to create 'confidence' intervals.

3.9 Discussion

In this chapter we analysed numbers of sandeels in seal scat, illustrated how we moved from a Poisson GLM to a ZINB GAM, and subsequently reverted to the NB GAM. Our motivation to continue to the ZINB GAM was curiosity as to why the negative binomial distribution was needed. We argued that the zeros consisted of true zeros and false zeros, which could have been an argument in favour of beginning the statistical analysis with zero inflated models. However, estimated parameters of the ZINB GAM indicated that the probability of measuring false zeros is 0, which means that the ZINB GAM becomes an NB GAM. So we did a lot of work for nothing. However, we now know that the negative binomial GAM is capable of fitting the zeros. The negative binomial distribution is needed because the data contains large variations.

The outline of the steps is given in Figure 3.1. Did we follow the most sensible approach? Would a different roadmap have been more efficient? Having written the chapter, we think it can be done more logically, but it is always easy to conclude that afterwards. In a perfect world, the choice of which distribution to apply could be made before seeing the data, prior to initiating the analysis. It would not be difficult to argue that an observed number of zeros consists of true and false zeros; thus this is a justification for applying zero inflated models. But beginning with zero inflated models may be daunting; an ordinary Poisson GLM is simpler, faster, and easier to understand. Perhaps a practical approach is to apply a Poisson GLM and inspect the residuals for potential problems. Then, depending on the cause of the overdispersion, choose any of the potential extensions (Figure 3.13).

The 'opportunistic' aspect of these data is also a concern. We are aware of the difficulties related to sampling this type of data. The current data set is unique, as it covers a long time span, and the lead scientists were involved throughout the study period, ensuring that sampling protocols were not changed. We demonstrated seasonality in the data and that sampling did not always take place in the same months. We dealt with this by analysing data from only certain months. However, there are years in which no sampling took place. The biggest change in the temporal smoother seemed to coincide with a period of no available data, and that is worrying. Caution is needed in interpretation of the trends.

In this chapter we used the package `VGAM` to estimate the smoothers of a GAM with a zero inflated distribution. Although this is an interesting package with many useful functions, it is difficult to work with at times. In Chapter 9, WinBUGS is used to estimate a smoother in a GAM with a zero inflated distribution. WinBUGS takes considerably more time to set up, but once running properly it is easier to work with. But gaps in time stay gaps!

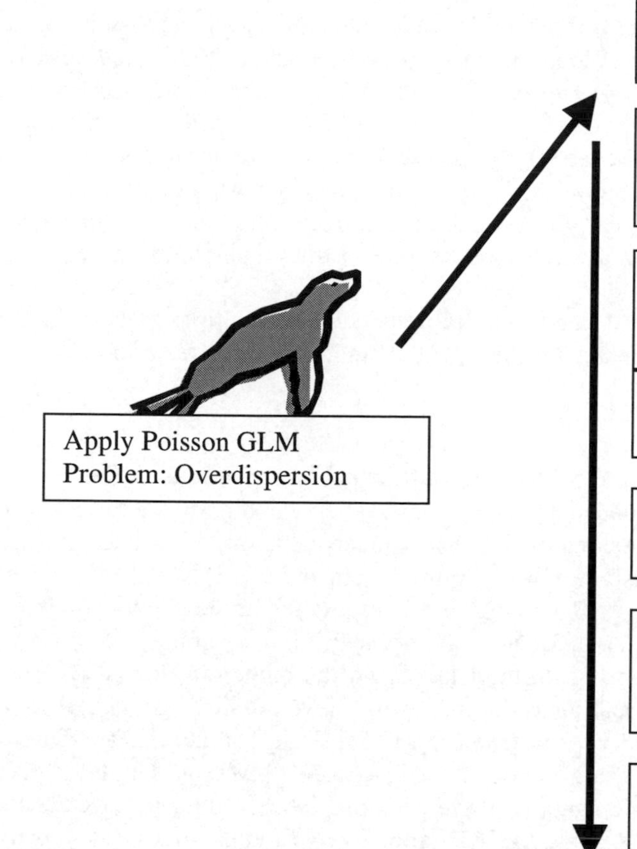

Plot residuals versus each covariate in the model. If patterns are apparent, apply a GAM.

Plot residuals versus each covariate not in the model. If patterns are apparent, add them to the GLM and refit the Poisson GLM.

Make a coplot of residuals. If patterns are apparent, add interactions and refit the Poisson GLM.

Check for outliers in the Pearson residuals. Inspect Cook's distance values.

Check for zero inflation in the data exploration. Is there a justification for zero inflated models?

Make an auto-correlation plot or variogram of the Pearson residuals (in case of time series or spatial data). This means GLMM or GAMM.

Plot residuals versus fitted values. Do we have the appropriate distribution and the appropriate link function? Consider the negative binomial distribution.

Choose one, or more, of the potential solutions and refit the GLM or GAM.

Figure 3.13. Improved roadmap for analysing overdispersed data.

3.10 What to present in a paper?

In the introduction of a paper on the seal scat study, you need to include the underlying questions and explain why the data were collected. In the biological part of the Methods section, explain how the data were collected. Referees may object to the 'sampling by opportunity' nature of the data, although those who work in the field will not necessarily be unaware of such issues. If referees are harsh regarding the time gaps in the data, results can be labelled as 'exploratory.' We strongly suggest that you are honest about the limitations of the data and include a paragraph in the discussion stating that results should be interpreted with caution. Also, the unbalanced design with respect to month requires some remarks in the discussion, but, by focussing on only the summer months, this should not be a serious issue.

In describing the statistics in the Methods section, explain how the optimal model was arrived at. In this case we would say that we started with a Poisson GLM and obtained non-linear residual patterns and overdispersion. We were suspicious of the ZINB GAM results in `VGAM` as there were convergence issues.

Cite R and any packages used. This information is available via the `citation()` function. To cite packages, use the results of `citation("VGAM")`, `citation("mgcv")`, `citation("lattice")`, and `citation("gstat")`. Include statements such as, 'We used the packages `VGAM` (Yee and Wild, 1996), `mgcv` (Wood 2006), `lattice` (Sakar, 2008), and `gstat` (Pebesma, 2004) from the software package R (R Development Core Team, 2011) to fit GAMs and zero inflated GAMs and to verify independence.'

In the Results section, present the NB GAM model `M4`. There is no need to show results of the other models, but explain the process of selecting it. This may include the discussion of the zeros and the zero inflated models. Include a statement that you conducted a model validation, and that there were no patterns in the residuals. It may be an option to present one of the graphs in the paper or as online supplemental material.

In the Discussion section, address potential dependence problems such as multiple scats from the same seal. Add cautionary remarks. Finally, explain the shape of the smoothers in terms of biology.

Acknowledgements

The original 1986-89 study was funded by the (then) Department of Agricultural Fisheries for Scotland. Sampling during the 1990s was funded by a Stevenson Scholarship, two Marie Curie awards, and the Carnegie Trust for the Universities of Scotland. Subsequent work has been funded from a variety of sources. Graham Pierce also received funding from the EU ANIMATE project (MEXC-CT-2006-042337). We are grateful to Paul Thompson, the late Peter Boyle, George Dunnet, Phil Hammond, Mike Harris, Tony Hawkins, John Hislop, and John Prime for support and advice; Tommy and Colin Mainland for providing transport to the island; and the numerous people who assisted with sample collection and processing and analysis, including Steve Adams, Antony Bishop, Ron Bisset, Adam Bratt, Ken Brown, Lois Calder, Avril Conacher, Tom Craig, Ian Diack, Jane Diack, Paul Doyle, Gaelle Ecobichon, Bill Edwards, Sara Frears, Beatriz Gimenez, Maxine Grisley, Mike Harding, Lee Hastie, Hassan Heba, Gema Hernandez, Steve Hoskin, Ishbel Hunter, Nick Jacob, Amanda Kear, Verena Keller, Linda Key, Kit Kovak, Steven Land, Evelyn Lauda, Sue Lewis, Andy Lucas, Gayle Mackay, Alasdair Miller, Nicola Murray, Cecile Nielsen, Fiona Read, Annette Ross, Gill Robertson, Nikki Rowberry, Angeles Santos, Laura Seaton, Arnor Sigfusson, Gabrielle Stowasser, Dominic Tollit, Sawai Wanghongsa, Jon Watt, Gerard Wijnsma, and Cynthia Yau.

4 GLMM and zero inflated Poisson GLMM applied to 2-way nested marmot data

Alain F Zuur, Kathleen R Nuckolls, Anatoly A Saveliev, and Elena N Ieno

4.1 Introduction

In this chapter, we analyse data on the number of female offspring produced by a female yellow-bellied marmot (*Marmota flaviventris*) during its lifespan. Understanding the variables that influence reproductive success provides insight into the evolution of particular traits, and can suggest new hypotheses about the adaptive significance of traits or behaviours (Clutton-Brock, 1988).

Marmots are widely distributed in mountainous environments of the western United States, and are most often found above 2,000 meters (Frase and Hoffmann, 1980). This habitat subjects marmots to environmental extremes that are likely to exert selection pressure on populations. The high elevations have a short growing season and frequent bouts of stormy weather, and marmots must produce young sufficiently early in the year for both young and adults to gain weight for the coming winter. Marmots have adapted to survive the long winter by reducing energy demands through an extended hibernation period (Armitage, 2003).

Living in a colony confers certain benefits. Members of the group may benefit from hearing the alarm calls of others, and reproductive success may be greater in larger groups. There are also costs; young females in groups with older females are often reproductively suppressed (Armitage, 1991). Marmots live in groups structured around an association of closely related mothers, daughters, and sisters. These female groups are called matrilines, and one or more matriline may share a suitable open meadow to form the colony.

It is common for species that live in groups to show social behaviours that may have adaptive functions, and marmots are no exception. A number of agonistic (conflict) and amicable (cohesive) behaviours are typical of marmots. Interactions among adult females within a matriline are typically amicable, while inter-matriline behaviours are primarily agonistic. Interactions between females and yearlings are harder to predict (Armitage, 1991). Not all yearling females remain in their natal colony. About half of all yearling females, and nearly all yearling males, disperse to new areas. This has the potential to release them from reproductive suppression, but it also reduces chances of survival and their lifetime reproductive success (Oli and Armitage, 2008).

The data analysed in this chapter are part of a larger data set explored in the PhD thesis of Kathleen Nuckolls. The data were collected by Kenneth Armitage and colleagues between 1963 and 2004 in the East River Valley near the Rocky Mountain Biological Laboratory in Gunnison County, Colorado, USA. Marmots were live-trapped each year of the study and age, sex, and reproductive status were recorded. The first time each marmot was trapped, it was fitted with numbered ear tags for identification. Animals were marked with dye for identification from a distance. Each female was monitored annually until she disappeared from the study area. Some females were recorded once and some as many as 14 times. For more information on the methods and the biology of marmots, see Armitage (1991).

The young were trapped as soon as possible after they first emerged from the natal burrow when they are about a month old and just being weaned. This is the most efficient way of estimating the number of young in a litter.

Variables that may explain the number of female offspring fell into four general categories: variables intrinsic to the female, environmental variables, demographic variables, and behavioural variables (Table 4.1). Variables intrinsic to the individual marmot include age, resident status, and whether the female reproduced the previous year. Demographic variables take into account the identity and number of members in the colony. Behaviour variables include interactions with age and sex classes within the matriline and with members of competing matrilines in the colony.

Underlying question

The underlying ecological question dealt with in this chapter is simple: What factors (intrinsic, environmental, demographic, or behavioural) affect reproductive success across the lifespan of a female marmot?

Table 4.1. Variables and terminology used.

	Name	Description of Data
Identifier		
	ANIMAL	Unique identification code for each Marmot
Response variable		
	YOUNG	Number of female young produced in the year
Intrinsic to female		
	AGE	Age of the marmot in years
	RESIDENT	Whether marmot was born into its current colony or immigrated into it
	REPLAST	Whether the female reproduced the previous year
Environmental		
	COLONY	Numeric code for colony and satellite sites
	ELEV	Elevation of colony in meters
	FSNOWF	Julian day of first snowfall >1 inch in fall prior to reproduction
	LSNOWF	Julian day of last snowfall >1 inch in spring prior to weaning
	FSNOWC	Julian day of first snow cover >1 inch in fall prior to reproduction
	LSNOWC	Julian day of last snow cover >1 inch in spring prior to weaning
	LWINT1	1st snowfall in autumn to last snowfall in following spring
	LWINT2	1st snow cover in autumn to last snow cover in following spring
	GROWSEAS1	Last snow cover in spring to first snow cover the next autumn
	GROWSEAS2	Last 24°F record in spring to first subsequent 24°F record
	SEPMAYPRECIP	Precip. during hibernation prior to reproduction, inches of water
	JUNEPRECIP	Inches of precip. in June in the year prior to reproduction
	JULYPRECIP	Inches of precip. in July in the year prior to reproduction
	AUGPRECIP	Inches of precip. in August in the year prior to reproduction
	JUNJULPRECIP	Inches of precip. in June and July summer prior to reproduction
	JULAUGPRECIP	Inches of precip. in July and Aug summer prior to reproduction
	JUNAUGPRECIP	Inches of precip. June to August summer prior to reproduction
	TEMPSEPTNOV	Mean Temp. (°F) Sept.–Nov., fall prior to reproduction
	TEMPJUNAUG	Mean Temp. (°F) June–August, summer prior to reproduction
	YEAR	Four digit study year

Table 4.1 continued. Variables and terminology used in this chapter.

	Name	Description of Data
Demographic		
	MYEARL	Number of male yearlings present from previous year's litter
	FYEARL	Number of female yearlings present from previous year's litter
	MATRILINE	Number of related females present
	OTHFEM	Number of unrelated females present in the colony
	MALEPRES	Whether there was a male observed
	NEWMALE	Whether the male observed was new to the site
	YOUNGAD	Number of younger adult matriline females present
	OFFSPYOUNG	Number of pups produced by younger matriline adults
	OLDAD	Number of older adult matriline females present
	OFFSPOLD	Number of pups produced by older matriline adults
	SAMEAD	Number of matriline females the same age as the subject
	OFFSSAME	Number of pups produced by matriline females the same age
Behaviour		
	MYOUNGAM	Rate of amicable interactions with related male young
	FYOUNGAM	Rate of amicable interactions with related female young
	MYEARLAM	Rate of amicable interactions with related male yearlings
	FYEARLAM	Rate of amicable interactions with related female yearlings
	FADAM	Rate of amicable interactions with related female adults
	MYEARLAG	Rate of agonistic interactions with related male yearlings
	FYEARLAG	Rate of agonistic interactions with related female yearlings
	FADAG	Rate of agonistic interactions with related female adults
	MYEARLAMNK	Rate of amicable interactions with unrelated male yearlings
	FYEARLAMNK	Rate of amicable interactions with unrelated female yearlings
	FADAMNK	Rate of amicable interactions with unrelated female adults
	MYEARLAGNK	Rate of agonistic interactions with unrelated male yearlings
	FYEARLAGNK	Rate of agonistic interactions with unrelated female yearlings
	FADAGNK	Rate of agonistic interactions with unrelated female adults
	MATDOM	Space Use Overlap with females within the matriline
	ADFEMDOM	Space Use Overlap with unrelated females

4.2 Data exploration and visualisation

We apply data exploration to check for zero inflation, outliers, and collinearity and to assess types of relationships (e.g. linear versus non-linear).

4.2.1 Import the data and get a first impression

Let us first import the data from the text file.

```
> Marmots <- read.table(file = "Marmots.txt", header = TRUE)
> names(Marmots)

 [1] "ANIMAL"      "YEAR"       "AGE"        "REPLAST"
 [5] "YOUNG"       "MYEARL"     "FYEARL"     "MATRILINE"
 [9] "OTHFEM"      "MALEPRES"   "NEWMALE"    "YOUNGAD"
[13] "OFFSPYOUNG"  "OLDAD"      "OFFSPOLD"   "SAMEAD"
[17] "OFFSSAME"    "X2YR"       "GPUP"       "MYOUNGAM"
[21] "FYOUNGAM"    "MYEARLAM"   "FYEARLAM"   "FADAM"
```

```
[25] "MYEARLAG"      "FYEARLAG"      "FADAG"         "MYEARLAMNK"
[29] "FYEARLAMNK"    "FADAMNK"       "MYEARLAGNK"    "FYEARLAGNK"
[33] "FADAGNK"       "COLONY"        "ELEV"          "RESIDENT"
[37] "FSNOWF"        "LSNOWF"        "FSNOWC"        "LSNOWC"
[41] "LWINT1"        "LWINT2"        "GROWSEAS1"     "GROWSEAS2"
[45] "SEPMAYPRECIP"  "JUNEPRECIP"    "JULYPRECIP"    "AUGPRECIP"
[49] "JUNJULPRECIP"  "JULAUGPRECIP"  "JUNAUGPRECIP"  "TEMPSEPTNOV"
[53] "TEMPJUNAUG"    "MATDOM"        "ADFEMDOM"
```

The variable YOUNG is the response variable, and most other variables are covariates. We do not show the results of the str(Marmots) command, as its output consumes two pages, but the variables COLONY and ANIMAL are coded numerically, and we convert them to categorical variables using:

```
> Marmots$fCOLONY <- factor(Marmots$COLONY)
> Marmots$fANIMAL <- factor(Marmots$ANIMAL)
```

4.2.2 Zero inflation in the response variable

Figure 4.1 contains the frequency plot for the response variable YOUNG, the number of female offspring. It contains many zeros. There are 507 animals with 0 young, a considerably higher number than the group with 1 young (113), 2 young (126), etc. This is a clear indication that the data are zero inflated. The R code to make Figure 4.1 is simple, and is given below. The table function determines the frequencies and the type = "h" draws the vertical lines.

```
> plot(table(Marmots$YOUNG), type = "h", xlab = "Number of young",
        ylab = "Numbers per class")
```

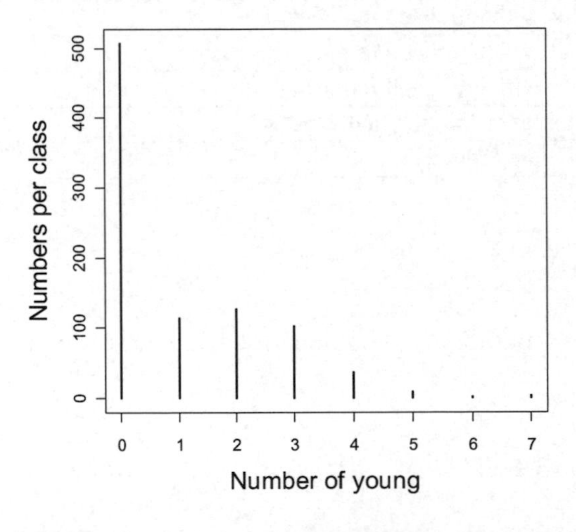

Figure 4.1. Frequencies of specific numbers of young per female; 507 animals bore 0 young, 113 had 1, 126 animals had 2, etc.

The first question we ask is why we have so many zeros. Chapters 2 and 3 contain summaries of the concept of false zeros and true zeros. False zeros are due to a poor experimental design (e.g. insufficient sampling period) or inadequate observing (counting zero young whereas, in reality, young exist). We do not imply that ecologists who collected the data did a poor job, but it may have been difficult at times to count the marmots due to behaviour of the animals, weather conditions, etc. True zeros are zeros occurring because the habitat is not favourable (e.g. harsh winter). In this case, the zeros are both false and true. Because we work with counts, we should use the Poisson or negative binomial distribution, and this means that we are in the world of zero inflated Poisson (ZIP) or zero inflated negative binomial (ZINB) models. These models were discussed in detail in Chapters 2 and 3. We recommend that you familiarise yourself with the information in these chapters before continuing. Based on the range of the observed values (from 0 to 7), we expect that a Poisson distribution will suffice.

During initial data exploration we noticed that if there were no males present the number of young was always 0. If there are no males present, the probability of non-zero young is 0. Austin and Meyers (1996) classified such observations as 'naughty noughts,' and argued that these should be removed from the data before doing the analysis. Given that the statistical population under study is the number of young, and not the number of tagged females, we agree and remove these 59 observations. This also reduces some of the numerical problems we encounter in the simulation techniques discussed later in this chapter. Note that in analysis of other aspects of the data, these 59 observations may not be classified as naughty zeros.

Male presence is coded as TRUE or FALSE, which makes removing rows with FALSE simple:

```
> Marmots2 <- Marmots[Marmots$MALEPRES, ]
```

4.2.3 Number of missing values

A quick view of the original data file shows a large number of missing values, and we would like to know the percent of missing values per variable. This can be calculated as:

```
> Marmots.X <- Marmots2[ ,c(
    "fANIMAL",         "fCOLONY",         "AGE",
    "MATRILINE",       "OTHFEM",          "MALEPRES",
    "NEWMALE",         "YOUNGAD",         "OFFSPYOUNG",
    "OLDAD",           "OFFSPOLD",        "SAMEAD",
    "OFFSSAME",        "MYOUNGAM",        "FYOUNGAM",
    "MYEARLAM",        "FYEARLAM",        "FADAM",
    "MYEARLAG",        "FYEARLAG",        "FADAG",
    "MYEARLAMNK",      "FYEARLAMNK",      "FADAMNK",
    "MYEARLAGNK",      "FYEARLAGNK",      "FADAGNK",
    "ELEV",            "RESIDENT",        "FSNOWF",
    "LSNOWF",          "FSNOWC",          "LSNOWC",
    "LWINT1",          "LWINT2",          "GROWSEAS1",
    "GROWSEAS2",       "SEPMAYPRECIP",    "JUNEPRECIP",
    "JULYPRECIP",      "AUGPRECIP",       "JUNJULPRECIP",
    "JULAUGPRECIP",    "JUNAUGPRECIP",    "TEMPSEPTNOV",
    "TEMPJUNAUG",      "MATDOM",          "ADFEMDOM")]

> Info.na <- colSums(is.na(Marmots.X)) / nrow(Marmots.X)
> print(Info.na, digits = 3)

     fANIMAL      fCOLONY          AGE    MATRILINE       OTHFEM
       0.000        0.000        0.010        0.018        0.061
    MALEPRES      NEWMALE      YOUNGAD   OFFSPYOUNG        OLDAD
       0.000        0.000        0.001        0.001        0.001
    OFFSPOLD       SAMEAD     OFFSSAME     MYOUNGAM     FYOUNGAM
       0.002        0.000        0.002        0.833        0.833
    MYEARLAM     FYEARLAM        FADAM     MYEARLAG     FYEARLAG
       0.849        0.830        0.698        0.849        0.829
       FADAG   MYEARLAMNK   FYEARLAMNK      FADAMNK   MYEARLAGNK
       0.698        0.870        0.841        0.645        0.865
  FYEARLAGNK      FADAGNK         ELEV     RESIDENT       FSNOWF
       0.838        0.645        0.000        0.000        0.000
      LSNOWF       FSNOWC       LSNOWC       LWINT1       LWINT2
       0.000        0.000        0.000        0.000        0.000
   GROWSEAS1    GROWSEAS2 SEPMAYPRECIP   JUNEPRECIP   JULYPRECIP
       0.000        0.000        0.000        0.000        0.000
   AUGPRECIP JUNJULPRECIP JULAUGPRECIP JUNAUGPRECIP  TEMPSEPTNOV
       0.000        0.000        0.000        0.000        0.000
  TEMPJUNAUG       MATDOM     ADFEMDOM
       0.000        0.821        0.786
```

The data frame `Marmot.X` contains all explanatory variables, and `Info.na` gives the missing values as a percent. It is important to realise that most statistical techniques will remove an entire row (one age reading) of data if there is a missing value in that row. Deciding what defines 'many' missing values is subjective, but we do not want to drop 85% of the data because one or more variables has 85% missing values. In this case, it is easy to decide what to do. There are sixteen variables with noticeably more missing values than the others, and we will discard them from further analyses. These are `MYOUNGAM`, `FYOUNGAM`, `MYEARLAM`, `FYEARLAM`, `FADAM`, `MYEARLAG`, `FYEARLAG`, `FADAG`, `MYEARLAMNK`, `FYEARLAMNK`, `FADAMNK`, `MYEARLAGNK`, `FYEARLAGNK`, `FADAGNK`, `MATDOM`, and `ADFEMDOM`.

4.2.4 Outliers

The next step in the data exploration is determining whether there are outliers in the response variable YOUNG and the numerical explanatory variables. Useful tools for this are boxplots and Cleveland dotplots. We do not present the dotplot of `YOUNG` here, as the frequency plot in Figure 4.1 gives an impression of the spread. There are no clear outliers in the response variable or in the explanatory variables (Figure 4.2). The multi-panel Cleveland dotplot was created with the `dotplot` function from the `lattice` package. There are several methods of using `dotplot`, but the simplest is to provide it with a matrix containing the data in columns. The `scales` argument ensures that each panel has its own x and y range. The -1, -2, -6, and -7 remove some of the covariates.

```
> library(lattice)
> Z <- as.matrix(Marmots.X[ , c(-1, -2, -6, -7)])
> dotplot(Z ,groups = FALSE, col = 1, cex  = 0.5, pch = 16,
          scales = list(x = list(relation = "free"),
                        y = list(relation = "free"), draw = FALSE),
          xlab = "Value of the variable",
          ylab = "Order of the data from text file")
```

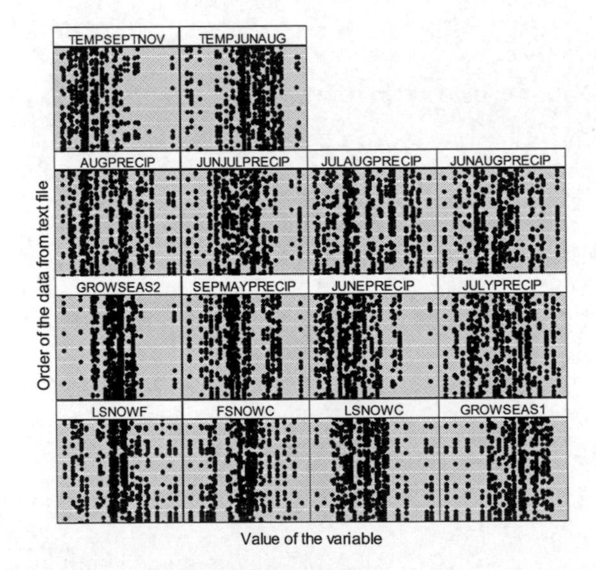

Figure 4.2. Cleveland dotplots of the explanatory variables. There are no explanatory variables with extremely small or extremely large values.

4.2.5 Collinearity

Zuur et al. (2010) provided a detailed explanation of collinearity (correlated explanatory variables) and discussed the use of multipanel scatterplots, Pearson correlation coefficients, variance inflation factors (VIF), and generalised variance inflation factors (GVIF). Before reading on, we suggest that you familiarise yourself with VIFs and GVIFs.

In the marmot study, we have a great many explanatory variables, and determining VIFs using all these covariates is a major challenge. Because the explanatory variables can be divided into four groups: intrinsic to individual marmot, environmental, demographic, and behavioural, we will investigate collinearity per group.

The environmental group

The VIF function cannot be applied to all 15 environmental explanatory variables. This is an indication that there is 100% collinearity in the explanatory variables (one variable being a linear combination of other variables). Dropping JUNJULPRECIP and JULAUGPRECIP solves this problem. The remaining set of explanatory variables still demonstrates strong collinearity, as can be seen from the estimated VIF values.

```
> source(file = "HighstatLib.R")
> Marmots.X <- Marmots2[,c(
      "ELEV",          "FSNOWF",         "LSNOWF",         "FSNOWC",
      "LSNOWC",        "LWINT1",         "LWINT2",         "GROWSEAS1",
      "GROWSEAS2",     "SEPMAYPRECIP",   "JUNEPRECIP",     "JULYPRECIP",
      "AUGPRECIP",     "JUNJULPRECIP",   "JULAUGPRECIP",   "JUNAUGPRECIP",
      "TEMPSEPTNOV",   "TEMPJUNAUG")]
> corvif(Marmots.X[ , c(-14, -15)])

Variance inflation factors
                      GVIF
  ELEV            1.062392
  FSNOWF          4.995907
  LSNOWF          4.891540
  FSNOWC        120.097729
  LSNOWC        108.083927
  LWINT1          6.257708
  LWINT2        286.112461
  GROWSEAS1       6.413001
  GROWSEAS2       1.860775
  SEPMAYPRECIP    1.993273
  JUNEPRECIP      6.463727
  JULYPRECIP      8.511789
  AUGPRECIP       7.953277
  JUNAUGPRECIP   21.207554
  TEMPSEPTNOV     2.025825
  TEMPJUNAUG      1.928070
```

The `source` function is used to load our VIF code into the `HighStatLib.R` file (available from the book website at www.highstat.com). The data frame `Marmots.X` contains the environmental explanatory variables, and the `corvif` function is only applied to the selected covariates. The estimated VIF values indicate that there is serious collinearity, and therefore we drop LWINT2. The process of calculating VIF values and dropping the variable with the highest VIF is continued until all VIFs are smaller than 3. The variables JUNAUGPRECIP, LWINT1, and GROWSEAS1 are dropped, in that order. If any of the remaining variables are significant in the statistical analysis, we must mention in the discussion that the significant variable may not be *the* driving variable, but that due to collinearity, one of the omitted variables may have an impact on the number of young.

The selected explanatory variables are stored in a data frame `X.env` using the R command:

```
> X.Env <- Marmots2[ ,c("ELEV", "FSNOWF", "LSNOWF", "FSNOWC",
```

```
        "LSNOWC", "GROWSEAS2", "SEPMAYPRECIP", "JUNEPRECIP",
        "JULYPRECIP", "AUGPRECIP", "TEMPSEPTNOV", "TEMPJUNAUG")]
```

The demographic explanatory variables

Using the same process as above, we drop MATRILINE and OFFSPOLD. The VIF indicates that we should drop OLDAD, but, for biological reasons, we instead remove OFFSPOLD, as it is collinear with OLDAD. The remaining variables are combined with:

```
> X.Demog <- Marmots2[,c( "MYEARL", "FYEARL", "OTHFEM", "MALEPRES",
                          "NEWMALE", "YOUNGAD", "OFFSPYOUNG",
                          "OLDAD", "SAMEAD", "OFFSSAME")]
```

The intrinsic explanatory variables

The VIF values are all smaller than 3, and the data are combined using:

```
> X.Intrin <- Marmots2[,c("AGE", "RESIDENT", "REPLAST")]
```

The behaviour explanatory variables

Because of the large number of missing values, these variables were dropped (see Subsection 4.2.3).

All selected explanatory variables

All selected explanatory variables can be combined using:

```
> Xsel <- cbind(X.Intrin, X.Env, X.Demog)
```

When applying VIFs on the selected variables, they are all smaller than 3, implying that they exhibit no serious collinearity. Instead of using VIF values, it is also possible to apply principal component analysis on groups of the explanatory variables, and use the extracted axes as new covariates. The problem with this approach lies in comprehending what such an axis means in terms of biology. Note that we still have 25 covariates!

4.2.6 Visualisation of relationships between number of young and the covariates

Before applying any statistical models, we want to visualise the relationships between the response variable and each explanatory variable. We start with the categorical explanatory variables `colony` and `animal`, as we expect correlation between the number of female offspring of an individual animal and between the number of offspring of animals from the same colony. These are the two covariates that will later be used for random effects (Figure 4.3). A strong correlation among observations of a single animal or colony results in boxplots that are different from one another. In this case, all `COLONY` boxplots show similarities, and the boxplots for `ANIMAL` are inconclusive. The R code to produce Figure 4.3 is simple and is given below. The `par` function prepares a graphic window with two panels, each containing a boxplot. It is also possible to make a multi-panel boxplot, but it does not provide a clearer picture.

```
> par(mfrow = c(1, 2))
> boxplot(YOUNG ~ fCOLONY, data = Marmots2, xlab = "Colony",
          ylab = "Number of young")
> boxplot(YOUNG ~ fANIMAL, data = Marmots2, xlab = "Animal",
          ylab = "Number of young")
```

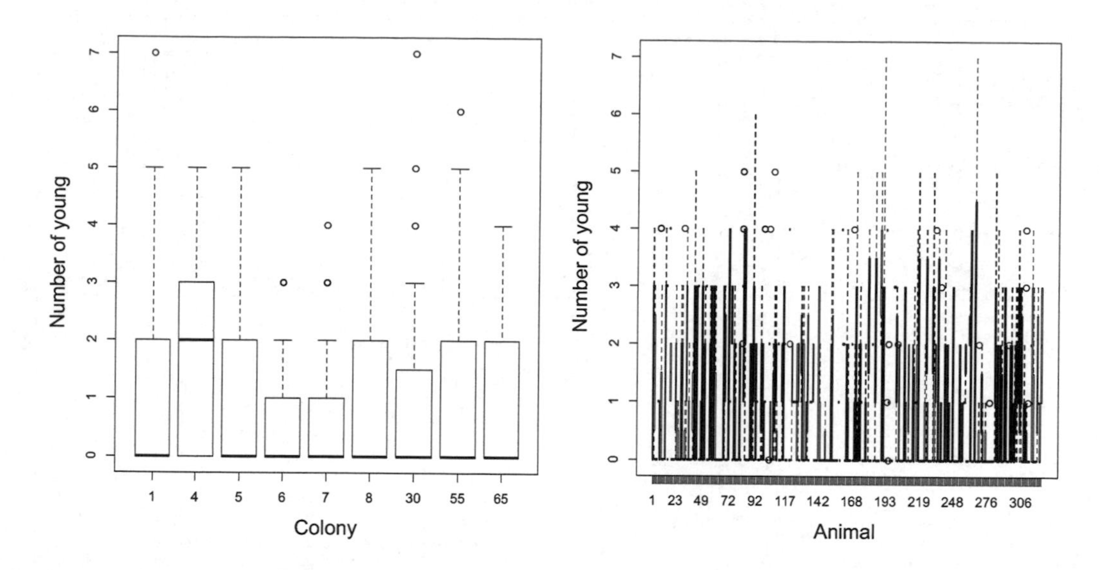

Figure 4.3. Left: Boxplot of the number of young, conditional on colony. Right: Boxplot of the number of young, conditional on animal.

Figure 4.4 shows boxplots of the number of young conditional on `REPLAST`, `RESIDENT`, `MALEPRES`, and `NEWMALE`. We use the original `MARMOT` object to show why we decided to remove 59 rows. There is no clear pattern in any of these graphs except for the `MALEPRES` boxplot, which shows that if there are no males present, as is the case for 59 observations, the number of young is zero. There is no situation where we have a non-zero number of young with no males present. Common sense will argue that we have zero probability of a non-zero number of young, if there are no males present.

R code to make Figure 4.4 is complicated. First, we need to create three variables, `Y4`, `X4`, and `ID4`. `X4` contains the four concatenated variables `REPLAST`, `RESIDENT`, `MALEPRES`, and `NEWMALE`, and `ID4` indicates to which variable an observation belongs. Finally, `Y4` contains the variable `YOUNG` repeated four times. The multi-panel boxplot is made with the `bwplot` function from the `lattice` package.

```
> library(lattice)
> Y4 <- rep(Marmots$YOUNG, 4)
> X4 <- with(Marmots, c(REPLAST, RESIDENT, MALEPRES, NEWMALE)
> ID4 <- rep(c("REPLAST", "RESIDENT", "MALEPRES", "NEWMALE"),
             each = nrow(Marmots))
> bwplot(Y4 ~ X4 | ID4, horizontal = FALSE,
    ylab = "Number of young",
    xlab = "Levels of categorical variables",
    par.settings = list(
      box.rectangle = list(col = 1),
      box.umbrella  = list(col = 1),
      plot.symbol   = list(cex = .5, col = 1)))
```

Figure 4.5 shows the response variable `YOUNG` plotted versus all continuous explanatory variables. There seems to be a non-linear age effect, and also a MYEARL and OFFSPYYOUNG effect. The statistical analysis presented later will tell us whether this is indeed the case.

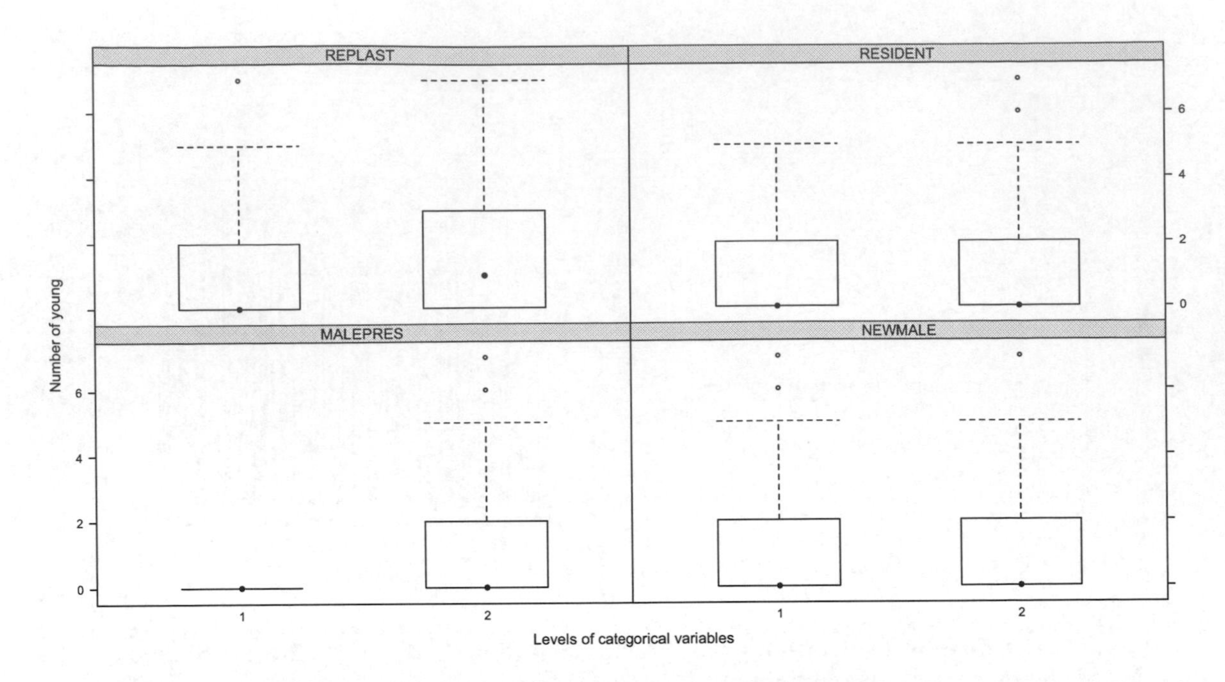

Figure 4.4. All categorical explanatory variables (except `colony` and `animal`) plotted versus the number of young. No males present means that there are no young.

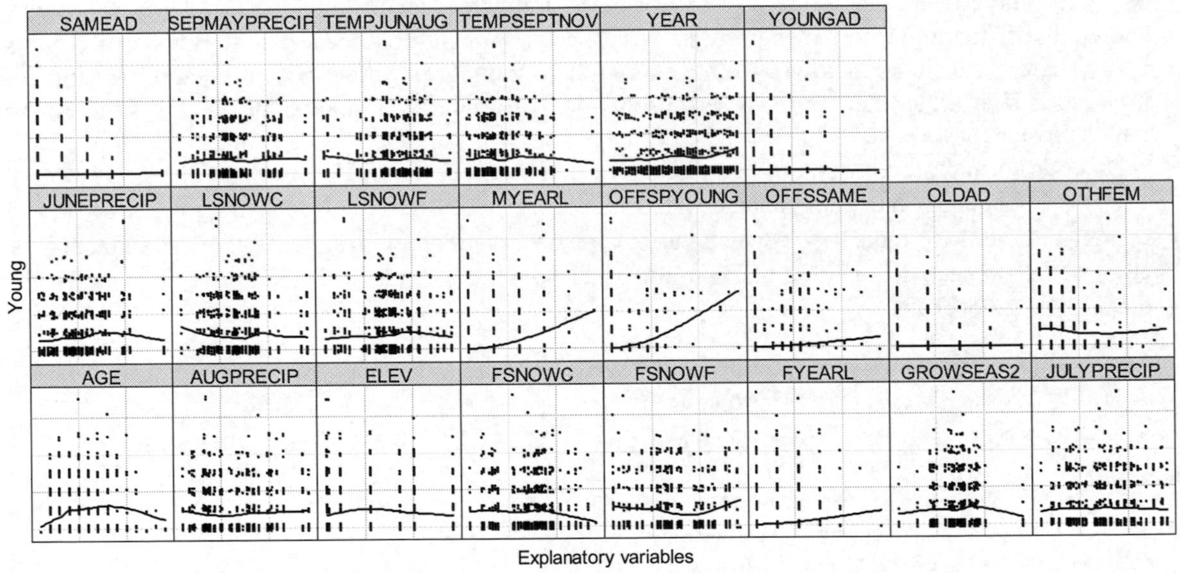

Figure 4.5. All continuous explanatory variables plotted versus the number of young. A LOESS smoother was added to aid visual interpretation.

The R code for this figure is given below. The `xyplot` makes the multiplanel scatterplot of `YOUNG` against each covariate. First, we need to put all selected covariates into a single long vector, `Xvec`. Note that we add `YEAR`. We had not included it in the data exploration up to this point, but there may be reasons based on biology to include it in the analysis. The problem is that age and year are collinear to a certain level, although all VIF values are smaller than 3. The variable `YOUNG` is repeated 22 times, and is also stored in the vector `Yvec`. Finally, `IDvec` informs the `xyplot` function which observations are from the same variable. The options in the `scales` and `panel` functions are cosmetic to make an attractive graph. Change them and see what happens.

```
> library(lattice)
> X22 <- Marmots2[,c(
     "YEAR", "AGE", "MYEARL", "FYEARL", "OTHFEM", "YOUNGAD",
     "OFFSPYOUNG", "OLDAD", "SAMEAD", "OFFSSAME", "ELEV", "FSNOWF",
     "LSNOWF", "FSNOWC", "LSNOWC", "GROWSEAS2", "SEPMAYPRECIP",
     "JUNEPRECIP", "JULYPRECIP", "AUGPRECIP", "TEMPSEPTNOV",
     "TEMPJUNAUG")]
> Xvec   <- as.vector(as.matrix(X22))
> Yvec   <- rep(Marmots2$YOUNG, 22)
> IDvec <- rep(colnames(X21), each = 896)
> xyplot(jitter(Yvec) ~ Xvec | IDvec ,
      type = "s", pch = 16, col = 1, cex = 0.2,
      scales = list(x = list(relation = "free"),
                    y = list(relation = "same"), draw = FALSE),
      xlab  = "Explanatory variables", ylab = "Young",
      panel = function(x, y) {
       panel.grid(h = -1, v = 2)
       panel.xyplot(x, y, col = 1, pch = 16, cex = 0.2)
       panel.loess(x, y, span = 0.5, col = 1, lwd = 2) })
```

Finally we present some basic information on the number of colonies using the `table` function.

```
> table(Marmots2$fCOLONY)

  1   4   5   6   7   8  30  55  65
176  68 118  31  50  39  46 252  57
```

There are 9 colonies, and the `table` function shows how many observations we have per colony (how often was it sampled), which is between 31 (colony 6) and 252 (colony 55); thus the data are highly unbalanced. It is interesting to look at the number of observations per marmot, for example for colony number 5 we have:

```
> table(Marmots2$ANIMAL[Marmots2$fCOLONY == "5"])
```

Results are not shown here, but some animals were observed only once and others eight times.

4.3 What makes this a difficult analysis?

There are multiple issues that make the statistical analysis of the marmot data a challenging exercise:

1. There are 25 covariates in addition to colony and animal, and a certain degree of relationship between these variables may be expected. We dealt with linear collinearity using VIF values, but there may be non-linear relationships present. In addition, we have limited prior knowledge in the sense that we are not able to come up with 10 – 15 potential models and apply the techniques described in Burnham and Anderson (2002).
2. We have 2-way nested data, multiple observations per marmot and multiple marmots per colony (Figure 4.6). It is likely that the number of young of marmots per female within a colony is more similar than the number per females of different colonies. Also, observations of the same marmot are likely to be correlated. A marmot in good condition living in a good habitat patch may produce consistently large litters year after year. Alternatively, because conditions at high

altitudes are harsh, marmots may not readily recover from reproduction over the course of the short growing season. This may lead to a negative correlation of successive years; a large number of young produced in one year may lead to a small number the following year. Using colony and animal nested in colony means that we allow for correlation among observations of the same animal and also between observations from the same colony. In more advanced models, we could try to include the auto-regressive correlation structure, so that sequential observations are more similar than are those further apart. That will not be done here, as the issues are already complex, and there are relatively few observations per animal.

3. The response variable `YOUNG` is likely to be zero inflated. In the previous section we argued that we have true and false zeros, and therefore zero inflated Poisson GLM (ZIP) or zero inflated negative binomial GLM (ZINB) models need to be applied. Due to the small range of the response variable we expect that a ZIP will be adequate.
4. The relationship between the number of young and the prime variable of interest, age of the female marmot, may be non-linear.

This amounts to arguments sufficient to make this a demanding analysis.

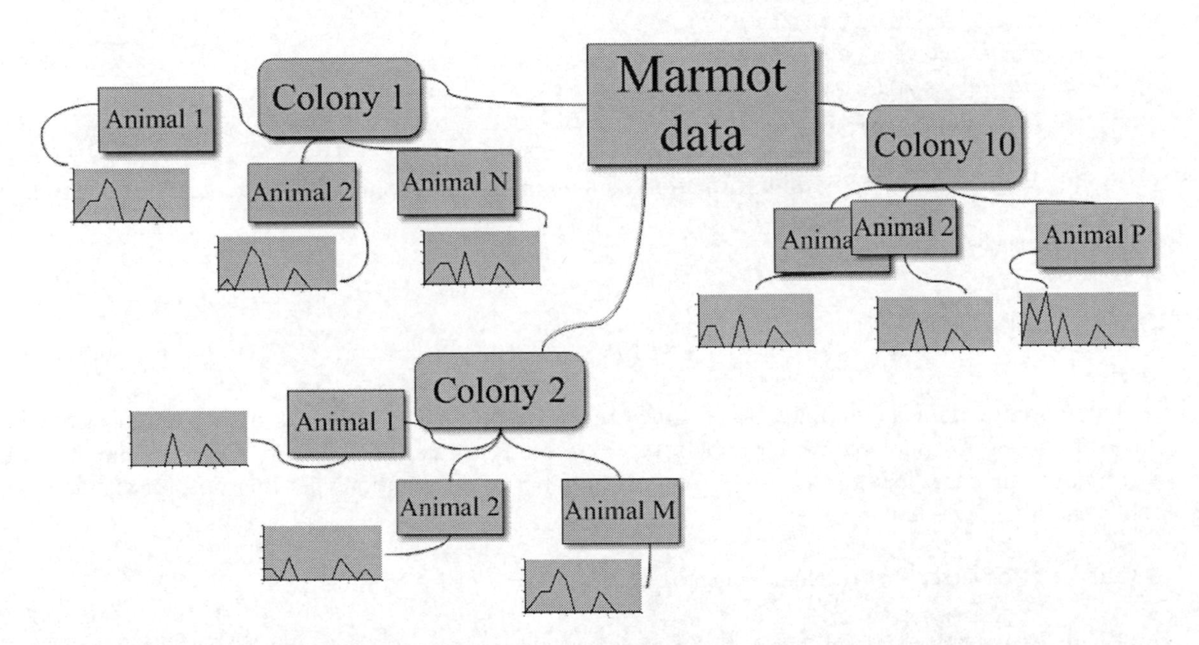

Figure 4.6. Layout of the marmot data. The full data set consists of 10 colonies with 9 used in these analyses, and from each colony multiple marmots have been observed. The number of young produced by an individual female marmot at different ages was counted. The data are 2-way nested: multiple observations per marmot, and `animal` is nested in `colony`.

4.4 Which statistical techniques and software to apply?

The problems sketched in Section 4.3, plus the information revealed by the data exploration, indicate that we must deal with zero inflated models for nested data. As there may be non-linear relationships between response variables and covariates, we may need smoothing techniques such as generalised additive modelling (GAM). However, it is most likely that a GAM with 20 – 25 smoothers will have numerical estimation problems. It may be more effective to use generalised linear modelling (GLM) techniques, as these are more numerically stable. To deal with non-linear patterns we consider quadratic terms. Due to the presence of zero inflation and the nested design of the data, it seems sensible to apply zero inflated Poisson GLM with random effects for `animal` and `colony`. This makes it a ZIP GLMM for 2-way nested data.

We will use Markov Chain Monte Carlo (MCMC) techniques. Therefore, we recommend that you are fully familiar with Chapters 1 and 2 before continuing.

4.5 Short description of the statistical methodology

We ended the previous section with the decision to apply a ZIP GLMM for 2-way nested data. Such a model is complicated, and, to clarify, we will begin by explaining the Gaussian linear mixed effects model for 2-way data and discuss its matrix notation and induced correlation structure. We will then give similar information for a Poisson GLM and GLMM using the same data, and finally present the models in the zero inflated context.

4.5.1 Viewing it (wrongly) as a Gaussian linear mixed effects model

To explain the matrix notation, we keep things simple and only use `age` as a covariate. Recall that there are 9 colonies containing from 16 to 73 animals. From each animal we have multiple observations (counts of the number of young). To reduce complexity of the mathematical notation, we assume for the moment that there are 9 colonies, 3 animals per colony, and each animal is sampled twice (i.e. its young are counted in two separate years). Once we have introduced the matrix notation, we will describe generalization of the notation for our unbalanced data set.

We define y_{ijk} as the number of young at observation k of animal j in colony i. The indices i, j, and k range from 1 to 9, 1 to 3, and 1 to 2, respectively. A linear mixed effects model, using `age` as fixed covariate and `animal` and `colony` as random effects, is given by:

$$\begin{aligned}
&y_{ijk} = \alpha + \beta \times Age_{ijk} + b_i + b_{ij} + \varepsilon_{ijk} \\
&b_i \sim N(0, \sigma^2_{colony}) \\
&b_{ij} \sim N(0, \sigma^2_{animal}) \\
&\varepsilon_{ijk} \sim N(0, \sigma^2) \\
&i = 1,...,9, \quad j = 1,...3, \quad k = 1,2
\end{aligned} \tag{4.1}$$

The term $\alpha + \beta \times \text{Age}_{ijk}$ is the fixed part of the model and is of prime interest, as it shows whether there is an age effect. The intercept b_i allows for a random variation around the intercept α for each colony, and b_{ij} does the same, but for each animal. Thus there is a single overall intercept α, and the term b_i allows for random variation around the intercept α for each colony and b_{ij} for random variation around the intercept α for each animal within a colony. Finally, the error term ε_{ijk} is 'real' noise. The random intercepts b_i, b_{ij}, and ε_{ijk} are assumed to be independent.

Instead of the notation in Equation (4.1), we use matrix notation, as it will help to explain the underlying correlation structure for the observed data. Matrix notation is not something that every reader may be familiar with, but all you need to know is that:

$$\begin{pmatrix} a & b \\ c & d \end{pmatrix} \times \begin{pmatrix} e \\ f \end{pmatrix} = \begin{pmatrix} a \times e + b \times f \\ c \times e + d \times f \end{pmatrix}$$

This is pure notation, nothing threatening so far. Using this type of notation, we can write the linear mixed effects model as:

$$\begin{pmatrix} y_{11} \\ y_{12} \\ y_{21} \\ y_{22} \\ y_{31} \\ y_{32} \end{pmatrix}_i = \begin{pmatrix} 1 & Age_{11} \\ 1 & Age_{12} \\ 1 & Age_{21} \\ 1 & Age_{22} \\ 1 & Age_{31} \\ 1 & Age_{32} \end{pmatrix}_i \times \begin{pmatrix} \alpha \\ \beta \end{pmatrix} + \begin{pmatrix} 1 & 1 & 0 & 0 \\ 1 & 1 & 0 & 0 \\ 1 & 0 & 1 & 0 \\ 1 & 0 & 1 & 0 \\ 1 & 0 & 0 & 1 \\ 1 & 0 & 0 & 1 \end{pmatrix}_i \times \begin{pmatrix} b \\ b_1 \\ b_2 \\ b_3 \end{pmatrix}_i + \begin{pmatrix} \varepsilon_{11} \\ \varepsilon_{12} \\ \varepsilon_{21} \\ \varepsilon_{22} \\ \varepsilon_{31} \\ \varepsilon_{32} \end{pmatrix}_i$$

The small index i at the lower right of each vector and matrix refers to COLONY. This looks intimidating, but what we have done is write, for a particular colony, i, the equation $y_{ijk} = \alpha + \beta \times \text{Age}_{ijk} + b_i + b_{ij} + \varepsilon_{ijk}$ for each observation. Note that the index i is attached to nearly every vector or matrix. The reason we use this matrix notation is simple; it allows us to write the linear mixed effects model for 2-way nested data in a general matrix notation of the form:

$$\mathbf{y}_i = \mathbf{X}_i \times \beta + \mathbf{Z}_i \times \mathbf{b}_i + \mathbf{e}_i \tag{4.2}$$

This is a famous notation in linear mixed effects modelling literature, the Laird and Ware notation (Laird and Ware, 1982). It is the general notation for a linear mixed effects model. Once we have written a model in this form, we can apply general results, as we will demonstrate.

In the expression in Equation (4.2), the vector $\mathbf{y}_i$ contains all the data from colony i, $\mathbf{X}_i$ contains all the fixed effects, and $\boldsymbol{\beta}$ has the regression parameters of interest. The $\mathbf{Z}_i$ and $\mathbf{b}_i$ implement the random effects. The use of bold face $\mathbf{b}_i$ in Equation (4.2) and the italicized b_i in Equation (4.1) may be confusing. The $\mathbf{b}_i$ contains multiple b_{ij}s as well as the b_i from Equation (4.1). The same holds for the $\boldsymbol{\beta}$ in Equation (4.2); it contains the α and β from Equation (4.1).

Having written the model in matrix notation, we also need to write an expression for the variances of the error terms:

$$\begin{aligned} \mathbf{b}_i &\sim N(0, \mathbf{D}) \\ \boldsymbol{\varepsilon}_i &\sim N(0, \boldsymbol{\Sigma}_i) \end{aligned} \tag{4.3}$$

It is an option to assume that the random intercepts are independent, and therefore:

$$\mathbf{D} = \begin{pmatrix} \sigma^2_{colony} & 0 & 0 & 0 \\ 0 & \sigma^2_{animal} & 0 & 0 \\ 0 & 0 & \sigma^2_{animal} & 0 \\ 0 & 0 & 0 & \sigma^2_{animal} \end{pmatrix}$$

Recall that the first element of $\mathbf{b}_i$ contains the random intercept b_i, which is normally distributed with mean 0 and variance σ^2_{colony}. Similarly, the random intercept b_{ij}, reflecting differences among animals within a colony, is normally distributed, with mean 0 and variance σ^2_{animal}. This is exactly what the notation $\mathbf{b}_i \sim N(0, \mathbf{D})$ and the definition of $\mathbf{D}$ tells us. As to the second expression in Equation (4.3), recall that $\boldsymbol{\varepsilon}_i$ contains the residuals from colony i. Its covariance matrix $\boldsymbol{\Sigma}_i$ is defined by:

$$\Sigma_i = \begin{pmatrix} \sigma^2 & 0 & 0 & 0 & 0 & 0 \\ 0 & \sigma^2 & 0 & 0 & 0 & 0 \\ 0 & 0 & \sigma^2 & 0 & 0 & 0 \\ 0 & 0 & 0 & \sigma^2 & 0 & 0 \\ 0 & 0 & 0 & 0 & \sigma^2 & 0 \\ 0 & 0 & 0 & 0 & 0 & \sigma^2 \end{pmatrix}$$

We assume that all colonies have the same residual variance, σ^2, and that the residuals from the same colony are independent. Note that this does not mean that the values of the response variable 'number of young,' are independent. On the contrary, as we will soon see. For longer time series, it may be an option to modify the definition of $\mathbf{\Sigma}$. For example, the (2,1) element is currently 0, and it represents the residual covariance between the two repeated observations of the first animal. The same is true of the (2,3) element for animal 2. It may be an option to use an auto-regressive correlation structure of order 1 to model residual correlation structures. For irregularly spaced data, any of the spatial correlation structures discussed in Chapter 5 in Pinheiro and Bates (2000) may be used. We will address later in which situations these would be used.

At this point, it is useful to consult Chapter 5 in Zuur et al. (2009a), which describes estimation of parameters, ML and REML estimation, model selection, and induced correlation structures. The latter means that the linear mixed effects model in Equations (4.2) and (4.3) define the following covariance structure on the observed data.

$$\text{cov}(\mathbf{y}_i) = \mathbf{Z}_i \times \mathbf{D} \times \mathbf{Z}_i^t + \mathbf{\Sigma}_i \tag{4.4}$$

Add the definitions of $\mathbf{Z}_i$, $\mathbf{D}$, and $\mathbf{\Sigma}$, convert the covariance to correlation, and the following expression for the correlation between two observations of a single animal is obtained.

$$\rho_{\text{within animal}} = \frac{\sigma^2_{animal} + \sigma^2_{colony}}{\sigma^2_{animal} + \sigma^2_{colony} + \sigma^2} \tag{4.5}$$

The correlation between observations of two animals within a colony is given by:

$$\rho_{\text{within colony}} = \frac{\sigma^2_{colony}}{\sigma^2_{animal} + \sigma^2_{colony} + \sigma^2} \tag{4.6}$$

We can write the correlation matrix for all observations in $\mathbf{y}_i$ as:

$$\begin{pmatrix}
1 & & & & & \\
\frac{\sigma_a^2+\sigma_c^2}{\sigma_a^2+\sigma_c^2+\sigma^2} & 1 & & & & \\
\frac{\sigma_c^2}{\sigma_a^2+\sigma_c^2+\sigma^2} & \frac{\sigma_a^2+\sigma_c^2}{\sigma_a^2+\sigma_c^2+\sigma^2} & 1 & & & \\
\frac{\sigma_c^2}{\sigma_a^2+\sigma_c^2+\sigma^2} & \frac{\sigma_c^2}{\sigma_a^2+\sigma_c^2+\sigma^2} & \frac{\sigma_a^2+\sigma_c^2}{\sigma_a^2+\sigma_c^2+\sigma^2} & 1 & & \\
\frac{\sigma_c^2}{\sigma_a^2+\sigma_c^2+\sigma^2} & \frac{\sigma_c^2}{\sigma_a^2+\sigma_c^2+\sigma^2} & \frac{\sigma_c^2}{\sigma_a^2+\sigma_c^2+\sigma^2} & \frac{\sigma_a^2+\sigma_c^2}{\sigma_a^2+\sigma_c^2+\sigma^2} & 1 & \\
\frac{\sigma_c^2}{\sigma_a^2+\sigma_c^2+\sigma^2} & \frac{\sigma_c^2}{\sigma_a^2+\sigma_c^2+\sigma^2} & \frac{\sigma_c^2}{\sigma_a^2+\sigma_c^2+\sigma^2} & \frac{\sigma_c^2}{\sigma_a^2+\sigma_c^2+\sigma^2} & \frac{\sigma_a^2+\sigma_c^2}{\sigma_a^2+\sigma_c^2+\sigma^2} & 1
\end{pmatrix}$$

To fit this onto the page, we use σ^2_a for σ^2_{animal} and σ^2_c for σ^2_{colony}. The upper triangular area of the correlation matrix is identical to the lower triangular area, as correlation matrices are symmetrical. The correlation between two observations of a single animal and between observations within a colony are the same for all colonies. Hence, the correlation matrix holds for all colonies.

Any extra residual correlation structure (e.g. the auto-regressive correlation) for time series observations of the same animal would influence the definition of $\mathbf{\Sigma}$, and therefore the σ^2 in the expression above. It is also possible to allow for heterogeneity, which would influence the definition of σ^2.

In our matrix notation, we assumed only 3 animals per colony and two repeated measurements (longitudinal, as they are over time). In reality, our data are highly unbalanced. The matrix notation is flexible in this respect; simply adjust the definitions of the design matrices $\mathbf{X}_i$ and $\mathbf{Z}_i$.

Unfortunately, our data set should not be analysed with a Gaussian distribution, as the response variable is a count. A Poisson or negative binomial distribution is more appropriate.

We continue with the Poisson generalised linear mixed models (GLMM).

4.5.2 Viewing the data (potentially wrongly) as a Poisson GLM or GLMM

Because the data are count data, we use a Poisson or negative binomial distribution. The negative binomial distribution allows for overdispersion, but because the range of the observed number of young is small, we do not expect major overdispersion. Hence we will focus on the Poisson distribution. A Poisson GLM for the marmot data is given by:

$$\begin{aligned} & Y_{ijk} \sim P(\mu_{ijk}) \\ & E(Y_{ijk}) = \mu_{ijk} \quad \text{and} \quad \text{var}(Y_{ijk}) = \mu_{ijk} \\ & \log(\mu_{ijk}) = \alpha + \beta_1 \times X_{1ijk} + \cdots + \beta_n \times X_{nijk} \end{aligned} \tag{4.7}$$

For example, a model with only age as covariate gives:

$$\log(\mu_{ijk}) = \alpha + \beta_1 \times Age_{ijk} \tag{4.8}$$

Note that this model does not contain the term ε_{ijk}. Elston et al. (2001) show how adding a residual term ε_{ijk} to the linear predictor allows for overdispersion (see also Chapter 2). This gives:

$$\begin{aligned} & \log(\mu_{ijk}) = \alpha + \beta_1 \times Age_{ijk} + \varepsilon_{ijk} \\ & \varepsilon_{ijk} \sim N(0, \sigma^2) \end{aligned} \tag{4.9}$$

However, we will not include the term ε_{ijk}, but continue with the log-link function in Equation (4.8). We can easily extend this model with a random intercept b_i for a `COLONY` effect, and b_{ij} for an `ANIMAL`-within-`COLONY` effect.

$$\begin{aligned} & Y_{ijk} \sim P(\mu_{ijk}) \\ & \log(\mu_{ijk}) = \alpha + \beta_1 \times Age_{ijk} + b_i + b_{ij} \\ & b_i \sim N(0, \sigma^2_{colony}) \quad \text{and} \quad b_{ij} \sim N(0, \sigma^2_{animal}) \end{aligned} \tag{4.10}$$

The first line states that Y_{ijk}, the number of young in year k of animal j in colony i, is Poisson distributed. In a more general context, we should write:

$$\log(\mu_{ijk}) = \alpha + \beta_1 \times X_{1ijk} + \cdots + \beta_n \times X_{nijk} + b_i + b_{ij} \tag{4.11}$$

The new addition is $b_i + b_{ij}$, which allows for the random intercept for COLONY, and a random intercept for ANIMAL nested in COLONY. We can also use matrix notation for the linear predictor function:

$$\log(\boldsymbol{\mu}_i) = \mathbf{X}_i \times \boldsymbol{\beta} + \mathbf{Z}_i \times \mathbf{b}_i,$$

where the second part contains the two random terms. The data are zero inflated, and this is likely to lead to overdispersed Poisson GLMMs. We therefore continue with zero inflated models.

4.5.3 Viewing the data (potentially wrong) as a zero inflated Poisson GLM

In Chapter 11 of Zuur et al. (2009a), the principle of ZIP models and true and false zeros was explained with a schematic similar to Figure 4.7. A short summary of the principle of ZIP models and true and false zeros is presented here.

In a ZIP model, two imaginary processes are the source of the zeros, a binary part and a count process. The zeros in the binary part are false zeros. The count process also contains zeros, but they are due to marmots being absent because of a harsh winter, lack of food, or unfavourable conditions of other covariates.

A binomial distribution is assumed for the binary part of the data and a Poisson distribution for the count data. Ghosh et al. (2003) showed that a whole range of distributions can be used for the count part of the data, but a Poisson distribution is the easiest. Alternatively, the negative binomial distribution can be used. However, because the range of the observed number of young is relatively small, we will use the Poisson distribution. Just as in Poisson GLM, we model the mean μ_i of the positive count data as:

$$\mu_i = e^{\alpha + \beta_1 \times X_{i1} + \cdots + \beta_q \times X_{iq}}$$

The probability of having a false zero, π_i, is modelled as:

$$\pi_i = \frac{e^{\nu + \gamma_1 \times X_{i1} + \cdots + \gamma_q \times X_{iq}}}{1 + e^{\nu + \gamma_1 \times X_{i1} + \cdots + \gamma_q \times X_{iq}}}$$

These two equations can also be written as:

$$\log(\mu_i) = \mathbf{X}_i \times \beta$$
$$\text{logit}(\pi_i) = \mathbf{X}_i \times \gamma$$

The mean and variance of a ZIP are given by:

$$E(Y_{ijk}) = \mu_{ijk} \times (1 - \pi_{ijk})$$
$$\text{var}(Y_{ijk}) = (1 - \pi_{ijk}) \times (\mu_{ijk} + \pi_{ijk} \times \mu_{ijk}^2)$$

The marmot data are nested; hence we have to extend the ZIP model to allow for a COLONY effect and an ANIMAL nested in COLONY effect.

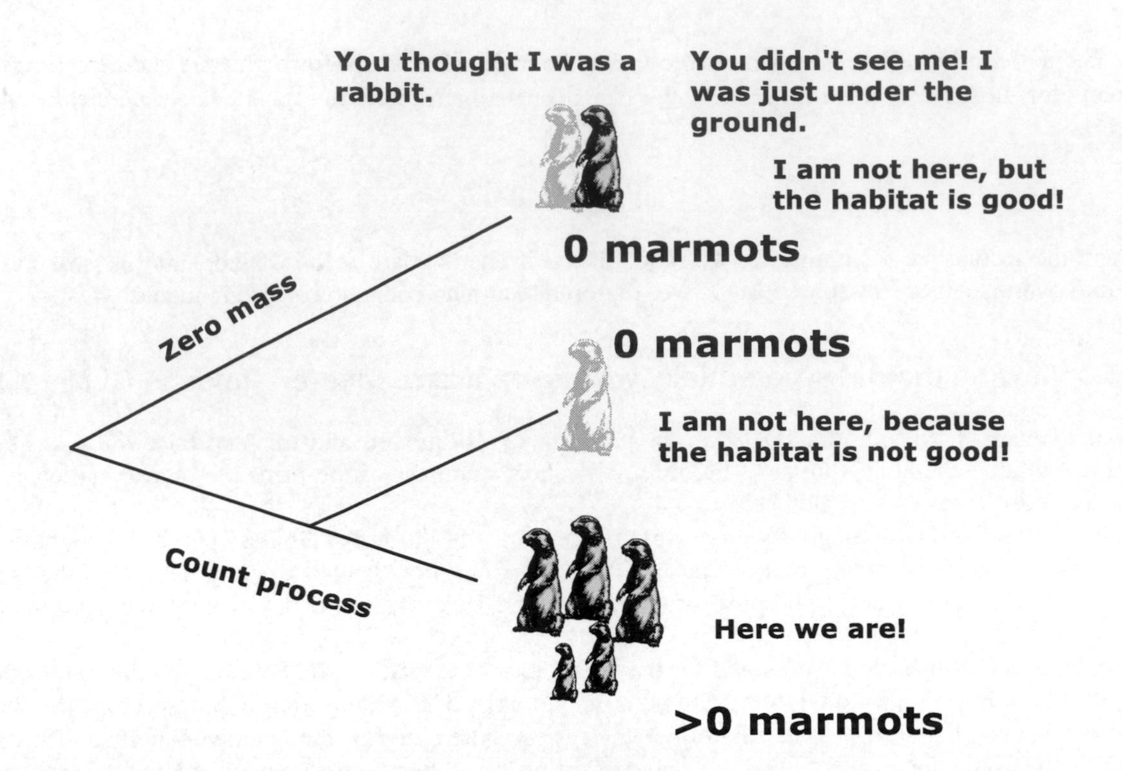

Figure 4.7. Schematic overview of a zero inflated Poisson GLM. Zero counts can be a part of the counting process, or occur because of poor, or difficult, sampling. The presence of false zeros is not a *requirement* for the application of ZIP models; see Chapter 10 for a discussion.

4.5.4 Viewing the data (possibly correct) as a 2-way nested ZIP GLMM

In the previous subsections, we discussed the Gaussian linear mixed effects model for 2-way data, the GLMM with a Poisson distribution, and the ZIP GLM. Now we need to combine them, as our data requires a ZIP for 2-way nested data. The underlying model is given by:

$$
\begin{aligned}
&Y_{ijk} \sim ZIP(\pi_{ijk}, \mu_{ijk}) \\
&E(Y_{ijk}) = \mu_{ijk} \times (1 - \pi_{ijk}) \quad \text{and} \quad \text{var}(Y_{ijk}) = (1 - \pi_{ijk}) \times (\mu_{ijk} + \pi_{ijk} \times \mu_{ijk}^2) \\
&\text{logit}(\pi_{ijk}) = \alpha + \gamma_1 \times X_{1ijk} + \cdots + \gamma_n \times X_{nijk} + d_i + d_{ij} \\
&\log(\mu_{ijk}) = \alpha + \beta_1 \times X_{1ijk} + \cdots + \beta_n \times X_{nijk} + b_i + b_{ij}
\end{aligned}
$$

This looks like a tortuous set of equations, but it is merely a combination of the models discussed so far. The first line states that Y_{ijk} follows a zero inflated Poisson distribution with parameters π_{ijk} and μ_{ijk}. The π_{ijk} is the probability of measuring a false zero, and μ_{ijk} the expected value for the count process for observation k of animal j in colony i. These are modelled as functions of the covariates and the random intercepts for `COLONY` and `ANIMAL` nested in `COLONY`. Hence, we allow for the situation in which covariates have differing effects on the binary process and the count process. This means that we must estimate a double number of parameters, including two sets of variances for the random terms.

We can simplify the expressions for the logit and log functions using the matrix notation:

$$
\begin{aligned}
\text{logit}(\pi_i) &= \mathbf{X}_i \times \gamma + \mathbf{Z}_i \times \mathbf{d}_i \\
\log(\mu_i) &= \mathbf{X}_i \times \beta + \mathbf{Z}_i \times \mathbf{b}_i
\end{aligned}
$$

If this looks intimidating, you can use the first notation. We also need to record the assumptions of the random terms:

$$b_i \sim N(0, \sigma^2_{1,colony}) \quad \text{and} \quad b_{ij} \sim N(0, \sigma^2_{1,animal})$$
$$d_i \sim N(0, \sigma^2_{2,colony}) \quad \text{and} \quad d_{ij} \sim N(0, \sigma^2_{2,animal})$$

We assume that all random terms are independent of one another. Again, this is merely a repetition of the Gaussian linear mixed effects model.

4.6 Dealing with 25 covariates in a 2-way nested ZIP GLMM

We started the data exploration with a large number of covariates but dropped some, based on variance inflation factors. However, we still have 25 covariates, in addition to `COLONY` and `ANIMAL`, which far exceeds the number appropriate to the sample size and complexity of the methodology GLMM and ZIP GLMM are numerical estimation techniques that require heavy computation. The data exploration showed no strong ecological patterns.

Model selection is the process of finding the optimal set of covariates, and there is a heated debate in the literature on how, and how not, to achieve this. In some scientific fields, classical model selection is routinely applied using backward or forward selection based on criteria such as the Akaike Information Criteria (AIC). In other fields such an approach is condemned and will almost certainly lead to results being rejected for publication. An alternative approach is the Information Theoretic (IT) approach (Burnham and Anderson, 2002), in which a set of models, typically 10 – 15, is specified. Although model comparison is also based on the AIC (or better, AIC weights), the difference is that we enumerate and present all potentially interesting and important models, so one would typically present the results of a series of models. It is possible to obtain an average model.

Another approach is to apply model selection to the interaction terms and retain the main terms in the model, even if they are not significant. Yet another option is to do no model selection and simply present the results of the model that you chose at the outset. In all approaches it is important that you deal with collinearity before fitting the model. Useful references are Anderson et al. (2000), Burnham and Anderson (2002), Johnson and Omland (2004), Whittingham et al. (2006), Murtaugh (2009), and Mundry and Nunn (2009).

Within a Bayesian context, variable selection can be applied, see, for example, Chapter 11 in Ntzoufras (2009).

The IT approach is rapidly gaining popularity in a broad range of ecological fields. It dictates that we thoroughly assess the data and formulate a range of plausible ecological hypotheses with associated statistical models. As an example of this approach, see the 10 – 12 models formulated in Reed et al. (2011). Formulating the underlying biological questions and associated models presented in that paper was a lengthy process. However, for the marmot data we have no idea which models to choose. Previous independent analyses have shown many of the 25 covariates to be important, but we have limited knowledge of which are the most important or how they might interact. Thus we are unable to specify 10 – 15 models *a priori*, as required for the IT approach. A model with 25 covariates is not an easy place to start, so we have a major obstacle. A ZIP GLMM for 2-way nested data is a complicated model. Later we will show MCMC code and run WinBUGS. Too many parameters relative to the sample size are likely to result in chains that do not mix, correlated parameters, and extremely long computing time. Weak ecological signals result in poorly mixed chains. A single model may need to run overnight, imagine the computing time required for a forward or backward selection with 25 covariates!

Because this book deals with statistics and not biology, at this point we decided to re-assess the aim of this chapter. We initially intended to use the marmot data to illustrate how to do a ZIP GLMM and present a detailed analysis and results. Because of the extensive computing time and the issues with model selection we will use a single model and demonstrate the steps of analysis. We will show how to fit a ZIP GLMM for 2-way nested data in WinBUGS and discuss relevant aspects. When you are familiar with the methodology, you can use it for your own data. If you face similar model selection problems, you have three options: keep the model simple, formulate 10 – 15 straightforward models and apply the IT approach, or apply a model selection. Figure 4.8 shows a flowchart of the layout of this chapter.

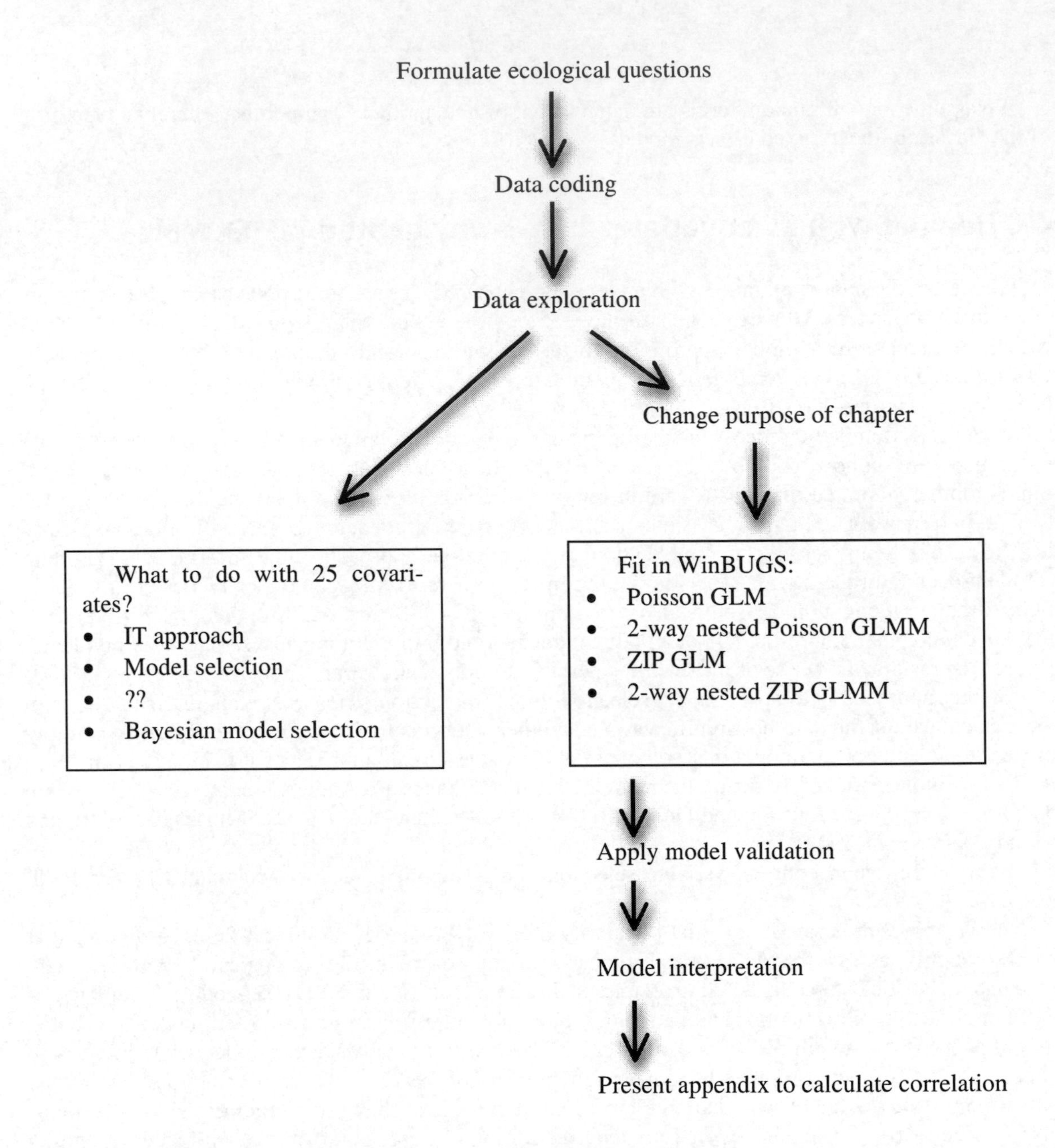

Figure 4.8. Flowchart of information presented in this chapter.

Model selection

Readers strongly opposed to model selection may want to skip the remainder of this section.

To apply a forward model selection approach means that our analysis is becoming a data phishing, data snooping, cherry picking, data dredging exercise. There is no way to avoid this for these data, other than choosing the IT approach, and we must label the results as 'exploratory in nature.' If you are happy with this, as many field ecologists will be, since it is often the only feasible approach, read on.

Later, we will discuss the R and WinBUGS code for the MCMC method to estimate the parameters of the nested ZIP models. Because the MCMC algorithm requires substantial computing time, we suggest doing a forward, rather than backward, selection. A backward selection beginning with all 25 covariates is impractical to fit. We have additional complicating factors for model selection in that we have two equations, the logit link function and the log link function. Also some covariates are fitted as quadratic terms and others as linear terms.

We begin by addressing the second and less demanding issue. We mentioned earlier that instead of fitting a GAM, we can use quadratic terms to model non-linear patterns. However, covariates with a limited number of unique values (e.g. `MYEARL`, `YOUNGAD`, etc.) are better fitted as linear terms.

Now for the more difficult problem: We have two equations in which a covariate can enter the model, the log link and the logit link. In a forward selection, we can add each of the covariates to the model in either the linear predictor for the log or in the logit link function, calculate a selection criterion, and select the covariate with the best selection criterion for the second round. Hence a possible approach is to run the following two loops:

Step 1 of forward selection

For each covariate fit the following model:

$$\begin{aligned}\text{logit}(\pi_{ijk}) &= \alpha + \gamma_1 \times \text{Covariate}_{ijk} + d_i + d_{ij} \\ \log(\mu_{ijk}) &= \alpha + b_i + b_{ij}\end{aligned}$$

For each covariate fit the following model:

$$\begin{aligned}\text{logit}(\pi_{ijk}) &= \alpha + d_i + d_{ij} \\ \log(\mu_{ijk}) &= \alpha + \beta_1 \times \text{Covariate}_{ijk} + b_i + b_{ij}\end{aligned}$$

This means that we need to run 2 X 25 models. From each set, we choose the best model, as judged by the Deviance Information Criterion (DIC), but recall the critical remarks in Section 2.10 on the use of DIC in mixed effect models. The covariate that corresponds to the best model is selected and used in the second round. Suppose this is `REPLAST` for the Poisson part. The second round of the forward selection is then:

Step 2 of forward selection

For each covariate fit the following model:

$$\begin{aligned}\text{logit}(\pi_{ijk}) &= \alpha + \gamma_1 \times \text{Covariate}_{ijk} + d_i + d_{ij} \\ \log(\mu_{ijk}) &= \alpha + \beta_1 \times \text{REPLAST}_{ijk} + b_i + b_{ij}\end{aligned}$$

For each covariate fit the following model:

$$\begin{aligned}\text{logit}(\pi_{ijk}) &= \alpha + d_i + d_{ij} + \eta_{ijk} \\ \log(\mu_{ijk}) &= \alpha + \beta_1 \times \text{REPLAST}_{ijk} + \beta_2 \times \text{Covariate}_{ijk} + b_i + b_{ij}\end{aligned}$$

This process is then repeated until the DIC indicates that no further terms should be added. Be prepared to let your computer run for a couple of nights!

Another approach is to use only an intercept in the logistic link function and add covariates only to the log link function. A justification for this is that anything beyond an intercept in the logistic link function may cause non-mixed chains. This depends on the sample size, strength of the ecological signals, amount of unbalanced data, etc. For your own data, using covariates and random effects in both the logistic and log link function may work.

4.7 WinBUGS and R code for Poisson GLM, GLMM, and ZIP models

In this section we present WinBUGS and R code for a Poisson GLM, 2-way nested Poisson GLMM, and an ordinary ZIP GLM without random effects. We begin with code for 2-way nested ZIP GLMMs. It is relatively easy to adjust the code presented in this section to accommodate a negative binomial distribution. See also Ntzoufras (2009).

One of our philosophies is 'seeing is believing.' If we perform a complicated task; for example program a ZIP model in WinBUGS, and it gives results similar to those produced by existing functions in R, we naively take it for granted that our WinBUGS code is correct. This, of course, assumes that the existing R function is correct. It would be exceedingly coincidental if both approaches were wrong. Therefore, throughout this section, we present the essential elements of our WinBUGS code, and when possible, also show results of the frequentist approach.

4.7.1 WinBUGS code for a Poisson GLM

To illustrate the coding and results we quasi-arbitrarily select the covariates `AGE`, `MYEARL`, `SAMEAD`, `OFFSSAME`, and `JUNEPRECIP`. Quasi-arbitrary refers to initial GLM analysis indications that these covariates may have some influence compared to the others.

Suppose that we want to fit a Poisson GLM in which we model `YOUNG` as a function of these covariates. Such a model is given by:

$$
\begin{aligned}
&Y_{ijk} \sim P(\mu_{ijk}) \\
&\log(\mu_{ijk}) = \beta_1 + \beta_2 \times AGE_{ijk} + \beta_3 \times MYEARL_{ijk} + \beta_4 \times SAMEAD_{ijk} + \\
&\qquad \beta_5 \times OFFSSAME_{ijk} + \beta_6 \times JUNEPRECIP_{ijk}
\end{aligned}
$$

Y_{ijk} is the kth observation of animal j of colony i, and *AGE* is the corresponding value of `AGE`. The same holds for the other covariates. We have seen R and WinBUGS code for a Poisson GLM in Chapter 2. However, when building a zero inflated Poisson GLMM for 2-way nested data it is best to start from scratch. Therefore, we will repeat the code for a Poisson GLM and, step-by-step, extend this to a 2-way nested ZIP GLMM. Modelling a 2-way nested ZIP GLMM from scratch is bound to produce many errors in WinBUGS.

First we need to specify an object that contains the response variable and the required covariates. To keep the WinBUGS code 'simple' we delete any missing values with the `na.exclude` function.

```
> Marmots3 <- data.frame(YOUNG      = Marmots2$YOUNG,
                         AGE        = Marmots2$AGE,
                         MYEARL     = Marmots2$MYEARL,
                         SAMEAD     = Marmots2$SAMEAD,
                         OFFSSAME   = Marmots2$OFFSSAME,
                         JUNEPRECIP = Marmots2$JUNEPRECIP,
                         COLONY     = Marmots2$COLONY,
                         ANIMAL     = Marmots2$ANIMAL)
> Marmots4 <- na.exclude(Marmots3)
```

Note that applying the na.exclude command on the Marmots2 object would remove most data, as this object contains variables with many missing values. We prefer to make a new data frame, collect all variables that we will need later, and remove the missing values (2) from the object Marmots3. The final object is called Marmots4.

Initial WinBUGS analysis shows a correlation of approximately 0.8 between the estimated intercept and some of the parameters. This is because covariates such as AGE, OFFSAME, and JUNEPRECIP have values that are far from 0. High correlation between estimated parameters may cause poor mixing, and a standard solution in statistics for removing the correlation is to centre (subtract the mean) or standardise (subtract the mean and divide by the standard deviation) these covariates. We use standardisation:

```
> Marmots4$AGE.C        <- with(Marmots4,
                            (AGE - mean(AGE)) / sd(AGE))
> Marmots4$OFFSSAME.C <- with(Marmots4,
                            (OFFSSAME - mean(OFFSSAME)) / sd(OFFSSAME))
> Marmots4$JUNEPRECIP.C <- with(Marmots4,
                         (JUNEPRECIP - mean(JUNEPRECIP)) / sd(JUNEPRECIP))
```

Next we define the object win.data1. It contains the relevant data for the GLM in WinBUGS, although it contains some double code. We could skip the Marmots3 and Marmots4 steps and do it in a single step, but we prefer easy coding.

```
> win.data1 <- list(YOUNG       = Marmots4$YOUNG,
                    AGE         = Marmots4$AGE.C,
                    MYEARL      = Marmots4$MYEARL,
                    SAMEAD      = Marmots4$SAMEAD,
                    OFFSSAME    = Marmots4$OFFSSAME.C,
                    JUNEPRECIP  = Marmots4$JUNEPRECIP.C,
                    N           = nrow(Marmots4))
```

The WinBUGS model code is:

```
> sink("GLM.txt")
cat("
model {
  for (i in 1:N) {
    #Poisson part
    YOUNG[i]   ~  dpois(mu[i])
    log(mu[i]) <- max(-20, min(20, eta.mu[i]))
    eta.mu[i]  <- aP[1]  +
                  aP[2] * AGE[i] +
                  aP[3] * MYEARL[i] +
                  aP[4] * SAMEAD[i] +
                  aP[5] * OFFSSAME[i] +
                  aP[6] * JUNEPRECIP[i]
    L[i] <- YOUNG[i] * log(mu[i]) - mu[i]  - loggam(YOUNG[i] + 1)
    }
  dev <- -2 * sum(L[1:N])
  #Priors for regression parameters
  for (i in 1:6) { aP[i] ~ dnorm(0.0, 0.001) }
  }
",fill = TRUE)
sink()
```

The `sink` command writes the code between `model {}` in a text file called `GLM.txt`. The code is similar to that used in Chapter 2. `YOUNG[i]` is Poisson distributed with mean `mu[i]`, and `eta.mu[i]` is the predictor function. To avoid numerical errors resulting from taking the exponential of large values, we use the `max` construction. If this causes trouble (i.e. odd mixing of chains), adjust the diffuse priors for the regression parameters `aP[i]`.

We add code to calculate the likelihood, `L[i]`, and the deviance, `dev`. This is not necessary for a Poisson GLM, as WinBUGS gives this information. However, for the ZIP GLMM, the DIC quoted by WinBUGS is incorrect, and we need to calculate it ourselves. Seeing how the deviance is calculated for a simple Poisson GLM makes it easier to understand how it works for the ZIP and ZIP GLMM models presented later in this chapter.

We need to initialise the parameters and tell WinBUGS from which parameters to save the chains. We will only extract the chains from the regression parameters and the deviance, `dev`. For the initial values we sample from a normal distribution. It would perhaps make more sense to use the estimated regression coefficients from a GLM obtained with the `glm` function as initial values, possibly with some noise added. We will do this later for the ZIP GLM.

```
> inits1  <- function () { list(aP = rnorm(6, 0, 0.01))}
> params1 <- c("aP", "dev")
```

We now have all the tools in place to run the MCMC. We use 3 chains each with 100,000 iterations, and the burn-in is 5,000. The thinning rate is 50, which means that 5,700 iterations are generated for the posterior of each parameter. Finally we set the directory on our computer that contains WinBUGS, load the `R2WinBUGS` package, and run WinBUGS from within R.

```
> nc <- 3          #Number of chains
> ni <- 100000     #Number of draws from posterior
> nb <-   5000     #Number of draws to discard as burn-in
> nt <-    500     #Thinning rate
> library(R2WinBUGS)
> MyWinBugsDir <- "C:/WinBUGS14/"
> out1 <- bugs(data = win.data1, inits = inits1,
               parameters = params1, model = "GLM.txt",
               n.thin = nt, n.chains = nc, n.burnin = nb,
               n.iter = ni, debug = TRUE,
               bugs.directory = MyWinBugsDir)
```

Mixing of the chains is reasonable, with minor auto-correlation. We suggest expanding the thinning rate by a factor of 10, and therefore increasing the number of draws as well, in order to obtain a reasonable number of draws for the posteriors. Results are:

```
> print(out1, digits = 3)
```

```
Inference for Bugs model at "GLM.txt", fit using WinBUGS,
 3 chains, each with 1e+05 iterations (first 5000 discarded), n.thin = 50
 n.sims = 5700 iterations saved
             mean    sd     2.5%      25%      50%      75%    97.5%  Rhat n.eff
aP[1]      -0.016 0.051   -0.117   -0.051   -0.016    0.019    0.082 1.002  1500
aP[2]       0.184 0.034    0.118    0.161    0.184    0.207    0.248 1.001  5700
aP[3]       0.224 0.034    0.157    0.202    0.224    0.247    0.292 1.003   820
aP[4]      -0.205 0.064   -0.332   -0.248   -0.205   -0.159   -0.082 1.004   580
aP[5]       0.257 0.040    0.178    0.230    0.257    0.284    0.335 1.003   840
aP[6]       0.141 0.034    0.075    0.118    0.140    0.164    0.206 1.001  5700
dev      2401.030 3.425 2396.000 2398.000 2400.000 2403.000 2409.000 1.001  5700
deviance 2401.030 3.425 2396.000 2398.000 2400.000 2403.000 2409.000 1.001  5700

For each parameter, n.eff is a crude measure of effective sample size,
and Rhat is the potential scale reduction factor (at convergence, Rhat=1).
```

```
DIC info (using the rule, pD = Dbar-Dhat)
pD = 6.0 and DIC = 2407.0
DIC is an estimate of expected predictive error (lower deviance is better).
```

Note that our calculated deviance, `dev`, is identical to the deviance given by WinBUGS, which means that our likelihood function has been implemented correctly. We will return to deviances and DIC later in this chapter.

We compare the results, especially the estimated regression coefficients, to those obtained with the `glm` function.

```
> M1 <- glm(YOUNG ~ AGE.C + MYEARL + SAMEAD + OFFSSAME.C +
            JUNEPRECIP.C, data = Marmots4, family = poisson)
> summary(M1)

               Estimate Std. Error z value Pr(>|z|)
(Intercept)    -0.01310    0.05054  -0.259  0.79549
AGE.C           0.18337    0.03327   5.512 3.54e-08
MYEARL          0.22585    0.03391   6.660 2.74e-11
SAMEAD         -0.20584    0.06494  -3.170  0.00153
OFFSSAME.C      0.25890    0.04017   6.446 1.15e-10
JUNEPRECIP.C    0.14125    0.03340   4.229 2.35e-05

(Dispersion parameter for poisson family taken to be 1)
    Null deviance: 1541.1  on 826  degrees of freedom
Residual deviance: 1386.2  on 821  degrees of freedom
AIC: 2407
```

The estimated regression parameters obtained by the `glm` function are similar to the mean posterior values, as are the standard errors. Note that the GLM is overdispersed, as can be seen from the ratio of the residual deviance to the residual degrees of freedom. We could also obtain this information from the MCMC by calculating and extracting Pearson residuals (see Chapter 2). We expect that adding random effects and/or zero inflation will eliminate the overdispersion. Note that explained deviance is $(1541.1 - 1386.2) / 1541.1 = 0.10$. Hence these 5 covariates only explain 10% of the variation in the number of young.

As part of any model selection one should plot the (Pearson) residuals versus fitted values and the residuals versus each covariate. We should also assess the correlation between the estimated parameters. In WinBUGS we can simply make scatterplots of the simulated values of each parameter (Figure 4.9). We do not want to see clear patterns in these scatterplots. Note that the intercept and slope for `SAMEAD` (`aP[1]` and `aP[4]`) are correlated. This is also true of `SAMEAD` and `OFFSAME`. Standardising `SAMEAD` may remove these correlations. When we make this graph for the original (unstandardised) variables there are more of these high correlations, and standardising removes them.

The R code to make Figure 4.9 follows. The `panel.cor` function is a modified version of the `panel.cor` function in the `pairs` help file and is in the `HighstatLib.R` file sourced earlier. We used greyscale for the points, as it uses less ink.

```
> pairs(out1$sims.list$aP, lower.panel = panel.cor,
     upper.panel =  function(x, y) points(x, y, pch = 16, cex = 0.2,
                                          col = gray(0.5)),
     labels = c("Intercept", "Age", "MYEARL", "SAMEAD", "OFFSAME",
               "JUNEPRECIP"))
```

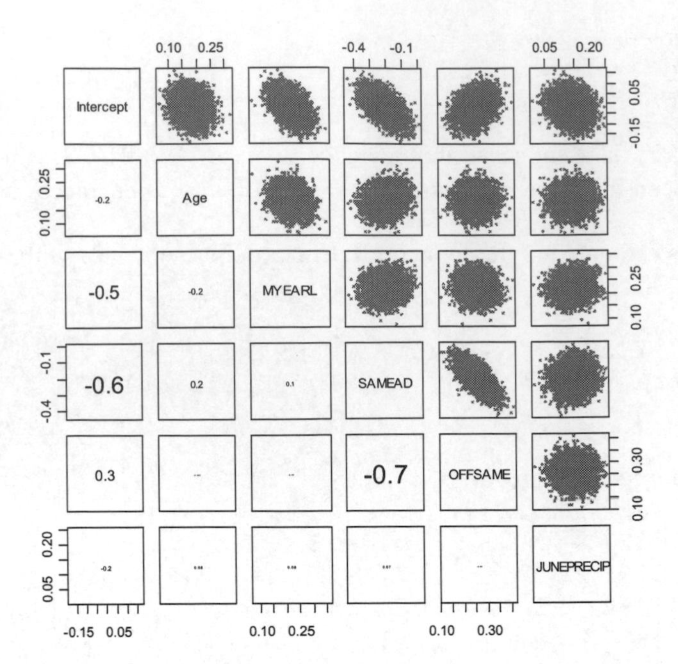

Figure 4.9. Multi-panel scatterplot of the MCMC simulated values for each regression parameter.

4.7.2 WinBUGS code for 1-way and 2-way nested Poisson GLMMs

In this subsection we discuss code for 2-way nested Poisson GLMMs. The underlying model is given by:

$$
\begin{aligned}
& Y_{ijk} \sim P(\mu_{ijk}) \\
& \log(\mu_{ijk}) = \beta_1 + \beta_2 \times AGE_{ijk} + \beta_3 \times MYEARL_{ijk} + \beta_4 \times SAMEAD_{ijk} + \\
& \qquad \beta_5 \times OFFSSAME_{ijk} + \beta_6 \times JUNEPRECIP_{ijk} + b_i + b_{ij} \\
& b_i \sim N(0, \sigma^2_{Colony}) \\
& b_{ij} \sim N(0, \sigma^2_{Animal})
\end{aligned}
$$

We again use standardised variables for `AGE`, `OFFSAME`, and `JUNEPRECIP`. We use the function `glmer` from the `lme4` package to fit the 2-way Poisson GLMM. Before running the `glmer` function we define `COLONY` and `ANIMAL` as categorical variables.

```
> library(lme4)
> Marmots4$fCOLONY <- factor(Marmots4$COLONY)
> Marmots4$fANIMAL <- factor(Marmots4$ANIMAL)
> M2 <- glmer(YOUNG ~ AGE.C + MYEARL + SAMEAD + OFFSSAME.C +
                JUNEPRECIP.C + (1| fCOLONY / fANIMAL),
                data = Marmots4, family = poisson)
> summary(M2)
```

```
Random effects:
 Groups          Name        Variance   Std.Dev.
 fANIMAL:fCOLONY (Intercept) 2.8988e-01 5.3841e-01
 fCOLONY         (Intercept) 2.8369e-11 5.3263e-06

Number of obs: 827, groups: fANIMAL:fCOLONY, 279; fCOLONY, 9
Fixed effects:
```

```
                Estimate Std. Error z value Pr(>|z|)
(Intercept)     -0.13808    0.06724  -2.054 0.040017
AGE.C            0.21030    0.04190   5.019 5.19e-07
MYEARL           0.13393    0.03922   3.415 0.000637
SAMEAD          -0.21610    0.07547  -2.863 0.004193
OFFSSAME.C       0.26152    0.04653   5.620 1.91e-08
JUNEPRECIP.C     0.14331    0.03642   3.935 8.31e-05
```

The estimated value for the variance of the random effects for COLONY is small. This is not surprising, since the data exploration indicated that differences in the number of young among colonies were minimal. We could refit the model without the random intercept COLONY, or we could argue that the design of the study dictates that we retain the random intercepts COLONY and ANIMAL.

In the previous section we explained how to assess the model for overdispersion, and we repeat it here. We determined the residual sum of squares and divide it by the sample size minus 8 (6 regression parameters and 2 variances). The ratio 1.168 is sufficiently close to 1; hence there is no evidence of overdispersion.

```
> E2 <- resid(M2) #Pearson residuals are extracted
> Overdispersion <- sum(E2^2) / (nrow(Marmots4) - 6 - 2)
> Overdispersion

1.168022
```

The following code extracts the residuals and fitted values and plots them against one another. The resulting graph in Figure 4.10 shows no problems. The bands of dots are the observed values 0, 1, 2, etc.

```
> plot(fitted(M2), residuals(M2))
> MyData <- seq(0, 30, length = 101)
> F2 <- fitted(M2)
> tmp2 <- loess(E2 ~ F2)
> P2 <- predict(tmp2, newdata = MyData)
> lines(MyData, P2, col = 2, lwd = 2)
> abline(h = 0, col = "gray")
```

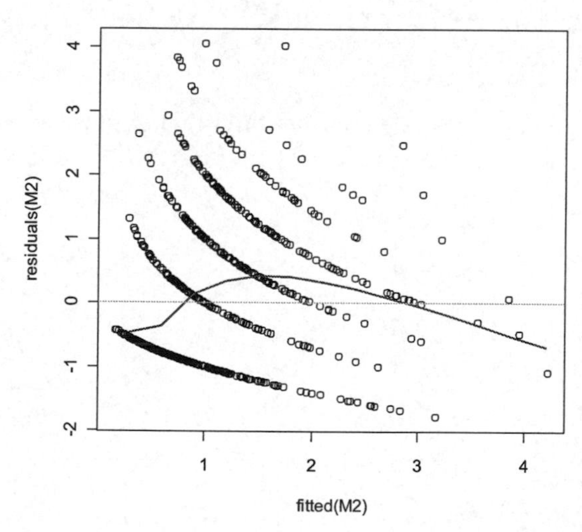

Figure 4.10. Fitted values versus residuals for model M2; the Poisson GLMM for 2-way nested data.

To fit the same model in WinBUGS, we first need to recode COLONY and ANIMAL as sequential numbers, beginning with 1. In the original data these were coded with non-sequential numbers. We use a simple loop that re-labels the COLONY with values of 1 through 9.

```
> AllColonies <- unique(Marmots4$COLONY)
> Marmots4$COLONYNEW <- vector(length= nrow(Marmots4))
> j <- 1
> for (i in AllColonies) {
   Marmots4$COLONYNEW[Marmots4$COLONY==i] <- j
   j <- j + 1  }
```

We do the same for ANIMAL. Each animal is assigned a unique sequential number. This allows for nested random effects. If we repeated some animal numbers we would model crossed random effects.

```
> AllAnimals <- unique(Marmots4$ANIMAL)
> Marmots4$ANIMALNEW <- vector(length= nrow(Marmots4))
> j <- 1
> for (i in AllAnimals) {
   Marmots4$ANIMALNEW[Marmots4$ANIMAL==i] <- j
   j <- j + 1  }
```

To convince yourself that this code only renames the colonies and animals, run the glmer function using COLONYNEW and ANIMALNEW as categorical variables in the nested random effects structure. You will get the same results. We also need to determine the number of unique animals:

```
> NANIMAL <- length(unique(Marmots4$ANIMALNEW))
```

We now have all variables required for the 2-way nested Poisson GLMM and can store them in a list that is called win.data2. We use normalised covariates for AGE, OFFSSAME, and JUNEPRECIP.

```
> win.data2 <- list(YOUNG      = Marmots4$YOUNG,
                    AGE        = Marmots4$AGE.C,
                    MYEARL     = Marmots4$MYEARL,
                    SAMEAD     = Marmots4$SAMEAD,
                    OFFSSAME   = Marmots4$OFFSSAME.C,
                    JUNEPRECIP = Marmots4$JUNEPRECIP.C,
                    N          = nrow(Marmots4),
                    NCOLONY    = 9,
                    COLONYNEW  = Marmots4$COLONYNEW,
                    ANIMALNEW  = Marmots4$ANIMALNEW,
                    NANIMAL    = NANIMAL)
```

The WinBUGS modelling code is a simple extension of the code presented in the previous section. The new notations, bP and bbP, are the random intercepts for COLONY and ANIMAL nested in COLONY. We use normal diffuse priors for them.

```
> sink("GLMM.txt")
  cat("
  model {
   for (i in 1:N) {
        #Poisson part
        YOUNG[i]   ~  dpois(mu[i])
        log(mu[i]) <- max(-20, min(20, eta.mu[i]))
        eta.mu[i]  <- aP[1]   +
                      aP[2] * AGE[i] +
                      aP[3] * MYEARL[i] +
                      aP[4] * SAMEAD[i] +
                      aP[5] * OFFSSAME[i] +
```

```
                        aP[6] * JUNEPRECIP[i] +
                        bP[COLONYNEW[i]] +
                        bbP[ANIMALNEW[i]]
        L[i] <- YOUNG[i] * log(mu[i]) - mu[i]  - loggam(YOUNG[i]+1)
      }
    dev <- -2 * sum(L[1:N])

    #Priors for regression parameters
    for (i in 1:6) { aP[i] ~ dnorm(0.0, 0.001) }
    for (i in 1:NCOLONY) { bP[i]  ~ dnorm(0.0,  tau.Colony) }
    for (i in 1:NANIMAL) { bbP[i] ~ dnorm(0.0,  tau.Animal) }

    #Priors for the random effects for colony
    tau.Colony <- 1 / (sigma.Colony * sigma.Colony)
    tau.Animal <- 1 / (sigma.Animal * sigma.Animal)
    sigma.Colony ~ dunif(0,10)
    sigma.Animal ~ dunif(0,10)
  }
",fill = TRUE)
sink()
```

The bP[COLONYNEW[i]] construction implements 9 random levels, because COLONYNEW has 9 values. The same was done with bbP[ANIMALNEW[i]]. Instead of this construction, we could have used two 'for' loops and bP[i] + bPP[i,j], but this requires slightly more coding, as all covariates and the response variable also have to be coded as 2-dimensional matrices.

We code the log-likelihood and deviance, with initial values set as before. We use the estimated values of M2 as starting values, with a small amount of random noise added, for the regression parameters.

```
> inits2 <- function () {
   list(aP              = rnorm(6, fixef(M2), 0.01),
        bP              = rnorm(9, 0, 1),
        bbP             = rnorm(NANIMAL, 0, 1),
        sigma.Colony    = rlnorm(1, meanlog = 0, sdlog = 0.1),
        sigma.Animal    = rlnorm(1, meanlog = 0, sdlog = 0.1))}
```

Regression parameters, sigmas, and random effects are extracted.

```
> params2 <- c("aP", "sigma.Colony", "sigma.Animal", "bP", "bbP",
               "dev")
```

We use a slightly higher thinning rate:

```
> nc <- 3          #Number of chains
> ni <- 500000     #Number of draws from posterior
> nb <-  50000     #Number of draws to discard as burn-in
> nt <-    500     #Thinning rate
```

This gives 2,700 draws for each posterior. Mixing is adequate, but can be improved by multiplying ni, nb, and nt by 5 or 10. The MCMC is started with the following command. The output is presented below.

```
> library(R2WinBUGS)
> out2 <- bugs(data = win.data2, inits = inits2,
               parameters = params2, model = "GLMM.txt",
```

```
                n.thin = nt, n.chains = nc, n.burnin = nb,
                n.iter = ni, debug = TRUE,
                bugs.directory = MyWinBugsDir)
> print(out2, digits = 3)

Inference for Bugs model at "GLMM.txt", fit using WinBUGS,
 3 chains, each with 5e+05 iterations (first 50000 discarded), n.thin = 500
 n.sims = 2700 iterations saved
                mean     sd    2.5%      25%      50%      75%    97.5%  Rhat n.eff
aP[1]         -0.152  0.097  -0.357   -0.210   -0.147   -0.088    0.029 1.021   110
aP[2]          0.211  0.041   0.130    0.183    0.211    0.239    0.293 1.001  2700
aP[3]          0.131  0.041   0.049    0.105    0.131    0.159    0.210 1.008   270
aP[4]         -0.219  0.075  -0.364   -0.269   -0.218   -0.168   -0.075 1.013   160
aP[5]          0.260  0.046   0.169    0.229    0.259    0.292    0.350 1.008   260
aP[6]          0.144  0.036   0.073    0.120    0.145    0.169    0.214 1.004   930
sigma.Colony   0.161  0.117   0.006    0.071    0.144    0.227    0.429 1.002  2700
sigma.Animal   0.534  0.067   0.410    0.485    0.533    0.579    0.670 1.066    42
...
dev         2198.310 24.328 2154.000 2181.000 2199.000 2215.000 2245.000 1.057    42
deviance    2198.310 24.328 2154.000 2181.000 2199.000 2215.000 2245.000 1.057    42

For each parameter, n.eff is a crude measure of effective sample size,
and Rhat is the potential scale reduction factor (at convergence, Rhat=1).
DIC info (using the rule, pD = Dbar-Dhat)
pD = 104.3 and DIC = 2302.7
DIC is an estimate of expected predictive error (lower deviance is better).
```

A pairplot (not shown here) showed that all correlations among the parameters are < 0.3, except `aP[4]` and `aP[5]` which were -0.6. Perhaps we should also standardise the two remaining covariates.

To assess whether there is overdispersion we should extend the code by calculating Pearson residuals inside the MCMC code and extracting these residuals. Take the sum of squares and divide by the number of observations minus the number of parameters (6 regression parameters + 2 variances). This ratio should be approximately 1. We will provide the code to do this for the 2-way ZIP GLMM in the following section.

Other model validation steps consist of plotting the residuals versus each covariate and making histograms or QQ-plots of the Pearson residuals and random effects for both `bP` and `bbP`, although we lack a sufficient number of random effects for `COLONY`, as there are only 9. We will demonstrate this for the 2-way ZIP GLMM in the next section.

We also need to evaluate mixing of the chains. Because of the highly unbalanced design, not all chains for the residuals mixed well. However, results of the `glmer` approach and the WinBUGS approach are remarkably similar. The only difference is that the estimated variance for the `COLONY` effect is close to 0 in `glmer`, whereas in WinBUGS it is 0.161, although the standard error is large.

In addition to comparing the estimated parameters obtained by `glm`, `glmer`, and WinBUGS, it is useful to plot the results side by side. For this we use the `coefplot2` function from the `coefplot2` package (Volker and Su, 2011). This package is not in CRAN and you will need to download it using:

```
> install.packages("coefplot2",repos="http://r-forge.r-project.org")
```

Estimated parameters and 95% confidence intervals are plotted in Figure 4.11. Note that the estimation approaches produce nearly identical estimated regression parameters. The following code was used to create the figure:

```
> library(coefplot2)
> coefplot2(list("glm" = M1,
                 "glm WinBUGS" = out1,
                 "glmm with glmer" = M2,
```

```
            "glmm with WinBUGS" = out2),
      merge.names = FALSE, intercept=TRUE,
      legend = TRUE, xlim = c(-0.5, 0.5),
      legend.x = "right")
```

In this section we presented code for 2-way nested Poisson GLMMs. The code can easily be modified for 1-way nested data by dropping one of the random effect terms.

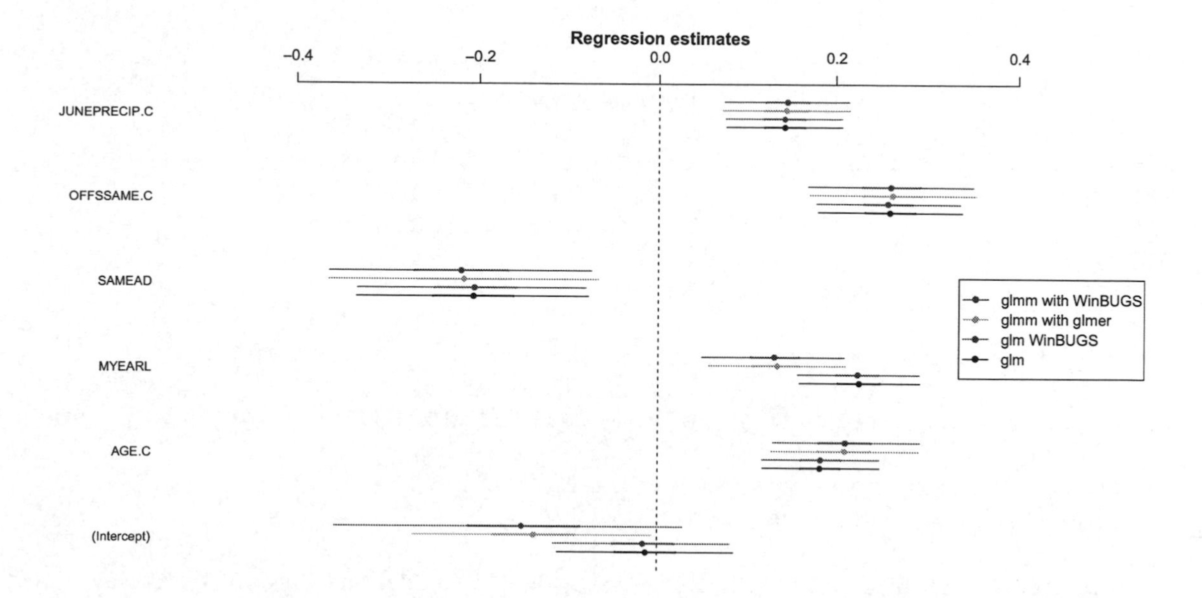

Figure 4.11. Estimated regression parameters for the Poisson GLM (using `glm` and WinbUGS) and 2-way nested Poisson GLMM (using `glmer` and WinBUGS). For WinBUGS results the mean of the posterior of each parameter is shown. The limits along the horizontal axis were set from -0.5 to 0.5. The original graph uses colours but these were set to greyscale during printing. The order of the points/lines corresponds to the order of the designations in the legend. The top lines are for the GLMM with WinBUGS and the bottom lines for the `glm` function.

4.7.3 WinBUGS code for ZIP GLM

Before showing code for the 2-way nested ZIP GLMM, we will fit a ZIP GLM and ignore the random effects. We have two reasons for this step. First, we are working on building up the WinBUGS code for the 2-way nested ZIP GLMM. The second reason will become clear when we compare all models later in this chapter. The ZIP GLM is of the form:

$$
\begin{aligned}
&Y_{ijk} \sim ZIP(\mu_{ijk,} \pi) \\
&\log(\mu_{ijk}) = \beta_1 + \beta_2 \times AGE_{ijk} + \beta_3 \times MYEARL_{ijk} + \beta_4 \times SAMEAD_{ijk} + \\
&\qquad \beta_5 \times OFFSSAME_{ijk} + \beta_6 \times JUNEPRECIP_{ijk} \\
&\text{logit}(\pi) = \gamma
\end{aligned}
$$

In principle we could extend the ZIP presented above by allowing the probability of false zeros, π, to be a function of covariates as well. However, with such a model it is more difficult to obtain well mixed chains in WinBUGS, especially when random effects are added. The ZIP GLM can be fitted in R with the `zeroinfl` function from the `pscl` package (Zeileis et al., 2008):

```
> library(pscl)
> M3 <- zeroinfl(YOUNG ~ AGE.C + MYEARL + SAMEAD + OFFSSAME.C +
                 JUNEPRECIP.C | 1,
                 data = Marmots4, dist = "poisson", link = "logit")
```

The notation `| 1` ensures that the logistic part of the model contains only an intercept. The numerical output is:

```
> summary(M3)

Count model coefficients (poisson with log link):
              Estimate Std. Error z value Pr(>|z|)
(Intercept)    0.52447    0.06569   7.984 1.42e-15
AGE.C          0.17566    0.04340   4.047 5.18e-05
MYEARL         0.13600    0.03701   3.674 0.000238
SAMEAD        -0.19784    0.07667  -2.581 0.009865
OFFSSAME.C     0.20830    0.04746   4.389 1.14e-05
JUNEPRECIP.C   0.12361    0.03999   3.091 0.001993

Zero-inflation model coefficients (binomial with logit link):
            Estimate Std. Error z value Pr(>|z|)
(Intercept)  -0.4116     0.1079  -3.814 0.000137

Number of iterations in BFGS optimization: 14
Log-likelihood: -1100 on 7 Df
```

Differences between the estimated parameters of the ZIP model and those of the 2-way nested Poisson GLMM are small, but we will plot coefficients obtained by all techniques in a single graph presented later. The estimated value for the intercept in the logistic function is -0.412. This means that the probability of a false zero is exp(-0.412) / (1 + exp(-0.412)) = 0.398. A large negative value for this intercept would have meant that the probability of false zeros is 0, in which case the ZIP GLM would become a Poisson GLM, but this is clearly not the case.

To assess whether there is overdispersion, we extract the Pearson residuals and calculate the ratio of the residual sum of squares of the Pearson residuals to the residual degrees of freedom.

```
> E3 <- resid(M3, type = "pearson")
> Dispersion3 <- sum(E3^2) / (nrow(Marmots4) - 7)
> Dispersion3

0.9868699
```

There is no evidence of overdispersion. The ZIP can also be fitted in WinBUGS with the following code. Recall that this code was also used in Chapter 2. We need the following variables:

```
> win.data3 <- list(YOUNG      = Marmots4$YOUNG,
                    AGE        = Marmots4$AGE.C,
                    MYEARL     = Marmots4$MYEARL,
                    SAMEAD     = Marmots4$SAMEAD,
                    OFFSSAME   = Marmots4$OFFSSAME.C,
                    JUNEPRECIP = Marmots4$JUNEPRECIP.C,
                    N          = nrow(Marmots4))
```

The modelling code is as follows.

```
> sink("ZIPGLM.txt")
```

```
   cat("
   model {
    for (i in 1:N) {
         #Logistic part
         W[i] ~ dbern(psi.min1)
         #Poisson part
         YOUNG[i]  ~  dpois(mu.eff[i])
         mu.eff[i]  <- W[i] * mu[i]
         log(mu[i]) <- max(-20, min(20, eta.mu[i]))
         eta.mu[i]  <- aP[1]  +
                       aP[2] * AGE[i] +
                       aP[3] * MYEARL[i] +
                       aP[4] * SAMEAD[i] +
                       aP[5] * OFFSSAME[i] +
                       aP[6] * JUNEPRECIP[i]
         EZip[i]   <- mu[i] * (1 - psi)
         VarZip[i] <- (1 - psi) * (mu[i] + psi * pow(mu[i], 2))
         PRES[i] <- (YOUNG[i] - EZip[i]) / sqrt(VarZip[i])

         #log-likelihood
         lfd0[i] <- log(psi + (1 - psi) * exp(-mu[i]))
         lfd1[i] <- log(1 - psi) + YOUNG[i] * log(mu[i]) - mu[i]  -
                    loggam(YOUNG[i]+1)
         L[i]    <- (1 - W[i]) * lfd0[i] + W[i] * lfd1[i]
       }
      dev <- -2 * sum(L[1:N])
      psi.min1 <- min(0.99999, max(0.00001,(1 - psi)))
      eta.psi <- aB
      logit(psi) <- max(-20, min(20, eta.psi))

      #Priors for regression parameters
      for (i in 1:6) { aP[i] ~ dnorm(0.0, 0.001) }
      aB ~ dnorm(0, 0.001)
   }
",fill = TRUE)
sink()
```

The Poisson part of the ZIP is identical to that of the Poisson GLM; we use the same link function and the same priors. The logistic link function contains only an intercept, denoted by `aB`, and it uses a diffuse normal prior. Our usual security steps are in place to ensure we do not experience a fatal crash as a consequence of taking the exponential of a large number. The ZIP is a mixture of a binomial and Poisson distribution, as explained in Chapter 2. We also calculate Pearson residuals (`PRES`), and the underlying equation for these was given in Chapter 11 in Zuur et al. (2009a). Finally, we obtain the fitted values (`EZip`), variance (`VarZip`), and the likelihood function (denoted by `L[i]`).

We need to define `W` as a vector or zeros (`YOUNG` = 0) and ones (`YOUNG` > 0).

```
> W <- Marmots4$YOUNG
> W[Marmots4$YOUNG > 0] <- 1
```

Initial values are as follows. We use the estimated regression coefficients of the 2-way Poisson GLMM as initial values for the regression parameters in the Poisson part of the ZIP and added a small amount of noise.

```
> inits3 <- function () {
             list(aP = rnorm(6, fixef(M2), 0.01),
```

```
                aB = rnorm(1, 0, 0.01),
                W = W)}
```

We want to extract information from aP, aB, and the deviance, dev. If this were the final model, we would also extract the residuals and fitted values. We will do this in the next section.

```
> params3 <- c("aP", "aB",  "dev")
```

We use values similar to those used previously for the number of chains, draws, burn-in, and thinning rate. If this were the final model, we would expand the latter three by a factor 10.

```
> nc <- 3          #Number of chains
> ni <- 100000     #Number of draws from posterior (for each chain)
> nb <-  10000     #Number of draws to discard as burn-in
> nt <-    100     #Thinning rate
```

WinBUGS is run as before with the bugs command, and we call the resulting object outzip. Mixing of the chains is shows only minor auto-correlation. The numerical results are obtained with the print command.

```
> print(outzip, digits = 3)

Inference for Bugs model at "ZIPGLM.txt", fit using WinBUGS,
 3 chains, each with 1e+05 iterations (first 10000 discarded), n.thin = 100
 n.sims = 2700 iterations saved
             mean     sd     2.5%      25%      50%      75%    97.5%  Rhat n.eff
aP[1]       0.521  0.066    0.390    0.476    0.521    0.565    0.648 1.002  1100
aP[2]       0.175  0.043    0.090    0.145    0.177    0.203    0.257 1.002  1800
aP[3]       0.135  0.037    0.064    0.110    0.136    0.160    0.208 1.001  2700
aP[4]      -0.203  0.076   -0.357   -0.254   -0.201   -0.151   -0.061 1.001  2700
aP[5]       0.210  0.047    0.120    0.179    0.211    0.240    0.303 1.001  2700
aP[6]       0.124  0.039    0.049    0.098    0.124    0.151    0.199 1.001  2700
aB         -0.426  0.109   -0.645   -0.501   -0.425   -0.351   -0.220 1.010   220
dev      2501.140 26.039 2450.000 2484.000 2501.000 2518.000 2553.000 1.002  1200
deviance 1583.461 45.498 1497.475 1552.000 1583.000 1614.000 1675.000 1.007   310

For each parameter, n.eff is a crude measure of effective sample size,
and Rhat is the potential scale reduction factor (at convergence, Rhat=1).

DIC info (using the rule, pD = var(deviance)/2)
pD = 1029.0 and DIC = 2612.4
DIC is an estimate of expected predictive error (lower deviance is better).
```

Note that the posterior mean values obtained by WinBUGS are close to the estimated values of zeroinfl. The only difference is between the deviance calculated by our code (dev) and the reported deviance. With frequentist statistics, we have tools such as comparison of deviances (for nested models) and AIC at our disposal for comparing models. With Bayesian statistics, one can use the DIC (Spiegelhalter et al., 2002). Let $D(\boldsymbol{\theta})$ be defined as -2 log($f(\mathbf{y}|\boldsymbol{\theta})$), where $\boldsymbol{\theta}$ contains the model parameters, and $f(\mathbf{y}|\boldsymbol{\theta})$ is the likelihood. $D(\boldsymbol{\theta})$ is calculated for each realisation of the MCMC chain. Let $\bar{D}$ be the average of $D(\boldsymbol{\theta})$. The effective number of parameters, p_D, is calculated as the posterior mean deviance minus the deviance evaluated at the posterior mean:

$$p_D = \bar{D} - D(\bar{\theta})$$

where $\bar{\theta}$ is the average of all realisations of $\boldsymbol{\theta}$. Then, analogous to AIC, the DIC is defined as

$$DIC = D(\bar{\theta}) + 2 \times p_D = \bar{D} + p_D$$

Models with a small DIC are preferred. The problem arises in that the DIC and the deviance reported above are based on a Poisson distribution and not on a ZIP distribution. Hence we need to calculate the DIC ourselves (note that this was not a problem for the Poisson GLM and 2-way nested Poisson GLMM). To calculate the DIC, we take the average deviance from WinBUGS, but, when the WinBUGS code has finished, we also need to calculate the deviance for the optimal parameters. The average DIC value ($\bar{D}$) is given by:

```
> outzip$mean$dev
```

```
2501.14
```

This is the same value as shown in the printed output. Next we need $D(\bar{\theta})$, and for this we substitute the mean posterior values in the model and calculate the likelihood. The R code for this follows. First we calculate π (denoted by `psi`) and define `L` as a vector. The loop with the code for `eta.mu`, `mu`, and the log likelihood `L` was copied from the MCMC code above. It is not necessary to use `mu` and `eta.mu` as vectors. The mean posterior is substituted for the parameter value, and `Marmot4$` is added to every covariate name. We also changed the WinBUGS function `loggam` to the R function `lfactorial`.

```
> psi     <- exp(outzip $mean$aB) / (1 + exp(outzip$mean$aB))
> L       <- vector(length = nrow(Marmots4))
> for (i in 1: nrow(Marmots4)){
      eta.mu <- outzip$mean$aP[1]  +
                outzip$mean$aP[2] * Marmots4$AGE[i] +
                outzip$mean$aP[3] * Marmots4$MYEARL[i] +
                outzip$mean$aP[4] * Marmots4$SAMEAD[i] +
                outzip$mean$aP[5] * Marmots4$OFFSSAME[i] +
                outzip$mean$aP[6] * Marmots4$JUNEPRECIP[i]
      mu <- exp(eta.mu)

      #log-likelihood
      lfd0 <- log(psi + (1 - psi) * exp(-mu))
      lfd1 <- log(1 - psi) + Marmots4$YOUNG[i] * log(mu)  -
              mu - lfactorial(Marmots4$YOUNG[i])
      L[i] <- (1 - W[i]) * lfd0 + W[i] * lfd1
    }
> devMeanTheta <- -2 * sum(L)
```

The value of `L` is the log-likelihood, which we hope is close to -1100, as this was the value calculated by `zeroinfl`. To check it, type:

```
> devMeanTheta
```

```
2199.991
```

It seems that our code correctly calculates the likelihood value, as the deviance is -2 times the log-likelihood value. Using the formula for pD we can calculate the DIC as:

```
> pD <-  outzip$mean$dev - devMeanTheta
> DIC <- outzip$mean$dev + pD
> DIC
```

```
2802.289
```

We will compare the DIC values of all models later, but recall the critical comments in Section 2.10 on the use of DIC for comparing models with random effects. If this were the final model, a complete model validation should be applied (see the following section).

4.7.4 WinBUGS code for 2-way nested ZIP Poisson GLMMs

In the previous subsections we presented code for a 2-way nested Poisson GLMM and a ZIP GLM. In this subsection we will combine the two models to create a 2-way nested ZIP GLMM model. Such a model cannot be fitted in the `lme4` package, so we do not have a reference model. The underlying mathematical model is given by:

$$
\begin{aligned}
&Y_{ijk} \sim ZIP(\mu_{ijk,}\pi) \\
&\log(\mu_{ijk}) = \beta_1 + \beta_2 \times AGE_{ijk} + \beta_3 \times MYEARL_{ijk} + \beta_4 \times SAMEAD_{ijk} + \\
&\qquad \beta_5 \times OFFSSAME_{ijk} + \beta_6 \times JUNEPRECIP_{ijk} + b_i + b_{ij} \\
&\text{logit}(\pi) = \gamma \\
&b_i \sim N(0, \sigma^2_{Colony}) \\
&b_{ij} \sim N(0, \sigma^2_{Animal})
\end{aligned}
$$

In principle we could extend this model by allowing the probability of false zeros, π, to be a function of random effects (colony and animal) and the covariates as well. However, such a model is more difficult to fit (i.e. obtain well mixed chains). The WinBUGS code follows. The object `win.data2` from the 2-way Poisson GLMM (Subsection 4.7.2) can be used.

The WinBUGS model code is a combination of the ZIP code and the 2-way Poisson GLMM from the previous subsections. We also calculated Pearson residuals and fitted values following the equations presented in Chapter 11 in Zuur et al. (2009a). As explained in the previous subsection, the DIC presented by WinBUGS is not the DIC of a ZIP model, and therefore we need to calculate the correct deviance.

```
> sink("ZIPGLMM.txt")
  cat("
  model {
  for (i in 1:N) {
        #Logistic part
        W[i] ~ dbern(psi.min1)

        #Poisson part
        YOUNG[i]   ~  dpois(mu.eff[i])
        mu.eff[i]   <- W[i] * mu[i]
        log(mu[i]) <- max(-20, min(20, eta.mu[i]))
        eta.mu[i]   <- aP[1]   +
                       aP[2] * AGE[i] +
                       aP[3] * MYEARL[i] +
                       aP[4] * SAMEAD[i] +
                       aP[5] * OFFSSAME[i] +
                       aP[6] * JUNEPRECIP[i] +
                       bP[COLONYNEW[i]] +
                       bbP[ANIMALNEW[i]]
        EZip[i]    <- mu[i] * (1 - psi)
        VarZip[i] <- (1 - psi) * (mu[i] + psi * pow(mu[i], 2))
        PRES[i]    <- (YOUNG[i] - EZip[i]) / sqrt(VarZip[i])
```

```
        #log-likelihood
        lfd0[i] <- log(psi + (1 - psi) * exp(-mu[i]))
        lfd1[i] <- log(1 - psi) + YOUNG[i] * log(mu[i])   - mu[i]   -
                   loggam(YOUNG[i]+1)
        L[i]    <- (1 - W[i]) * lfd0[i] + W[i] * lfd1[i]
     }
    dev <- -2 * sum(L[1:N])

    psi.min1 <- min(0.99999, max(0.00001,(1 - psi)))
    eta.psi <- aB
    logit(psi) <- max(-20, min(20, eta.psi))

    #Priors for regression parameters
    for (i in 1:6) { aP[i] ~ dnorm(0.0, 0.001) }
    aB ~ dnorm(0, 0.001)
    for (i in 1:NCOLONY) { bP[i] ~ dnorm(0.0,  tau.Colony) }
    for (i in 1:NANIMAL) { bbP[i] ~ dnorm(0.0,  tau.Animal) }

    #Priors for the random effects for colony
    tau.Colony <- 1 / (sigma.Colony * sigma.Colony)
    tau.Animal <- 1 / (sigma.Animal * sigma.Animal)
    sigma.Colony ~ dunif(0,10)
    sigma.Animal ~ dunif(0,10)
  }
",fill = TRUE)
sink()
```

This is extensive coding, but consists of a combination of code from previous subsections. Our main objective in presenting WinBUGS for all intermediate models is that it should now be relatively easy to detect the structure of the code. If you skipped some of the previous subsections and don't understand parts of this code, now is the time to read them.

The initial values are set as before. The variable `W` has values 0 and 1. If `YOUNG` is greater than 0, `W` is set to 1; otherwise `W` is set to 0.

```
> W <- Marmots4$YOUNG
> W[Marmots4$YOUNG > 0] <- 1
> inits4 <- function () {
   list(aP           = rnorm(6, fixef(M2), 0.01),
        aB           = rnorm(1, 0, 0.01),
        bP           = rnorm(9, 0, 1),
        bbP          = rnorm(NANIMAL, 0, 1),
        sigma.Colony = rlnorm(1, meanlog = 0, sdlog = 0.1),
        sigma.Animal = rlnorm(1, meanlog = 0, sdlog = 0.1),
        W            = W)}
```

We store the draws of the posteriors of the regression parameters, variances, random effects, deviance, Pearson residuals, and fitted values.

```
> params4 <- c("aP", "aB", "sigma.Colony", "sigma.Animal", "bP",
               "bbP", "dev", "PRES", "EZip")
```

Initial MCMC runs show that we need a large burn-in and a high thinning rate. The values shown below ensure that we have proper mixing. The number of draws for each posterior is 3,000.

```
> nc <- 3         #Number of chains
```

```
> ni <- 1000000  #Number of draws from posterior (for each chain)
> nb <-   500000  #Number of draws to discard as burn-in
> nt <-      500  #Thinning rate
```

The MCMC is run with the code:

```
> out4 <- bugs(data = win.data2, inits = inits4,
               parameters = params4, model = "ZIPGLMM.txt",
               n.thin = nt, n.chains = nc, n.burnin = nb,
               n.iter = ni, debug = TRUE,
               bugs.directory = MyWinBugsDir)
```

Calculation time was a full night on a fast computer. We present a detailed model validation in the next section.

4.8 Validating the 2-way nested ZIP GLMM

We begin by presenting the numerical output of the 2-way ZIP GLMM fitted in Subsection 4.7.4.

```
> print(out4, digits = 3)

Inference for Bugs model at "ZIPGLMM.txt", fit using WinBUGS,
 3 chains, each with 1e+06 iterations (first 5e+05 discarded), n.thin = 500
 n.sims = 3000 iterations saved

                 mean     sd     2.5%      25%      50%      75%    97.5%  Rhat n.eff
aP[1]           0.481  0.090    0.293    0.427    0.483    0.540    0.644 1.003   950
aP[2]           0.184  0.046    0.092    0.154    0.185    0.215    0.271 1.002  3000
aP[3]           0.133  0.039    0.056    0.108    0.133    0.159    0.207 1.001  3000
aP[4]          -0.190  0.081   -0.352   -0.245   -0.188   -0.134   -0.031 1.001  3000
aP[5]           0.202  0.049    0.109    0.168    0.203    0.236    0.298 1.001  3000
aP[6]           0.127  0.041    0.049    0.100    0.126    0.156    0.208 1.001  3000
aB             -0.466  0.120   -0.712   -0.543   -0.463   -0.383   -0.244 1.004   810
sigma.Colony    0.098  0.078    0.005    0.040    0.081    0.136    0.290 1.006   540
sigma.Animal    0.117  0.082    0.007    0.052    0.101    0.168    0.306 1.006   370
...
dev          2490.322 27.468 2434.000 2472.000 2491.000 2509.000 2542.000 1.002  1200
...
deviance     1589.932 46.380 1502.000 1558.000 1589.000 1620.000 1685.000 1.002  1700

For each parameter, n.eff is a crude measure of effective sample size,
and Rhat is the potential scale reduction factor (at convergence, Rhat=1).

DIC info (using the rule, pD = var(deviance)/2)
pD = 1074.9 and DIC = 2664.9
DIC is an estimate of expected predictive error (lower deviance is better).
```

The mean and sign of the prior distribution for the Poisson regression parameters are similar to those obtained by the 2-way Poisson GLMM, so that is encouraging. The standard deviations are also similar. Thus the biological conclusions for the count part of the model are similar to that of the 2-way Poisson GLMM.

The posterior mean of σ_{Animal} is smaller than that of the 2-way Poisson GLMM, and its standard deviation is relatively large. Both models indicate that σ_{Colony} is close to 0.

The main point of interest of the 2-way ZIP GLMM is the parameter π, which is the probability of a false zero. It is modelled by logit(π) = aB = -0.466. A large negative value for aB would have meant that π is close to 0, in which case the 2-way ZIP GLMM becomes a 2-way Poisson GLMM. The 95% credible interval for aB is from -0.712 to -0.244, which does not contain 0. This means that the zero inflated part of the model is doing its job. A histogram of the π values is presented in Figure 4.12. The code to create this figure is:

```
> eta <- out4$sims.list$aB
> pi1 <- exp(eta) / (1 + exp(eta))
> hist(pi1, breaks = 50)
```

The posterior samples are stored in the `sims.list$aB` object, and the inverse of the logit link function is used to arrive at the π values. The mean of the probability of false zeros is 0.386.

The 2-way ZIP GLMM is not overdispersed, as can be inferred by the ratio of the sum of squared Pearson residuals to the residual degrees of freedom:

```
> sum(out4$mean$PRES^2) / (nrow(Marmots4) -9)

0.982
```

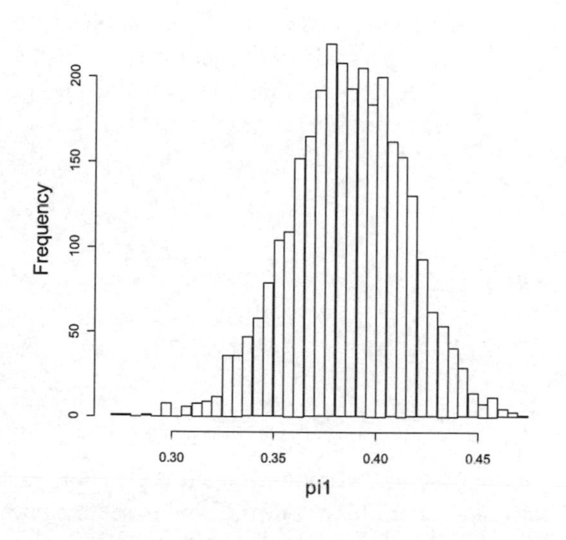

Figure 4.12. Histogram of the transformed posterior of aB. The value along the x-axis is π, the probability of a false zero.

The ratio is close to 1, which is good. As part of the model validation we should also plot the Pearson residuals in `out4$mean$PRES` versus each covariate in the model and each covariate not included in the model. Any patterns present in these graphs would suggest that quadratic terms, splines, or smoothers should be used. Figure 4.13 shows the Pearson residuals plotted versus each covariate in the model. We add a LOESS smoother to aid visual interpretation. The shape of the smoother for `AGE` suggests that a quadratic term should be added to the 2-way ZIP GLMM. Upon closer investigation, the `AGE` smoother is found to be significant, but it only explains about 3% of the variation in the residuals.

R code to produce Figure 4.13 is:

```
> E4    <- rep(out4$mean$PRES, 5)
> Xall  <- with(Marmots4, c(AGE.C, MYEARL, SAMEAD, OFFSSAME.C,
                             JUNEPRECIP.C))
> IDall <- rep(c("AGE.C", "MYEARL", "SAMEAD", "OFFSSAME.C",
                 "JUNEPRECIP.C"), each = nrow(Marmots4))
> xyplot(E4 ~ Xall | IDall,  xlab = "Covariate",
         ylab = "Pearson residuals",
         strip = strip.custom(bg = 'white',
         par.strip.text = list(cex = 0.8)),
         scales = list(x = list(relation = "free"),
                       y = list(relation = "same"), draw = TRUE),
         panel = function(x, y,...) {
           panel.points(x, y, col = 1, cex = 0.3, pch = 16)
           panel.loess(x, y, span = 0.7, lwd = 3, col = 1)
           panel.abline(h = 0, lty = 2,  col = 1)})
```

This is familiar code by now and is not explained here.

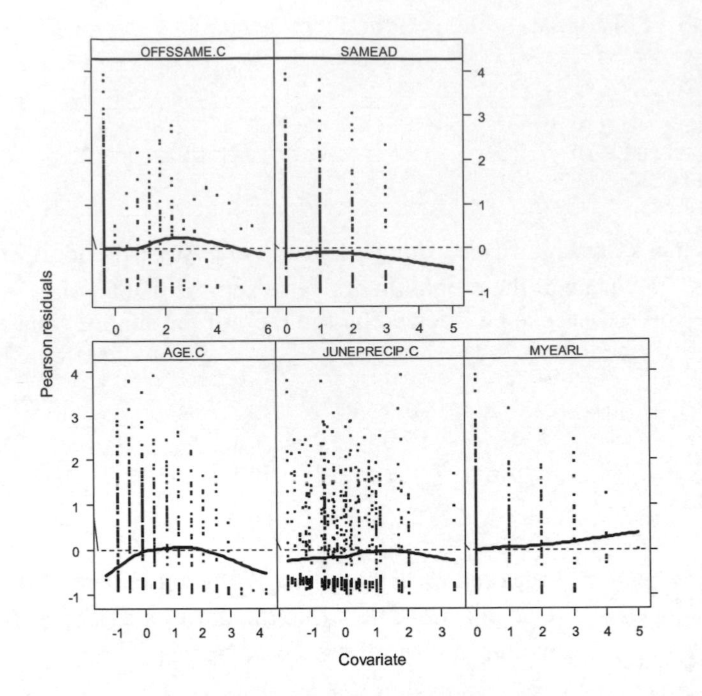

Figure 4.13. Pearson residuals plotted versus each covariate used in the model. A LOESS smoother with a span of 0.7 was added to aid visual interpretation.

We also plot the draws of the posterior distribution for each parameter versus the others (Figure 4.14). Except for SAMEAD and OFFSAME there are no high correlations. Perhaps the covariate SAMEAD should have been centred as well. R code to make the pairplot follows. First we use the cbind function to bind the draws of the relevant variables and the colnames function to label the columns so that the pairplot shows the variable names.

```
> Z <- cbind(out4$sims.list$aP,
             out4$sims.list$aB,
             out4$sims.list$sigma.Colony,
             out4$sims.list$sigma.Animal)
> colnames(Z) <- c("Intercept", "AGE", "MYEARL", "SAMEAD",
                   "OFFSSAME", "JUNEPRECIP", "aB", "Sigma.C",
                   "Sigma.A")
> pairs(Z, lower.panel = panel.cor, cex.labels = 1.3)
```

Finally we discuss mixing of the chains. The code below plots the chains for each regression parameter of the Poisson part of the model. We did not reproduce them here, but the chains seem to be mixing well (no auto-correlation). For smaller burn-in values there was no mixing at the beginning of the chains.

```
> for (i in 1:6) {
   par(mfrow = c(1,1))
   Chains <- out4$sims.array[,, paste("aP", "[", i, "]", sep = "")]
   plot(Chains[,1], type = "l", col = 1,
        main = paste("beta", i, sep = ""), ylab = "")
   lines(Chains[,2], col = 2)
   lines(Chains[,3], col = 3)
   win.graph()  #quartz() on Mac OS
```

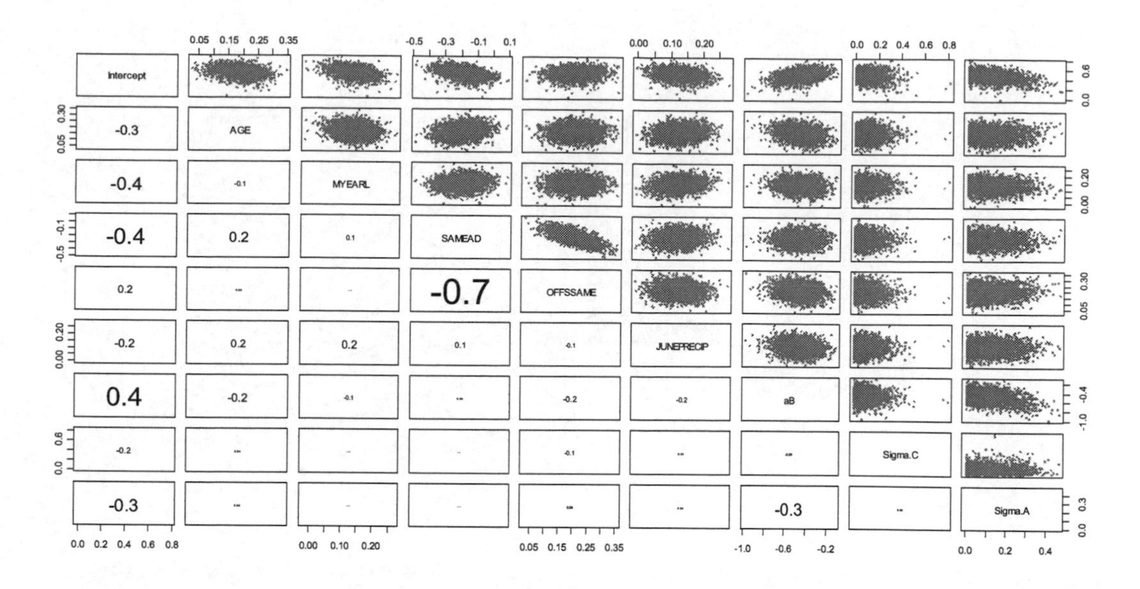

Figure 4.14. Multi-panel scatteplot of the posterior values for each variable.

Auto-correlation functions (ACF) of the chains are another essential part of the model validation process. Except for the intercept in the log-link function, the auto-correlation in the chains is low. The ACFs (not presented here) are created with the following code. The `out4$sims.array` object contains the posterior draws in the original order and can therefore be used for auto-correlation graphs. The loop is used to access the chains for each parameter, and the `paste` command is used to create variable names of the form `aP[1]`, `aP[2]`, etc.

```
> for (i in 1:6) {
   par(mfrow = c(1,1))
   Chains <- out4$sims.array[, , paste("aP", "[", i ,"]", sep = "")]
   plot(Chains[,1], type = "l", col = 1,
        main = paste("beta", i, sep = ""), ylab = "")
   lines(Chains[,2], col = 2)
   lines(Chains[,3], col = 3)
   win.graph()   #quartz() on Mac OS
}
```

The code can easily be adjusted so that the draws of the other parameters are presented. For example, for the intercept in the binary part of the model we can use `out4$sims.array[,,"aB"]`.

In addition to the advanced model validation graphs we plot the sorted residuals and plot residuals versus fitted values (Figure 4.15). These graphs do not show problems. R code to make them is:

```
> par(mfrow = c(1, 2))
> plot(sort(out4$mean$PRES), type = "h", xlab = "Sorted order",
       ylab = "Pearson residuals")
> plot(x= out4$mean$EZip, y = out4$mean$PRES,
       ylab = "Pearson residuals", xlab = "Fitted values")
```

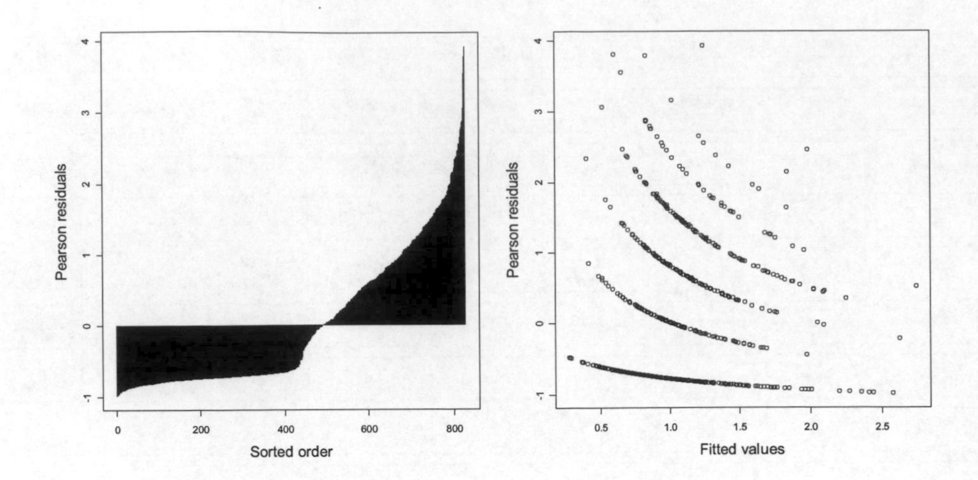

Figure 4.15. Left: Sorted residuals (small to large). Right: Residuals versus fitted values.

Finally, we assess normality of the random effects b_i and b_{ij}. Because there are only 9 colonies, histograms or QQ-plots of the COLONY random effects, b_i, would have little value, but for the random effects ANIMAL nested in COLONY we make QQ-plots for b_{ij} (Figure 4.16). There seems to be a certain amount of non-normality in these random intercepts. Making the graphs requires simple coding:

```
> qqnorm(out4$mean$bbP, main = "Random intercepts for animal")
```

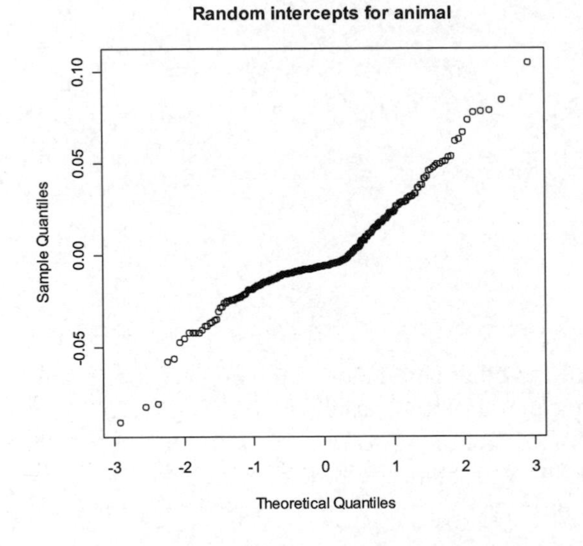

Figure 4.16. QQ-plots for the Poisson random intercepts for the ANIMAL effect.

Calculation for the DIC is similar to that with the ZIP GLM:

```
> MeanDeviance4 <- out4$mean$dev
> psi     <- exp(out4$mean$aB) / (1 + exp(out4$mean$aB))
> L       <- vector(length = nrow(Marmots4))
> for (i in 1: nrow(Marmots4)){
         eta.mu <- out4$mean$aP[1]  +
                   out4$mean$aP[2]  * Marmots4$AGE.C[i] +
                   out4$mean$aP[3]  * Marmots4$MYEARL[i] +
                   out4$mean$aP[4]  * Marmots4$SAMEAD[i] +
                   out4$mean$aP[5]  * Marmots4$OFFSSAME.C[i] +
                   out4$mean$aP[6]  * Marmots4$JUNEPRECIP.C[i]
```

```
        mu <- exp(eta.mu)

        #log-likelihood
        lfd0 <- log(psi + (1 - psi) * exp(-mu))
        lfd1 <- log(1 - psi) + Marmots4$YOUNG[i] * log(mu)   -
                       mu - lfactorial(Marmots4$YOUNG[i])
        L[i]    <- (1 - W[i]) * lfd0 + W[i] * lfd1
     }
> devMeanTheta4 <- -2 * sum(L)
```

We now have the quantities necessary to calculate the DIC.

```
> pD4 <-  out4$mean$dev - devMeanTheta4
> DIC4 <- out4$mean$dev + pD4
> DIC4

2802.289
```

This is higher than we found with the 2-way Poisson GLMM, and we will address its implications in the discussion section of this chapter.

4.9 Interpretation of the 2-way nested ZIP GLMM model

To fully understand the results of the 2-way nested ZIP GLMM we record the estimated model and sketch its fitted values. The estimated model is given by:

$$
\begin{aligned}
& Y_{ijk} \sim ZIP(\mu_{ijk,} \pi) \\
& \log(\mu_{ijk}) = 0.48 + 0.18 \times AGE_{ijk} + 0.13 \times MYEARL_{ijk} - 0.19 \times SAMEAD_{ijk} + \\
& \qquad 0.20 \times OFFSSAME_{ijk} + 0.13 \times JUNEPRECIP_{ijk} + b_i + b_{ij} \\
& \text{logit}(\pi) = -0.47 \\
& b_i \sim N(0, 0.10^2) \\
& b_{ij} \sim N(0, 0.12^2)
\end{aligned}
$$

Modelling the counts shows a positive `AGE` effect, a positive effect of the number of male yearlings present from previous year's litter (`MYYEARL`), a negative effect of the number of matriline females of the same age as the subject (`SAMEAD`), a positive effect of the number of pups produced by matriline females the same age (`OFFSAME`) and a positive effect of precipitation in June (`JUNEPRECIP`). No 95% credible interval of these parameters contains 0, so all terms are important. There is a small amount of random variation around the intercept due to `COLONY` and `ANIMAL`. The probability of a false 0 is 0.386.

Instead of looking at the value and sign of the regression parameters we can sketch the fitted values. This is a simple exercise and follows the code presented in the Appendix of Zuur et al. (2009a). We show it for one covariate and leave the other covariates as an exercise for the reader. First we create values for age:

```
> MyAge <- seq(min(Marmots4$AGE.C), max(Marmots4$AGE.C),
              length = 100)
```

This gives 100 age values within the range of observed (standardised) values. We then calculate the mean value (the median or any other value will do) of other covariates and calculate predicted values:

```
> eta.age <- out3$mean$aP[1]  +
             out3$mean$aP[2]  * MyAge +
             out3$mean$aP[3]  * mean(Marmots4$MYEARL) +
             out3$mean$aP[4]  * mean(Marmots4$SAMEAD) +
             out3$mean$aP[5]  * mean(Marmots4$OFFSSAME.C) +
             out3$mean$aP[6]  * mean(Marmots4$JUNEPRECIP.C)
> mu.age <- exp(eta.age)
```

We can now print `MyAge` versus `mu.age` to visualise the effect of `AGE` on the predicted values of `YOUNG` given average values of the other covariates. We repeat these three commands for each covariate, concatenate all the vectors, and make a multi-panel scatterplot (Figure 4.17).

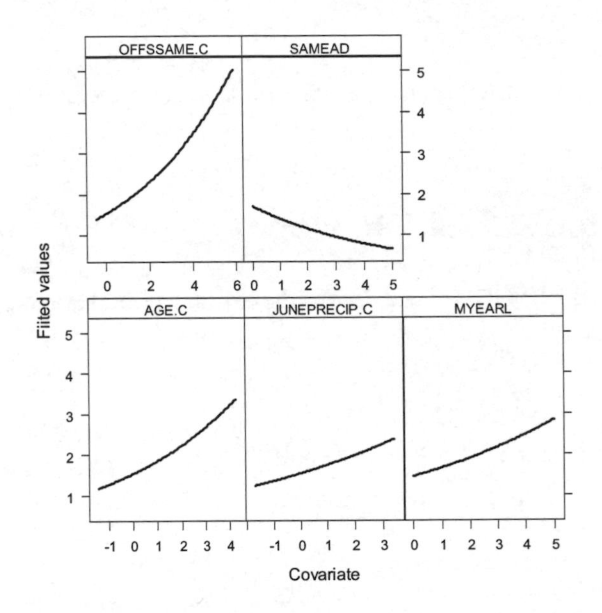

Figure 4.17. Multi-panel scatterplot depicting the effect of each covariate given average values of the other covariates.

4.10 Discussion

During the analysis of the marmot data we encountered four main obstacles: a large number of collinear variables, zero inflation, nested data, and weak ecological patterns.

Collinearity was dealt with by computing variance inflation factors to identify collinear variables and dropping some of them. However, this results in problems identifying which covariates are driving the system, those that we kept or the collinear variables that we dropped. We can't resolve this question without conducting a controlled experiment. This is unrealistic for marmots, as one will not be able to create laboratory conditions to mimic those in the field. After dropping the collinear variables, 25 covariates remained, far too many for the amount of data and the model complexity. We might have utilized the Information Theory approach and defined 10 – 15 models *a priori* for comparison. However, we have little knowledge of the processes driving marmot reproduction. An alternative would have been to apply a model selection using, for example, the DIC, but computing would be lengthy and the literature has recently contained criticism of model selection. Instead we demonstrated how techniques such as GLM, ZIP, and GLMM can be fitted in WinBUGS and presented interpretation of the output. We quasi-arbitrarily chose a model with 5 covariates, based on initial Poisson GLM analyses. These were likely the covariates that would be selected in a backwards selection, but we can't be sure, and this would not be an appropriate strategy for finding the optimal set of covariates in a genuine analysis. We also chose these covariates because they resulted in minimal removal of rows due to missing values. If a different set of covariates had been used, the outcome would likely be different, since fewer observations would be included in the analysis.

To demonstrate a 2-way nested ZIP GLMM, we began with a Poisson GLM, proceeded to a 2-way nested Poisson GLMM and the ZIP, and finally coded a 2-way nested ZIP GLMM. The spread of the response variable was small, so there was no need for negative binomial models. The DIC values for all 4 models are given below. To present this information in a paper you would also need to include the deviances and pD values for each method.

```
                          DIC
Poisson GLM           2407.017
2-way Poisson GLMM 2304.025
ZIP GLM               2802.289
2-way ZIP GLMM        2806.547
```

Only the Poisson GLM was overdispersed and is therefore not a candidate model. The DIC difference between the ZIP GLM and the 2-way nested ZIP GLMM is negligible. The validation of all models indicated that any of the remaining 3 models is a candidate model to present. Based on the DIC values, the 2-way nested Poisson GLMM seems to perform better than the 2-way nested ZIP GLMM, but there are serious criticisms of the use of DIC for mixed models.

Estimated parameters produced by all techniques are presented in Figure 4.18. Note that the intercept is different for all zero inflated models.

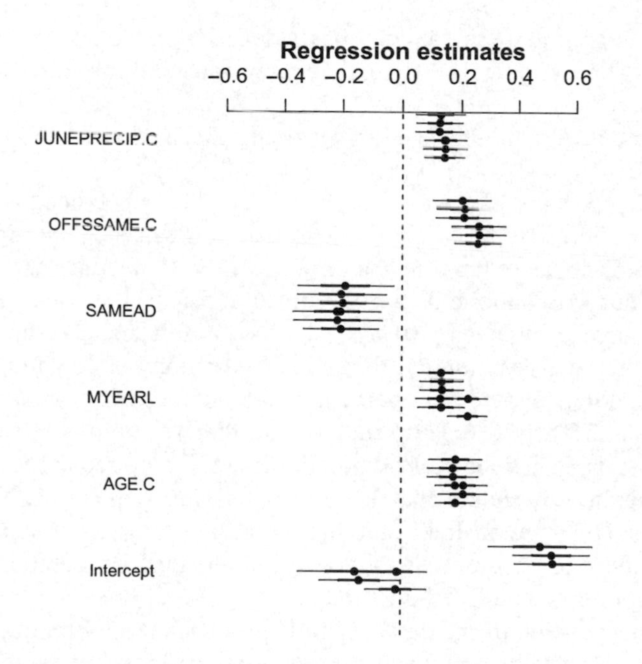

Figure 4.18. Regression parameters in the log link function obtained by all methods. From bottom to top each line represents: Poisson GLM with `glm` function, Poisson GLM in WinBUGS, 2-way nested Poisson GLMM in `glmer`, 2-way nested Poisson GLMM with WinBUGS, ZIP GLM with `zeroinfl`, and 2-way nested ZIP GLMM with WinBUGS.

Which model should we present? Why is the 2-way nested Poisson GLMM better that the ZIP GLM or ZIP GLMM, as indicated by the DIC? To address the latter question: Extensive zeros may mean that there is greater between-animal variation, especially if the zeros are observed for the same animal. This means that the high between-animal variation in the Poisson GLMM may be due to the large number of zeros. For these data and this model, the random effects for `ANIMAL` are able to capture the effect of the zeros in the number of young. The ZIP models explicitly model the zeros, and, when both components are present, the random effect `ANIMAL` and the probability of false zeros are likely to compete for the zeros. This is most likely the reason that the variance for the random effect `ANIMAL` in the 2-way nested ZIP is close to 0. Thus we have a wide variation between animals due to the zeros. If we would have had a large variation between animals due to the counts, the variance for the random effect `ANIMAL` would have been more important in the 2-way nested ZIP GLMM.

This also establishes which model we should present. For this specific data, the 2-way nested Poisson GLMM is adequate, but for other data sets we may need the 2-way nested ZIP GLMM. We can argue that the design of the study requires the random effects `COLONY` and `ANIMAL`, and, because we have zero inflation, we need the 2-way nested ZIP GLMM. This is an *a priori* choice.

We have two approaches to modelling the data (random effects vs. zero inflation), and it is important that we select the correct one. In Chapter 2 we simulated zero inflated data and found that any other model gave biased parameters. We need to choose the right model for the problem, and the zero inflation models with random effects `COLONY` and `ANIMAL` seem to make more sense for these data.

Using a different set of covariates could reduce the data set by 30 – 40% due to missing values, and the results and interpretation may be different.

The fact that the ecological patterns are weak is something we can't resolve. The scatterplots in the data exploration section showed that it would be a difficult task to find any significant effects. In cases of weak ecological signals it is even more important to deal with even the smallest amount of collinearity.

A full analysis of all data and an ecological discussion of the results are presented in the PhD thesis of Kathleen Nuckolls.

4.11 What to present in a paper?

To write a paper presenting these data, the large number of covariates must be dealt with first, and the IT approach seems a sensible solution. If you do not object to model selection, and the journal you will submit to is not opposed to model selection, the forward selection using DIC is an option. So let us assume you resolve this issue.

Whether you can write an ecological paper around these data obviously depends on the results. But let us assume that the outcome makes ecological sense and is of sufficient relevance to present. What should you say about the statistics?

Start with an Introduction, in which you describe the motivation for doing the fieldwork. Formulate your underlying questions.

In the Methods section, start by describing the steps of the data exploration and how you dealt with collinearity. You can cite Zuur et al. (2010). Some referees may believe that you can move straight from typing in data to the statistical analysis. Emphasise the four major problems with the data; it may buy you empathy from referees. At this point it is important that you make clear to the reader and referees that the data are zero inflated, otherwise, referees will ask why you didn't apply ordinary linear regression or a GLM with a Poisson or negative binomial distribution. Mention biased parameters and overdispersion if zero inflation is ignored. Provide the percent of the number of zeros in the response variable. Also explain that the data are nested (or correlated), and that you must use zero inflated generalised linear mixed effects models. It may be worthwhile to present the nested aspect of the data in a flowchart. State the model applied. It should match your underlying hypothesis. The Bayesian approach applied in this chapter is well documented in the literature; hence there is no need to give any algorithms. You need to report the type of prior distributions used, the burn-in, number of draws for the posterior, and a brief summary of how you fitted the model.

In the Results section, show a graphic presentation of the data. A multi-panel scatterplot produced with the `xyplot` function from the `lattice` package will do. A referee or reader that can visualize your data is a referee or reader that is more likely to understand your data. Summarise the model selection results and present the final model. Describe the model validation and performance of the MCMC iteration process (i.e. mixing of the chains). Try to visualize what your optimal model tells you.

In the Discussion mention the collinearity problem and that it is not possible to identify which covariate is driving the system.

Appendix A: Correlation between observations in GLMMs*

The following technical appendix presents theory and R code to calculate the correlation among observations of the same animal and observations from the same colony.

In Section 4.5 we explained how the random effects in a linear mixed effects model work their way to correlation among the observed data. Recall that the linear mixed effect model with the random effect `COLONY`, and a random effect `ANIMAL` nested within `COLONY`, led to a system in which:

- There was correlation among observations of an individual animal.
- There was a (different) correlation among observations of different animals from the same colony.
- Observations from different colonies were independent.

In a GLMM, similar rules apply, although they are considerably more complicated, and not many sources are available in which it is explained how a GLMM imposes correlation structures on raw data.

We begin with the correlation in Poisson GLMMs for 1-way nested data and proceed to that for 2-way nested data. This is repeated for binomial GLMMs. Finally, we combine the two and show the induced correlations for ZIP GLMMs for 1-way and 2-way data.

A.1 Correlations in a Poisson GLMM for 1-way nested data*

Suppose we have a Poisson GLM of the form:

$$Y_{ij} \sim P(\mu_{ij})$$
$$\mu_{ij} = e^{\eta_{ij}+b_i}$$

Y_{ij} is the observed number of young for observation j in colony i, the predictor function η_{ij} contains the intercept and all covariate effects, and b_i is the random intercept for colony. It follows a normal distribution with mean 0 and variance σ^2_{colony}. We now seek an expression for the correlation between two observations from the same colony, cor(Y_{ij}, Y_{im}). As part of this expression, we need the equations for E(Y_{ij}) and var(Y_{ij}). For the expectation, we have:

$$E(Y_{ij}) = e^{\eta_{ij}} \times E(e^{b_i}) = e^{\eta_{ij}} \times e^{\frac{\sigma^2_{colony}}{2}}$$

The last step is based on the log-normal distribution (see, for example, Johnson et al., 1994). For the variance the starting point is:

$$\text{var}(Y_{ij}) = V_{b_i}\left(E(Y_{is} \mid b_i)\right) + E_{b_i}\left(V(Y_{is} \mid b_i)\right)$$

This may look intimidating, but it is a standard expression for the variance. Working out the expression for the variance gives:

$$
\begin{aligned}
\text{var}(Y_{ij}) &= \text{var}_{b_i}\left(e^{\eta_{ij}+b_i}\right) + E_{b_i}\left(e^{\eta_{ij}+b_i}\right) \\
&= e^{2\eta_{ij}} \times \text{var}_{b_i}\left(e^{b_i}\right) + e^{\eta_{ij}} \times E_{b_i}\left(e^{b_i}\right) \\
&= e^{2\eta_{ij}} \times (e^{\sigma^2_{colony}} - 1) \times e^{\sigma^2_{colony}} + e^{\eta_{ij}} \times e^{\frac{\sigma^2_{colony}}{2}} \\
&= e^{\eta_{ij}} \times \left(e^{\eta_{ij}} \times (e^{\sigma^2_{colony}} - 1) \times e^{\sigma^2_{colony}} + e^{\frac{\sigma^2_{colony}}{2}} \right)
\end{aligned}
\tag{A.1}
$$

The first step is based on the general rule that var($c \times X$) = $c^2 \times$ var(X) for any constant c, and the second step uses the variance for a log-normal distribution (see page 212 in Johnson et al., 1994). We now have the ingredients required for calculating the covariance between Y_{ij} and Y_{im}:

$$
\text{cov}(Y_{ij}, Y_{im}) = E_{b_i}\left[E[Y_{ij} \times Y_{im} \mid b_i]\right] - E_{b_i}\left[E[Y_{ij} \mid b_i]\right] \times E_{b_i}\left[E[Y_{im} \mid b_i]\right] \tag{A.2}
$$

This is one of the standard expressions for the covariance. There are three components to this equation. The first component can easily be found by substituting the equations for Y_{ij} and Y_{im}:

$$
E_{b_i}\left[E[Y_{ij} \times Y_{im} \mid b_i]\right] = E_{b_i}\left[e^{\eta_{ij}+b_i} \times e^{\eta_{im}+b_i}\right] = e^{\eta_{ij}+\eta_{im}} E_{b_i}\left[e^{2 \times b_i}\right] = e^{\eta_{ij}+\eta_{im}} \times e^{2 \times \sigma^2_{colony}}
$$

The final step is based on the rule that if b_i is normally distributed with mean 0 and variance σ^2_{colony}, $2 \times b_i$ is normally distributed with mean 0 and variance $4 \times \sigma^2_{\text{colony}}$. This is used in the log-normal distribution. For the second component in Equation (A.2) we have:

$$
E_{b_i}\left[E[Y_{ij} \mid b_i]\right] = E_{b_i}\left[e^{\eta_{ij}+b_i}\right] = e^{\eta_{ij}} \times E_{b_i}\left[e^{b_i}\right] = e^{\eta_{ij}} \times e^{\frac{\sigma^2_{colony}}{2}}
$$

A similar expression is obtained for the third term. As a result, we can write the covariance as:

$$
\text{cov}(Y_{ij}, Y_{im}) = e^{\eta_{ij}+\eta_{im}} \times e^{2 \times \sigma^2_{colony}} - e^{\eta_{ij}+\eta_{im}} \times e^{\sigma^2_{colony}} = e^{\eta_{ij}+\eta_{im}} \times \left(e^{\sigma^2_{colony}} - 1\right) \times e^{\sigma^2_{colony}} \tag{A.3}
$$

This is the same expression as on page 333 in Ntzoufras (2009). The correlation is then defined as:

$$
\text{cor}(Y_{ij}, Y_{im}) = \frac{\text{cov}(Y_{ij}, Y_{im})}{\sqrt{\text{var}(Y_{ij})} \times \sqrt{\text{var}(Y_{im})}}
$$

The expression for the variance was given in Equation (A.1), and substituting this expression for the variance into the equation above is a straightforward exercise. A problem with the equation for the correlation coefficient is that it depends on the covariates via η_{ij} and η_{im}. To calculate correlation coefficients, we need to choose particular values for the covariates (e.g. mean values), which then gives values for η_{ij} and η_{im}. Ntzoufras (2009) gives code for including these calculations inside the WinBUGS code. A practical approach is to calculate the correlation using the mean values of the posterior distributions when the MCMC algorithm is finished.

A.2 Correlations in a Poisson GLMM for 2-way nested data*

We continue with a model that contains a random intercept b_i for colony as well as a random intercept b_{ij} for ANIMAL within COLONY. Suppose we have a model of the form:

$$Y_{ijk} \sim P(\mu_{ijk})$$
$$\mu_{ijk} = e^{\eta_{ijk} + b_i + b_{ij}}$$

Y_{ijk} is the observed number of young at observation k of animal j in colony i; η_{ijk} contains the intercept and all covariates; b_i is the random intercept for COLONY following a normal distribution with mean 0 and variance σ^2_{colony}; and b_{ij} is the random intercept allowing for differences per animal within a colony, and is normally distributed with mean 0 and variance σ^2_{animal}. The expression for the correlation between two observations of the same animal is cor(Y_{ijk}, Y_{ijm}), and two observations of different animals from the same colony is cor(Y_{ijk}, Y_{ilm}).

Just as before, the starting point is an expression for E(Y_{ijk}) and var(Y_{ijk}).

$$E[Y_{ijk}] = e^{\eta_{ijk}} \times E[e^{b_i + b_{ij}}] = e^{\eta_{ijk}} \times e^{\frac{\sigma^2_{colony} + \sigma^2_{animal}}{2}}$$

Because b_i and b_{ij} are assumed independent, $b_i + b_{ij}$ is normally distributed with mean 0 and variance $\sigma^2_{colony} + \sigma^2_{animal}$. This new variance is then used in the log-normal distribution. It is now a matter of repeating all equations from the previous subsection using this new variance. The starting point for the variance is similar to before:

$$\text{var}(Y_{ijk}) = V_\theta\left(E(Y_{ijk} \mid \theta)\right) + E_\theta\left(V(Y_{ijk} \mid \theta)\right)$$

To simplify notation, we use $\theta = (b_i, b_{ij})$. Following the same steps as in the previous subsection, we get:

$$\begin{aligned}
\text{var}(Y_{ijk}) &= \text{var}_\theta\left(E(Y_{ijk} \mid \theta)\right) + E_\theta\left(V(Y_{ijk} \mid \theta)\right) \\
&= \text{var}_\theta\left(e^{\eta_{ijk} + b_i + b_{ij}}\right) + E_\theta\left(e^{\eta_{ijk} + b_i + b_{ij}}\right) \\
&= e^{2 \times \eta_{ijk}} \times \text{var}_\theta\left(e^{b_i + b_{ij}}\right) + e^{\eta_{ijk}} \times E_\theta\left(e^{b_i + b_{ij}}\right) \\
&= e^{2 \times \eta_{ijk}} \times \left(e^{\sigma^2_{colony} + \sigma^2_{animal}} - 1\right) \times e^{\sigma^2_{colony} + \sigma^2_{animal}} + e^{\eta_{ijk}} \times e^{\frac{\sigma^2_{colony} + \sigma^2_{animal}}{2}} \\
&= e^{\eta_{ijk}} \times \left(e^{\eta_{ijk}} \times (e^{\sigma^2_{colony} + \sigma^2_{animal}} - 1) \times e^{\sigma^2_{colony} + \sigma^2_{animal}} + e^{\frac{\sigma^2_{colony} + \sigma^2_{animal}}{2}}\right)
\end{aligned}$$

The same can be done for the covariance, which produces approximately 4 pages of equations, with the end result given by:

$$\text{cov}(Y_{ijk}, Y_{ijm}) = e^{2 \times \eta_{ijk}} \times \left(e^{\sigma^2_{colony} + \sigma^2_{animal}} - 1\right) \times e^{\sigma^2_{colony} + \sigma^2_{animal}}$$

The covariance can be converted into a correlation by dividing it by the product of the square roots of var(Y_{ijk}) and var(Y_{ijm}).

Using similar algebra, the covariance between two observations of different animals from the same colony is given by:

$$\text{cov}(Y_{ijk}, Y_{ilm}) = e^{\eta_{ijk} + \eta_{ilm}} \times \left(e^{2 \times \sigma^2_{colony} + \sigma^2_{animal}} - e^{\sigma^2_{colony} + \sigma^2_{animal}} \right)$$

Divide by the product of the square roots of var(Y_{ijk}) and var(Y_{ilm}) to obtain the correlation.

A.3 Correlations in a binomial GLMM for 1-way nested data*

Due to the log-link function, deriving correlations between two observations of the same animal, and from different animals from the same colony, was relatively easy for the Poisson GLMM. It was based on definitions of the variance and the log-normal distribution plus basic algebra. For binomial GLMMs the situation is more complicated. Suppose Y_{ij} is a binary variable, e.g. presence or absence of young at observation *j* of colony *i*. In the first instance we again ignore the `ANIMAL` effect. The binomial GLMM with a random intercept for `COLONY` is given by:

$$Y_{ij} \sim B(\pi_{ij}, 1)$$
$$\pi_{ij} = \frac{e^{\eta_{ij} + b_i}}{1 + e^{\eta_{ij} + b_i}}$$

We follow the same steps as for the Poisson GLM; that is, finding expressions for E(Y_{ij}), var(Y_{ij}), cov(Y_{ij},Y_{ik}), and cor(Y_{ij},Y_{ik}). Let us start with the expectation:

$$E[Y_{ij}] = E[\frac{e^{\eta_{ij} + b_i}}{1 + e^{\eta_{ij} + b_i}}]$$

There is no convenient way to remove the η_{ij} from the expectation and use the log-normal distribution as we did with the Poisson GLMM. Goldstein et al. (2002) discuss three options to address this problem:

- Linearization with a Taylor expansion.
- Simulation.
- Approximation with a Gaussian model.

Browne et al. (2005) provide a comparison of these approaches. We follow the simulation approach, as it is relatively easy to implement and can be applied to the ZIP equations that will be presented later. However, before doing this we must again derive equations for the mean, variance, covariance, and correlation. The variance is given by:

$$\begin{aligned} \text{var}(Y_{ij}) &= \text{var}_{b_i}\left(E(Y_{ij} \mid b_i)\right) + E_{b_i}\left(\text{var}(Y_{ij} \mid b_i)\right) \\ &= \text{var}_{b_i}\left(\frac{e^{\eta_{ij} + b_i}}{1 + e^{\eta_{ij} + b_i}}\right) + E_{b_i}\left(\pi_{ij} \times (1 - \pi_{ij})\right) \end{aligned} \quad \text{(A.4)}$$

Again, we cannot further simplify this equation. The covariance between Y_{ij} and Y_{im} is given by:

$$\begin{aligned} \text{cov}(Y_{ij}, Y_{im}) &= E_{b_i}\left[E[Y_{ij} \times Y_{im} \mid b_i]\right] - E_{b_i}\left[E[Y_{ij} \mid b_i]\right] \times E_{b_i}\left[E[Y_{im} \mid b_i]\right] \\ &= E_{b_i}\left[\frac{e^{\eta_{ij} + \eta_{im} + 2b_i}}{(1 + e^{\eta_{ij} + b_i}) \times (1 + e^{\eta_{im} + b_i})}\right] - E_{b_i}\left[\frac{e^{\eta_{ij} + b_i}}{1 + e^{\eta_{ij} + b_i}}\right] \times E_{b_i}\left[\frac{e^{\eta_{im} + b_i}}{1 + e^{\eta_{im} + b_i}}\right] \end{aligned} \quad \text{(A.5)}$$

The simulation method as described in Goldstein et al. (2002) and Browne et al. (2005) consists of the following steps:

1. Fit the model and estimate the parameters, including σ^2_{colony}.
2. Sample a large number (N) of values from N(0, σ^2_{colony}) and call them b_i^*. Note that we are using the sample estimate of the variance. Goldstein et al. (2002) proposed N = 5000 values.
3. Choose values for the covariates; hence η_{ij} has a fixed value. Using the logit link function, calculate π_{ij} values for each b_i^* and call the resulting values π_{ij}^*. Calculate all terms in the covariance function.
4. Calculate the variance according to Equation (A.4).
5. Calculate the covariance, using Equation (A.5), and the correlation.

It may help to see these equations in action with R code and data. The first two lines of the following R code convert the number of young to a binary variable (presence/absence). A binomial GLMM is applied using the same covariates as before. For the time being we only use `COLONY` as random effect. The `summary(M4)` command shows that the estimated variance for the colony effect is 0.0290, and it also gives the estimated values for the intercept and slopes.

```
> Marmots4$YOUNG01 <- Marmots4$YOUNG
> Marmots4$YOUNG01[Marmots4$YOUNG01 > 0] <- 1
> M4 <- glmer(YOUNG01 ~ AGE.C + MYEARL + SAMEAD + OFFSSAME.C +
                 JUNEPRECIP.C + (1|COLONY),
              family = binomial, data = Marmots4)
> summary(M4)

Random effects:
 Groups Name        Variance Std.Dev.
 COLONY (Intercept) 0.029073 0.17051
Number of obs: 827, groups: COLONY, 9

Fixed effects:
              Estimate Std. Error z value Pr(>|z|)
(Intercept)   -0.13761    0.11954  -1.151 0.249647
AGE.C          0.30627    0.08015   3.821 0.000133
MYEARL         0.38251    0.09754   3.922 8.80e-05
SAMEAD        -0.27904    0.11834  -2.358 0.018378
OFFSSAME.C     0.45590    0.09791   4.656 3.22e-06
JUNEPRECIP.C   0.20080    0.07371   2.724 0.006447
```

Next we set all covariates to their mean value (we could have used any value), allowing only `AGE` to change, and calculate the predictor function.

```
> Sigma2Colony <- 0.029073
> Betas <- fixef(M4)
> eta.ij <- Betas[1] +
            Betas[2] *  mean(Marmots4$AGE.C) +
            Betas[3] *  mean(Marmots4$MYEARL) +
            Betas[4] *  mean(Marmots4$SAMEAD) +
            Betas[5] *  mean(Marmots4$OFFSSAME.C) +
            Betas[6] *  mean(Marmots4$JUNEPRECIP.C)
> eta.im <- Betas[1] +
            Betas[2] *  min(Marmots4$AGE.C) +
            Betas[3] *  mean(Marmots4$MYEARL) +
            Betas[4] *  mean(Marmots4$SAMEAD) +
            Betas[5] *  mean(Marmots4$OFFSSAME.C) +
```

```
            Betas[6] *  mean(Marmots4$JUNEPRECIP.C)
```

These commands define a fixed value for the predictor functions η_{ij} and η_{ik}. We arbitrarily used the mean and the minimum of AGE; all other covariates remained constant. The following two blocks of code define two functions, one for the variance and one for the covariance following Equations (A.4) and (A.5).

```
#Function for the variance
> MyVariance <- function(eta.ij, sigma_bi, N) {
     bi.star <- rnorm(N, 0, sigma_bi)
     pi.ij   <- exp(eta.ij + bi.star) / (1 + exp(eta.ij + bi.star))
     Var_bi  <- var(pi.ij)
     Exp_bi  <- mean(pi.ij * (1 - pi.ij))
     VarY    <- Var_bi + Exp_bi
     VarY}
#Function for the covariance
> MyCov <- function(eta.ij, eta.im, sigma_bi, N) {
     bi.star <- rnorm(N, 0,sigma_bi)
     pi.ij   <- exp(eta.ij + bi.star) / (1 + exp(eta.ij + bi.star))
     pi.im   <- exp(eta.im + bi.star) / (1 + exp(eta.im + bi.star))
     Cov     <- mean(pi.ij * pi.im, na.rm=TRUE) -
                mean(pi.ij, na.rm = TRUE) * mean(pi.im, na.rm = TRUE)
     Cov}
```

The remainder of the code is a matter of calling these functions with the appropriate arguments. In this case, the correlation between two observations from the same colony, where the two objects are the average and the minimum of AGE, was 0.0066. This is an extremely low value and reflects minimal COLONY effect in the data. We also ignored the ANIMAL effect, so this was purely a coding exercise:

```
> V.ij     <- MyVariance(eta.ij, sqrt(Sigma2Colony), 5000)
> V.im     <- MyVariance(eta.im, sqrt(Sigma2Colony), 5000)
> Cov.i.jm <- MyCov(eta.ij, eta.im, sqrt(Sigma2Colony), 5000)
> Cor.i.jm <- Cov.i.jm / (sqrt(V.ij) * sqrt(V.im))
> Cor.i.jm

0.006615143
```

In the next subsection we extend the 1-way binomial GLMM to a 2-way binomial GLMM and discuss the induced correlation.

A.4 Correlations in a GLMM binomial model for 2-way nested data*

The binomial GLMM with a random intercept for COLONY and for ANIMAL nested within COLONY is given by:

$$Y_{ijk} \sim B(\pi_{ijk}, 1)$$
$$\pi_{ijk} = \frac{e^{\eta_{ijk} + b_i + b_{ij}}}{1 + e^{\eta_{ijk} + b_i + b_{ij}}}$$

We follow the same steps as for the 1-way binomial GLMM. Let us start with the expectation, which is given by:

$$E[Y_{ijk}] = E[\frac{e^{\eta_{ijk}+b_i+b_{ij}}}{1+e^{\eta_{ijk}+b_i+b_{ik}}}]$$

Again, no simplification is possible. For the variance, we have:

$$\begin{aligned}\text{var}(Y_{ijk}) &= \text{var}_\theta\left(E(Y_{ijk} \mid b_i)\right) + E_\theta\left(\text{var}(Y_{ijk} \mid b_i)\right) \\ &= \text{var}_\theta\left(\frac{e^{\eta_{ijk}+b_i+b_{ij}}}{1+e^{\eta_{ijk}+b_i+b_{ij}}}\right) + E_\theta\left(\pi_{ijk} \times (1-\pi_{ijk})\right)\end{aligned}$$

We give an expression for the correlation between two observations of an individual animal and for the correlation between observations of two different animals from the same colony. In the first case, correlations between observations of the animal j of colony i, we have:

$$\begin{aligned}\text{cov}(Y_{ijk}, Y_{ijm}) &= E_\theta\left[E[Y_{ijk} \times Y_{ijm} \mid \theta]\right] - E_\theta\left[E[Y_{ijk} \mid \theta]\right] \times E_\theta\left[E[Y_{ijm} \mid \theta]\right] \\ &= E_\theta\left[\frac{e^{\eta_{ijk}+\eta_{ijm}+2b_i+2b_{ij}}}{(1+e^{\eta_{ijk}+b_i+b_{ij}}) \times (1+e^{\eta_{ijm}+b_i+b_{ij}})}\right] - E_\theta\left[\frac{e^{\eta_{ijk}+b_i+b_{ij}}}{1+e^{\eta_{ijk}+b_i+b_{ij}}}\right] \times E_\theta\left[\frac{e^{\eta_{ijm}+b_i+b_{ij}}}{1+e^{\eta_{ijm}+b_i+b_{ij}}}\right]\end{aligned}$$

Conversion to a correlation coefficient is as before. We used similar R code for the simulation:

```
> M5 <- glmer(YOUNG01 ~ AGE.C + MYEARL + SAMEAD + OFFSSAME.C +
                        JUNEPRECIP.C + (1 | fCOLONY / fANIMAL),
                        family = binomial, data = Marmots4)

> MyVar2 <- function(eta.ijk, sigma_bi, sigma_bij, N) {
        bi.star    <- rnorm(N, 0, sigma_bi)
        bij.star   <- rnorm(N, 0, sigma_bij)
        pi.ijk     <- exp(eta.ijk + bi.star + bij.star) /
                           (1 + exp(eta.ijk + bi.star + bij.star))
        Var_theta <- var(pi.ijk)
        Exp_theta <- mean(pi.ijk * (1 - pi.ijk))
        VarY       <- Var_theta + Exp_theta
        VarY }
> MyCov.ij <- function(eta.ijk, eta.ijm, sigma_bi, sigma_bij, N) {
        bi.star <- rnorm(N, 0, sigma_bi)
        bij.star <- rnorm(N, 0, sigma_bij)
        pi.ijk <- exp(eta.ijk + bi.star + bij.star) /
                   (1 + exp(eta.ijk + bi.star + bij.star))
        pi.ijm <- exp(eta.ijm + bi.star + bij.star) /
                  (1 + exp(eta.ijm + bi.star + bij.star))
        Cov <- mean(pi.ijk * pi.ijm, na.rm = TRUE) -
               mean(pi.ijk, na.rm = TRUE) *
               mean(pi.ijm, na.rm = TRUE)
        Cov }

> Sigma2Colony <- 0.00000000000794   #Taken from summary(M5)
> Sigma2Animal <- 0.15616            #Taken from summary(M5)
> Betas <- fixef(M5)
> eta.ijk <- Betas[1] +
             Betas[2]  *  mean(Marmots4$AGE.C) +
             Betas[3]  *  mean(Marmots4$MYEARL) +
             Betas[4]  *  mean(Marmots4$SAMEAD) +
```

```
               Betas[5] *  mean(Marmots4$OFFSSAME.C) +
               Betas[6] *  mean(Marmots4$JUNEPRECIP.C)
> eta.ijm <- Betas[1] +
               Betas[2] *  min(Marmots4$AGE.C) +
               Betas[3] *  mean(Marmots4$MYEARL) +
               Betas[4] *  mean(Marmots4$SAMEAD) +
               Betas[5] *  mean(Marmots4$OFFSSAME.C) +
               Betas[6] *  mean(Marmots4$JUNEPRECIP.C)

> V.ijk <- MyVar2(eta.ijk, sqrt(Sigma2Colony), sqrt(Sigma2Animal),
                  5000)
> V.ijm <- MyVar2(eta.ilm, sqrt(Sigma2Colony), sqrt(Sigma2Animal),
                  5000)
> Cov <- MyCov.ij(eta.ijk, eta.ijm, sqrt(Sigma2Colony),
                  sqrt(Sigma2Animal), 5000)
> Cor <- Cov / (sqrt(V.ijk) * sqrt(V.ijm))
> Cor

0.0338
```

The object `M5` contains the estimated intercept and slopes and the variances for the `COLONY` and `ANIMAL` within `COLONY` effect. These are used to create fixed values for η_{ijk} and η_{ijm} in the simulation process.

For the covariance between observations of different animals from the same colony, we can write:

$$\begin{aligned}\text{cov}(Y_{ijk}, Y_{ilm}) &= E_\theta\left[E[Y_{ijk} \times Y_{ilm} \mid \theta]\right] - E_\theta\left[E[Y_{ijk} \mid \theta]\right] \times E_\theta\left[E[Y_{ilm} \mid \theta]\right] \\ &= E_\theta\left[\frac{e^{\eta_{ijk}+\eta_{ilm}+2b_i+b_{ij}+b_{il}}}{(1+e^{\eta_{ijk}+b_i+b_{ij}}) \times (1+e^{\eta_{ilm}+b_i+b_{il}})}\right] - E_\theta\left[\frac{e^{\eta_{ijk}+b_i+b_{ij}}}{1+e^{\eta_{ijk}+b_i+b_{ij}}}\right] \times E_\theta\left[\frac{e^{\eta_{ilm}+b_i+b_{il}}}{1+e^{\eta_{ilm}+b_i+b_{il}}}\right]\end{aligned}$$

We leave it as an exercise for the reader to write the R code.

A.5 Correlations in a ZIP model for 1-way nested data*

The underlying model for a ZIP model for 1-way nested data was given in Section 4.5 and is repeated below.

$$\begin{aligned}&Y_{ij} \sim ZIP(\pi_{ij}, \mu_{ij}) \\ &E(Y_{ij}) = \mu_{ij} \times (1 - \pi_{ij}) \quad \text{and} \quad \text{var}(Y_{ij}) = (1-\pi_{ij}) \times (\mu_{ij} + \pi_{ij} \times \mu_{ij}^2) \\ &\text{logit}(\pi_{ij}) = \eta_{ij} + d_i \\ &\log(\mu_{ij}) = \xi_{ij} + b_i\end{aligned}$$

This means that we have the following equation for the variance:

$$\begin{aligned}\text{var}(Y_{ij}) &= \text{var}_\theta\left(E(Y_{ij} \mid \theta)\right) + E_\theta\left(\text{var}(Y_{ijk} \mid \theta)\right) \\ &= \text{var}_\theta\left(\mu_{ij} \times (1 - \pi_{ij})\right) + E_\theta\left((1-\pi_{ijk}) \times (\mu_{ij} + \pi_{ij} \times \mu_{ij}^2)\right)\end{aligned}$$

where $\theta = (b_i, d_i)$, which are the two random effect terms. The covariance between Y_{ij} and Y_{im} is given by:

$$\operatorname{cov}(Y_{ij}, Y_{im}) = E_\theta\left[E[Y_{ij} \times Y_{im} \mid \theta]\right] - E_\theta\left[E[Y_{ij} \mid \theta]\right] \times E_{b_i}\left[E[Y_{im} \mid \theta]\right]$$
$$= E_\theta\left[\mu_{ij} \times (1-\pi_{ij}) \times \mu_{ik} \times (1-\pi_{ik})\right] - E_\theta\left[\mu_{ij} \times (1-\pi_{ij})\right] \times E_\theta\left[\mu_{ik} \times (1-\pi_{ik})\right]$$

No results or R code are presented. We will do this for the ZIP model for 2-way nested data.

A.6 Correlations in a ZIP model for 2-way nested data*

The underlying model for a 2-way nested ZIP model was presented in Section 4.5, and is repeated below.

$$Y_{ijk} \sim ZIP(\pi_{ijk}, \mu_{ijk})$$
$$E(Y_{ijk}) = \mu_{ijk} \times (1-\pi_{ijk}) \quad \text{and} \quad \operatorname{var}(Y_{ijk}) = (1-\pi_{ijk}) \times (\mu_{ijk} + \pi_{ijk} \times \mu_{ijk}^2)$$
$$\operatorname{logit}(\pi_{ijk}) = \eta_{ijk} + d_i + d_{ij}$$
$$\log(\mu_{ijk}) = \xi_{ijk} + b_i + b_{ij}$$

The variance is given by:

$$\operatorname{var}(Y_{ijk}) = \operatorname{var}_\theta\left(E(Y_{ijk} \mid \theta)\right) + E_\theta\left(\operatorname{var}(Y_{ijk} \mid \theta)\right)$$
$$= \operatorname{var}_\theta\left(\mu_{ijk} \times (1-\pi_{ijk})\right) + E_\theta\left((1-\pi_{ijk}) \times (\mu_{ijk} + \pi_{ijk} \times \mu_{ijk}^2)\right)$$

where $\theta = (b_i, b_{ij}, d_i, d_{ij})$, and the covariance between two observations of animal j of colony i is given by:

$$\operatorname{cov}(Y_{ijk}, Y_{ijm}) = E_\theta\left[E[Y_{ijk} \times Y_{ijm} \mid \theta]\right] - E_\theta\left[E[Y_{ijk} \mid \theta]\right] \times E_\theta\left[E[Y_{ijm} \mid \theta]\right]$$
$$= E_\theta\left[\mu_{ijk} \times (1-\pi_{ijk}) \times \mu_{ijm} \times (1-\pi_{ijm})\right] - E_\theta\left[\mu_{ijk} \times (1-\pi_{ijk})\right] \times E_\theta\left[\mu_{ijm} \times (1-\pi_{ijm})\right]$$

The covariance between observations i of animals j and l from colony is:

$$\operatorname{cov}(Y_{ijk}, Y_{ilm}) = E_\theta\left[E[Y_{ijk} \times Y_{ilm} \mid \theta]\right] - E_\theta\left[E[Y_{ijk} \mid \theta]\right] \times E_\theta\left[E[Y_{ilm} \mid \theta]\right]$$
$$= E_\theta\left[\mu_{ijk} \times (1-\pi_{ijk}) \times \mu_{ilm} \times (1-\pi_{ilm})\right] - E_\theta\left[\mu_{ijk} \times (1-\pi_{ijk})\right] \times E_\theta\left[\mu_{ilm} \times (1-\pi_{ilm})\right]$$

R code and results for the optimal model are given in Section A.8.

A.7 Non-technical summary of the appendix

In this appendix we have shown how, in a Poisson and binomial GLMM, random effects induce correlations. For the Poisson GLMM this process is fairly simple, as it depends on characteristics of the variance and the log-normal distribution, and the final results are analytical expressions for the correlation between multiple observations of an individual animal and between observations of different animals from the same colony. For binomial GLMMs no simple formula exists that allows us to find an analytical expression for the correlations; therefore a simulation approach described in Goldstein et al. (2002) was applied to estimate the correlations. R code and results for a simple model were shown.

We included expressions for the ZIP for 1-way and 2-way nested data. Results and R code are presented in the next subsection.

A.8 Example of correlations for the 2-way nested ZIP GLMM*

In the previous subsection, we gave the underlying mathematics for calculating correlations between multiple observations of a single animal and for observations of different animals from the same colony. Here we present the code for a 2-way nested ZIP model.

The equations derived were for a general 2-way nested ZIP GLMM with covariates in the log and logistic link functions. Such a model is specified by:

$$YOUNG_{ijk} \sim ZIP(\mu_{ijk}, \pi_{ijk})$$
$$\log(\mu_{ijk}) = \eta_{ijk} + b_i + b_{ij}$$
$$\text{logit}(\pi_{ijk}) = \xi_{ijk} + d_i + d_{ij}$$

The η_{ijk} and ξ_{ijk} are predictor functions and contain an intercept and covariates. In order to demonstrate the correlation equations, we can either extend our 2-way nested ZIP GLMM with random effects or use the existing model and set some of the terms to 0. We will opt to do the latter. We set the variances of d_i and d_{ij} to 0 and use only an intercept in the predictor function ξ_{ijk}. We present only the code to calculate the variance between two observations of a single animal and leave the correlation between observations of different animals from the same colony as an exercise for the reader.

First we need to choose values for the two predictor functions. For ξ_{ijk} this is easy, as it only contains an intercept. For the log link function we set all covariates to fixed values, but let one covariate change, as before. The following R function calculates the variance based on the 2-way ZIP model.

```
> ZIPVar <- function(xi.ijk, eta.ijk, sigma_bi, sigma_bij, sigma_di,
                     sigma_dij, N) {
     bi.star  <- rnorm(N, 0, sigma_bi)
     bij.star <- rnorm(N, 0, sigma_bij)
     di.star  <- rnorm(N, 0, sigma_di)
     dij.star <- rnorm(N, 0, sigma_dij)
      pi.ijk <- exp(eta.ijk + di.star + dij.star) /
                  (1 + exp(eta.ijk + di.star + dij.star))
      mu.ijk <- exp(xi.ijk + bi.star + bij.star)
      Var_theta <- var(mu.ijk * (1-pi.ijk))
      Exp_theta <- mean((1 - pi.ijk) * (mu.ijk + pi.ijk * mu.ijk^2))
      VarY <- Var_theta + Exp_theta
      VarY }
```

For the covariance we have:

```
> ZIPCov <- function(eta.ijk, eta.ijm, xi.ijk, xi.ijm, sigma_bi,
                     sigma_bij, sigma_di, sigma_dij, N) {
        bi.star  <- rnorm(N, 0,sigma_bi)
        bij.star <- rnorm(N, 0,sigma_bij)
        di.star  <- rnorm(N, 0,sigma_di)
        dij.star <- rnorm(N, 0,sigma_dij)
        pi.ijk <- exp(eta.ijk + di.star + dij.star) /
                   (1 + exp(eta.ijk + di.star + dij.star))
        pi.ijm <- exp(eta.ijm + di.star + dij.star) /
                   (1 + exp(eta.ijm + di.star + dij.star))
        mu.ijk <- exp(xi.ijk + bi.star + bij.star)
        mu.ijm <- exp(xi.ijm + bi.star + bij.star)
        Cov <- mean(mu.ijk * (1 - pi.ijk) * mu.ijm *
                    (1-pi.ijm), na.rm=TRUE) -
               mean(mu.ijk * (1 - pi.ijk), na.rm=TRUE) *
```

```
         mean(mu.ijm * (1 - pi.ijm), na.rm=TRUE)
    Cov }
```

This code follows the equations for the variance and the covariance between two observations of a single animal. Now extract the estimated variances of the random effects:

```
> Bin.Sigma2Colony <- 0 #Colony variance in logistic link
> Bin.Sigma2Animal <- 0 #Animal variance in logistic link
> Poi.Sigma2Colony <- outzip$mean$sigma.Colony^2
> Poi.Sigma2Animal <- outzip$mean$sigma.Animal^2
```

Define the two predictor functions `xi.ijk`, `eta.ijk`, `xi.ijm` and `eta.ijm` similar way as before.

```
> xi.ijk <- outzip$mean$aB
> xi.ijm <- outzip$mean$aB
```

We will use the same values for `eta.ijk` and `eta.ijm` as before. We are now ready to run the functions:

```
> V.ijk <- ZIPVar(xi.ijk, eta.ijk, sqrt(Poi.Sigma2Colony),
                  sqrt(Poi.Sigma2Animal), sqrt(Bin.Sigma2Colony),
                  sqrt(Bin.Sigma2Animal),5000)
> V.ijm <- ZIPVar(xi.ijm, eta.ijm, sqrt(Poi.Sigma2Colony),
                  sqrt(Poi.Sigma2Animal), sqrt(Bin.Sigma2Colony),
                  sqrt(Bin.Sigma2Animal),5000)
> Cov.km <- ZIPCov(eta.ijk, eta.ijm, xi.ijk, xi.ijm,
                   sqrt(Poi.Sigma2Colony), sqrt(Poi.Sigma2Animal),
                   sqrt(Bin.Sigma2Colony), sqrt(Bin.Sigma2Animal),
                   5000)
> Cor.km <- Cov.km / (sqrt(V.ijk) * sqrt(V.ijm))
> Cor.km

0.007236424
```

The correlation between two observations of a single animal, based on these covariate values, is low. The results are not unexpected, as the random effects play only a small role in the model (see Discussion). Obviously, the correlation may change depending on the values of the covariates. One way of obtaining insight into how the correlation coefficient depends on the covariates is by changing the covariates. The following code puts the R code above in a loop and uses different values for `AGE`.

```
> MyAge <- seq(from = min(Marmots4$AGE.C), to = max(Marmots4$AGE.C),
               length = 50)
> CorGrid.km <- matrix(nrow = 50, ncol = 50)
> for (k in 1:50){
      for (m in k:50){
        eta.ijk <- Betas[1] +
            Betas[2] *  MyAge[k] +
            Betas[3] *  mean(Marmots4$MYEARL) +
            Betas[4] *  mean(Marmots4$SAMEAD) +
            Betas[5] *  mean(Marmots4$OFFSSAME.C) +
            Betas[6] *  mean(Marmots4$JUNEPRECIP.C)
        eta.ijm <- Betas[1] +
            Betas[2] *  MyAge[m] +
            Betas[3] *  mean(Marmots4$MYEARL) +
```

```
        Betas[4] *  mean(Marmots4$SAMEAD) +
        Betas[5] *  mean(Marmots4$OFFSSAME.C) +
        Betas[6] *  mean(Marmots4$JUNEPRECIP.C)
    V.ijk <- ZIPVar(xi.ijk, eta.ijk, sqrt(Poi.Sigma2Colony),
                   sqrt(Poi.Sigma2Animal),
                   sqrt(Bin.Sigma2Colony),
                   sqrt(Bin.Sigma2Animal), 5000)

    V.ijm <- ZIPVar(xi.ijm, eta.ijm, sqrt(Poi.Sigma2Colony),
                   sqrt(Poi.Sigma2Animal),
                   sqrt(Bin.Sigma2Colony),
                   sqrt(Bin.Sigma2Animal), 5000)

    Cov.km <- ZIPCov(eta.ijk, eta.ijm, xi.ijk, xi.ijm,
                   sqrt(Poi.Sigma2Colony),
                   sqrt(Poi.Sigma2Animal),
                   sqrt(Bin.Sigma2Colony),
                   sqrt(Bin.Sigma2Animal), 5000)
    CorGrid.km[k,m] <- Cov.km / (sqrt(V.ijk) * sqrt(V.ijm))
    CorGrid.km[m,k] <- CorGrid.km[k,m] }}
```

It looks complicated, but it is only a matter of importing different AGE values, and for each combination a correlation coefficient is calculated. Instead of printing the results, we can plot the estimated correlation coefficients versus AGE using the persp function, which makes a 3-dimensional graph. The results are given in Figure 4.19.

```
> persp(MyAge, MyAge, CorGrid.km, zlim = c(0, 0.06), theta = 140,
    phi = 20, expand = 0.5, col = "lightblue", ltheta = 120,
    shade = 0.75, ticktype = "detailed", xlab = "k",
    ylab = "m", zlab = ")
```

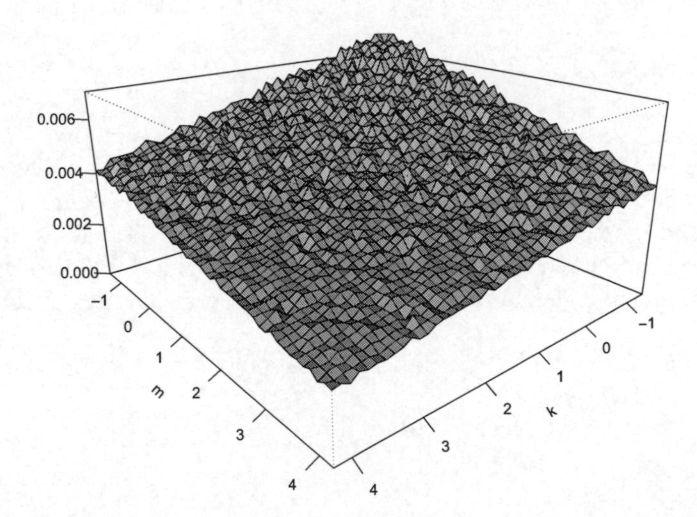

Figure 4.19. Visualisation of the correlation between two observations, *m* and *k*, of the same animal.

5 Two-stage GAMM applied to zero inflated Common Murre density data

Alain F Zuur, Neil Dawson, Mary A Bishop, Kathy Kuletz, Anatoly A Saveliev, and Elena N Ieno

5.1 Introduction

This chapter uses a dataset from surveys of seabirds carried out in Prince William Sound (PWS), on the northern Gulf of Alaska. The full data set consists of observations of seven key species made during surveys conducted outside the breeding season. We analyse the data on the Common Murre, *Uria aalge*, also known as the Common Guillemot.

The data was collected during eight cruises in November of 2007 and 2008 and January and March of 2008 and 2009 (Table 5.1). These were not repeat surveys and were often opportunistic, so areas covered and kilometres travelled were not consistent among surveys. However, the sampling effort per site was the same.

Survey methods followed those of United States Fish and Wildlife Service, recording all seabirds encountered in a 300 m wide transect. All surveys were carried out by the same observer, with the exception of March 2009, from the same boat. Surveys were conducted while travelling on a direct course at a constant speed under 12 knots and only when wind, wave height, and visibility allowed for all birds to be detected. Detection was assumed to be close to 100% and consistent across the 300 m wide survey band.

There is little survey work done outside the breeding season, and a greater understanding of factors affecting seabird distribution at these times would be of value to fisheries managers, tourism operators, spill response teams, and to further research. The chief question addressed in this analysis is which of the environmental variables can be used to describe the density of Common Murre in PWS.

5.2 Sampling

Cruises followed survey tracks, also called transects, divided into segments of 1 km. Density of Common Murre per km^2 was calculated for each kilometre of survey track. As an example, Figure 5.1 shows the position of the 1 km segments (or sites) for the cruise Herringmarch08, in March 2008. The 419 observations from this cruise were distributed over 16 transects. To visualise these 16 transects, we could have used colours in Figure 5.1 but instead present a multi-panel scatterplot (Figure 5.2). The left graph in this figure shows the 16 transects. It is interesting to zoom in to visualize the individual transects, as in the right graph in Figure 5.2. Note that transects are not always on a straight line, and sometimes there is a gap between sequential sites. The 1 km segments are the dots in Figure 5.1 and Figure 5.2, and we have a bird density value for each site.

Table 5.1. Months and years in which cruises were carried out.

Cruise label	Month	Year
Herringnov07	November	2007
Whalesnov07	November	2007
Whalesjan08	January	2008
Herringmarch08	March	2008
Herringnov08	November	2008
Whalesjan2009	January	2009
March09whales	March	2009
HerringMarch09	March	2009

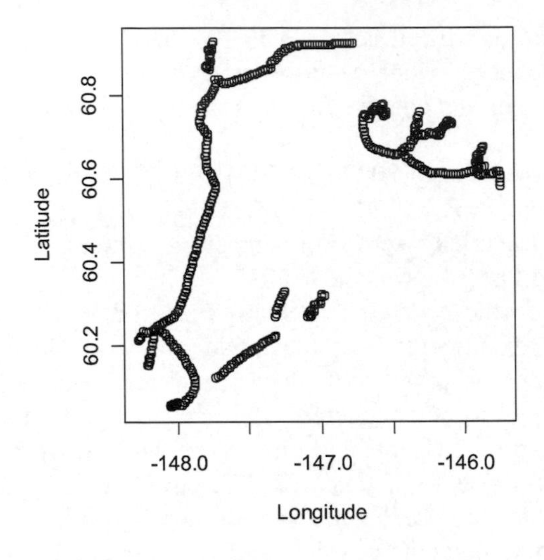

Figure 5.1. Survey positions in cruise Herringmarch08, March 2008.

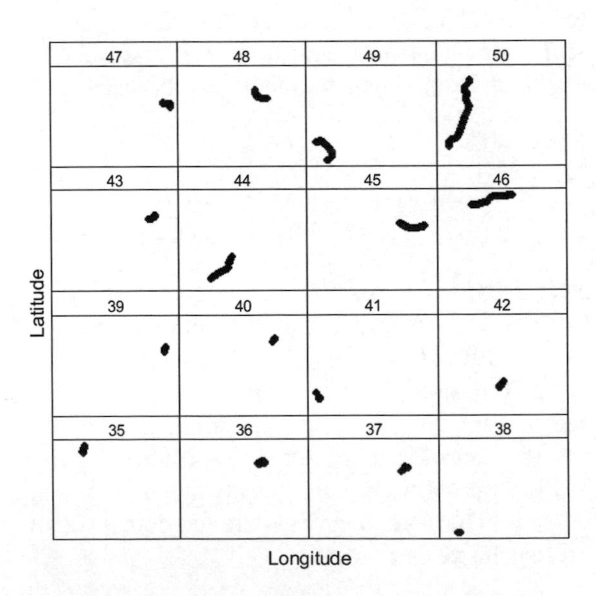

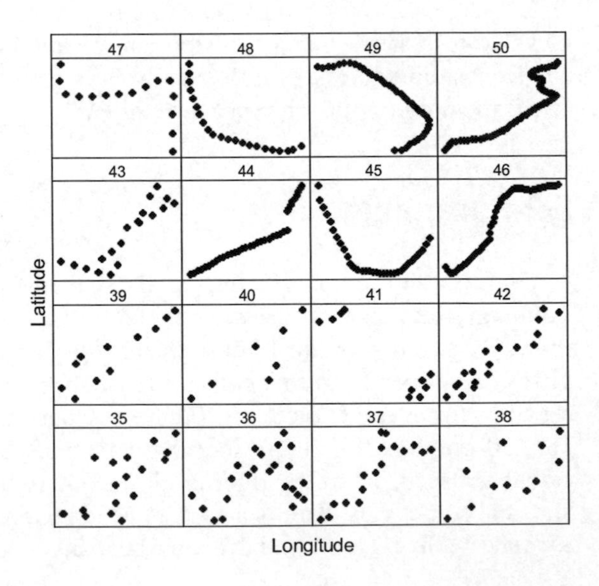

Figure 5.2. Survey positions of cruise Herringmarch08 in March 2008. In both graphs each panel represents a transect. The points in a transect are 1 km segments. In the left graph, the horizontal and vertical ranges for all panels are the same. In the right graph each panel has a unique horizontal and vertical range.

Figure 5.3 shows the geographic position of the sites for all cruises. Because sampling took place in transects following a roughly circular pattern, we do not have coverage of the entire area and must be cautious in selecting a spatial correlation structure for the analysis.

The surveyed locations were not consistent for all cruises, which is potentially a serious problem. Figure 5.4 shows the position of the sites for each winter, and in Figure 5.5 we added the month information to the graph. Although the survey covers roughly the same area in both winters, there are differences in the routes of the cruises within a winter. In both winters, the March data is more complete in terms of comparable spatial resolution. If we fail to take this discrepancy into consideration, we may identify a habitat effect that is essentially due to neglecting to survey a particular segment of the study area in a given month. The apparent habitat effect could, in reality, be a month effect, which may represent weather conditions.

The best, and perhaps only, way to solve this problem is to truncate the data and ensure that like is compared with like, but this leaves only a small amount of data to analyse. Our options are to:

- Not analyse the data.
- Truncate the data so that roughly the same areas are compared. The problem with this approach is that, any choice of which data to omit is subjective and open to criticism.
- Analyse all the data and view the analysis with a pragmatic approach. After all, the design was ad-hoc and on a limited budget and did not reflect optimal survey design. This approach necessitates caution with the interpretation of the results. A significant habitat variable may be due to differences in sampling location over time. We would need to investigate this after obtaining the results. This may lead a reviewer to criticise the results and biological interpretation of the analysis applied to all data. On the positive side, the results of an analysis of the full data may show interesting patterns and have important implications, possibly providing an argument to use in applying for a larger budget to carry out more surveys in the area, making it worth the effort of analysing all data.

In this chapter we will focus on the statistical aspects of the analysis: How do we analyse zero inflated density data that includes spatial correlation? Results, with respect to biological interpretation, must be viewed pragmatically.

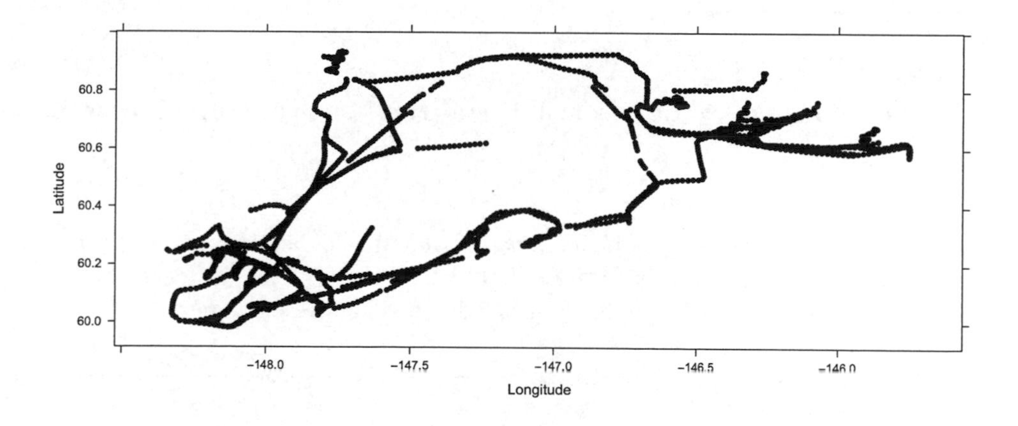

Figure 5.3. Position of all survey sites.

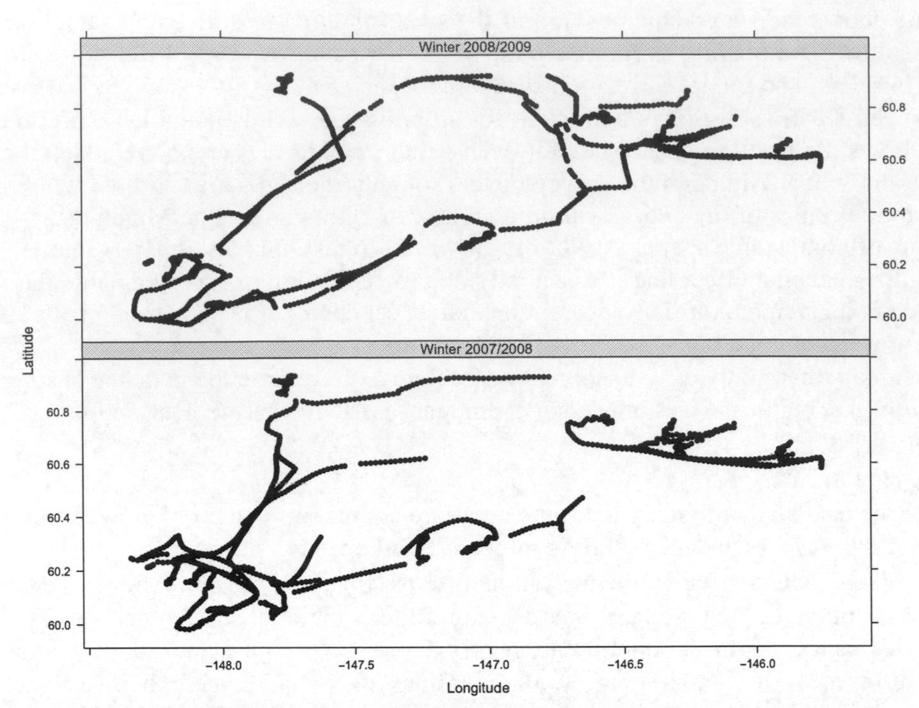

Figure 5.4. Position of survey sites in each winter.

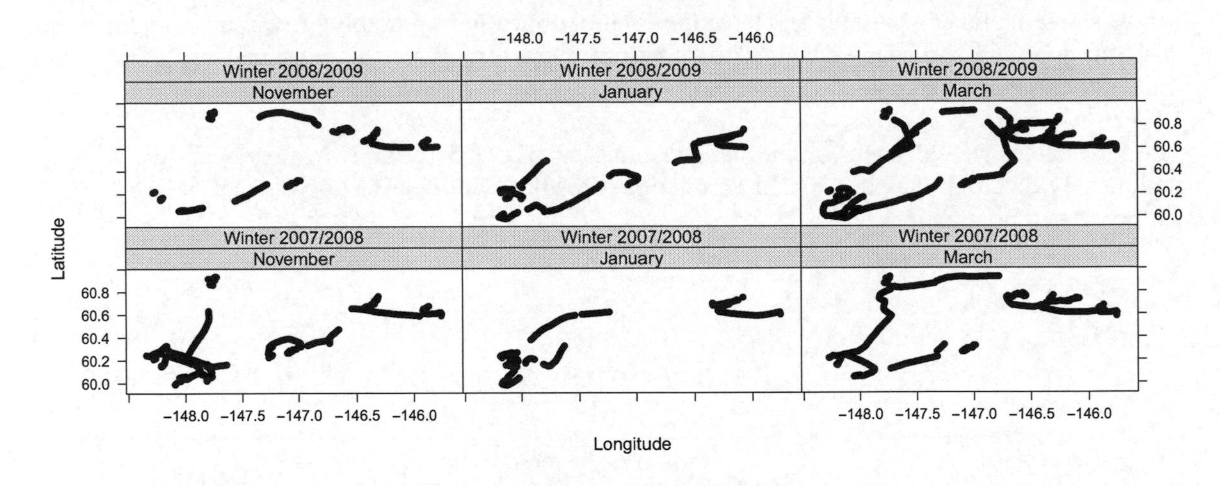

Figure 5.5. Position of the survey sites per month and year. March 2009 and November 2007 included two cruises.

The following R code is used to import the data, do some basic data preparation, check the input with the `names` and `str` functions, and make Figure 5.3. Code to make Figure 5.1, Figure 5.2, Figure 5.4 and Figure 5.5 is similar to that for Figure 5.3 and is not presented here.

```
> CM <- read.table(file = "CommonMurre.txt", header = TRUE)
> CM$fOceancatnom <- factor(CM$Oceancatnom)
> CM$fWinter      <- factor(CM$Winter)
> CM$fSubstrate   <- factor(CM$Substrate)
> CM$fDepth       <- factor(CM$Depth_nominal)
> CM$fCruise      <- factor(CM$Cruise)
> CM$fTransect    <- factor(CM$Transect)
> names(CM)          #Results not printed here
```

```
> str(CM)                    #Results not printed here
> library(lattice)
> xyplot(Midpointlat ~ Midpointlong, xlab = "Longitude", col = 1,
         ylab = "Latitude", aspect = "iso", pch = 16, data = CM)
```

The `aspect = "iso"` argument in the `xyplot` function ensures that units along the vertical and horizontal axes are identical.

5.3 Covariates

A large number of covariates are available, most taken from a GIS database. The covariates were calculated for each kilometre of survey track and include month, marine habitat, distance to shore, depth, substrate, wave exposure, slope of seabed, and sea surface temperature. In more detail:

- Depth in metres was taken from the Alaska Ocean Observing System grid of bathymetry for PWS, modelled to 500 m resolution.
- Location of coastal kelp beds was obtained from a separate project, and distance to them in metres was calculated using GIS.
- 'Slope' is the angle in degrees of seabed derived from GIS calculations based on the AOOS bathymetry grid.
- The ocean categories are: Inside Bay, Mouth of Bay, Passage, Mouth of Passage, and Open Water.
- Wave exposure of the nearest point of land and location of eelgrass and kelp beds were obtained from a shoreline mapping project. For wave exposure, areas were categorized as exposed/semi-exposed, semi-protected, and protected.
- Substrate defines the substrate type around the nearest shoreline. It comprises 3 categories: rock, rock and sediment, and sediment.
- Sea surface temperature in degrees Celsius was obtained from the NASA Giovanni website, which gives monthly averages obtained by satellite for points in a c. 10 km grid. This provides 233 points in and around PWS.

The covariates that are used in this chapter are given in Table 5.2.

Table 5.2. Covariates used in the analyses. The column "Category" indicates whether a covariate is fitted as a categorical variable or as a continuous variable.

Covariate	Category	Short description
Cruise	Factor	Identifier for cruise.
Transect	Factor	Identifier for transect within a cruise.
Midpointlat	Continuous	Latitude
Midpointlong	Continuous	Longitude
Oceancatnom	Factor	
Distancetoshore	Continuous	Metres
Exposure	Factor	
Depth	Continuous	Depth in metres
Slope	Continuous	Degrees
Disteelgrass	Continuous	Metres
Distkelps	Continuous	Metres
Substrate	Factor	
Monthlyavesst	Continuous	Degrees Celsius
Monthlyavetempmidsound	Continuous	Degrees Celsius

5.4 Data exploration

Before attempting statistical analysis, data exploration should be carried out to look for outliers, collinearity, zero inflation, and the types of relationships between birds and covariates. We should also look at the geographical position of the survey sites and visualise how measurements were taken over time, as was done in Section 5.2. We will follow the protocol sketched in Zuur et al. (2010).

5.4.1 Potential outliers in the bird data

To assess the presence of potential outliers, we create a Cleveland dotplot. This is a graph in which the observed values are plotted along the horizontal axis, and the index number, following the order of the data as imported from the text file, is along the vertical axis (Figure 5.6). We are interested in whether there are points extending to the right or left side of the graph, as these are observations that are larger or smaller, respectively, than the majority of observations. Note that some observations show atypically high numbers of Common Murre. At this stage, we do not label such points as 'outliers.' We simply note their presence and will investigate their role in the statistical analysis.

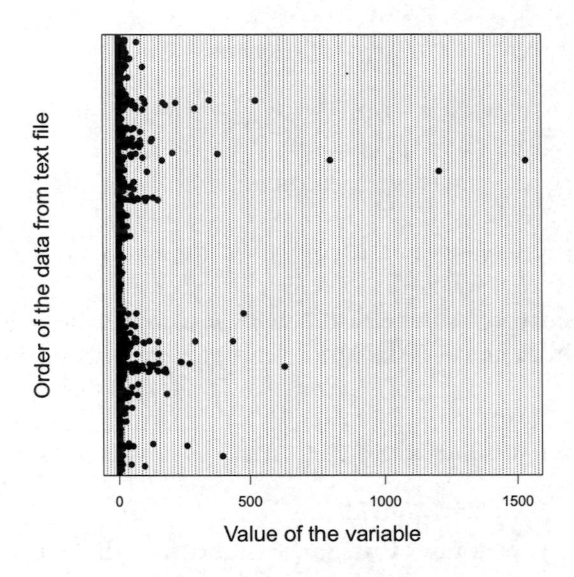

Figure 5.6. Cleveland dotplot of the Common Murre data. The horizontal axis shows the observed values and the vertical axis the order of the data. Note that several observations show considerably higher counts than the majority.

An alternative approach would have been to make a boxplot, but this graphics tool tends to label a certain number of observations as outliers even if they do not differ substantially from the majority of observations. The Cleveland dotplot has the advantage of giving more detailed information on the data. The R code to make Figure 5.6 is:

```
> dotchart(CM$Comu, col = 1, cex  = 1, pch = 16,
        xlab = "Value of the variable",
        ylab = "Order of the data from text file")
```

5.4.2 Zero inflation of the bird data

Figure 5.6 shows another potential problem. Although it is not apparent in the Cleveland dotplot, there seems to be a large number of zeros. Figure 5.7 shows a frequency plot. The *x*-axis contains the observed values and the *y*-axis shows how often each value was measured. The graph indicates that the data for this species are zero inflated. It was created with the R code:

```
> plot(table(CM$Comu), type = "h",
       ylab = "Frequency", xlab = "Observed values")
```

The issue arising with zero inflation is that the normal, Poisson, negative binomial, and binomial distributions would expect considerably fewer zeros. Applying generalised linear models (GLM) or generalised additive models (GAM) using one of these distributions on zero inflated data is likely to produce biased parameters.

In addition to frequency plots, we can calculate the percent of zero observations, length, and the number of non-zero observations for Common Murre using the following R code:

```
> ZEROS        <- 100 * sum(CM$Comu ==0 ) / length(CM$Comu)
> LENGTH       <- length(CM$Comu)
> Z            <- cbind(ZEROS, LENGTH, (100 - ZEROS) * LENGTH/100)
> colnames(Z) <- c("%0", "Length", "#non-0")
> rownames(Z) <- c("Common Murre")
> Z

                    %0 Length #non-0
Common Murre 67.10469    2283    751
```

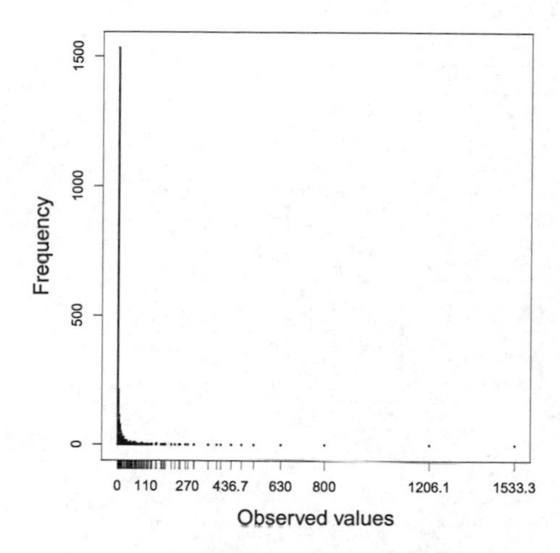

Figure 5.7. Frequency plot for the Common Murre data. The horizontal axis shows the observed values, and the vertical axis shows how often a given value was observed. Note that 0 is the most frequently observed value.

The data contain 67.1% zeros. Later we will apply a so-called hurdle model. In such a model we perform two analyses, with the second being on the presence-only data; all observations with a density of 0 are dropped from the analysis. Problems will arise if there are only a few non-zero observations. The consensus in statistical data analysis is to have at least 30 observations per covariate in regression-type models. In our models, we will have at least 5 categorical covariates and two smoothing functions. We have 751 non-zero observations, which means we can safely perform the analysis on the non-zero data.

5.4.3 Outliers in the covariates

We also produce Cleveland dotplots for the covariates (Figure 5.8) and include the categorical and the continuous covariates in the graph. The information provided by Cleveland dotplots of categorical covariates is limited. They only show the number of observations per class, which gives an indication of whether the data are balanced with respect to a categorical covariate.

The lower left panel in Figure 5.8 represents the covariate `fCruise`. The data used here comprise 8 cruises. The Cleveland dotplot for distance to shore, slope, and distance to eelgrass indicate that there are no observations with noticeably larger or smaller values. We have no observations from isolated locations (See the latitude and longitude midpoint values). Although `depth` is a continuous variable, most observations were made at shallow or deep locations; there are only a few observations in depth mid-ranges. This is an argument in favour of converting `depth` to a categorical variable with two or three classes.

Also note that we have few observations of the first class of `substrate`. The Cleveland dotplot for the two temperature covariates shows only a few unique values, which will cause problems with techniques such as generalised additive modelling, and at some sites the temperature is different from the majority of the observations. The following R code was used to make Figure 5.8:

```
> MyCovariates <- c("fCruise", "fTransect", "Midpointlat",
        "Midpointlong", "Oceancatnom", "Winter", "Distancetoshore",
        "Depth", "Slope", "Disteelgrass", "Distkelps", "Substrate",
        "Monthlyavesst", "Monthlyavetempmidsound")
> CM.X <- CM[, MyCovariates]
> dotplot(as.matrix(CM.X), groups = FALSE, col = 1, cex  = 0.5,
    strip = strip.custom(bg = 'white',
            par.strip.text = list(cex = 1.0)),
    scales = list(x = list(relation = "free"),
                  y = list(relation = "free"), draw = FALSE),
    xlab = "Value of the variable", pch = 16,
    ylab = "Order of the data from text file")
```

The variable `MyCovariates` contains all the names of the covariates for the `dotplot`, and the function `dotplot` from the `lattice` package draws the Cleveland dotplots.

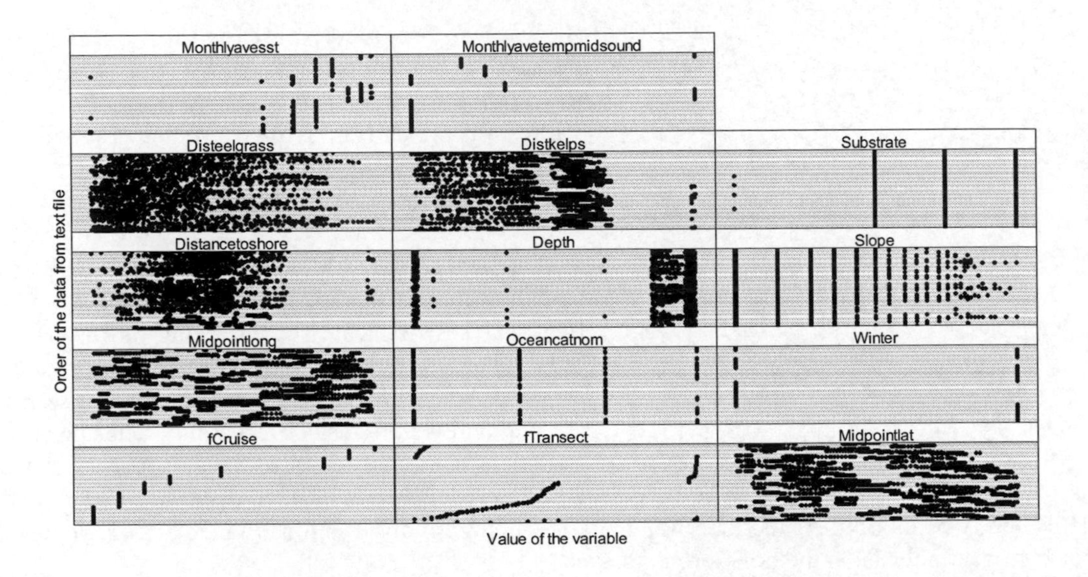

Figure 5.8. Cleveland dotplots for the covariates. The horizontal axes show the values of the covariates and the vertical axes the order of the data.

5.4.4 Collinearity of the covariates

Another important step in data exploration is assessing whether the covariates are collinear (i.e. correlated with one another). Collinearity inflates the *p*-values in linear regression and related techniques and also causes problems with model selection. If weak ecological patterns are expected, it is important to deal with even the smallest amount of collinearity before doing the analysis, as it is likely to mask such covariate effects.

Useful tools to detect collinearity are pairplots (multi-panel scatterplots) and variance inflation factors (VIF). Pairplots capture only 2-way relationships, whereas VIFs detect high-dimensional collinearity.

Figure 5.9 shows the pairplot for the continuous covariates. As can be expected, the two temperature covariates are highly collinear, and only one of them should be used. We will keep `Monthlyavesst`. Distance to kelp beds and latitude are also highly collinear.

`Distancetoshore` and `disteelgrass` have a Pearson correlation of 0.5, and for `depth` and `slope` it is -0.4. Normally, we would label correlations in this range as moderate. But for such a large data set (2283 observations) we can consider it to be strong.

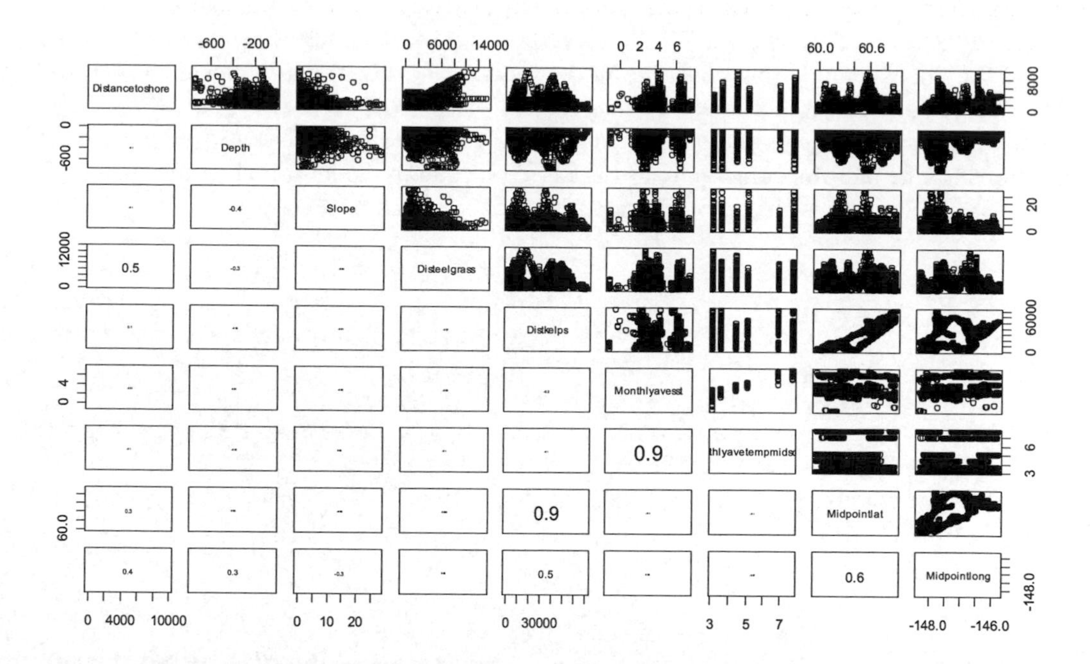

Figure 5.9. Multipanel scatterplot (pairplot) for all continuous covariates. The lower diagonal panels contain estimated Pearson correlation coefficients with the size of the font proportional to the value of the estimated correlation coefficient.

The estimated VIF values are given below. Values lower than 3 indicate that there is no substantial collinearity.

```
Variance inflation factors
                         GVIF
  Distancetoshore   1.680072
  Depth             1.717803
  Slope             1.434426
  Disteelgrass      2.603872
  Distkelps         1.335501
  Monthlyavesst    11.830847
  fExp              2.375947
  fTime            15.147221
  fOceancatnom      2.784125
  fSubstrate        1.101168
```

The VIFs of `Time` and `Monthlyavesst` are both high, probably due to correlation between these variables, which is confirmed by a simple boxplot of `Monthlyavesst` conditional on `fTime`.

Dropping the SST data (`Monthlyavesst`) and re-calculating VIF values gives a new set of VIF values (not presented here) that are all smaller than 3. However, `Oceancatnom` and `Disteel-grass` still have relatively high VIF values (between 2.5 and 3), which means that standard errors in regression models are multiplied by a factor 1.7 compared to the situation in which there is no collinearity. Dropping one of them reduces the VIFs to approximately 2. We therefore decide to drop `Disteelgrass`.

To visually assess whether there is collinearity between continuous and categorical variables, we can make conditional boxplots. Figure 5.10 is a boxplot of `Monthlyavesst` conditional on `Month` and `Year`. It shows a clear difference among the months. This means that we should not use covariates such as `Monthlyavesst` and `Winter` or `Monthlyavesst` and `Month` in the same model.

Figure 5.10 also shows that we basically have two survey periods: November 2007, January 2008 and March 2008 versus November 2008, January 2009 and March 2009. The variable `Winter` represents this. We could include this covariate in the models and test for a winter effect. However, the variable `fCruise` contains the same information.

The R code for the pairplot and the first set of VIF values follows. The `source` command loads our functions for the pairplot construction; it is based on code in the help files of the `pairs` function, and the `corvif` function is also in our library file `HighstatLib.R`.

```
> source(file = "HighstatLib.R") #Available from the first author
> pairs(CM[,c("Distancetoshore", "Depth", "Slope", "Disteelgrass",
            "Distkelps", "Monthlyavesst","Monthlyavetempmidsound",
            "Midpointlat","Midpointlong")],
      lower.panel = panel.cor)            #Figure 5.9
> Z <- CM[,c("Distancetoshore", "Depth", "Slope", "Disteelgrass",
            "Distkelps", "Monthlyavesst", "fExp", "fTime",
           "fOceancatnom", "fSubstrate")]
> corvif(Z)
```

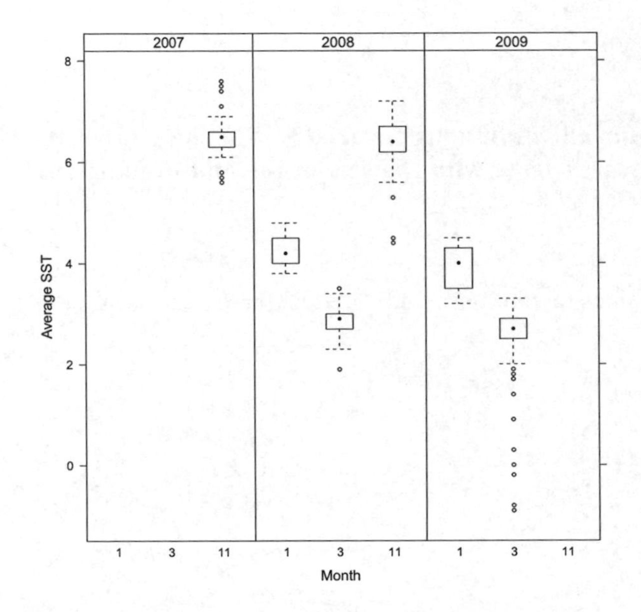

Figure 5.10. Boxplot of `Monthlyavesst` conditional on `Month` and `Year`. The width of a boxplot is proportional to the sample size in that month.

The boxplot in Figure 5.10 was created using the code:

```
> bwplot(Monthlyavesst ~ factor(Month) | factor(Year),
     strip = strip.custom(bg = 'white'),
     cex = .5, layout = c(3, 1),
     data = CM, xlab = "Month", ylab = "Average SST",
     par.settings = list(
        box.rectangle = list(col = 1),
        box.umbrella  = list(col = 1),
        plot.symbol   = list(cex = .5, col = 1)))
```

During the calculations of the VIF values, we note problems caused by the highly unbalanced `Substrate`, as can be seen from the `table` command:

```
> table(CM$Substrate)

Manmade    Rock    Rockandsediment     Sediment
     4     452              1252          575
```

There are only four observations in the Manmade category. It may be better to remove these four observations to avoid numerical problems in the models that will be applied later. This is done with the following code:

```
> CM2 <- CM[CM$Substrate != "Manmade",]
> CM2$fSubstrate <- factor(CM2$Substrate)
```

5.4.5 Visualisation of relationships between birds and covariates

In the next step of data exploration, we visualise relationships between the bird data and the covariates. The bird data are zero inflated, which makes it difficult to visualise patterns using all the data. In later chapters we will apply a hurdle model, which means that we first analyse the data as presence/absence data, and, in a second step, analyse the presence-only data. It makes sense to structure this part of the data exploration in the same way. We first convert the data to presence/absence data and create multi-panel scatterplots (Figure 5.11). To aid visual interpretation, we fit a binomial GAM. However, this model does not take into account spatial dependence; hence the smoother should be interpreted with care. Note that these graphs are only presented to obtain an overview of the data; they do not represent the outcome of a final analysis.

Some of the covariates seem to have an effect on the Common Murre data. Monthlyavesst apparently reflects seasonal patterns, and the probability of encountering Common Murre decreases with slope and increases with depth up to a certain level, after which it decreases.

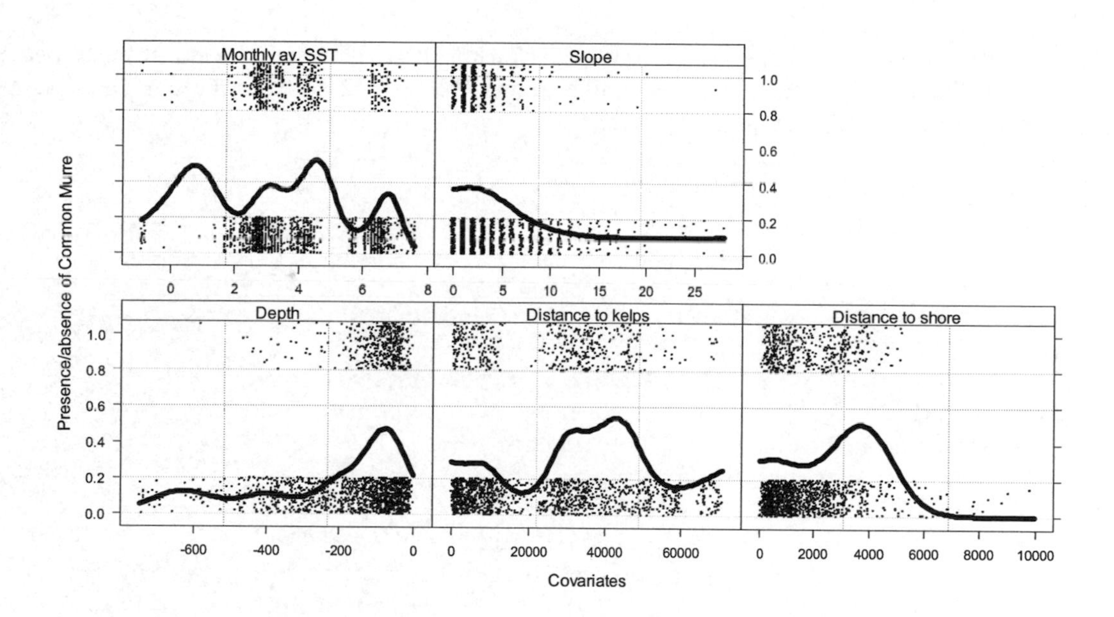

Figure 5.11. Presence/absence of Common Murre versus each continuous covariate. A certain amount of jittering (adding random noise to the data) was applied in order to distinguish a single value from multiple observations having the same value (both will be shown as a single dot in the graph without jittering). A binomial GAM was added to aid visual interpretation.

The R code to make Figure 5.11 may be intimidating. First we copy and paste the Common Murre data five times. These are converted to presence/absence data (Birds01). Then the five continuous covariates are concatenated. AllID contains the five covariate names, each repeated a large number of times. The xyplot applies a binomial GAM with cross-validation to estimate the optimal amount of smoothing. The predict function is used to sketch the fitted line.

```
> Birds   <- rep(CM2$Comu, 5)
> Birds01 <- Birds
> Birds01[Birds > 0 ] <- 1
> XNames <- rep(c("Distance to kelps", "Distance to shore",
                  "Depth", "Slope", "Monthly av. SST"))
> AllID <- rep(XNames, each = length(CM2$Comu))
> AllX <- c(CM2$Distkelps, CM2$Distancetoshore,
            CM2$Depth, CM2$Slope, CM2$Monthlyavesst)
> library(lattice)
> library(mgcv)
> xyplot(Birds01 ~ AllX | AllID, col = 1,
    strip = function(bg='white', ...)
            strip.default(bg='white', ...),
    scales = list(alternating = T, x = list(relation = "free"),
                  y = list(relation = "same")),
    xlab = "Covariates", ylab = "Presence/absence of Common Murre",
    panel = function(x, y){
      panel.grid(h = -1, v = 2)
      panel.points(jitter(x), abs(jitter(y)), col = 1, cex = 0.2)
      tmp <- gam(y ~ s(x), family = binomial)
      DF1 <- data.frame(x = seq(min(x, na.rm = TRUE),
                                max(x, na.rm = TRUE), length = 100))
      P1 <- predict(tmp, newdata = DF1, type = "response")
      panel.lines(DF1$x, P1, col = 1, lwd = 5) })
```

We also produce these graphs for the non-zero density data, and there are no obvious patterns (Figure 5.12). The R code for this graph is similar to that above. CM3 contains the same data as CM2, except that all zero densities are omitted.

```
> CM3    <- CM2[CM2$Comu > 0,]
> Birds  <- rep(CM3$Comu, 5 )
> XNames <- rep(c("Distance to kelps", "Distance to shore",
                  "Depth", "Slope", "Monthly av. SST"))
> AllID <- rep(XNames, each = length(CM3$Comu))
> AllX  <- c(CM3$Distkelps, CM3$Distancetoshor,
             CM3$Depth, CM3$Slope, CM3$Monthlyavesst)
> xyplot(Birds ~ AllX | AllID, col = 1,
         strip = function(bg = 'white', ...)
                 strip.default(bg='white', ...),
         scales = list(alternating = T,
                       x = list(relation = "free"),
                       y = list(relation = "same", log = TRUE)),
         xlab = "Covariates", ylab = "Positive counts of birds",
         panel = function(x, y){
            panel.grid(h =- 1, v = 2)
            panel.points(x, y, col = 1, cex = 0.5)
            tmp <- gam(y ~ s(x), family = Gamma(link = "log"))
            DF1 <- data.frame(x = seq(min(x, na.rm = TRUE),
                                      max(x, na.rm = TRUE),
                                      length = 100))
            P1 <- predict(tmp, newdata = DF1, type = "response")
            panel.lines(DF1$x, P1, col = 1, lwd = 5) })
```

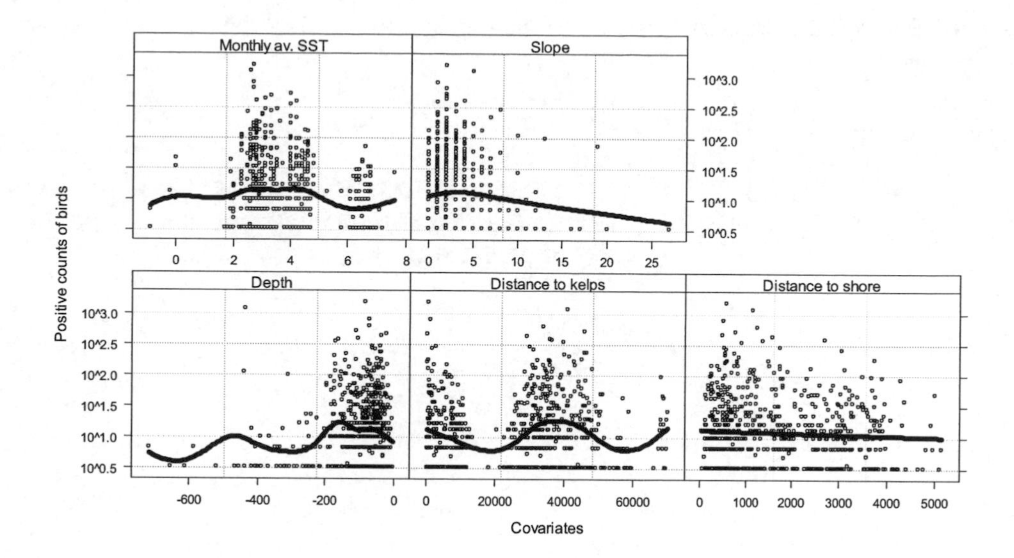

Figure 5.12. Non-zero densities for Common Murre versus each covariate. To aid visual interpretation, a gamma GAM with log-link function is added, and we use a logarithmic scale for the vertical axes.

5.4.6 Cruise and transect effects

Counts were made on 8 cruises, with each cruise divided into transects. It is likely that observations from the same cruise are more similar than observations from different cruises. Also, it is likely that observations from the same transect are more similar than observations from different transects in the same cruise. Figure 5.13 is a boxplot of the log-transformed positive (zeros were dropped) Common Murre data, conditional on `fCruise`. We use the log-transformation to enhance the clarity of the graph. We will not apply a transformation in the actual analysis. Note that there are only marginal differences among the cruises.

Figure 5.14 contains boxplots of the same data, conditional on `Transect` within `Cruise`. Note the vertical displacement of the boxplots (some do not overlap), indicating that there is a `Transect` effect. This means that a possible analytic approach would be to apply generalised linear mixed effects models (GLMM) or generalised additive mixed effects models (GAMM) using `Transect` nested in `Cruise` as random intercepts. Such an approach would allow for correlation among observations from the same transect and a different correlation among observations from the same cruise, but it assumes that there is no correlation between observations from different cruises. The R code to create these two figures is:

```
> par(mar = c(8, 4, 3, 3))
> boxplot(log(Comu) ~ fCruise, varwidth = TRUE,
          data = CM3, cex = 0.5, las = 3)
> bwplot(log(Comu) ~ fTransect | fCruise,
   strip = strip.custom(bg = 'white'),
   cex = .5, data = CM3, xlab = "Transect", ylab = "Bird density",
   par.settings = list(
      box.rectangle = list(col = 1),
      box.umbrella  = list(col = 1),
      plot.symbol   = list(cex = .5, col = 1)),
      scales = list(alternating = T,
                x = list(relation = "free"),
                y = list(relation = "same")))
```

The first boxplot uses the ordinary `boxplot` function, and Figure 5.14 uses the `bwplot` from the `lattice` package.

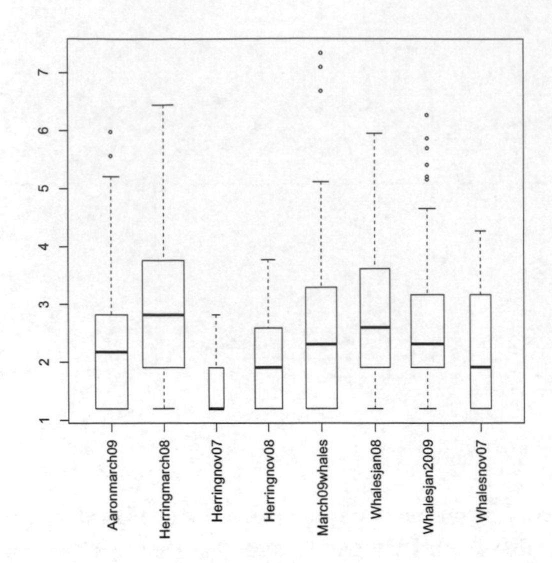

Figure 5.13. Boxplot of the log-transformed positive Common Murre data conditional on `Cruise`.

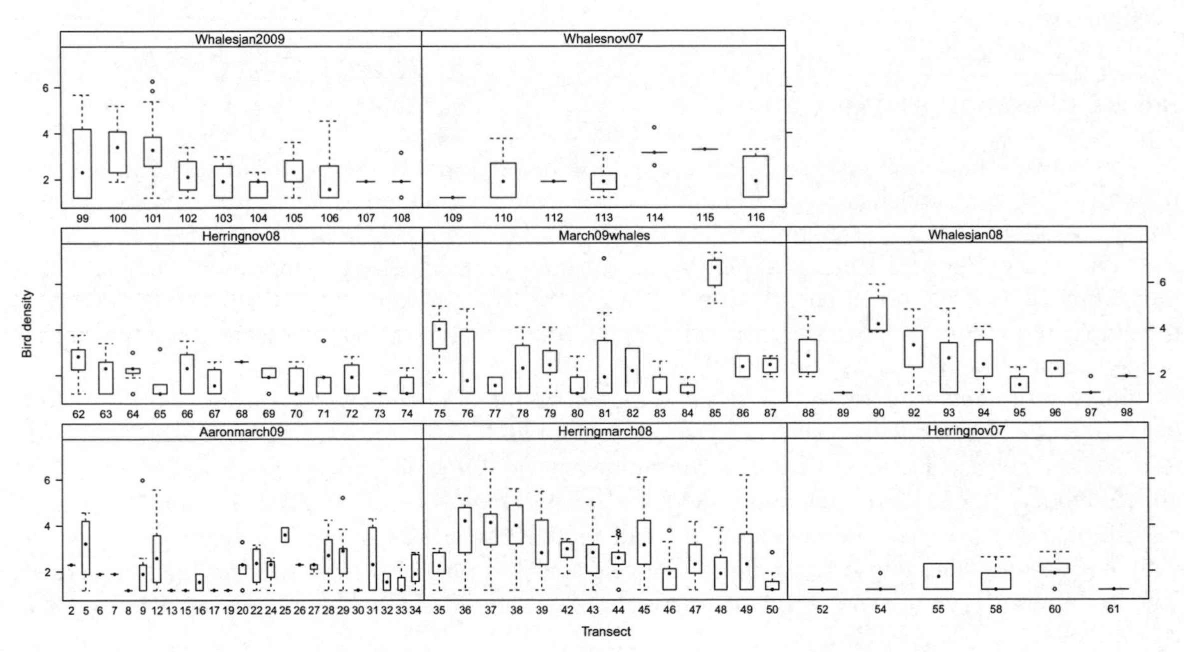

Figure 5.14. Boxplots of the log-transformed positive Common Murre data conditional on `Transect` within `Cruise`.

5.4.7 Summary of the data exploration

The data exploration showed that the bird data are clearly zero inflated. There are no obvious outliers in the covariates. The two temperature covariables are highly collinear and one should be dropped (It does not matter which one). Also, `Monthlyavesst` is collinear with the covariate `Time`, which represents the month and year. Because we are interested in changes over time, we will drop `Monthlyavesst`. There is some minor collinearity left, and we drop `Disteelgrass` and `Slope`. Table 5.3 shows a list of covariates to be used in the analysis and those dropped. The spatial coordinates are not used together with the covariates, as most covariates are correlated in space.

Scatterplots for the presence or absence of birds and the presence-only data shows that weak patterns can be expected for the former, and very weak patterns, if any, for the presence-only data. This is additional grounds for dealing with even the smallest amount of collinearity. There is a clear transect effect. The covariate `Substrate` is unbalanced and we remove the `Manmade` class, which has only four observations.

Figure 5.5 shows that the geographic positions of the survey sites differed among cruises. For example, in January 2008, the south-eastern part of the survey area was not monitored, whereas in March 2009 it was included. If densities are high in the south-eastern part of the survey area, the density may show an increase for March 2009, but in reality this is due to a difference in the location surveyed.

Focusing on the underlying biological question of how bird distribution within PWS changes relative to environmental variables in the non-breeding seasons, we will use a model of the form:

$$Birds = f(\text{All selected covariates})$$

This model is affected by the changes over time of surveyed locations. Thus a significant covariate effect may be due to differences in position of the surveys.

If it were not for the nested structure of the data, the model could be a simple linear regression model. Some of the covariates will represent spatial information such as distance to the shore and distance to kelp beds (which changes over time), and we will also use the temporal covariate `Time`, the month and year the count was made. Again, care is needed in the interpretation of the temporal covariate. It is not the trend in the data, but merely the temporal effect on the bird data while taking into account all other selected covariates.

Table 5.3. Covariates used in the analyses. The categorical variable `Time` defines the month and year.

Covariate	Category	Status
`Cruise`	Factor	Discussed later
`Transect`	Factor	Used as random intercept
`Midpointlat`	Continuous	**Dropped**
`Midpointlong`	Continuous	**Dropped**
`Oceancatnom`	Factor	Used
`Winter`	Factor	**Dropped**
`Distancetoshore`	Continuous	Used
`Exposure`	Factor	Used
`Depth`	Factor	Used
`Slope`	Continuous	**Dropped**
`Disteelgrass`	Continuous	**Dropped**
`Distkelps`	Continuous	Used
`Substrate`	Factor	Used
`Monthlyavesst`	Continuous	**Dropped**
`Monthlyavetempmidsound`	Continuous	**Dropped**
`Time`	Factor	Used

5.5 GAMM for zero inflated and correlated densities

In this section we discuss the models used to answer the underlying biological question that was formulated in Section 5.1.

5.5.1 Brainstorming

The bird data are clearly zero inflated, and that means that regression type techniques such as generalised linear models (GLM) and generalised additive models (GAM) are likely to produce biased estimated parameters and smoothers (See Chapter 2 for a discussion and simulation example).

In Chapters 2 and 3 we applied mixture models such as zero inflated Poisson models and zero inflated negative binomial models. For the mixture models we introduced the concept of true zeros and false zeros. A false zero is a zero due to, for example, an insufficient sampling period or an observer not spotting a bird. In both cases a 0 is recorded but birds were present. The ecologists involved in this case study assured us that all birds within the 300-meter observation range were detected, eliminating the possibility of false zeros. As explained in Chapter 10 we can still apply mixture models, but instead we will ask ourselves the following questions:

1. Which covariates drive the presence or absence of Common Murre?
2. When the birds are present, which covariates drive their numbers?

This means that we can apply a hurdle (or two-stage) model. In such models, we first analyse the data as presence/absence, and in the second step the presence-only data are analysed. For the presence/absence data we select a binomial generalised additive mixed model (GAMM) with `Transect` as random intercept. This ensures that we impose a correlation structure on observations from the same transect.

Our plan for the presence-only data is to use a GAMM with a gamma distribution, as we are working with density data, and `Transect` is used as random intercept. Again, the use of the random intercept ensures that observations from the same transect are correlated. Hence, we will apply the following two models:

$$
\begin{aligned}
&Y_{ijk}^{01} \sim B(\pi_{ijk}) \\
&\text{logit}(\pi_{ijk}) = \alpha + f(Slope) + Time + \cdots + Transect_{ij} \\
&Transect_{ij} \sim N(0, \sigma^2_{1,Transect})
\end{aligned}
$$

and

$$
\begin{aligned}
&Y_{ijk}^{1+} \sim Gamma(\mu_{ijk}, \nu) \\
&\log(\mu_{ijk}) = \alpha + f(Slope) + Time + \cdots + Transect_{ij} \\
&Transect_{ij} \sim N(0, \sigma^2_{2,Transect})
\end{aligned}
$$

The index i refers to cruise, j to transect within cruise, and k to observations within transect j and cruise i. A variable that is gamma distributed with mean μ and parameter ν has variance μ^2/ν, allowing for a quadratic variance structure. A gamma distribution is useful for density data (continuous) and is zero truncated, i.e. the response variable cannot take the value of 0, which makes it useful for the second step of the hurdle model. The continuous covariates are modelled using splines with the optimal amount of smoothing determined by cross-validation (Wood 2006).

If we assume that the random intercepts for the binary part and the presence-only part are independent, the hurdle model can be analysed in two steps (Boucher and Guillén, 2009). This means that we can use existing software to model the two-stage GAMM with random effects. On the other hand, if we want to apply mixture models, we need to program a substantial amount of code. First a GAM with random effects must be created in, for example, WinBUGS, and this must be extended such that a binomial model is used for the false zeros versus the densities and the gamma distribution is used for the non-zero data. The code for the GAM part is discussed in Chapter 9, and the inclusion of random effects is explained in Chapter 2. Replacing the Poisson distribution by a gamma distribution in the count part of a mixture model is simple. The advantage of programming such a zero inflated gamma (ZIG) smoothing model is that we can easily include a more advanced correlation structure (see also Chapters 6 and 7).

We may wish to include a random effect `fCruise` in the models. The drawback of this approach is that `fCruise` also represents time. Recall from Table 5.1 that, with the exception of two cruises, each cruise represents a different month and year combination. Using `fCruise` as a random effect and `Time` as a categorical variable means that these two terms will 'fight' for the same information. Since one of our primary questions is whether there are changes over time, we will use `Time` as a fixed covariate and refrain from using `fCruise` as a random intercept.

5.5.2 Four model selection approaches

We would like to know which covariates influence Common Murre density, and this leads to model selection. There is debate in the scientific community over how this should be done, and some approaches are more prevalent in certain scientific fields. Four are discussed (see also Chapter 4):

1. The classical approach is to use the Akaike Information Criterion (AIC; Akaike 1973) along with a model selection strategy, such as backward selection, to find the optimal model.
2. It is also possible to use hypothesis testing to drop each non-significant term in turn until all covariates are significant.

Both these options are heavily criticised (Whittingham et al., 2006).

3. Yet another approach is to ensure that we have a set of non-collinear covariates and present the full model, with no covariates removed. This means that we present a model in which some of the covariates may be non-significant.
4. A fourth approach is the Information-Theoretic (IT) approach discussed in Burnham and Anderson (2002). They advocate no model selection and instead define 15 to 20 models *a priori,* which are then compared to one another using Akaike weights. Instead of presenting only one model and labelling it as the best model, the IT approach leads to a series of competing models that are all presented.

The Information-Theoretic approach will present the fewest obstacles to acceptance for publication. The authors of this book prefer to present the full model without prior model selection. The crucial condition is that there be no collinearity. This is most practical for models with fewer than 10 covariates. Some of the examples discussed in Burnham and Anderson (2002) have more than 100 covariates. Here we will present results (and the process) of the full model approach. Potential models for the Information-Theoretic approach are presented in the Discussion.

5.6 Results of the full model approach

5.6.1 Results of the presence/absence GAMM

A binomial GAMM with following logistic link function is applied to the presence/absence data.

$$\begin{aligned}\text{logit}(\pi_{ijk}) &= \alpha + f(\Delta Shore_{ijk}) + f(\Delta Kelps_{ijk}) + fExposure_{ijk} + fDepth_{ijk} + fSubstrate_{ijk} + \\ &\quad fTime_{ijk} + Transect_{ij} \\ Transect_{ij} &\sim N(0, \sigma^2_{Transect})\end{aligned}$$

To fit this model in R, we need the following code:

```
> CM2$fExposure <- factor(CM2$Exp_nominal)
> CM2$fTime <- factor(paste(CM2$Month, CM2$Year, sep = "."),
           levels = c("11.2007", "1.2008", "3.2008",
                      "11.2008", "1.2009", "3.2009"),
           labels = c("Nov07", "Jan08", "Mar08",
                      "Nov08", "Jan09", "Mar09"))
> CM2$Comu01 <- CM2$Comu
> CM2$Comu01[CM2$Comu>0] <- 1
> library(mgcv)
> M1.Comu <- gamm(Comu01 ~ s(Distancetoshore) + s(Distkelps) +
                 fDepth + fOceancatnom + fSubstrate + fExposure +
                 fTime, random = list(fTransect =~ 1),
                 family = binomial, data = CM2)
```

The first line defines `Exposure` as a factor, so that we can eliminate the more cumbersome `factor(Exp_nominal)` in the remainder of the code. We also need to make a new covariate `fTime`, representing the month and year. The variable `Comu01` contains the presence/absence data, and the `gamm` function from the `mgcv` package is used for the estimation of the model. The numerical output for this model is:

```
> anova(M1.Comu$gam)

Parametric Terms:
             df      F  p-value
fDepth        3 17.855 1.88e-11
fOceancatnom  3  2.433 0.063229
fSubstrate    2  1.483 0.227214
fExposure     2  1.111 0.329265
fTime         5  5.026 0.000139

Approximate significance of smooth terms:
                     edf Ref.df     F  p-value
s(Distancetoshore) 3.429  3.429 2.915   0.0268
s(Distkelps)       5.229  5.229 6.926 1.25e-06
```

To investigate potential collinearity in the model, we drop each covariate in turn and note that the effects of the remaining covariates stay roughly the same, except when `fDepth` is removed. In this case the smoother for `Distancetoshore` became highly significant. This means that there is some collinearity between these two covariates, although a scatterplot is non-conclusive.

Only the covariates `fDepth` and `Distkelps` are highly significant. It should also be noted that *p*-values obtained by a GAM are approximate; thus a covariate with a *p*-value of 0.02 should not be labelled as a strong effect. Wood (2006) suggested that there is some evidence that *p*-values ranging from 0.01 to 0.05 in GAM should be doubled. Therefore, one should not put too much emphasis on the `Ocean` category and `Distancetoshore`.

Figure 5.15 shows the estimated smoothers for `Distancetoshore` and `Distkelps`, of which only the latter is of interest, as it is highly significant. The shape of these smoothers indicate that the probability of observing Common Murre decreases when the distance to the shore is greater than 4000 m, although the effect is weak, and it is reduced at distance to kelp of around 2000 m. The R code for the graph with the two smoothers follows. The `scale = FALSE` argument allows each panel to have its own vertical scale.

```
> par(mfrow = c(1, 2))
> plot(M1.Comu$gam, scale = FALSE, select = c(1))
> plot(M1.Comu$gam, scale = FALSE, select = c(2))
```

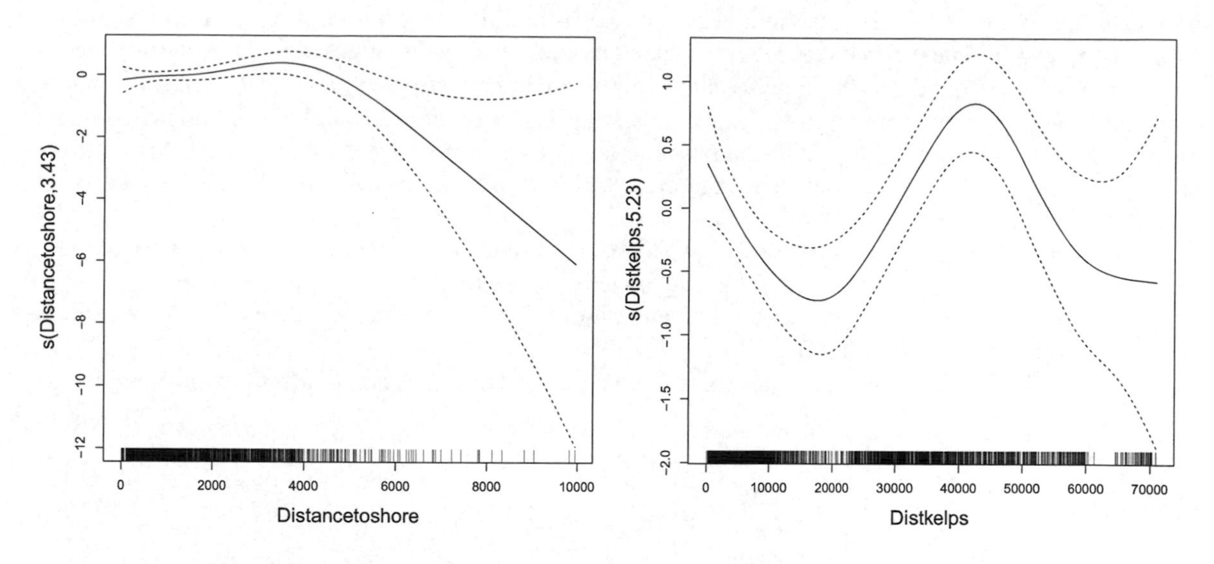

Figure 5.15. Estimated smoothers for distance to shore and distance to kelp.

5.6.2 More detailed output for the binomial GAMM

More detailed information on the categorical variables is given below and indicates that the probability of encountering Common Murre is significantly higher at depth levels 2, 3, and 4 as compared to level 1. A similar conclusion is obtained when `Depth` is used as a smoother instead of as a categorical variable.

```
> summary(M1.Comu$gam)

Parametric coefficients:
                          Estimate Std. Error t value Pr(>|t|)
(Intercept)               -2.79076    0.37625  -7.417 1.68e-13
fDepth2                    0.95033    0.19613   4.845 1.35e-06
fDepth3                    1.20592    0.18332   6.578 5.90e-11
fDepth4                    0.61146    0.20045   3.050 0.002312
fOceancatnom2              0.21413    0.22677   0.944 0.345145
fOceancatnom3              0.35594    0.24114   1.476 0.140075
fOceancatnom4             -0.35892    0.27069  -1.326 0.184983
fSubstrateRockandsediment  0.16730    0.14759   1.134 0.257112
fSubstrateSediment         0.30203    0.17541   1.722 0.085222
fExposure2                 0.15550    0.16635   0.935 0.350024
fExposure3                -0.08181    0.26203  -0.312 0.754901
fTimeJan08                 1.16525    0.49828   2.339 0.019445
fTimeMar08                 1.64755    0.43744   3.766 0.000170
fTimeNov08                 1.09142    0.46530   2.346 0.019082
fTimeJan09                 1.65175    0.49431   3.342 0.000847
fTimeMar09                 0.39043    0.36532   1.069 0.285310
```

The estimated regression parameters for `fDepth2`, `fDepth3`, and `fDepth4` indicate these levels relative to the baseline level, which is depth category 1. In case of numerical problems it may be an option to change the baseline level, especially if it has only a few non-zero values. This is not only true for GAMMS, but also for linear regression models.

It is possible to visualise the effects of categorical variables in the model, and this may show the effect of `Depth` in the presence of the other covariates. We will create such a graph only for the significant covariates Depth and Time (Figure 5.16). The left panel illustrates the effect fDepth. As noted above, category 1 is the baseline; hence the horizontal line without confidence intervals represents fDepth = 1. The other levels are represented by their estimated values, and 95% confidence intervals are drawn around the estimated value using the standard errors obtained from the summary command. Note that no correction for multiple testing has been carried out, so we should compare only the levels and standard errors to the baseline level. However, if we do a correction for multiple comparison of the four levels, confidence intervals will increase. In this case, levels 2, 3, and 4 are not significantly different from one another.

With respect to the fTime effect in Figure 5.16, note that the probability of observing Common Murre was lowest in November 2007. This may have been due to surveying of different areas.

The summary(M1B.Comu$lme) command shows that the estimated values for the variances are $\sigma^2_{Transect} = 1.072^2$ and $\sigma^2_{\varepsilon} = 0.931^2$.

The R code to make Figure 5.16 uses the all.terms = TRUE argument in the plot function:

```
> par(mfrow = c(1, 2))
> plot(M1.Comu$gam, scale = FALSE, all.terms = TRUE, select = c(3))
> plot(M1.Comu$gam, scale = FALSE, all.terms = TRUE, select = c(7))
```

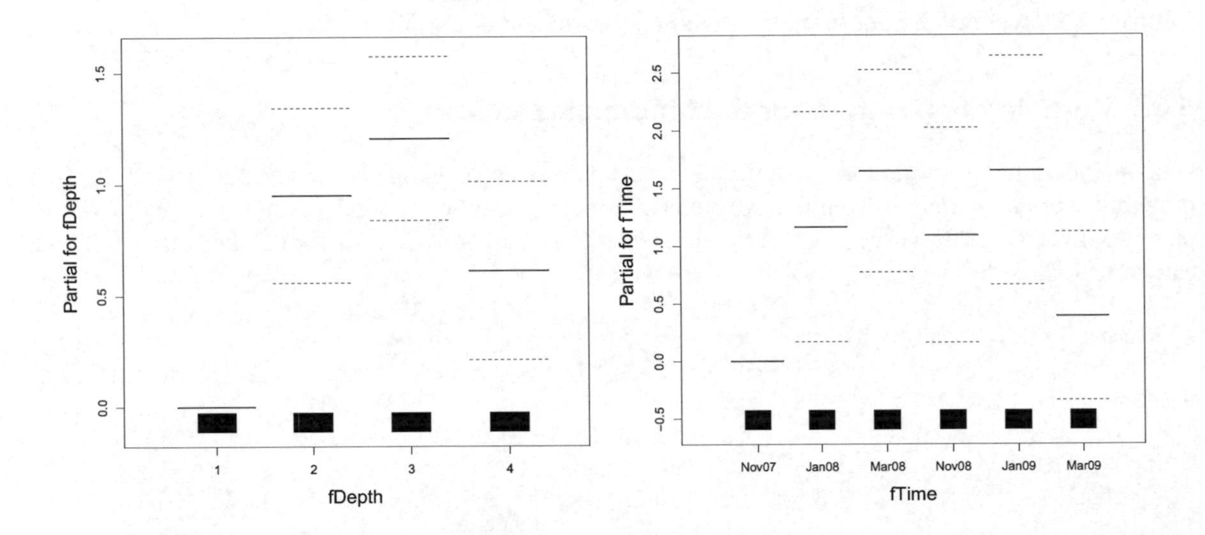

Figure 5.16. Effect of Depth and Time. Depth level 1 and November 2007 are used as baseline levels; thus there are no confidence bands.

5.6.3 Post-hoc testing

So far, we have concentrated on identifying covariates that are significantly related to the presence/absence of Common Murre. We applied a model and assessed the significance of the covariates. With a smoother in a GAM or a continuous covariate in a linear regression model, one can easily infer its effect from the shape of the smoother or the sign of the regression coefficient. With categorical variables having more than two levels, it is more difficult to determine the term's effects. When the analysis is complete, we may be faced with the question of which levels of a categorical variable differ from each other. This should be one of the underlying biological questions. There is a difference in asking 'Is there a depth effect?' and 'What is the depth effect?' For the first question, we can simply look at the numerical output of the GAMM and quote a single *p*-value for depth. For the second question, we need to do some post-hoc testing. Consensus is that unplanned post-hoc testing is bad science or data phishing. Useful references are Sokal and Rohlf (1995) and Quinn and Keough (2002), and an overview is given in Ruxton and Beauchamp (2008). One of many possible approaches is the Bonferonni correction. To carry out this process for Depth, we first run the GAMM four times, each time with a different baseline. The altered baselines are as follows:

```
> CM2$fDepth.1 <- factor(CM2$fDepth, levels = c("1", "2", "3", "4"))
> CM2$fDepth.2 <- factor(CM2$fDepth, levels = c("2", "1", "3", "4"))
> CM2$fDepth.3 <- factor(CM2$fDepth, levels = c("3", "1", "2", "4"))
> CM2$fDepth.4 <- factor(CM2$fDepth, levels = c("4", "1", "2", "3"))
```

Instead of fDepth in the GAMM, we use fDepth.1 and the summary command extracts the relevant output. In the previous subsection we saw the output when fDepth.1 is used. For fDepth.2 it is:

```
fDepth.21              -0.95033    0.19613  -4.845 1.35e-06
fDepth.23               0.25559    0.16676   1.533 0.125502
fDepth.24              -0.33887    0.18350  -1.847 0.064924
```

For fDepth.3 it is:

```
fDepth.31              -1.20592    0.18332  -6.578 5.90e-11
fDepth.32              -0.25559    0.16676  -1.533 0.125502
fDepth.34              -0.59446    0.13835  -4.297 1.81e-05
```

Finally, for fDepth.4 it is:

```
fDepth.41              -0.61146    0.20045  -3.050 0.002312
fDepth.42               0.33887    0.18350   1.847 0.064924
fDepth.43               0.59446    0.13835   4.297 1.81e-05
```

We have included all *p*-values in Table 5.4. Unfortunately, each time we conduct a test at the 5% significance level, there is a 5% chance that we make a type 1 error. By doing this a large number of times, in this case $4 \times (4 - 1) / 2 = 6$ times, the chance of error rapidly multiplies. A simple but controversial solution is to apply the Bonferonni correction. Let $N = k \times (k - 1) / 2$ be the number of comparisons, where k is the number of levels. In our case, $k = 4$ and $N = 6$. Using the Bonferonni correction we would not test at the 5% level but at the $5 / N$ % level, which is 0.0083 for our data. This results in Depth levels 2, 3, and 4 being significantly different from level 1, and level 4 significantly different from 1 and 3.

Alternative tests discussed in the literature include Scheffe's Procedure and Tukey's Honestly Significant Difference Test. However, it should be stressed that post-hoc tests should be conducted taking into consideration the underlying biological questions; hence they should not be carried out only because Depth shows significance.

Table 5.4. *P*-values obtained by using different baselines in the GAMM.

	1	2	3	4
1		**1.35e-06**	**5.90e-11**	0.002312
2	**1.35e-06**		0.125502	0.064924
3	**5.90e-11**	0.125502		**1.81e-05**
4	0.002312	0.064924	**1.81e-05**	

5.5.4 Validation of the binomial GAMM

As a component of model validation, we inspect the residuals for spatial dependence. The inclusion of a random intercept for `Transect` in the GAMM imposes a dependence structure among observations from the same transect, but this dependence is the same for any two observations from the same transect, whatever the distance between these observations. It may well be that sites in proximity are more similar than sites further separated. This can be investigated by taking Pearson residuals to see whether they exhibit spatial patterning. Figure 5.17 shows a bubble plot of the Pearson residuals for the binomial GAMM. These residuals are plotted versus their position. Grey dots are positive residuals and black dots are negative residuals. The dot size is proportional to the absolute value of the residual. We do not want to see clustering of residuals of the same sign. Note that for some transects all observations are under- or over-fitted.

The R code for the bubble plot uses the function `bubble` from the `gstat` package. This is not part of the base installation of R, so you will need to install it. Similar code was presented and explained in Chapter 7 of Zuur et al. (2009a). We converted the latitude and longitude values to Universal Transverse Mercato (UTM) coordinates using the `rgdal` package. These steps will be demonstrated in more detail in Chapter 6.

```
#See Chapter 6 for explanation of rgdal code
#project from Lat/Lon to UTM zone 6
> library(rgdal)
> prj.UTM <-
   "+proj=utm +zone=6 +ellps=WGS84 +datum=WGS84 +units=m +no_defs"
> p.UTM  <- project(as.matrix(CM2[,c("Midpointlong",
                                     "Midpointlat")]), prj.UTM)
> CM2$X <- p.UTM[,1] / 1000
> CM2$Y <- p.UTM[,2] / 1000
#End of "See Chapter 6"

> library(gstat)
> E1     <- resid(M1.Comu$gam, type = "p")
> mydata <- data.frame(E1, CM2$Comu01, CM2$X, CM2$Y)
> coordinates(mydata) <- c("CM2.X", "CM2.Y")
> bubble(mydata, "E1", col = c("black", "grey"), main = "Residuals",
         xlab = "X-coordinates", ylab = "Y-coordinates")
```

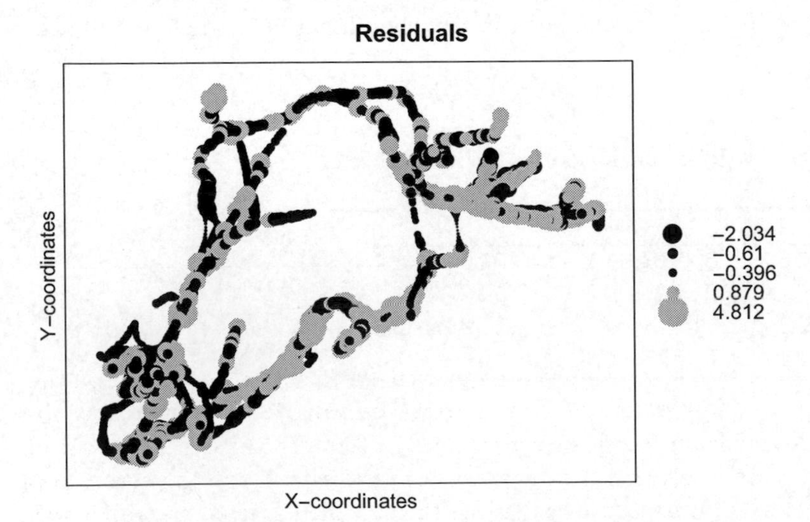

Figure 5.17. Bubble plot of the Pearson residuals from model M1.Comu. Grey dots are positive residuals and black dots are negative residuals. The dot size is proportional to the absolute value of the residual. We do not want to see clustering of residuals of the same sign.

The bubble plot is a subjective visualisation tool, and a sample variogram of the Pearson residuals gives more objective information (Figure 5.18A). Independence of residuals is indicated by a band of horizontal points, as is the case with our data. Hence we can assume independence of the residuals, at least on the presented scale. The code to make Figure 5.18A also uses a function from the gstat package:

```
> E1      <- resid(M1.Comu$lme, type = "n")
> Vario1 <- variogram(E1 ~ 1, mydata)
> plot(Vario1, col= 1 , pch = 16)
```

The function variogram uses all residuals, irrespective of Transect, to calculate the sample variogram, so residuals from sites in different transects are used. There are no definable patterns in the sample variogram, which is satisfactory.

It is also possible to use the Variogram function from the nlme package, but it will only accept combinations of sites within transects, because gamm uses a random effect, and therefore the correlation is imposed within the transects. This variogram (Figure 5.18B) seems to show a clear pattern. Figure 5.19A gives the number of sites per transect. The majority of transects consist of less than 20 sites.

At this point it is useful to note transect length, although we should have done this in the data exploration section. Figure 5.19B shows the maximum distance between two points for each transect, which represents the distance covered during surveying. The results in Figure 5.19B show that the majority of transects cover only a short distance, less than 20 km. This means that we may disregard values of the sample variogram in Figure 5.18B for which the distance is greater than 20 km. The R code to make Figure 5.18 and Figure 5.19A is:

```
> V1 <- Variogram(M1.Comu$lme, form =~ X + Y | fTransect,
                  robust = TRUE, resType = "normalized", data = CM2)
> plot(V1, smooth = FALSE, pch = 16)
> plot(table(CM2$fTransect),  type = "h",
       xlab = "Transects", ylab = "Frequency")
```

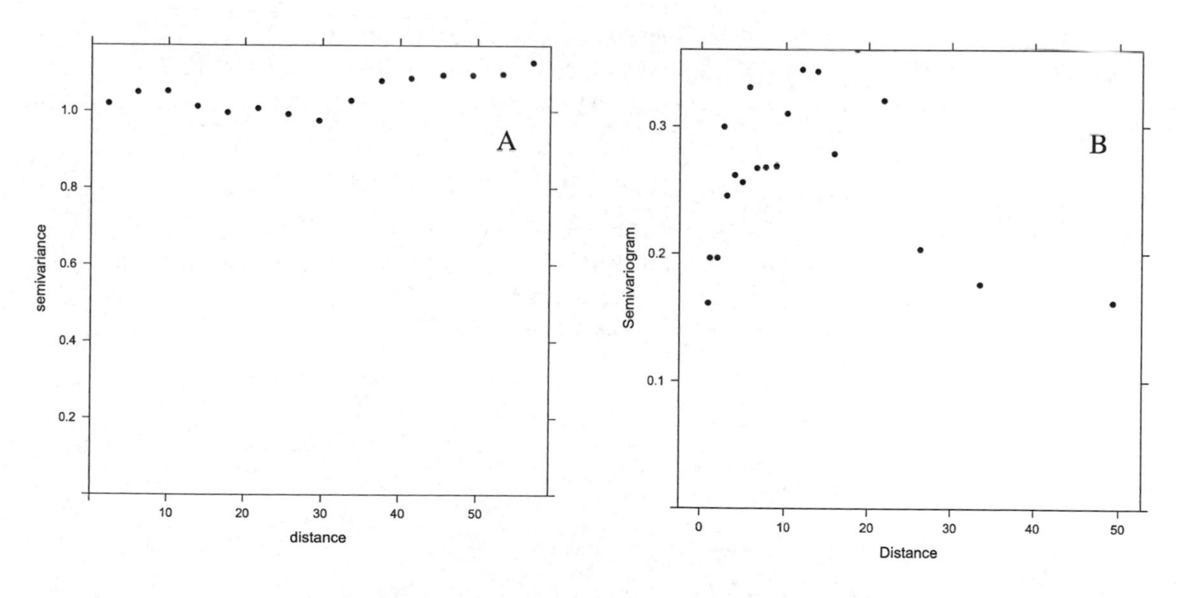

Figure 5.18. A: Sample variogram obtained by taking the normalized Pearson residuals and the geographic coordinates for each site, irrespective of Transect. The sample variogram was obtained with the Variogram function from the gstat package. B: Sample variogram obtained by taking the normalized Pearson residuals and the geographic coordinates for each site. Combinations of residuals within a transect are used to calculate the sample variogram, which was obtained with the Variogram function from the nlme package.

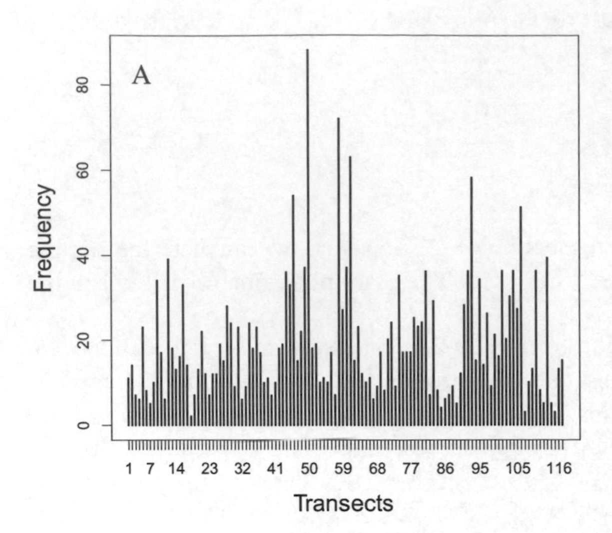

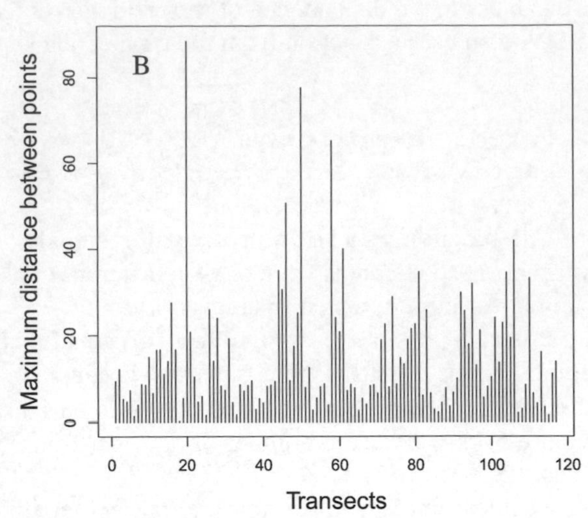

Figure 5.19. A: Number of observations per transect. B: Maximum distance between two points within each transect.

The R code to make Figure 5.19B is based on simple geometry and is not presented here. The sample variograms in Figure 5.18A and Figure 5.18B lead to concern about potential dependence structure, and we attempted to apply more advanced correlation structures in the `gamm` function without success. However, given that the number of observations per transect is small, and that there is a wide variation in transect length, this is not surprising.

5.5.5 Analysis of the presence-only data using a gamma GAMM

In the second step of the hurdle model, the zeros are dropped from the analysis, and the positive density data is analysed with a GAMM using a gamma distribution. Note that we have 751 non-zero observations of Common Murre. The R code to extract the presence-only data and to fit a gamma GAMM is:

```
> CM3 <- CM2[CM2$Comu > 0,]
> M2.Comu <- gamm(Comu ~ s(Distancetoshore) + s(Distkelps) + fDepth+
                  fExposure + fSubstrate + fTime + fOceancatnom,
                  random = list(fTransect =~ 1),
                  family = Gamma(link = "log"), data = CM3)
```

The numerical results are:

```
Parametric Terms:
             df     F  p-value
fDepth        3 0.658 0.578357
fExposure     2 3.374 0.034785
fSubstrate    2 0.526 0.591028
fTime         5 4.310 0.000717
fOceancatnom  3 1.376 0.248990
```

```
Approximate significance of smooth terms:
                       edf Ref.df      F  p-value
s(Distancetoshore) 1.000  1.000 1.276 0.258931
s(Distkelps)       4.442  4.442 4.758 0.000517
```

Note that there is a significant effect of `Distkelps`, with the effect of this covariate similar to that found earlier; i.e., lower densities at around 2000 m (Figure 5.20). In this case, `Depth` is not significant, but the categorical variable `Time` is significant. The estimated values of the variances are $\sigma^2_{Transect} = 0.650^2$ and $\sigma^2_{\varepsilon} = 1.184^2$. There is no clear residual spatial correlation. R code to extract the numerical information, plot the smoothers, and validate the model is similar to that used in Subsection 5.5.4 and is not presented here.

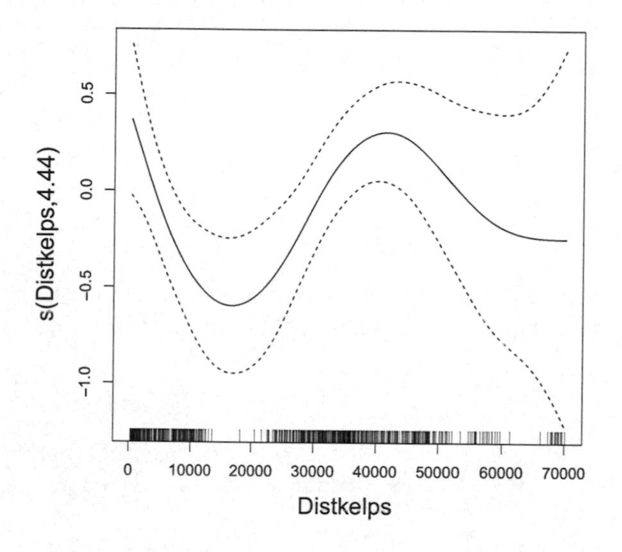

Figure 5.20. Partial effect of `Distkelps` obtained by a GAMM applied to the presence-only data.

5.7 Fitting a Gamma GAMM with CAR correlation in WinBUGS**

In previous chapters we demonstrated zero inflated GLMs, to be more specific, mixture models. In the previous sections of this chapter we applied two-stage models and, when analysing the density data, were able to use off-the-shelf software such as GAMM with a binomial distribution and a GAMM with a Gamma distribution. We used the function `gamm` from the `mgcv` package. Correlation among observations was dealt with by introducing a random effect, `Transect`. But what if:

1. We want to include a more complicated spatial correlation structure?
2. We want to analyse the counts of the birds rather than the density?
3. We think that there are false zeros and want to apply a zero inflated Poisson (or negative binomial) generalised additive model with a spatial correlation structure?

In Chapters 6 and 7 we will apply zero inflated Poisson GLM models with a conditional correlation structure (CAR) on the residuals. These methods will be implemented in WinBUGS. The CAR correlation can be easily applied on the binary part of the GAM that we used in this section, and, provided that we know how to fit a Gamma GAM in WinBUGS, we can do the same for the presence-only data. We will therefore summarize fitting a Gamma GLM in WinBUGS.

The density data were calculated by converting the number of birds observed in the 150 m range on the left side and the 150 m range on the right side of the boat (300 m in total) over a distance of 1 km to numbers per km^2. Hence all observed bird counts were multiplied by 10/3. To analyse the count data, we convert the density data back to numbers per 300 m by multiplying the densities by 3/10 and rounding the values to the nearest integer. We can then use a Poisson or negative binomial distribution. There is no need to use an offset, as sampling effort per site is the same.

In previous chapters we showed how a ZIP GLM can be applied in WinBUGS. In this section we use smoothers, as some of the covariates have a clear non-linear effect. In Chapter 9 we show how a GAM can be fitted in WinBUGS, and we will also implement a ZIP GAM.

The material presented in the remainder of this section may look daunting. We strongly suggest that, before continuing, you familiarize yourself with the material presented in Chapters 6 – 9.

5.7.1 Fitting a Gamma GLM in WinBUGS**

To fit a Gamma GAM with a CAR residual correlation in WinBUGS we need to be able to fit a Gamma GLM in WinBUGS. We will do this using only one covariate. Extending it to multiple covariates is simple, but may require more computing time. We will fit the following model in WinBUGS:

$$Y_i \sim Gamma(\mu_i, \tau)$$
$$E(Y_i) = \mu_i \quad \text{and} \quad \text{var}(Y_i) = \frac{\mu_i^2}{\tau}$$
$$\log(\mu_i) = \alpha + fExposure_i$$

Note that this is a mathematical notation that does not necessarily correspond to the coding in statistical software. The general notation for a gamma distribution and its density function in R is as follows (see the help file of `dgamma`):

$$Y_i \sim Gamma(\text{scale}, \text{shape})$$
$$E(Y_i) = \text{shape} \times \text{scale} \quad \text{and} \quad \text{var}(Y_i) = \text{shape} \times \text{scale}^2$$
$$f(Y_i; Y_i > 0) = \frac{1}{\text{scale}^{\text{shape}} \times \Gamma(\text{shape})} \times Y_i^{\text{shape}-1} \times e^{-\frac{Y_i}{\text{scale}}}$$

Depending on the software, the order of the shape and scale parameters may change. The `dgamma` function in R also uses a rate parameter, which is defined as 1 / scale. It is not the case that μ_i is equal to the shape, and τ is not equal to the scale. We will see that WinBUGS uses a slightly different parameterisation.

We first fit the gamma GLM in R to verify the results of WinBUGS. The R code for a gamma GLM in R is:

```
> MG1 <- glm(Comu ~ fExposure, family = Gamma(link="log"), data=CM3)
```

We use a log link to model the mean of Y_i (density of Common Murre for observation i). It is a challenge to determine how the `glm` function deals with the shape and scale parameters. From the help file of `Gamma` (which is used by the `family` function inside the `glm` function) and the coding of this function (type `Gamma` in R without any brackets to see the code that defines the AIC), it seems that the scale parameter is modelled as: `scale = mu * disp`, where `disp` is a function of the residuals. The `shape` parameter is modelled as `1 / disp`. Using this parameterization we get E(Y_i) = μ_i and var(Y_i) = μ_i^2 / τ, although the τ is not estimated with maximum likelihood. The μ_i in our equation is not necessarily the same as the `mu` in the R code. The bottom line is that the scale is modelled as a function of covariates.

The output of the GLM is:

```
> summary(MG1)
```

```
               Estimate Std. Error t value Pr(>|t|)
(Intercept)      2.6034     0.1782  14.608  < 2e-16
fExposure2       1.0647     0.2124   5.012 6.74e-07
fExposure3       1.0858     0.3045   3.565 0.000386

(Dispersion parameter for gamma family taken to be 6.097853)
    Null deviance: 1478.1  on 747  degrees of freedom
Residual deviance: 1343.8  on 745  degrees of freedom
AIC: 6516.7
```

The model is overdispersed, but that is irrelevant for the current presentation; at this stage we only want WinBUGS code that gives the same results. For WinBUGS we first need to bundle the required data. The covariate `Exposure` is a factor with three levels, and we use the `model.matrix` function to obtain the two dummy variables for levels 2 and 3.

```
> DesMat   <- model.matrix(~ fExposure , data = CM3)
> win.data <- list(Comu  = CM3$Comu,
                   Exp2  = DesMat[, "fExposure2"],
                   Exp3  = DesMat[, "fExposure3"],
                   N     = nrow(CM3))
```

Now we need the modelling code. The WinBUGS manual states that we need `dgamm(r, mu)`, and the matching density function is:

$$Y_i \sim dgamma(\mathrm{r}, \mathrm{mu})$$

$$f(Y_i; Y_i > 0) = \frac{\mathrm{mu}^r \times Y_i^{r-1}}{\Gamma(\mathrm{r})} \times e^{-\mathrm{mu} \times Y_i}$$

So the r in the `dgamma` function in WinBUGS is the shape parameter in the `dgamma` function in R. The `mu` in the `dgamma` function in WinBUGS is 1 / scale. Again, the `mu` in the `dgamma` function in WinBUGSe is not the mean of Y_i. Summarising, the two `dgamma` functions use a slightly different parameterisation, but the density functions are identical. In order to get the same results as with the `glm` function in R we will model the scale parameter as a function of covariates. Note that the shape parameter will be different from that in R, as we will estimate it using MCMC. The following code fits the gamma GLM:

```
sink("modelGammaGLM.txt")
cat("
model{

 #Priors
 for (i in 1:3) { betal[i]  ~ dunif(-10, 10) }
 tau ~ dgamma( 0.01, 0.01 )

 #Likelihood
 for (i in 1:N) {
   Comu[i]    ~  dgamma(tau, TauDivScale[i])
   TauDivScale [i]  <- tau / mu[i]
   log(mu[i]) <- max(-20, min(20, eta[i]))
   eta[i]      <- betal[1]  + betal[2] * Exp2[i] + betal[3] * Exp3[i]
 }}
",fill = TRUE)
sink()
```

By using `dgamma(tau, OneDivScale[i])` and `OneDivScale[i] <- tau / mu[i]` we obtain the mean and variance of Y_i, as specified earlier in this subsection. It is simple and elegant. The rest of the code is only a matter of specifying diffuse priors, initial values, parameters to be extracted, and the number of iterations, chains, etc. Additional covariates can easily be added to `eta[i]`.

```
> inits <- function () {
   list( beta1 = rnorm(3, 0, 0.01),
         tau   = 1)  }
> params <- c("beta1", "tau")
> nc <- 3          #Number of chains
> ni <- 10000      #Number of draws from posterior (for each chain)
> nb <-  5000      #Number of draws to discard as burn-in
> nt <-    10      #Thinning rate
> MyWinBugsDir <- "C:/WinBUGS14/"
> library(R2WinBUGS)
> out <- bugs(data = win.data, inits = inits, parameters = params,
             model = "modelGammaGLM.txt", n.thin = nt, n.chains = nc,
             n.burnin = nb, n.iter = ni, debug = TRUE,
             bugs.directory = MyWinBugsDir)
> print(out, digits = 3)
```

The results are:

```
Inference for Bugs model at "modelGammaGLM.txt", fit using WinBUGS,
 3 chains, each with 10000 iterations (first 5000 discarded), n.thin = 10
 n.sims = 1500 iterations saved


             mean    sd     2.5%      25%      50%      75%    97.5% Rhat n.eff
beta1[1]    2.610 0.090    2.436    2.547    2.609    2.671    2.785 1.000  1500
beta1[2]    1.061 0.106    0.850    0.988    1.062    1.134    1.264 1.001  1500
beta1[3]    1.092 0.150    0.799    0.988    1.094    1.188    1.382 1.003   740
tau         0.673 0.030    0.619    0.652    0.672    0.694    0.730 1.000  1500
deviance 6494.067 2.911 6490.000 6492.000 6493.000 6496.000 6501.000 1.000  1500

For each parameter, n.eff is a crude measure of effective sample size,
and Rhat is the potential scale reduction factor (at convergence, Rhat=1).

DIC info (using the rule, pD = Dbar-Dhat)
pD = 4.0 and DIC = 6498.1
DIC is an estimate of expected predictive error (lower deviance is better).
```

Note that the estimated regression parameters are similar to those obtained by the `glm` function. More complicated models require a larger number of iterations, thinning rate, and burn-in. Other methods of modelling the scale and shape are possible.

5.7.2 Fitting a ZIP GAM on the Common Murre counts in WinBUGS**

To reduce computing time, we will fit the following ZIP GAM on the counts of Common Murre. The model can easily be extended with other covariates and smoothers; by adding them to the predictor functions and increasing the number of iterations, burn-in, and thinning rate.

$$
\begin{aligned}
&Y_i \sim ZIP(\mu_i, \pi_i)\\
&\log(\mu_i) = \alpha + fTime_i + s(\Delta Kelps_i)\\
&\text{logit}(\pi_i) = \gamma + fDepth_i + s(\Delta Kelps_i)
\end{aligned}
$$

The ZIP model without the smoothers is fitted with the code below. We first need to calculate the counts, and the `model.matrix` function is used to obtain the dummy variables.

```
> CM2$IComu <- round(CM2$Comu * 3 / 10)
> CM2$fTime <- factor(paste(CM2$Month, CM2$Year, sep = "."),
                 levels = c("11.2007", "1.2008", "3.2008", "11.2008",
                            "1.2009", "3.2009"),
                 labels = c("Nov07", "Jan08", "Mar08", "Nov08",
                            "Jan09", "Mar09"))
> DesMat <- model.matrix(~ fDepth + fTime , data = CM2)
```

Next we collect all the required variables:

```
> win.data <- list(Y     = CM2$IComu,
                  Dep2   = DesMat[, "fDepth2"],
                  Dep3   = DesMat[, "fDepth3"],
                  Dep4   = DesMat[, "fDepth4"],
                  TJan08 = DesMat[, "fTimeJan08"],
                  TMar08 = DesMat[, "fTimeMar08"],
                  TNov08 = DesMat[, "fTimeNov08"],
                  TJan09 = DesMat[, "fTimeJan09"],
                  TMar09 = DesMat[, "fTimeMar09"],
                  N      = nrow(CM2))
```

We are now ready to run the ZIP model without the smoothers. The code below was used in Chapter 2 and modified for the Common Murre data.

```
sink("zipgam.txt")
cat("
model{

  #Priors
  for (i in 1:4) { g[i] ~ dunif(-5, 5) }
  for (i in 1:6) { b[i] ~ dunif(-5, 5) }

  #Likelihood
  for (i in 1:N) {

    #Binary part
    W[i] ~ dbern(psi.min1[i])
    psi.min1[i]   <- 1 - psi[i]
    logit(psi[i]) <- max(-20, min(20, etaBin[i]))
    etaBin[i] <- g[1] + g[2] * Dep2[i] + g[3] * Dep3[i] +
                 g[4] * Dep4[i]

    #Count process
    Y[i] ~ dpois(mu.eff[i])
    mu.eff[i]  <- W[i] * mu[i]
    log(mu[i]) <- max(-20, min(20, eta[i]))
    eta[i]     <- b[1] + b[2] * TJan08[i] + b[3] * TMar08[i] +
                  b[4] * TNov08[i] + b[5] * TJan09[i] +
                  b[6] * TMar09[i]

    #Residuals
    EZip[i]   <- mu[i] * (1 - psi[i])
```

```
    VarZip[i] <- (1 - psi[i]) * (mu[i] + psi[i] * pow(mu[i], 2))
    PRES[i] <- (Y[i] - EZip[i]) / sqrt(VarZip[i])
  }}
",fill = TRUE)
sink()
```

We run the code with:

```
> W         <- CM2$IComu
> W[CM2$IComu > 0] <- 1
> inits <- function () {
   list( b = runif(6, -5, 5),
         g = runif(4, -5, 5),
         W = W) }
> params  <- c("PRES", "b",  "g")
> ThisRun <- "Test"
> if (ThisRun == "Test") {
    nc <- 3          #Number of chains
    ni <- 10000      #Number of draws from posterior (for each chain)
    nb <- 1000       #Number of draws to discard as burn-in
    nt <- 10         #Thinning rate
  } else {
    nc <- 3          #Number of chains
    ni <- 100000     #Number of draws from posterior (for each chain)
    nb <- 10000      #Number of draws to discard as burn-in
    nt <- 100        #Thinning rate
  }
> library(R2WinBUGS)
> MyWinBugsDir <- "C:/WinBUGS14/"
  #Start Gibbs sampler
> ZipGAM <- bugs(data = win.data, inits = inits,
                 parameters = params, model = "zipgam.txt",
                 n.thin = nt, n.chains = nc, n.burnin = nb,
                 n.iter = ni, debug = TRUE,
                 bugs.directory = MyWinBugsDir)
> print(ZipGAM, digits = 2)
```

Computing time is a few minutes and the chains are well mixed. Adding the smoothers is a matter of extending the predictor functions. First we need to define the knots and the splines. The code below is taken from Chapter 9, where it is fully explained. To save space we did not include the function `GetZ`, which is also presented in Chapter 9. We standardised the covariate `Distkelps`.

```
> CM2$DistKelps <- as.numeric(scale(CM2$Distkelps))
> num.knots <- 9
> Prob      <- seq(0, 1, length = (num.knots + 2))
> Prob1     <- Prob[-c(1,(num.knots+2))]
> Knots     <- quantile(unique(CM2$DistKelps), probs = Prob1)
> N         <- nrow(CM2)
> X         <- cbind(rep(1,N), CM2$DistKelps)
> Z         <- GetZ(CM2$DistKelps, Knots)
```

We need to extend the `win.data`:

```
> win.data <- list(Y    = CM2$IComu,
                   Dep2 = DesMat[, "fDepth2"],
```

```
               Dep3    = DesMat[, "fDepth3"],
               Dep4    = DesMat[, "fDepth4"],
               TJan08 = DesMat[, "fTimeJan08"],
               TMar08 = DesMat[, "fTimeMar08"],
               TNov08 = DesMat[, "fTimeNov08"],
               TJan09 = DesMat[, "fTimeJan09"],
               TMar09 = DesMat[, "fTimeMar09"],
               N       = nrow(CM2),
               NumKnots  = 9,
               X       = X,
               Z       = Z)
```

The modelling code follows. The basis of the code is presented in Chapter 9.

```
sink("zipgam2.txt")
cat("
model{

  #Priors
  for (i in 1:4) { g[i] ~ dunif(-5, 5) }
  for (i in 1:6) { b[i] ~ dunif(-5, 5) }
  betab ~ dunif(-5, 5)
  betap ~ dunif(-5, 5)
  for (i in 1:NumKnots) {
    sb[i] ~ dnorm(0, taub)
    sp[i] ~ dnorm(0, taup)
  }
  taub ~ dgamma(0.01, 0.01)
  taup ~ dgamma(0.01, 0.01)

  #Likelihood
  for (i in 1:N) {

    #Binary part
    W[i] ~ dbern(psi.min1[i])
    psi.min1[i]   <- 1 - psi[i]
    logit(psi[i]) <- max(-20, min(20, etaBin[i]))
    etaBin[i] <- g[1] + g[2] * Dep2[i] + g[3] * Dep3[i] +
                 g[4] * Dep4[i] + SmoothBin[i]
    SmoothBin[i] <- betab * X[i,2] +
                    sb[1] * Z[i,1] + sb[2] * Z[i, 2] +
                    sb[3] * Z[i,3] + sb[4] * Z[i, 4] +
                    sb[5] * Z[i,5] + sb[6] * Z[i, 6] +
                    sb[7] * Z[i,7] + sb[8] * Z[i, 8] +
                    sb[9] * Z[i,9]

    #Count process
    Y[i] ~ dpois(mu.eff[i])
    mu.eff[i]  <- W[i] * mu[i]
    log(mu[i]) <- max(-20, min(20, eta[i]))
    eta[i]     <- b[1] + b[2] * TJan08[i] + b[3] * TMar08[i] +
                  b[4] * TNov08[i] + b[5] * TJan09[i] +
                  b[6] * TMar09[i] + SmoothPois[i]
    SmoothPois[i] <- betap * X[i,2] +
                     sp[1] * Z[i,1] + sp[2] * Z[i, 2] +
```

```
                    sp[3] * Z[i,3] + sp[4] * Z[i, 4] +
                    sp[5] * Z[i,5] + sp[6] * Z[i, 6] +
                    sp[7] * Z[i,7] + sp[8] * Z[i, 8] +
                    sp[9] * Z[i,9]

    #Residuals
    EZip[i]   <- mu[i] * (1 - psi[i])
    VarZip[i] <- (1 - psi[i]) * (mu[i] + psi[i] * pow(mu[i], 2))
    PRES[i] <- (Y[i] - EZip[i]) / sqrt(VarZip[i])
  }
}
",fill = TRUE)
sink()
```

The initial values are set as:

```
> inits <- function () {
   list(
    b      = runif(6, -5, 5),
    g      = runif(4, -5, 5),
    W      = W,
    betab  = runif(1, -5, 5),
    betap  = runif(1, -5, 5),
    sb     = runif(9, -5, 5),
    sp     = runif(9, -5, 5),
    taub   = runif(1, 0.1, 0.5),
    taup   = runif(1, 0.1, 0.5))  }
> params <- c("PRES", "b",  "g", "SmoothPois", "SmoothBin")
```

This will take some time and may require increasing the number of iterations, burn-in, and thinning rate. Code to extract and print the smoothers and 95% credible intervals are given in Chapter 9.

5.7.3 Fitting a ZIP GAM with residual CAR in WinBUGS**

In Chapter 6 we show how a residual CAR correlation can be added to a ZIP GLM. Whether we use a GLM or a GAM is irrelevant. First we need to define which sites are neighbours. We can do this based on the shape of a sample variogram of Pearson residuals of the ZIP GAM without spatial correlation, or we can define sites within a certain range as neighbours. Some of the sample variograms suggest defining sites as neighbours if they are separated by less than 0.1 km. Such a definition would allow us to calculate a matrix **A** with elements 1 (neighbour) or 0 (non-neighbour). **A** can then be used to calculate the matrices **C** and **M** (Chapter 6). We then add a term `EPS[i]` to the Poisson link function, representing the spatially correlated noise:

```
eta[i]      <- b[1] + b[2] * TJan08[i] + b[3] * TMar08[i] +
               b[4] * TNov08[i] + b[5] * TJan09[i] +
               b[6] * TMar09[i] + SmoothPois[i] + EPS[i]
MeanEps[i] <- 0
```

Outside the loop for the likelihood we specify the CAR correlation for `EPS`:

```
EPS[1:N] ~ car.proper(MeanEps[], C[], adj[], num[], M[], tau, rho)
```

To make the chains more stable we can also use:

```
eta[i]      <- b[1] + b[2] * TJan08[i] + b[3] * TMar08[i] +
               b[4] * TNov08[i] + b[5] * TJan09[i] +
               b[6] * TMar09[i] + SmoothPois[i] + tau * EPS[i]
```

and

```
EPS[1:N] ~ car.proper(MeanEps[], C[], adj[], num[], M[], 1, rho)
```

Information on priors is provided in Chapter 6. We can also add a CAR noise term to the logistic link function. Computation may take several hours.

5.8 Discussion

We will refrain from a biological discussion, because the results of statistical analyses applied to the full data set are being prepared for publication at the time of writing.

Analysing ecological data sets becomes a challenge when a large number of covariates obtained from GIS are included. Collinearity is often a serious problem in such cases, and the choice of which covariates to use is subjective.

Another challenge with these data is to decide how to include a dependence structure. For our study, we used random effects. If you are brave and have a fast computer, try adding a CAR residual correlation.

We can analyse either density data or counts, although the counts is better (Poisson or negative binomial GLMs can deal with heterogeneity and fitted values are always positive). Using counts necessitates the use of WinBUGS or other software for MCMC.

In this chapter we presented the results of a model containing all the covariates, and no model selection was applied. To use an Information Theoretic approach the following 16 models can be used. We leave it as an exercise for the reader to implement these models, obtain Akaike weights, and find the most appropriate models.

- M_1: Only time. Is bird density driven by timing through the non-breeding season?
- M_2: Only SST. Is bird density related to avoidance of cold waters or to an abundance of fish at certain temperatures?
- M_3: Ocean category and time. Do the birds occupy different habitats at different stages of the season?
- M_4: Ocean category, slope, and time. Do slope and habitat reflect predictable preferred feeding areas where birds remain throughout the winter?
- M_5: Time and exposure. Are bird movements related to storms at certain times?
- M_6: Distance to the shore and time. Do birds maintain a given distance from the shore at certain times?
- M_7: Exposure, distance to the shore, and winter. Are storminess and proximity to shore factors differing by year?
- M_8: Latitude, longitude, and time. Is there an area that the birds prefer; close to the Gulf of Alaska or in a certain part of the sound?
- M_9: Latitude, longitude, ocean category, and SST. Do birds congregate in a geographical area related to ocean habitat and water temperature, which may, for example, control fish spawning?
- M_{10}: Distance to eelgrass and SST. Are these key factors in herring spawning during this period?
- M_{11}: SST and exposure. Do energy requirements of keeping warm and out of storms dictate distribution?
- M_{12}: SST and winter. Does water temperature drive bird density?
- M_{13}: SST and distance to the shore. Do the birds have a preference for a given proximity to shore influenced by water temperature?
- M_{14}: Slope and substrate. Do slope of seabed and substrate provide feeding opportunities which affect distribution?

- M_{15}: Ocean category and depth. Habitat and bathymetry may control where the birds are most able to acquire food.
- M_{16}: Distance to shore, slope, and substrate. Do these reflect feeding opportunities? Juvenile fish of many species are known to remain near shore, perhaps interacting with slope and seabed type.

5.9 What to present in a paper?

In presenting results of the analysis given here, the following points need to be adressed: First, we need to explain why the surveys were taken (Introduction) and how they were taken (Methods) and the underlying questions need to be outlined (Introduction).

In the Methods section we explain in more detail how a flying bird in Alaska becomes a number in a large spreadsheet. Present a graph of the area and the survey tracks. Effective communication of this to the reader is crucial.

In the statistical part in the Methods section, explain the steps in the data exploration. Keywords are zero inflation, collinearity, and non-linearity. This leads to a zero inflated GAMM. We need to justify why a hurdle model or mixture model is applied, which means discussing the nature of the zeros. We focused on analyses of the density data, but we could have used the counts. Discuss how to deal with model selection.

In the Results section, present the numerical output (e.g. the condensed output obtained with the `anova` command applied on the `$gam` object) and the graphs with smoothers. Present the results of the models for the binary part and for the non-zero data. Include the results of the model validation and consider presenting model validation graphs as online supplemental material. If the ZIP GAM with CAR correlation was used, discuss mixing of the chains, etc.

In the Discussion, focus on what the results mean in terms of biology. If a covariate such as distance to kelp is significant, but does not make immediate biological sense, try to explain why it is significant. Is it a proxy for something else?

Mention that, because of collinearity, the covariates in the model that show significance may not be those driving the system; the real effectors may be those dropped due to collinearity.

Finally, mention that the data were obtained by ship-of-opportunity, and changes over time may be due to differences in surveying positions.

Don't forget to cite R and the packages that were used.

Acknowledgement

We would like to thank Dave Janka who, as skipper of the MV Auklet, made it possible to survey PWS in winter; John Moran and Jeep Rice of National Marine Fisheries Service and Jan Straley of University of Alaska, Sitka for allowing us to share their vessel time. We are also grateful to additional observers Karen Brenneman and Aaron Lang. Jodi Harney of Coastal and Ocean Resources was helpful in making data from the recently completed Shorezone project available. Some analyses and depictions used in this paper were produced with the Giovanni online data system, developed and maintained by NASA GES DISC. The research described in this paper was supported by a grant from the Exxon Valdez Oil Spill Trustee Council. The findings and conclusions presented by the authors are their own and do not necessarily reflect the views or position of the Trustee Council.

6 Zero inflated spatially correlated coastal skate data

Alain F. Zuur, Federico Cortés, Andrés J. Jaureguizar, Raúl A. Guerrero, Elena N. Ieno, and Anatoly A. Saveliev

6.1 Introduction

In this chapter, zero inflated counts of three skate species, *Sympterygia bonapartii*, *Rioraja agassizi* and *Atlantoraja castelnaui*, will be analysed. The data were collected from the Rio de la Plata Basin, a major river basin of the Argentine coast. Chondrichthyan fishes (sharks, skates, rays, and chimaeras) use the Río de la Plata coastal region as a feeding, mating, egg laying, and nursery area. Skates are a ubiquitous group in the region (Massa et al., 2004) and, as benthic mesopredators, play an important functional role in structuring the benthic system (Myers et al., 2007).

Skates of the area are represented by species with a variation of life history traits (total length, age and length at maturity, and reproductive frequency), and it can be expected that environmental variables play a role in their distribution (for a review see Menni and Stehmann, 2000). The effects of these environmental factors have not been quantified. We will model the distribution and habitat preferences of three skate species using fishery survey data of the Rio de la Plata coastal region.

6.2 Importing the data and data coding

The following R code imports the data, stores it in the object `Skates` and prints a list of the variable names:

```
> Skates <- read.table(file = "Skates.txt", header = TRUE)
> names(Skates)

"Station"     "Year"          "Month"      "Day"          "Date"
"CumDate"     "DayNumber"     "Latitude"   "Longitude"    "Zone"
"Depth"       "Temperature"   "Salinity"   "BottomType"   "SweptArea"
"RA"          "AC"            "SB"
```

The species data are counts (number of animals) and are contained in the variables `RA` (*R. agassizi*), `AC` (*A. castelnaui*), and `SB` (*S. bonapartii*). The variable `SweptArea` is the area surveyed and represents the sampling effort per site. Sampling effort differed per site, so we need a generalized linear (or additive) model with a Poisson or negative binomial distribution in which the log of `SweptArea` is used as an offset variable. The variables `Year`, `Month`, `Latitude`, `Longitude`, `Depth`, `Temperature`, `Salinity`, and `BottomType` are self-explanatory. We will treat `Year`, `Month`, and `BottomType` as categorical variables with 4, 2, and 4 levels, respectively. We define them as such with the `factor` function:

```
> Skates$fYear  <- factor(Skates$Year)
> Skates$fMonth <- factor(Skates$Month)
> Skates$fBottomType <- factor(Skates$BottomType,
             levels = c(1, 2, 3, 4),
             labels = c("Mud", "Sand & mud", "Sand",
                   "Sand, shell, rest and/or tuff"))
```

The variable `DayNumber` is the day within the year, and `CumDay` is the number of days since 1 January 1998. In the statistical analysis we will need to include distances between sites. The study area is large; therefore we need to convert latitude and longitude to UTM coordinates. For this we need to know the UTM zone, and some `rgdal` magic does the conversion.

```
> library(rgdal)
> UTM21S <- "+proj=utm +zone=21 +south"
> UTM21S <- paste(UTM21S, " +ellps=WGS84",sep = "")
> UTM21S <- paste(UTM21S, " +datum=WGS84",sep = "")
> UTM21S <- paste(UTM21S, " +units=m +no_defs", sep="")
> coord.UTM21S <- project(as.matrix(Skates[,
                c("Longitude", "Latitude")]), UTM21S)
> Skates$X <- coord.UTM21S[,1]
> Skates$Y <- coord.UTM21S[,2]
```

The object `UTM21S` is a character string that contains:

```
"+proj=utm +zone=21+south +ellps=WGS84 +datum=WGS84 +units=m +no_defs"
```

The `project` function is exacting about the spaces in this string. The first argument is a 2-column matrix with coordinates. An explanation of the intimidating syntax of the character string in `UTM21S` can be found in the help file of the `project` function, which is an interface to the PROJ.4 library of projection functions for geographical position data. See also http://trac.osgeo.org/proj/ for details. The projected UTM coordinates are stored as `X` and `Y` in the `Skates` object.

6.3 Data exploration

We will discuss potential outliers, collinearity, spatial independence, relationships, and zero inflation, beginning with outliers.

6.3.1 Outliers

A multipanel Cleveland dotplot is presented in Figure 6.1. Note that surveying took place in November and December of four years. The data are balanced with respect to these two months. There are no sites with extremely large or small values of depth, temperature, or salinity. The species data seem to be zero inflated, something we will explore later. There are no obvious outliers in the explanatory variables or species data. The following R code was used to make Figure 6.1. The code has been used in other chapters and is not explained here.

```
> MyX <- c("Year", "Month", "CumDate", "DayNumber", "X", "Y",
        "Depth", "Temperature", "Salinity", "BottomType",
        "SweptArea", "RA", "AC", "SB")
> library(lattice)
> dotplot(as.matrix(Skates[, MyX]), groups = FALSE,
     strip = strip.custom(bg = 'white',
            par.strip.text = list(cex = 0.8)),
```

```
scales = list(x = list(relation = "free"),
              y = list(relation = "free"), draw = FALSE),
col=1, cex  = 0.5, pch = 16, layout = c(3, 5),
xlab = "Value of the variable",
ylab = "Order of the data from text file")
```

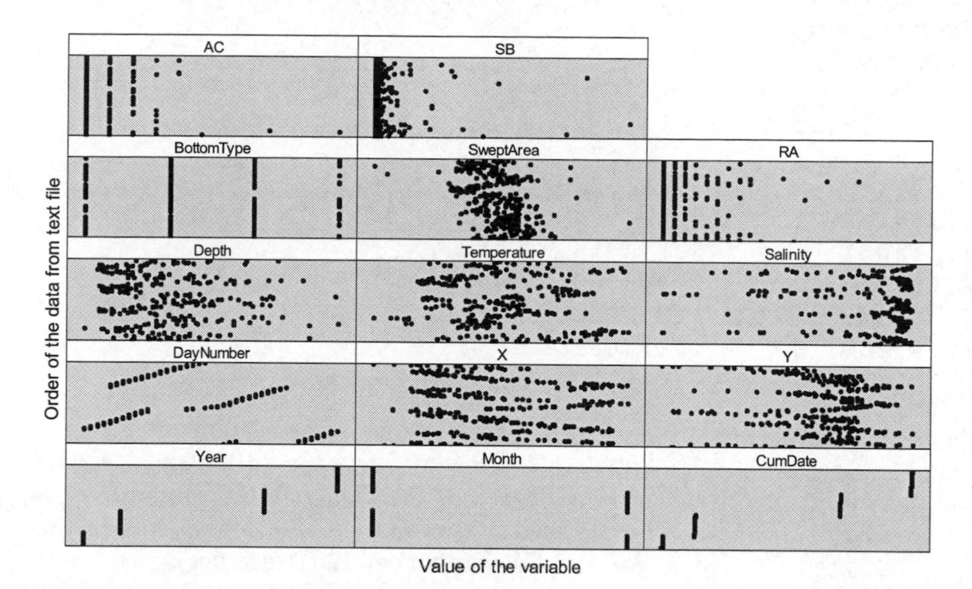

Figure 6.1. Cleveland dotplot for all variables. The horizontal axis represents the value of a variable and the vertical axis the order of the data as imported from the data file. Each panel corresponds to a particular variable.

6.3.2 Collinearity

Correlation between covariates causes inflated *p*-values and is called collinearity. To quantify the amount of collinearity we can calculate variance inflation factors (VIF) (see Chapter 4). The R code below calculates and prints the VIFs. A function for calculating VIF values is available in the file `HighstatLib.R`, which can be obtained from the first author of this book. The `car` package can also be used to calculate VIF values.

```
> source("HighstatLib.R")     #Contains functions for VIF
> MyX2 <- c("fYear", "fMonth", "Y", "X", "Depth", "Temperature",
        "Salinity", "fBottomType")
> corvif(Skates[, MyX2])

Variance inflation factors
                 GVIF Df GVIF^(1/2Df)
fYear       10.974011  3     1.490714
fMonth       9.592510  1     3.097178
X           11.628612  1     3.410075
Y            8.365349  1     2.892291
Depth        4.799223  1     2.190713
Temperature  5.489968  1     2.343068
Salinity     3.488660  1     1.867795
fBottomType  1.964894  3     1.119154
```

Note that the VIF values (GVIF) for `fYear` and `fMonth` are relatively high (3 is our cut-off level). Either `fYear` and `fMonth` are related, or each of them is related to one of more of the other covariates. Figure 6.2 shows a scatterplot of `fYear` and `fMonth`, Note that, except for 2003, surveys took place in only a single month. Any potential month effect may therefore be confounded with a year effect or vice versa. This means that we must refrain from using fMonth and fYear simultaneously.

Dropping fMonth reduces the VIF value for fYear, confirming that fYear and fMonth are indeed related, see the results below.

```
> MyX3 <- c("fYear", "X", "Y", "Depth", "Temperature", "Salinity",
            "fBottomType")
> corvif(Skates[, MyX3])

                 GVIF Df GVIF^(1/2Df)
fYear        1.835053  3     1.106475
X           11.504135  1     3.391775
Y            8.364165  1     2.892087
Depth        4.712021  1     2.170719
Temperature  5.489954  1     2.343065
Salinity     3.454965  1     1.858754
fBottomType  1.949450  3     1.117683
```

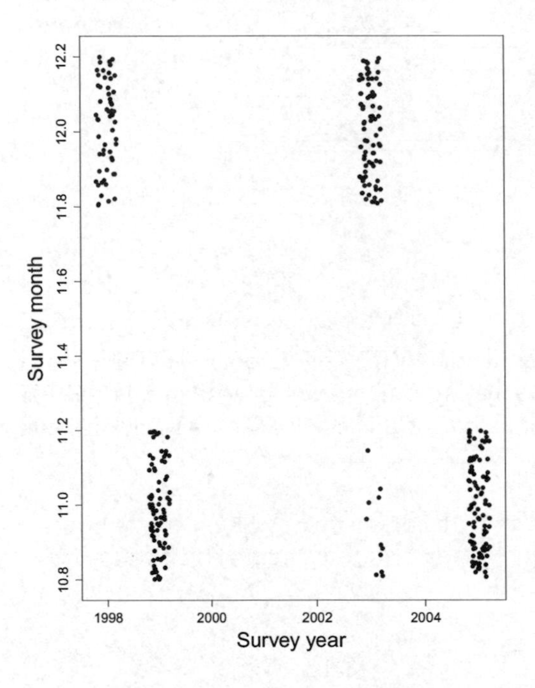

Figure 6.2. Plot of survey month versus survey year. A degree of jittering (adding random noise) was applied to spread the observations. Month was either 11 (November) or 12 (December).

The VIF values for the spatial coordinates are also high, probably because Temperature, Depth, and Salinity contain similar environmental information. If the VIF values for the spatial coordinates had shown a reduction after dropping fMonth, it would have indicated confounding between the area surveyed and the month of surveying. Fortunately, this is not the case. It is our intention to model the skate counts as a function of covariates such as Temperature, Salinity, Depth, fBottomType, and fYear, and check residual spatial correlation using the X and Y variables. We therefore drop the X and Y coordinates for the moment. When VIF values are calculated without the X and Y variables, Temperature and Depth still have high VIF values.

```
> MyX4 <- c("fYear", "Depth", "Temperature", "Salinity",
            "fBottomType")
> corvif(Skates[,MyX4])

                GVIF Df GVIF^(1/2Df)
fYear       1.748633  3     1.097614
Depth       3.187267  1     1.785292
Temperature 5.390796  1     2.321809
Salinity    2.580553  1     1.606410
fBottomType 1.259943  3     1.039262
```

The Pearson correlation between Depth and Temperature is -0.77, and therefore it seems sensible to drop one of them. We will drop Depth, as Temperature is more informative from a biological point of view. Recalculating VIF values shows that there is no obvious collinearity remaining.

```
> MyX5 <- c("fYear", "Temperature", "Salinity", "fBottomType")
> corvif(Skates[,MyX5])

Variance inflation factors
                GVIF Df GVIF^(1/2Df)
fYear       1.389014  3     1.056293
Temperature 2.849320  1     1.687993
Salinity    2.580375  1     1.606355
fBottomType 1.210090  3     1.032293
```

VIF values detect only linear relationships among covariates, and we recommend always combining this tool with scatterplots of covariates, as it provides information on potential non-linear relationships (Figure 6.3).

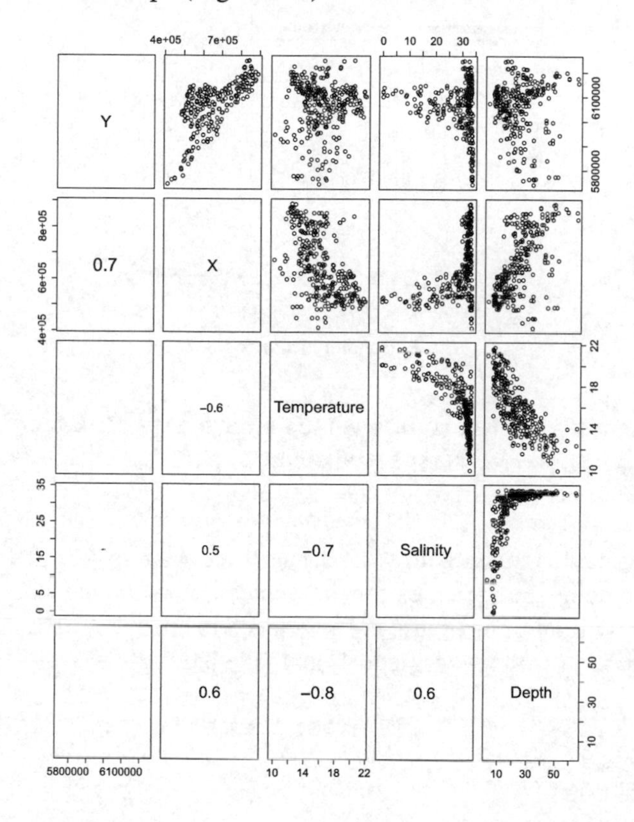

Figure 6.3. Multi-panel scatterplot. Lower diagonal elements are estimated Pearson correlation coefficients. Note that there is a strong correlation between temperature and salinity.

To confirm our original choices we use all continuous covariates, including those that we dropped based on high VIF values. It is better to double check and be 100% sure before removing covariates.

The reason for the high VIF values of `Depth` and `Temperature` becomes clear; there is a negative relationship. The further away from the coast to the east, the deeper the sites and the colder the water.

Note that there is still a relatively high correlation between `Temperature` and `Salinity`; the higher the temperature the lower the salinity. This corresponds to a VIF value of 2.849, which is close to our threshold of 3. The relationship between these two variables seems to be non-linear, and we will drop `Salinity` and continue the analysis with `Temperature`. This means that we are left with `fYear`, `Temperature`, and `fBottomType`. Boxplots (not presented here) of `Temperature` conditional on `fYear` and conditional on `BottomType` reveal no obvious strong collinearity. The problem with this set of covariates is the interpretation of `Year`. If it is significant in the models, we will need to determine what it represents.

The multi-panel scatterplot in Figure 6.3 is created with the following R code. The code assumes that the `HighstatLib.R` file has been sourced.

```
> MyX6 <- c("Y", "X", "Temperature", "Salinity", "Depth")
> pairs(Skates[, MyX6], lower.panel = panel.cor, cex.labels = 1)
```

6.3.3 Relationships

Scatterplots of the species data versus the explanatory variables should always be made. However, as we will see, the species data are zero inflated, and there is a difference in survey effort, which makes the interpretation of scatterplots difficult. For example, Figure 6.4A contains multi-panel scatterplots of the RA counts versus some of the covariates. A smoother was added to aid interpretation. There are no obvious patterns, but the graph is not informative, due to the difference in survey effort. An option is to plot the density: `RA/SweptArea`, but such a graph is not useful, as you will see if you attempt it. Because we are going to apply GLM models with a log link, plotting `log(RA/SweptArea+1)` versus the covariates makes more sense (Figure 6.4B). The LOESS smoother indicates some relationships, but zero inflation is causing trouble.

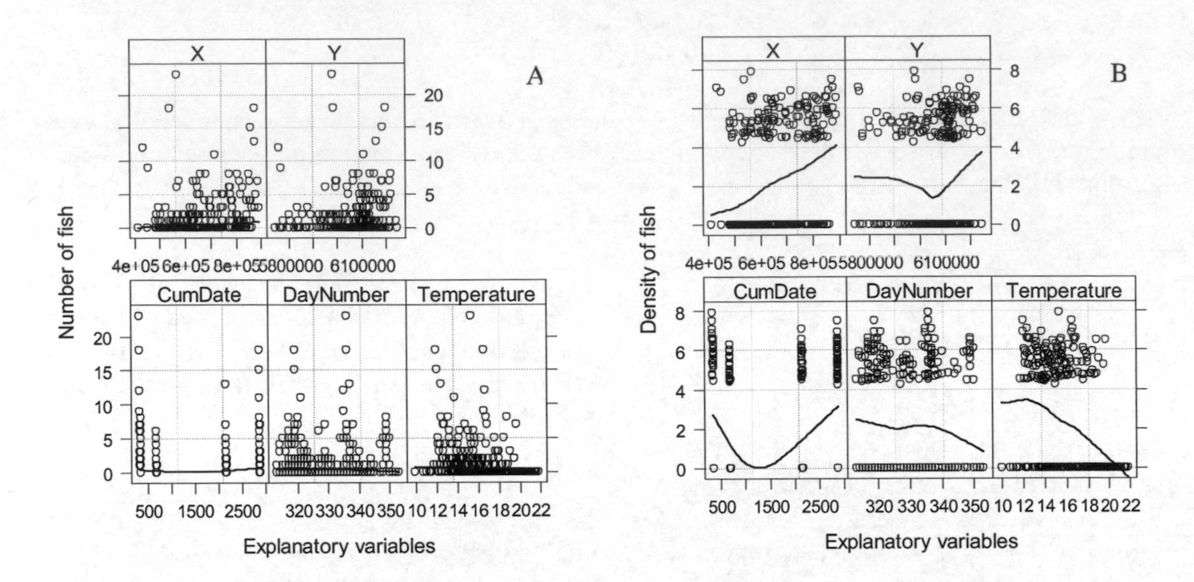

Figure 6.4. A: Multi-panel scatterplots representing the RA counts versus a covariate. B: Each panel represents a scatterplot of `log(RA/SweptArea+1)` versus a covariate.

R code to generate Figure 6.4A is intimidating and is given below. Creating three vectors `AllX`, `ID`, and `AllY` is essential. The vector `AllX` contains the variables for the *x*-axes, `ID` is the names of these variables, and `AllY` repeats the `RA` vector 5 times. For Figure 6.4B we need to change `AllY`: In the `rep` function divide `Skates$RA` by `Skates$SweptArea`, add 1, and take the log.

```
> MyX  <- c("CumDate", "DayNumber", "X", "Y", "Temperature")
> AllX <- as.vector(as.matrix(Skates[, MyX]))
> ID   <- rep(MyX, each = nrow(Skates))
> AllY <- rep(Skates$RA, length(MyX))
> library(lattice)
> xyplot(AllY ~ AllX|ID, col = 1, xlab = "Explanatory variables",
    ylab = "Number of fish",
    scales = list(alternating = T, x = list(relation = "free"),
                  y = list(relation = "same")),
    panel=function(x,y){
      panel.grid(h = -1, v = 2)
      panel.points(x, y, col = 1)
      panel.loess(x,  y, col = 1, lwd = 2)})
```

6.3.4 Geographical position of the sites

Figure 6.5A shows the positions of the sites. For presentations it may be an option to add the coastline. There is no land between any two sampling locations. The graph is created with the following R code:

```
> xyplot(Y ~ X, aspect = "iso", col = 1, pch = 16, data = Skates)
```

Putting information on the counts into this graph is also an option (Figure 6.5B). Note that the higher counts were taken in the northeast part of the study area. It may be an option to create a variogram of the species data, but they are zero inflated, making interpretation of the sample variogram difficult.

Figure 6.5B follows. The MyCex variable defines the size of a dot in the graph. The square root and multiplication factor of 4 is based on trial and error and was chosen such that the difference between low and high abundance is visually as clear as possible.

```
> MyCex <- sqrt(4 * Skates$RA / max(Skates$RA))
> xyplot(Y ~ X, aspect = "iso", pch = 16, cex = MyCex, col = 1,
      xlab = "X-coordinates", ylab = "Y-coordinates", data = Skates)
```

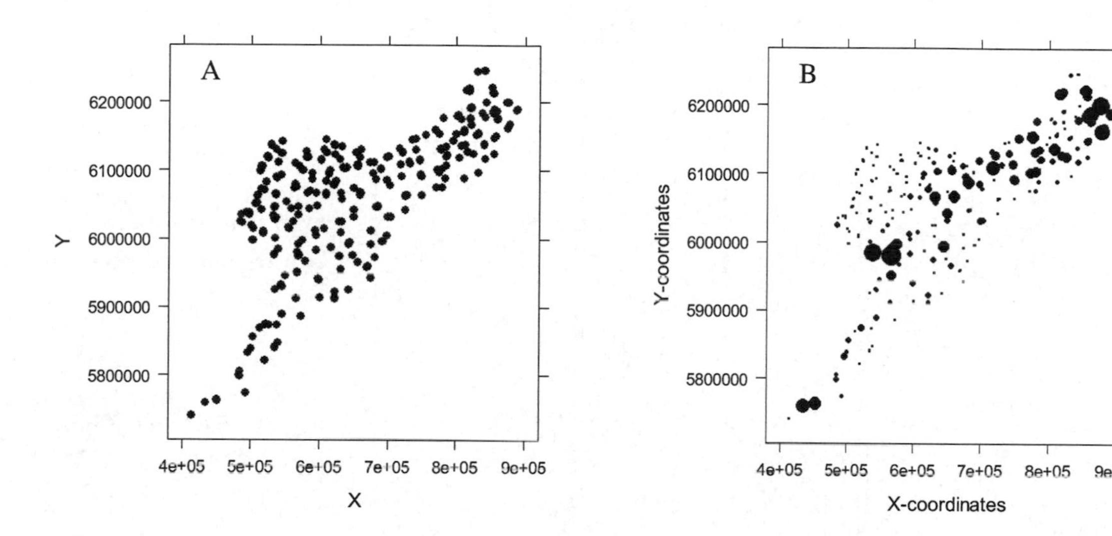

Figure 6.5. A: Position of the sites. B: Position of the sites. The size of a dot is proportional to the number of observed RA.

6.3.5 Zero inflation in the species data

Figure 6.6 contains Cleveland dotplots of the three species data. We presented these in Figure 6.1, and we plot them again, this time adding the scale along the horizontal axes. The RA counts range from 0 to 23, AC from 0 to 11, and SB from 0 to 190. Based on the range and variance of the data, our initial impression is that we may need a Poisson distribution for AC, either a Poisson or negative binomial distribution for RA, and a negative binomial distribution for SB.

The RA and AC variables are clearly zero inflated, as can be seen from the vertical band of observations of 0. We can also make frequency plots to get an impression of the zero inflation (Figure 6.7). It seems that all species are zero inflated. The percent of observations equal to 0 for each species is 60.54%, 74.49%, and 47.89% for RA, AC, and SB, respectively. We have 332 observations.

Why do we have so many zeros? In the marmot chapter and seal scat chapter we explained that an excessive number of zeros may be due to such things as the sampling period being too short, observer error, inadequate habitat conditions, etc. We made a distinction between true zeros and false zeros. We argue that we also have both true and false zeros in the skate data. A count of 0 may occur because an animal is not present due to unsuitable habitat (true zero), or perhaps it is present but was not counted (false zero) because it was not caught by the net. Hence if we apply zero inflated models, we should use mixture models such as zero inflated Poisson (ZIP) models and zero inflated negative binomial models (ZINB).

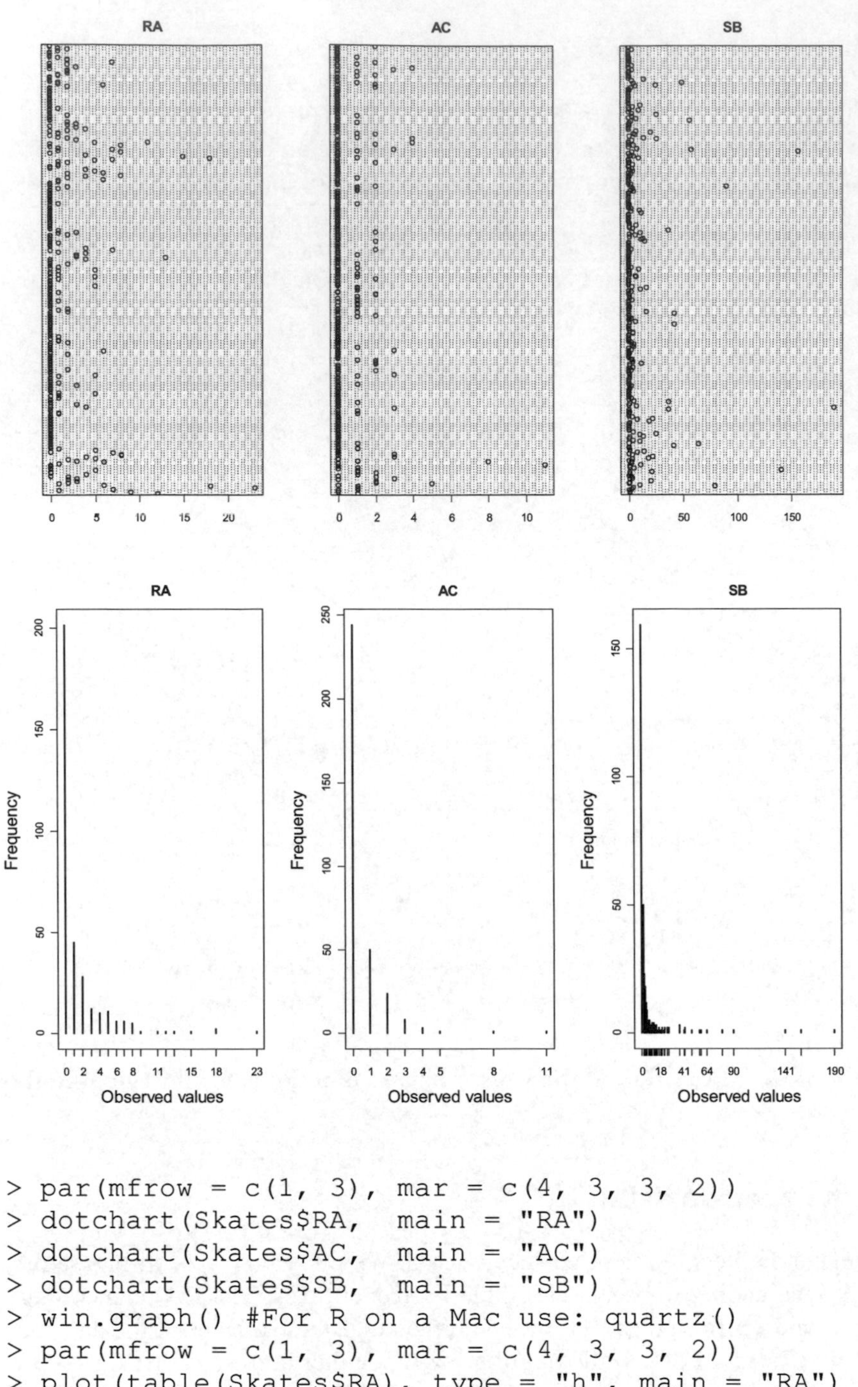

Figure 6.6. Cleveland dotplots for counts of the three species. The horizontal axis shows the observed value and the vertical axis the order of the data as imported from Excel.

Figure 6.7. Frequency plot for each species. Note that all species counts may be zero inflated.

R code to produce Figure 6.6 and Figure 6.7 is simple:

```
> par(mfrow = c(1, 3), mar = c(4, 3, 3, 2))
> dotchart(Skates$RA,  main = "RA")
> dotchart(Skates$AC,  main = "AC")
> dotchart(Skates$SB,  main = "SB")
> win.graph() #For R on a Mac use: quartz()
> par(mfrow = c(1, 3), mar = c(4, 3, 3, 2))
> plot(table(Skates$RA), type = "h", main = "RA")
> plot(table(Skates$AC), type = "h", main = "AC")
> plot(table(Skates$SB), type = "h", main = "SB")
```

6.3.6 What makes this a difficult analysis?

We have zero inflation and, potentially, spatial correlation and temporal correlation. We will begin by applying Poisson and negative binomial GLMs and also ZIP and ZINB models. If the residuals of these models are spatially or temporally correlated, we will need a Bayesian approach and must implement the models in WinBUGS (See also Chapter 7). It will be a challenge!

6.4 Zero inflated GLM applied to *R. agassizi* data

6.4.1 What is the starting point?

When modelling the counts of *R. agassizi*, we are faced with a series of questions: Which distribution do we need? Do we need a Poisson or negative binomial distribution? Why might a negative binomial distribution be needed? Should we include interactions? Are relationships linear or non-linear (i.e. do we need a GLM or a GAM)?

For the distribution, we will start with Poisson GLM and check for overdispersion. If there is overdispersion, we will inspect the residuals, and, if there are no clear patterns or outliers, we will apply a negative binomial GLM. If the negative binomial GLM is better than the Poisson GLM, as judged by a likelihood ratio test, we will try to determine why. Do we really need a negative binomial distribution, or is a zero inflated Poisson an option? In our opinion it is important to understand what is driving the increased overdispersion that is modelled by the negative binomial GLM. After all, a zero inflated Poisson GLM may be considered a simpler model than a negative binomial GLM.

We do not have *a priori* biological knowledge regarding potential interactions, so we will only use models with main terms and refrain from using interaction terms.

The question of whether we should use linear or non-linear relationships is misleading. The log link function in a GLM already imposes a non-linear relationship between the expected value of the species count and the covariates. What we need to ask ourselves is whether we need anything more complicated than parametric terms for the continuous covariates in the predictor function. Do we need quadratic functions, higher order polynomials, or smoothing functions? We will begin the analysis with a GLM and, if a plot of the Pearson residuals versus temperature shows any clear pattern, we will continue with a GAM.

We first model the counts of RA as a function of year, temperature, and bottom type. The log of the swept area ($LogSA_i$) is used as an offset variable. The mathematical formulation is:

$$
\begin{aligned}
& RA_i \sim Poisson(\mu_i) \\
& E(RA_i) = \text{var}(RA_i) = \mu_i \\
& \log(\mu_i) = \alpha + fYear_i + Temperature_i + fBottomType_i + offset(LogSA_i)
\end{aligned}
$$

RA_i represents the counts of *R. agassizi* for observation *i*. We simplified the notation for the predictor function by omitting the regression parameters before the covariate names. `fYear` and `fBottomType` are factors with 4 levels. Thus, in principle, there is one intercept and 7 regression parameters for the covariates (3 for `Year`, 1 for `Temperature`, and 3 for `fBottomType`), but it would be cumbersome to write this all out.

6.4.2 Poisson GLM applied to *R. agassizi* data

The R code to fit the Poisson GLM and calculate the dispersion is:

```
> Skates$LogSA <- log(Skates$SweptArea)
> M1 <- glm(RA ~ fYear + Temperature + fBottomType + offset(LogSA),
            family = poisson, data = Skates)
> E1 <- resid(M1, type = "pearson")
> Dispersion <- sum(E1^2) / M1$df.resid
> Dispersion

3.235325
```

Note that this model is overdispersed; thus it is not a good model. The residuals did not contain either outliers or residual patterns; therefore we apply a negative binomial GLM.

6.4.3 NB GLM applied to *R. agassizi* data

The negative binomial GLM is given by:

$$RA_i \sim NB(\mu_i, k)$$
$$E(RA_i) = \mu_i \quad \text{and} \quad \text{var}(RA_i) = \mu_i + \frac{\mu_i^2}{k} = \mu_i + \alpha \times \mu_i^2$$
$$\log(\mu_i) = \alpha + fYear_i + Temperature_i + fBottomType_i + offset(LogSA_i)$$

See Chapter 3 for a discussion on the notation for $\alpha = 1/k$. The chief difference of the Poisson GLM is the quadratic variance in the NB GLM. The NB GLM is fitted with the R code:

```
> library(MASS)
> M2 <- glm.nb(RA ~ fYear + Temperature + fBottomType +
                 offset(LogSA), data = Skates)
> E2 <- resid(M2, type = "pearson")
> Dispersion <- sum(E2^2) / M2$df.resid
> Dispersion

1.105699
```

This model is not overdispersed and is a good candidate model. Further numerical output is obtained by the `summary` command:

```
> summary(M2)

Coefficients:
                             Estimate Std. Error z value Pr(>|z|)
(Intercept)                  11.31425    0.86142  13.134  < 2e-16
fYear1999                    -2.48627    0.33561  -7.408 1.28e-13
fYear2003                    -1.60229    0.30659  -5.226 1.73e-07
fYear2005                    -1.23249    0.30029  -4.104 4.05e-05
Temperature                  -0.37448    0.04718  -7.937 2.07e-15
fBottomTypeSand & mud         0.80387    0.26835   2.996  0.00274
fBottomTypeSand               0.65271    0.28252   2.310  0.02087
fBottomTypeSand, shell, rest  0.78189    0.32568   2.401  0.01636

Dispersion parameter for Negative Binomial(0.5906) family taken to
be 1)
    Null deviance: 394.45  on 331  degrees of freedom
Residual deviance: 272.16  on 324  degrees of freedom
AIC: 909.09
Theta:  0.5906   Std. Err.: 0.0948    2 x log-likelihood:  -891.0860
```

`Temperature` is highly significant at the 5% level. A problem arises with the covariates `fYear` and `fBottomType` in that we have multiple *p*-values due to the individual levels. Information on the significance of these covariates can be obtained by fitting a full model and a series of nested models in which each term is dropped in turn. This can be done with the `drop1` function:

```
> drop1(M2, test = "Chi")

Single term deletions
Model: RA ~ fYear + Temperature + fBottomType + offset(LogSA)
```

```
              Df Deviance     AIC    LRT    Pr(Chi)
<none>            272.16 907.09
fYear          3  344.05 972.98 71.892 1.679e-15
Temperature    1  327.96 960.88 55.799 8.026e-14
fBottomType    3  281.86 910.79  9.705   0.02125
```

This shows a significant year effect, a significant temperature effect, and a weakly significant bottom type effect. Because *p*-values obtained by the likelihood ratio test are approximate, we refrain from stating that there is a significant bottom type effect. In applying the likelihood ratio tests, the `drop1` function keeps the parameter *k* in each sub-model fixed. This means that it will use the *k* from the full model in each of the three sub-models.

We are curious as to why the NB GLM is needed. The estimated value of *k* is 0.59 and the majority of the expected values μ_i are between 0 and 5; thus the term μ_i^2/k catches a considerable amount of the overdispersion. Is this because of non-linear effects, outliers in the response variable, correlation, zero inflation, or is the data just noisy? A plot of the Pearson residuals versus temperature does not show serious model misspecification, so there is no need to use a GAM. This also indicates that overdispersion is not due to a model misspecification of the covariates. There is also no evidence of outliers shown by the residual graphs. Based on the frequency plots in Figure 6.7 we suspect that it is the excessive number of zeros driving the overdispersion and the need for the NB GLM. We therefore apply zero inflated Poisson (ZIP) and zero inflated negative binomial (ZINB) GLMs and will see whether these models perform better.

6.4.4 Zero inflated Poisson GLM applied to R. agassizi data

The underlying theory for ZIP and ZINB GLM and GAM models was explained in Chapters 2 and 3, and is not repeated here. The ZIP GLM for *R. agassizi* is given by:

$$
\begin{aligned}
& RA_i \sim ZIP(\mu_i, \pi_i) \\
& E(RA_i) = \mu_i \times (1-\pi_i) \quad \text{and} \quad \text{var}(RA_i) = (1-\pi_i) \times (\mu_i + \pi_i \times \mu_I^2) \\
& \log(\mu_i) = \alpha_1 + fYear_i + Temperature_i + fBottomType_i + offset(LogSA_i) \\
& \text{logit}(\mu_i) = \alpha_2 + fYear_i + Temperature_i + fBottomType_i
\end{aligned}
$$

where π_i is the probability that observation RA_i is a false zero. The R code to fit the ZIP was discussed in the same two chapters and is:

```
> library(pscl)
> M3 <- zeroinfl(RA ~ fYear + Temperature + fBottomType +
                 offset(LogSA) | fYear + Temperature + fBottomType,
                 dist = "poisson", link = "logit", data = Skates)
> N  <- nrow(Skates)
> E3 <- resid(M3, type = "pearson")
> Dispersion <- sum(E3^2) / (N - 16)
> Dispersion

1.600723
```

The ZIP is slightly overdispersed. Its numerical output is:

```
Count model coefficients (poisson with log link):
                            Estimate Std. Error z value Pr(>|z|)
(Intercept)                 8.181886   0.628764  13.013  < 2e-16
fYear1999                  -1.558415   0.226738  -6.873 6.28e-12
fYear2003                  -1.009945   0.172956  -5.839 5.24e-09
```

```
fYear2005                          -0.825495   0.154425  -5.346 9.01e-08
Temperature                        -0.125252   0.040191  -3.116  0.00183
fBottomTypeSand & mud               0.370809   0.146936   2.524  0.01162
fBottomTypeSand                    -0.006113   0.191851  -0.032  0.97458
fBottomTypeSand, shell, rest        0.218000   0.193924   1.124  0.26095

Zero-inflation model coefficients (binomial with logit link):
                                 Estimate Std. Error z value Pr(>|z|)
(Intercept)                      -6.13316    1.48686  -4.125 3.71e-05
fYear1999                         2.05308    0.51395   3.995 6.48e-05
fYear2003                         1.13018    0.45918   2.461  0.01384
fYear2005                         0.48926    0.47437   1.031  0.30235
Temperature                       0.39265    0.08332   4.713 2.44e-06
fBottomTypeSand & mud            -0.98260    0.39567  -2.483  0.01301
fBottomTypeSand                  -1.16691    0.43887  -2.659  0.00784
fBottomTypeSand, shell, rest -1.43253    0.52971  -2.704  0.00684

Number of iterations in BFGS optimization: 22
Log-likelihood:  -486 on 16 Df
```

Because we obtained multiple *p*-values for the categorical variables, we apply a likelihood ratio test to measure the significance of each covariate in the ZIP GLM. This means that we must fit 6 sub-models. In each sub-model, we need to drop a term and apply a likelihood ratio test to compare the full model and the nested model. The coding is necessarily extensive:

```
#Drop fYear from the count model
> M3A <- zeroinfl(RA ~ Temperature + fBottomType +offset(LogSA) |
                       fYear + Temperature + fBottomType,
                  dist = "poisson", link = "logit", data = Skates)
#Drop Temperature from the count model
> M3B <- zeroinfl(RA ~ fYear + fBottomType +offset(LogSA) |
                       fYear + Temperature + fBottomType,
                  dist = "poisson", link = "logit", data = Skates)
#Drop fBottomType from the count model
> M3C <- zeroinfl(RA ~ fYear + Temperature + offset(LogSA) |
                       fYear + Temperature + fBottomType,
              dist = "poisson", link = "logit", data = Skates)
#Drop fYear from the binary model
> M3D <- zeroinfl(RA ~ fYear + Temperature + fBottomType +
                       offset(LogSA) | Temperature + fBottomType,
                  dist = "poisson", link = "logit", data = Skates)
#Drop Temperature from the binary model
> M3E <- zeroinfl(RA ~ fYear + Temperature + fBottomType +
                       offset(LogSA) | fYear +  fBottomType,
                  dist = "poisson", link = "logit", data = Skates)
#Drop fBottomType from the binary model
> M3F <- zeroinfl(RA ~ fYear + Temperature + fBottomType +
                       offset(LogSA) | fYear +  Temperature,
                  dist = "poisson", link = "logit", data = Skates)
```

We can write a function for the likelihood ratio test to condense the coding. The input for the function is the log-likelihood of the full and nested model and the degrees of freedom for the test. The output is the likelihood ratio statistic and corresponding *p*-value.

```
LikRatioTest <- function(L1, L2, df) {
 #L1 is full model, L2 is nested model
 #df is difference in parameters
 L <- 2 * (L1 - L2)
 L <- abs(as.numeric(L))
 p <- 1 - pchisq(L, df)
 round(c(L, p), digits = 4) }
```

We run the likelihood ratio test for each of the six sub-models to get an approximate p-value for each term in the model. Note that the covariates `fYear` and `fBottomtype` each consume 3 degrees of freedom.

```
> Z3A <- LikRatioTest(logLik(M3), logLik(M3A), 3)
> Z3B <- LikRatioTest(logLik(M3), logLik(M3B), 1)
> Z3C <- LikRatioTest(logLik(M3), logLik(M3C), 3)
> Z3D <- LikRatioTest(logLik(M3), logLik(M3D), 3)
> Z3E <- LikRatioTest(logLik(M3), logLik(M3E), 1)
> Z3F <- LikRatioTest(logLik(M3), logLik(M3F), 3)
```

The remainder of the code combines the output so that the results can be compared.

```
> Z <- rbind(Z3A, Z3B, Z3C, Z3D, Z3E, Z3F)
> rownames(Z) <- c("fYear in log link", "Temperature in log link",
                   "fBottomType in log link",
                   "fYear in logistic link",
                   "Temperature in logistic link",
                   "fBottomType in logistic link")
> colnames(Z) <- c("L", "p-value")
> round(Z,3)
                                    L    p-value
fYear in log link              75.709      0.000
Temperature in log link        10.458      0.001
fBottomType in log link        10.819      0.013
fYear in logistic link         14.669      0.002
Temperature in logistic link   25.355      0.000
fBottomType in logistic link   10.993      0.012
```

Note that `fBottomType` is only marginally significant. Due to the approximate nature of this test we consider it weakly significant at best. So what is the model telling us? To answer this question we examine the estimated parameters, or, better, make a graph of the fitted values (Figure 6.8). The left panel shows the fitted values for the count part of the ZIP and the right panel the probabilities of false zeros. According to the graphs for the count model, lower temperature means higher abundance of *R. agassizi*. The four lines in each panel represent the four bottom types. The lines for mud and sand are superimposed. The highest curve is always for sand and mud and the lowest curves are for mud and sand. Highest abundance was in 1998. The right panel shows that high temperature also means a higher chance of false zeros. The bottom type `mud` (top curve in each panel) has the highest probability of false zeros, but keep in mind that bottom type was only weakly significant.

Another way of presenting the results of the model is to plot year along the horizontal axis and the predicted counts along the vertical axis. Predictions can be made for discrete temperature values and the four bottom types. Such a graph puts more emphasis on changes over time. However, to obtain standard errors around the predicted values it may be better to fit the model in WinBUGS and do these predictions inside the simulation process. Generated draws can then be used to obtain a 95% point-wise confidence interval for the predicted values. (See also Chapter 9).

The R code to make Figure 6.8 follows. First we create a data frame `NewData` containing covariate values for which we want to make predictions. We will make predictions for all years, bottom types, and temperature (50 values between 9.9 and 21.75 degrees Celsius). We will use the average `LogSA` value for the offset. Predicted counts and probabilities for these covariate values are obtained with the `predict` function, and results are plotted with the `xyplot` function from the `lattice` package. The `groups` argument is used to put multiple lines in the same panel.

```
> NewData <- expand.grid(
                  fYear        = c("1998", "1999", "2003", "2005"),
                  Temperature = seq(9.9, 21.75, length = 50),
                  fBottomType = levels(Skates$fBottomType),
                  LogSA        = mean(Skates$LogSA))
> NewData$PCount <- predict(M3, newdata= NewData, type = "count" )
> NewData$PZero  <- predict(M3, newdata= NewData, type = "zero" )
> head(NewData,15)  #Results not shown here
> xyplot(PCount ~ Temperature | fYear, data = NewData,
        xlab = "Temperature", ylab = "Predicted counts", col =1,
        groups = fBottomType, layout = c(2,2), type = "l", lwd = 2)
> xyplot(PZero ~ Temperature | fYear,
      xlab = "Temperature", ylab = "Predicted probabilities",
      groups = fBottomType, layout = c(2,2),
        data = NewData, type = "l", lwd = 2, col =1)
```

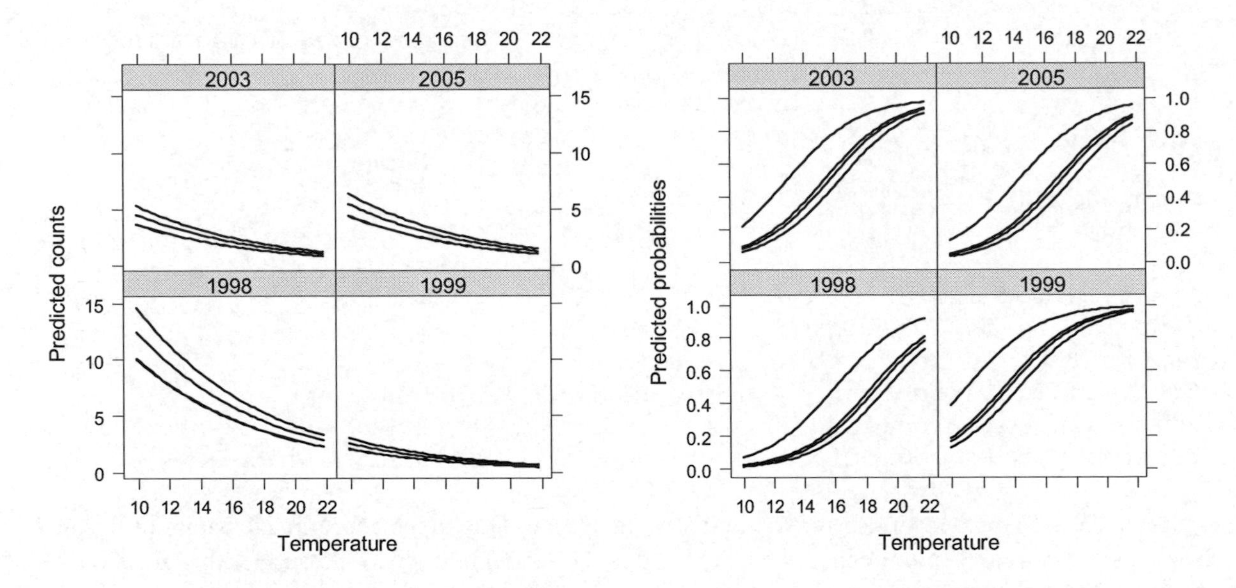

Figure 6.8. Left: Predicted values based on the count part of the ZIP model for *R. agassizi*. Right: Predicted probabilities of false zeros based on the binary part of the ZIP model applied on the *R. agassizi* data.

6.4.5 Zero inflated negative binomial GLM applied to *R. agassizi* data

The mathematical equation for the ZINB GLM is:

$$RA_i \sim ZINB(\mu_i, \pi_i, k)$$

$$E(RA_i) = \mu_i \times (1-\pi_i) \quad \text{and} \quad \text{var}(RA_i) = (1-\pi_i) \times \mu_i \times (1+\pi_i \times \mu_i + \frac{\mu_i^2}{k})$$

$$\log(\mu_i) = \alpha_1 + fYear_i + Temperature_i + fBottomType_i + offset(LogSA_i)$$

$$\text{logit}(\mu_i) = \alpha_2 + fYear_i + Temperature_i + fBottomType_i$$

The model is fitted with the R code:

```
> M4 <- zeroinfl(RA ~ fYear + Temperature + fBottomType +
                offset(LogSA) | fYear + Temperature + fBottomType,
                dist = "negbin", link = "logit", data = Skates)
> N  <- nrow(Skates)
> E4 <- resid(M4, type = "pearson")
> Dispersion <- sum(E4^2) / (N - 17)
> Dispersion

0.92076
```

The model is slightly underdispersed. The ZIP is nested in the ZINB, and comparison of these two models with a likelihood ratio test shows that the ZINB is significantly better (L = 119.82, df = 1, p < 0.001). The R code for this is:

```
> L    <- 2 * abs(as.numeric(logLik(M3) - logLik(M4)))
> Pval <- 1 - pchisq(L, 1)
> c(L, Pval)

119.8232   0.0000
```

So we should be focusing on the results of the ZINB GLM. The `summary` output for the ZINB GLM is:

```
> summary(M4)

Count model coefficients (negbin with log link):
                              Estimate Std. Error z value Pr(>|z|)
(Intercept)                    7.89489    1.11975   7.051 1.78e-12
fYear1999                     -2.16954    0.33802  -6.418 1.38e-10
fYear2003                     -1.56283    0.31492  -4.963 6.95e-07
fYear2005                     -0.92158    0.32378  -2.846  0.00442
Temperature                   -0.13507    0.06669  -2.025  0.04282
fBottomTypeSand & mud          0.71334    0.26628   2.679  0.00739
fBottomTypeSand                0.22955    0.30093   0.763  0.44557
fBottomTypeSand, shell, rest   0.39847    0.31030   1.284  0.19909
Log(theta)                    -0.25375    0.17361  -1.462  0.14383

Zero-inflation model coefficients (binomial with logit link):
                              Estimate Std. Error z value Pr(>|z|)
(Intercept)                   -56.6476    25.0318  -2.263   0.0236
fYear1999                       8.6284     4.1990   2.055   0.0399
fYear2003                       1.8572     2.0544   0.904   0.3660
fYear2005                       4.5597     2.7233   1.674   0.0941
Temperature                     3.0085     1.2964   2.321   0.0203
fBottomTypeSand & mud          -0.6741     1.4392  -0.468   0.6395
fBottomTypeSand                -1.9553     1.8848  -1.037   0.2995
fBottomTypeSand, shell, rest  -14.7846   798.8793  -0.019   0.9852
Theta = 0.7759
Number of iterations in BFGS optimization: 61
Log-likelihood: -426.1 on 17 Df
```

As with the ZIP Poisson GLM, we will need the likelihood ratio test to assess the significance of `fYear` and `fBottomType` based on the individual p-values.

If no regression parameter for the covariates in the logistic link function in the ZINB is significantly different from 0, and the intercept is a large negative number, π_i is close to 0 and the ZINB converts to a NB GLM. There are multiple means of investigating whether this is the case.

The complicated method is to apply a stepwise backward selection and drop all non-significant covariates. The disadvantages of model selection approaches have been discussed elsewhere in this book, and such a backward selection would convert our analysis into a data phishing exercise (or, alternatively, an 'exploratory analysis'). We will avoid this as far as is possible.

A simpler approach is to fit two ZINB GLMs. The first is `M4`, above, and the second uses only an intercept in the logistic link function (although the count part is the same as for `M4`). These two models can then be compared using a likelihood ratio test. If the models give approximately the same fit, and if the intercept of the sub-model is highly negative, we should use the NB GLM. The R code to fit the second model is:

```
> M4B <- zeroinfl(RA ~ fYear + Temperature + fBottomType +
                  offset(LogSA) | 1,
                  dist = "negbin", link = "logit", data = Skates)
```

The part of the model formula following ~ and preceding the | symbol are the covariates for the count part of the model, and the 1 after the | symbol means that only an intercept is used in the logistic link function.

The value of the log likelihood of `M4` is -426.10 and that of `M4B` is -445.54. The difference in parameters of the two models is 7; thus the likelihood ratio test indicates that some of the regression parameters in the logistic link function are significantly different from 0 at the 5% level ($L = 38.87$, $df = 7$, $p < 0.001$). Here is the required R code for the likelihood ratio test:

```
> L    <- 2 * abs(as.numeric(logLik(M4B) - logLik(M4)))
> Pval <- 1 - pchisq(L, 7)
> round(c(L, Pval), digits = 3)

38.875  0.000
```

The problem with this likelihood ratio test is that we should keep the parameter k in the sub-model `M4B` fixed. That is, k in the sub-model `M4B` should be the same k as in the full model `M4`. At present we have two different ks:

```
> c(M4$theta, M4B$theta)

0.7758826 0.6011039
```

The package `pscl` does not have a `drop1` function, nor does it easily allow for fixing the parameter k (called `theta` in the function `zeroinfl`) to a preset value. So, we need to determine how to fit a ZINB with the function `zeroinfl` and fix the dispersion parameter k. We also need this procedure to obtain a p-value for each term in the model. To use a fixed k in the `zeroinfl` function, we copy the code in `zeroinfl` to a new function called `Myzeroinfl` and reduce the number of parameters to be estimated by the optimization routine `optim` (called by `zeroinfl`) by 1. We drop the k estimation and set it to the fixed value from the full model. The file `Myzeroinfl` is available from the book website (www.highstat.com/book4.htm). The new function allows us to repeat the sub-model, but now with fixed a k value.

```
> M4C.fixed <- Myzeroinfl(RA ~ fYear + Temperature + fBottomType +
                 offset(LogSA) | 1 ,
                 dist = "negbin", link = "logit", data = Skates,
                 fixed.log.theta = log(M4$theta),
                 control = zeroinfl.control(trace = 0))
```

```
> L    <- 2 * abs(as.numeric(logLik(M4) - logLik(M4C.fixed)))
> Pval <- 1 - pchisq(L, 1)
> Out  <- c(L, Pval)
> round(Out, digits = 3)

39.227  0.000
```

Summarizing, we have refitted model M4B, ensuring that the dispersion parameter *k* from M4 was used and kept fixed in the estimation process. Results are stored in the object M4C, and the value of the likelihood ratio test statistic is slightly different from the first test (M4 versus M4B), although the conclusion is the same. Hence we need the intercept and perhaps some of the covariates in the logistic link function in the ZINB, and we should not use a NB GLM.

We can use the Myzeroinfl function to get a *p*-value for each term in the ZINB model. We drop each term in turn and apply the likelihood ratio test. As for the ZIP GLM we need to fit 6 sub-models and apply the likelihood ratio test.

```
> MyTheta <- M4$theta
#Drop fYear from the count model
> M4A <- Myzeroinfl(RA ~ Temperature + fBottomType + offset(LogSA) |
                         fYear + Temperature + fBottomType,
                    dist = "negbin", link = "logit", data = Skates,
                    fixed.log.theta = log(MyTheta))
#Drop Temperature from the count model
> M4B <- Myzeroinfl(RA ~ fYear  + fBottomType + offset(LogSA) |
                         fYear + Temperature + fBottomType,
                    dist = "negbin", link = "logit", data = Skates,
                    fixed.log.theta = log(MyTheta))
#Drop fBottomType from the count model
> my.control <- zeroinfl.control(trace = TRUE, method = "BFGS",
                                 EM = FALSE, reltol = 1e-5)
> M4C <- Myzeroinfl(RA ~ fYear + Temperature  + offset(LogSA) |
                         fYear + Temperature + fBottomType,
                    dist = "negbin", link = "logit", data = Skates,
                    fixed.log.theta = log(MyTheta),
                    control = my.control) #Numerical trouble
#Drop fYear from the binary model
> M4D <- Myzeroinfl(RA ~ fYear + Temperature + fBottomType +
                         offset(LogSA) | Temperature + fBottomType,
                    dist = "negbin", link = "logit", data = Skates,
                    fixed.log.theta = log(MyTheta))
#Drop Temperature from the binary model
> M4E <- Myzeroinfl(RA ~ fYear + Temperature + fBottomType +
                         offset(LogSA) | fYear +  fBottomType,
                    dist = "negbin", link = "logit", data = Skates,
                    fixed.log.theta = log(MyTheta))
#Drop fBottomType from the binary model
> M4F <- Myzeroinfl(RA ~ fYear + Temperature + fBottomType +
                         offset(LogSA) | fYear + Temperature,
                    dist = "negbin", link = "logit", data = Skates,
                    fixed.log.theta = log(MyTheta))
```

Applying the likelihood ratio test gives:

```
> Z4A <- LikRatioTest (logLik(M4), logLik(M4A), 3)
> Z4B <- LikRatioTest (logLik(M4), logLik(M4B), 1)
```

```
> Z4C <- LikRatioTest (logLik(M4), logLik(M4C), 3)
> Z4D <- LikRatioTest (logLik(M4), logLik(M4D), 3)
> Z4E <- LikRatioTest (logLik(M4), logLik(M4E), 1)
> Z4F <- LikRatioTest (logLik(M4), logLik(M4F), 3)
> Z <- rbind(Z4A, Z4B, Z4C, Z4D, Z4E, Z4F)
> rownames(Z) <- c("fYear in log link",
        "Temperature in log link", "fBottomType in log link"
        "fYear in logistic link",  "Temperature in logistic link",
        "fBottomType in logistic link")
> colnames(Z) <- c("L", "p-value")
> round(Z,3)
```

```
                                       L p-value
fYear in log link                 42.503   0.000
Temperature in log link          106.402   0.000
fBottomType in log link            7.311   0.063
fYear in logistic link             6.782   0.079
Temperature in logistic link      23.768   0.000
fBottomType in logistic link       2.696   0.441
```

This is basically the same process that the `drop1` function does for a linear regression model. The results of the likelihood ratio test show that `fBottomType` and `fYear` are not significant at the 5% level in the binary model, and `fBottomType` is weakly significant in the log link function. A model without `fBottomType` in the log link function showed numerical convergence problems and we changed the convergence criteria for this specific model. Alternatively the software can be modified so that it does not calculate a Hessian matrix, which is used for calculating standard errors.

The negative binomial zero inflated GLM is slightly underdispersed:

```
> p <- length(coef(M4)) + 1
> E4 <- resid(M4, type = "pearson")
> Dispersion <- sum(E4^2) / (N - p)
> Dispersion
```

```
0.92076
```

The term p is the number of regression parameters plus 1 (for the parameter k in the variance term). In principle, we can present this model, provided there are no residual patterns. The predicted counts and probabilities of false zeros are given in Figure 6.9. The graphs for the count model are nearly identical to those for the ZIP GLM. The binary part looks slightly different. It seems that the probability of false zeros increases rapidly at temperatures above 19 degrees Celsius.

Both the ZINB and ZIP are candidate models for presentation. The ZIP is slightly overdispersed and the ZINB is slightly underdispersed. Both models indicate that `fBottomType` is the least important covariate. It is difficult to choose between them, and techniques to compare ZIP and ZINB GLMs that were presented in Chapter 11 in Zuur et al. (2009a) could be applied.

We apply a model validation on the ZIP model as it is conceptually easier than on the ZINB GLM. We are also slightly suspicious of the shape of the logistic curves for the ZINB, as they are much steeper than in Figure 6.8.

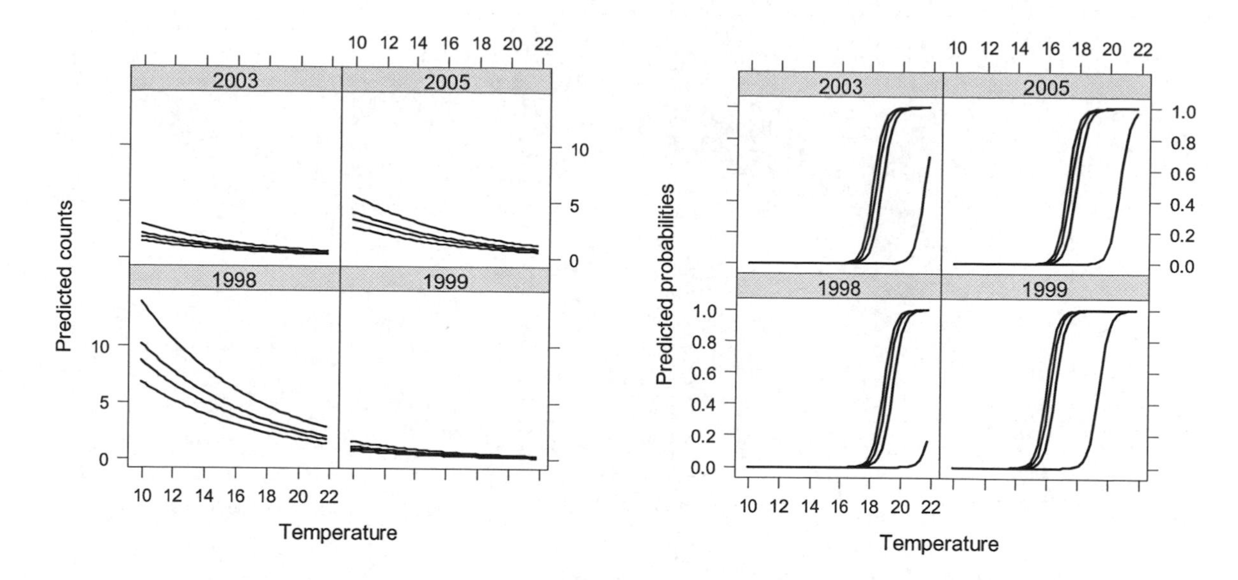

Figure 6.9. Left: Predicted values based on the count part of the ZINB model for *R. agassizi*. Right: Predicted probabilities of false zeros based on the binary part of the ZINB model applied to the *R. agassizi* data.

6.4.6 Model validation of the ZIP GLM applied to *R. agassizi* data

As a component of model validation, the Pearson residuals are plotted versus each covariate in the model as well as versus each covariate not included in the model. Results are not presented here, but none of these graphs indicate problems, so there is no need to use a more complicated model (e.g. by adding quadratic terms or smoothers).

The Pearson residuals should also be examined for spatial and temporal independence. The first step in this process is to plot the Pearson residuals versus spatial coordinates (Figure 6.10A). The size of the dots is proportional to the value of the Pearson residuals, and the colour indicates positive or negative residuals. This figure should look like the night sky; no patterns should be visible. Whether there are patterns present in Figure 6.10A is a subjective judgement; there may be a clustering of large residuals in the upper right area. A more objective tool is the sample variogram of the Pearson residuals (Figure 6.10B). Skates can swim up to 20 km per day, so we may expect correlation up to this distance. Correlation may also be due to excluded covariates acting on small or large scale distances. We therefore limit the distances to less than 50 km in the sample variogram. There is no clear indication of spatial correlation. The same conclusion is obtained when the sample variogram is applied to the Pearson residuals of the ZINB.

The variogram aggregates all combinations of sites, regardless of direction. If there are sea currents in a particular direction, perhaps the variogram of all site combinations in an east-west orientation will have a different shape than those of site combinations in a north-south orientation. This is called anisotropy. To investigate this, the variogram function can be instructed to calculate variograms per direction. These variograms are not presented here, but show no evidence of anisotropy.

The R code to produce Figure 6.10A follows. Pearson residuals are extracted, and the vector `MyCol` consists of black for negative residuals and grey for positive residuals. The `MyCex` vector defines the size of a dot and is proportional to the residuals. The selection of the square root and multiplication by 6 were obtained by trial and error. We chose values that resulted in a visually pleasing graph.

```
> E3 <- resid(M3, type = "pearson")
> MyCol <- vector(length = length(E3))
> MyCol[E3 <= 0] <- grey(0.1)
```

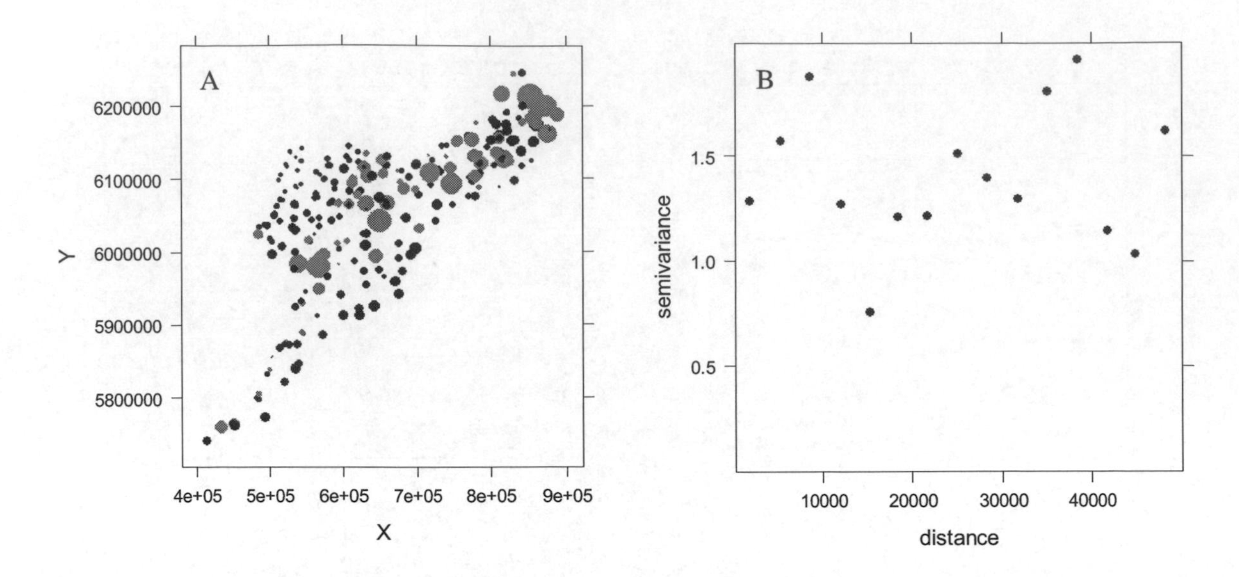

Figure 6.10. A: Plot of the Pearson residuals obtained from the ZIP GLM applied to the *R. agassizi* data. Dot size is proportional to the value of the residuals. Black dots represent negative residuals and grey dots positive residuals. B: Semi-variogram for the residuals of the ZIP GLM model applied to the *R. agassizi* data. Spatial coordinates are used to calculate the semi-variogram. The units along the horizontal axis are metres, and the axis represents 50 km.

```
> MyCol[E3  > 0] <- grey(0.5)
> MyCex <- 6 * abs(E3)/max(abs(E3))
> xyplot(Y ~ X, aspect = "iso", data = Skates, col = MyCol,
         pch = 16, cex = sqrt(MyCex))
```

For the sample variogram in Figure 6.10B, we use the `gstat` package.

```
> library(gstat)
> mydata <- data.frame(E8, Skates$X, Skates$Y)
> coordinates(mydata) <- c("Skates.X", "Skates.Y")
> Vario1 <- variogram(E8 ~ 1, mydata, cutoff = 50000)
> plot(Vario1, pch = 16, col =1)
```

The remaining question is whether there is temporal correlation in the residuals. Surveys took place in November or December of each year, and there may be correlation among observations recorded on different days within the same year. The following code produces the sample variogram in Figure 6.11. We will explain the graph with help of the R code. First we make a vector of ones:

```
> Skates$MyOnes <- rep(1, length(Skates$Y))
```

There are 332 observations in the data; thus `MyOnes` is a vector of length 332, and it only contains the value 1. Any value would have worked. The variable `CumDate` represents the survey day counted from 1 January 1998. The `variogram` function expects two variables with spatial coordinates; thus our dummy variable `MyOnes`. Because it has a consistent value, it does not influence the results of the sample variogram, which uses the variables `CumDate` and `MyOnes` as 'spatial' coordinates. Therefore, differences, expressed in days, among the survey dates are used as distances by the `variogram` function, and the cut-off level is chosen such that the first 10 days are used for all calculations.

```
> mydata <- data.frame(E3, Skates$CumDate, Skates$MyOnes)
> coordinates(mydata) <- c("Skates.CumDate", "Skates.MyOnes")
> Vario1 <- variogram(E3 ~ 1, mydata, cutoff = 10)
> plot(Vario1, pch = 16, col =1)
```

The shape of the sample variogram indicates that there is no temporal correlation within a 10-day period.

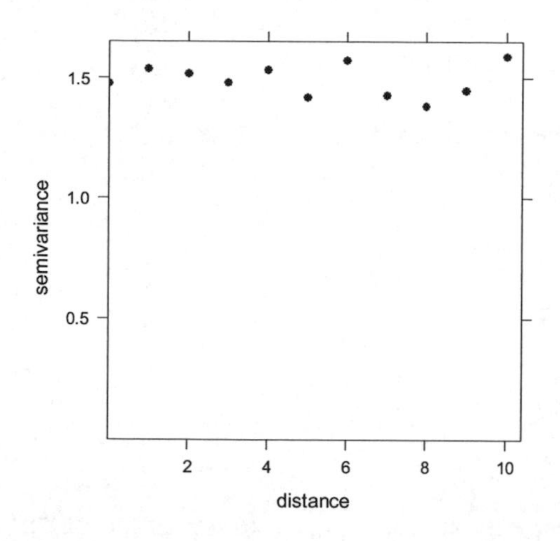

Figure 6.11. Semi-variogram for the residuals of the ZIP GLM model applied to the *R. agassizi* data. Observation date (`CumDate`) was used to create the semi-variogram. The units along the horizontal axis are days.

6.5 Zero inflated GLM applied to *A. castelnaui* data

We begin the analysis of the *A. castelnaui* data with a Poisson GLM and a ZIP GLM. The variable AC contains the counts of *A. castelnaui*. The Poisson GLM showed a dispersion of 2.12 (overdispersed), and the ZIP GLM had a dispersion of 1.11, so we will focus on the latter model. It is fitted in R with the code:

```
> M3 <- zeroinfl(AC ~ fYear + Temperature + fBottomType +
                 offset(LogSA) | fYear + Temperature + fBottomType,
                 dist = "poisson", link = "logit", data = Skates)
```

It is not wise to designate new data by a previously used object name; if we make a coding mistake, the object `M3` will contain results from *R. agassizi*. We do not want that! So it is better to close R and run all preparation code as before but with `AC` as response variable, so that `M3` is a new object. The estimated parameters, z-values, p-values, and likelihood ratio tests are:

```
> summary(M3)

Count model coefficients (poisson with log link):
                           Estimate Std. Error z value Pr(>|z|)
(Intercept)                 4.11753    1.00607   4.093 4.26e-05
fYear1999                  -0.47102    0.36140  -1.303   0.1925
fYear2003                  -0.77590    0.33107  -2.344   0.0191
fYear2005                  -0.31770    0.30698  -1.035   0.3007
Temperature                 0.11003    0.06158   1.787   0.0740
fBottomTypeSand & mud      -1.31205    0.32059  -4.093 4.27e-05
fBottomTypeSand            -0.78730    0.32232  -2.443   0.0146
fBottomTypeSand, shell, rest -0.43701  0.34760  -1.257   0.2087
```

```
Zero-inflation model coefficients (binomial with logit link):
                              Estimate Std. Error z value Pr(>|z|)
(Intercept)                    -8.3780     2.9478  -2.842 0.004481
fYear1999                       2.3007     0.8273   2.781 0.005420
fYear2003                       0.6458     0.7415   0.871 0.383800
fYear2005                       1.6199     0.7075   2.290 0.022049
Temperature                     0.6073     0.1844   3.293 0.000992
fBottomTypeSand & mud          -3.4134     0.9796  -3.485 0.000493
fBottomTypeSand                -3.1169     0.8854  -3.520 0.000431
fBottomTypeSand, shell, rest   -2.4932     0.7382  -3.377 0.000732
```

We apply a likelihood ratio test to obtain *p*-values for the covariates `fYear` and `fBottomType` in the log and logistic link functions. The results are:

```
                              Log likelihood p-value
fYear in log link                      6.150   0.104
Temperature in log link               97.889   0.000
fBottomType in log link               14.378   0.002
fYear in logistic link                 9.969   0.019
Temperature in logistic link          16.825   0.000
fBottomType in logistic link          29.642   0.000
```

Note that both the *z*-statistics and the likelihood ratio test indicate that `fYear` in the count model is not significant. It is weakly significant in the binary part. When the model has been fitted a model validation must be applied. A plot of the Pearson residuals versus the fitted values and each of the covariates shows no obvious problems (graphs are not presented). We also need to verify that the residuals do not contain temporal or spatial patterns, and the procedure used in the previous section is applied. A sample variogram of the Pearson residuals to check for spatial independence is presented in the left panel of Figure 6.12. The graph seems to suggest that there is no spatial correlation, but we are not 100% sure. Reducing the range of the *x*-axis by specifying a value for the `cutoff` option in the `variogram` function does not resolve whether there is spatial correlation. However, the `Cressie = TRUE` argument in the `variogram` function shows a clear pattern (Figure 6.12). This option uses the robust variogram estimate instead of the default classical method of moments estimate of the variogram. This is confusing; the Moran test for spatial correlation indicates that there is no significant spatial correlation. So we have the robust sample variogram signifying spatial correlation and the classical variogram estimate and spatial tests indicating no spatial correlation. In the following section we will explain how to include a spatial correlation structure in a GLM. As we will see, this correlation structure has various parameters, one of which measures the spatial correlation. If it is 0, there is no spatial correlation, or the model cannot capture the spatial correlation. When we apply this method to the *A. castelnaui* data, the spatial parameter is 0.

The sample variogram is created with the code below. The `variogram` function from the `gstat` package creates an object with distances, semi-variogram values, and the number of observations per distance band. We use elementary lattice code to create the multi-panel graph:

```
> library(gstat); library(lattice)
> E3 <- resid(M3, type = "pearson")
> mydata<-data.frame(E3, Skates$X, Skates$Y)
> coordinates(mydata) <- c("Skates.X", "Skates.Y")
> Vario1 <- variogram(E3 ~ 1, mydata)
> Vario2 <- variogram(E3 ~ 1, mydata, cressie = TRUE)
> Xall <- c(Vario1$dist,  Vario2$dist)
> Yall <- c(Vario1$gamma, Vario2$gamma)
> ID   <- rep(c("Classical variogram", "Robust variogram"),each=15)
> MyCex <- c(Vario1$np, Vario2$np)
> xyplot(Yall~Xall| ID, pch = 16, col = 1, cex = 2*MyCex/max(MyCex),
```

```
xlab = "Distance", ylab = "Sample variogram")
```

Why do we get such a difference between these two sample variograms? The reason may be that the classic variogram, as well as the tests for spatial correlation, assume normality, and this is not the case. We have more negative than positive residuals (Figure 6.13A and B).

There is no evidence of temporal correlation. Model interpretation can be done in the same way as in the previous section; that is, sketching the fitted values for different covariate values.

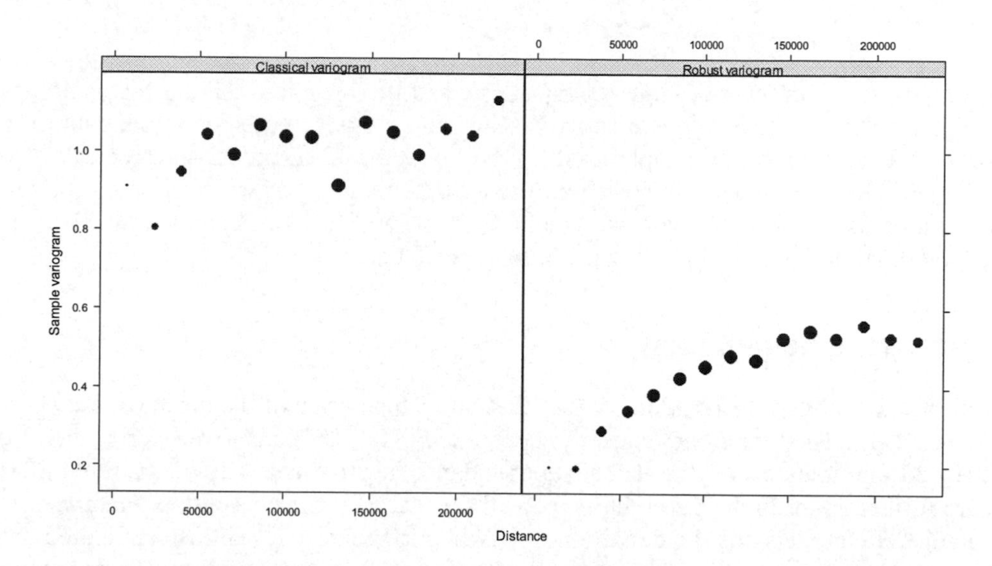

Figure 6.12. Sample variogram for the Pearson residuals of the ZIP model applied to *A. castelnaui.* The horizontal axis corresponds to 200 km. Default settings were used in the variogram function in the left panel, which is the classical method of moments estimate. In the right panel the robust variogram estimator is presented. The size of the points is proportional to the number of observations per distance band.

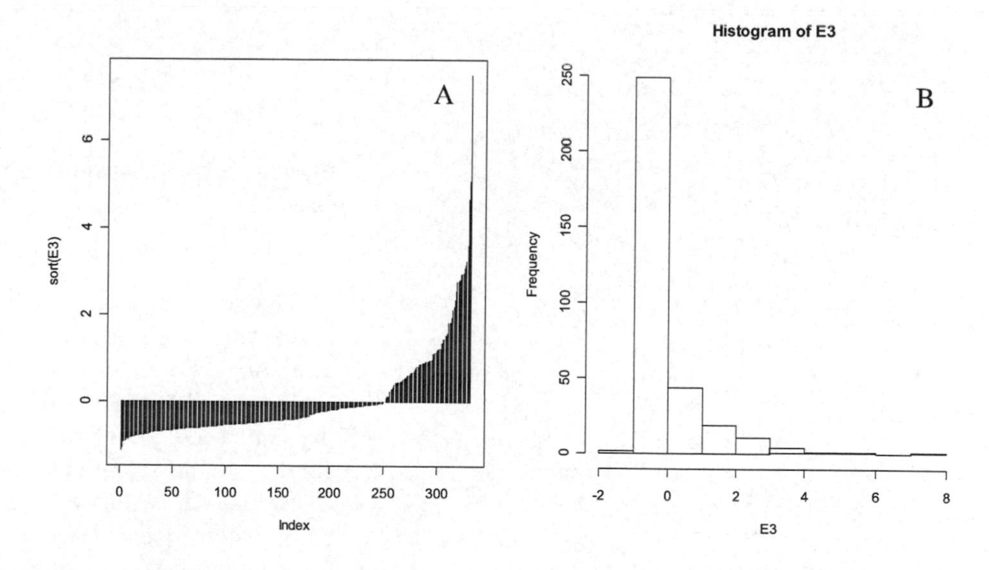

Figure 6.13. A: Sorted Pearson residuals. The x-axis is the index and the y-axis shows the value of the residual. Note that we have more negative than positive residuals. B: Histogram of Pearson residuals.

6.6 Adding spatial correlation to a GLM model

In the previous sections we analysed the *R. agassizi* and *A. castelnaui* data, and the final models were ZIP GLMs showing no clear residual spatial correlation. When analysing the *S. bonapartii* data in the next section, we will obtain a ZIP GLM in which the residuals are clearly spatially correlated. To solve this problem we will extend the GLM and ZIP GLM with a residual spatial correlation structure. Our aim is to use models of the form:

Abundance = function(Covariates + spatially correlated residuals)

where the covariates represent the large scale patterns and the residuals capture the small-scale spatially correlated patterns. There are various methods of doing this, but the conditional auto-regressive correlation (CAR) is the most frequently used. In the literature, CAR models are typically explained in the context of Gaussian models, but we will use a GLM model. We will follow the literature and explain CAR models in a Gaussian context, and in the next section we will show how this method can be adapted to GLM models and apply it to the *S. bonapartii* data.

6.6.1 Who are the neighbours?

In Figure 6.5 we plotted the positions of the sites, and a histogram of the distances between sites is given in Figure 6.14. The majority of sites are separated by 70 – 150 kms. However, some of the sites are less than 20 km apart. It may be that skate abundance at proximate sites is more similar than at sites that are further apart. In the next section we will see that this is the case for *S. bonapartii*.

To explain CAR models and the definition of neighbouring sites we will use simulated data. Suppose that we sample abundance at 10 sites. For illustration purposes we generate 10 sites with random spatial positioning (Figure 6.15A). The R code to make this graph is:

```
> set.seed(1234)
> x1 <- runif(10)
> y1 <- runif(10)
> plot(x = x1, y = y1, type = "n", xlab = "x", ylab = "y")
> text(x1, y1, seq(1:10))
```

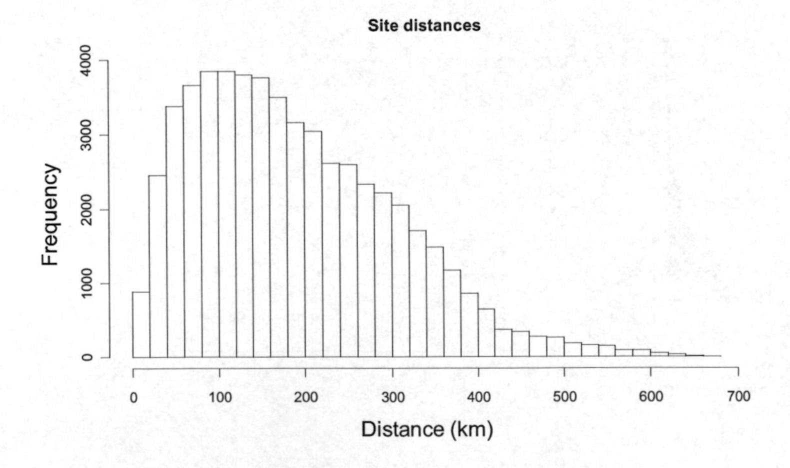

Figure 6.14. Distance between sites.

The *x*- and *y*-coordinates are drawn from a uniform distribution. The `set.seed(1234)` sets the random seed to 1234. If you use the same seed, your graphs and results will be identical to ours. Using the `dist` function, the distance between each two observations can be calculated. The `round` function presents the distances in 2 digits.

```
> Dist1 <- as.matrix(dist(cbind(x1, y1)))
> round(Dist1, digits = 2)
```

```
     1    2    3    4    5    6    7    8    9   10
1  0.00 0.53 0.64 0.56 0.85 0.55 0.42 0.44 0.75 0.61
2  0.53 0.00 0.26 0.38 0.35 0.29 0.67 0.48 0.36 0.33
3  0.64 0.26 0.00 0.64 0.25 0.56 0.60 0.38 0.11 0.11
4  0.56 0.38 0.64 0.00 0.67 0.09 0.88 0.76 0.74 0.70
5  0.85 0.35 0.25 0.67 0.00 0.59 0.85 0.63 0.22 0.35
6  0.55 0.29 0.56 0.09 0.59 0.00 0.84 0.70 0.65 0.62
7  0.42 0.67 0.60 0.88 0.85 0.84 0.00 0.22 0.66 0.51
8  0.44 0.48 0.38 0.76 0.63 0.70 0.22 0.00 0.44 0.28
9  0.75 0.36 0.11 0.74 0.22 0.65 0.66 0.44 0.00 0.16
10 0.61 0.33 0.11 0.70 0.35 0.62 0.51 0.28 0.16 0.00
```

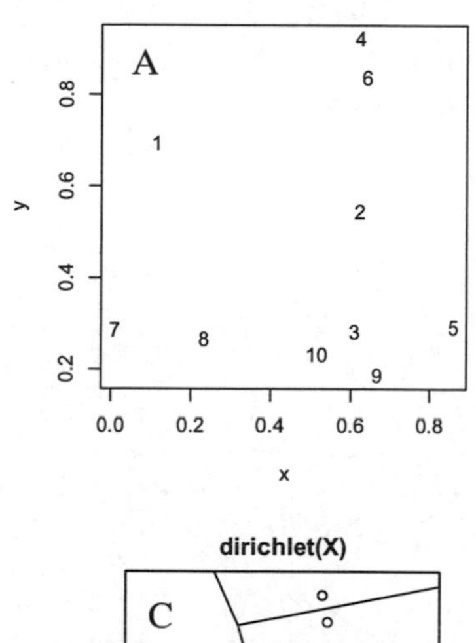

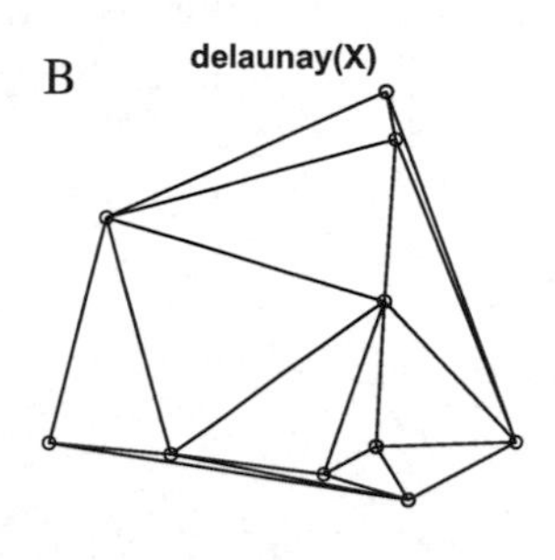

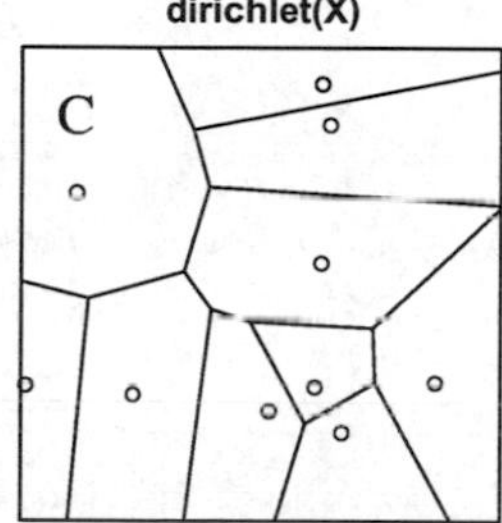

Figure 6.15. A: Ten random sites. B: Delaunay triangulation. Sites connected by a line are considered neighbours. C: Dirichlet tessellation. Each site has been allocated a certain area. Sites in areas sharing a common border are considered neighbours.

In ordinary linear regression models we assume independence. This means that the correlation between any two points is 0. This leads to a residual correlation matrix of the form:

$$\begin{pmatrix} 1 & 0 & \cdots & 0 & 0 \\ 0 & 1 & & & 0 \\ \vdots & & \ddots & & \vdots \\ 0 & & & 1 & 0 \\ 0 & 0 & \cdots & 0 & 1 \end{pmatrix}$$

In the example with 10 sites, the residual correlation matrix would be a 10×10 matrix with zeros everywhere except along the diagonal. To allow for correlation among observations, we can use a correlation function of the form:

$$cor(\varepsilon_s, \varepsilon_t) = \begin{cases} 1 & \text{if } s = t \\ h(\varepsilon_s, \varepsilon_t, \rho) & \text{else} \end{cases}$$

Assuming independence, we have $h(\varepsilon_s, \varepsilon_t, \rho) = 0$, if s is not equal to t, but allowing for dependence means that we need to choose a parametric structure for the correlation function $h(.)$, so that it is allowed to be different from 0. The easiest correlation function to explain is the exponential correlation function with no nugget. It is given by:

$$\gamma(s, \rho) = 1 - e^{-\frac{\Delta}{\rho}}$$

where ρ is the range and Δ the distance. The range is a parameter that we need to estimate using the observed data. Because we know the position of the sites, we also know the distance between them, Δ. Suppose that the range is 0.2. This means that there is relatively high correlation between sites separated by less than approximately 0.2. The variogram values are then:

```
> Range <- 0.2
> SampleVario <- 1 - exp(- Dist1 / Range)
> round(SampleVario, digits = 2)

      1    2    3    4    5    6    7    8    9   10
1  0.00 0.93 0.96 0.94 0.99 0.93 0.88 0.89 0.98 0.95
2  0.93 0.00 0.73 0.85 0.82 0.77 0.96 0.91 0.84 0.81
3  0.96 0.73 0.00 0.96 0.72 0.94 0.95 0.85 0.43 0.42
4  0.94 0.85 0.96 0.00 0.97 0.36 0.99 0.98 0.98 0.97
5  0.99 0.82 0.72 0.97 0.00 0.95 0.99 0.96 0.67 0.83
6  0.93 0.77 0.94 0.36 0.95 0.00 0.98 0.97 0.96 0.95
7  0.88 0.96 0.95 0.99 0.99 0.98 0.00 0.67 0.96 0.92
8  0.89 0.91 0.85 0.98 0.96 0.97 0.67 0.00 0.89 0.76
9  0.98 0.84 0.43 0.98 0.67 0.96 0.96 0.89 0.00 0.55
10 0.95 0.81 0.42 0.97 0.83 0.95 0.92 0.76 0.55 0.00
```

So, we pretend that we know the value of the range, we know the distance between sites, and the variogram values can be calculated using the exponential correlation formula. A value near 0 means that the quantities we are measuring at the two sites are correlated. The further from 0, the lower the correlation. Assuming a stationary process, we can use the rule

$$\text{variogram}(h) = \text{Var}(Y) - \text{Cov}(Y(x), Y(x + h))$$

to convert variogram values to a correlation coefficient. Correlation is defined as covariance divided by the variance of both terms:

$$\text{correlation}(Y(x), Y(x + h)) = 1 - \text{variogram}(h) / \text{var}(Y)$$

Using data with unit variance gives the correlation matrix:

```
> 1 - round(SampleVario, digits = 2)
```

	1	2	3	4	5	6	7	8	9	10
1	1.00	0.07	0.04	0.06	0.01	0.07	0.12	0.11	0.02	0.05
2	0.07	1.00	0.27	0.15	0.18	0.23	0.04	0.09	0.16	0.19
3	0.04	0.27	1.00	0.04	0.28	0.06	0.05	0.15	0.57	0.58
4	0.06	0.15	0.04	1.00	0.03	0.64	0.01	0.02	0.02	0.03
5	0.01	0.18	0.28	0.03	1.00	0.05	0.01	0.04	0.33	0.17
6	0.07	0.23	0.06	0.64	0.05	1.00	0.02	0.03	0.04	0.05
7	0.12	0.04	0.05	0.01	0.01	0.02	1.00	0.33	0.04	0.08
8	0.11	0.09	0.15	0.02	0.04	0.03	0.33	1.00	0.11	0.24
9	0.02	0.16	0.57	0.02	0.33	0.04	0.04	0.11	1.00	0.45
10	0.05	0.19	0.58	0.03	0.17	0.05	0.08	0.24	0.45	1.00

For large data sets, this correlation matrix becomes correspondingly large. To reduce computing time it may be an option to set some of the correlations equal to 0. One option is to set all correlations between non-neighbouring sites to 0. For this we need to know which sites are neighbours.

Figure 6.15B and Figure 6.15C show the same 10 sites, this time with a grid. In panel B we present a grid consisting of triangles. Creating such a triangular grid is a world in itself, and the interested reader may want to consult Wikipedia sites on Voronoi diagrams, Dirichlet tessellation, and Delaunay triangulation. The latter is used in Figure 6.15B, and the Dirichlet tessellation in 6.14C. The Delaunay triangulation is a common method of depicting a spatial neighbourhood. Connected points represent neighbouring sites. For example, for site 1, the neighbouring sites are 7, 8, 2, 6, and 4. By setting all correlations of non-neighbouring sites to 0, we obtain the following correlation matrix.

	1	2	3	4	5	6	7	8	9	10
1	1.00	0.07	0.00	0.06	0.00	0.07	0.12	0.11	0.00	0.00
2	0.07	1.00	0.27	0.00	0.18	0.23	0.00	0.09	0.00	0.19
3	0.00	0.27	1.00	0.00	0.28	0.00	0.00	0.00	0.57	0.58
4	0.06	0.00	0.00	1.00	0.03	0.64	0.00	0.00	0.00	0.00
5	0.00	0.18	0.28	0.03	1.00	0.05	0.00	0.00	0.33	0.00
6	0.07	0.23	0.00	0.64	0.05	1.00	0.00	0.00	0.00	0.00
7	0.12	0.00	0.00	0.00	0.00	0.00	1.00	0.33	0.04	0.00
8	0.11	0.09	0.00	0.00	0.00	0.00	0.33	1.00	0.00	0.24
9	0.00	0.00	0.57	0.00	0.33	0.00	0.00	0.00	1.00	0.45
10	0.00	0.19	0.58	0.00	0.00	0.00	0.00	0.24	0.45	1.00

Note that this matrix contains a high number of zeros. The more zeros we have in this correlation matrix, the easier it is for the numerical optimisation routines to estimate the parameters and standard errors. In the Dirichlet tessellation, sites sharing a common border are considered neighbours. To simplify the correlation matrix we can set all correlations to 0 for data collected in different years. For example, if sites 1 to 5 were surveyed in year 1 and sites 6 to 10 in year two, we could set all correlations between these two groups to 0. This would give a block-diagonal correlation structure:

	1	2	3	4	5	6	7	8	9	10
1	1.00	0.07	0.00	0.06	0.00	0.00	0.00	0.00	0.00	0.00
2	0.07	1.00	0.27	0.00	0.18	0.00	0.00	0.00	0.00	0.00
3	0.00	0.27	1.00	0.00	0.28	0.00	0.00	0.00	0.00	0.00
4	0.06	0.00	0.00	1.00	0.03	0.00	0.00	0.00	0.00	0.00
5	0.00	0.18	0.28	0.03	1.00	0.00	0.00	0.00	0.00	0.00
6	0.00	0.00	0.00	0.00	0.00	1.00	0.00	0.00	0.00	0.00
7	0.00	0.00	0.00	0.00	0.00	0.00	1.00	0.33	0.04	0.00
8	0.00	0.00	0.00	0.00	0.00	0.00	0.33	1.00	0.00	0.24
9	0.00	0.00	0.00	0.00	0.00	0.00	0.00	0.00	1.00	0.45
10	0.00	0.00	0.00	0.00	0.00	0.00	0.00	0.24	0.45	1.00

6.6.2 Neighbouring sites

To define which sites are neighbours we can use the Delaunay triangulation or the Dirichlet tessellation. We can define two sites as neighbours if they are connected by a line, or if they share a common border when the Dirichlet tessellation is used. To put this into mathematical notation we create a matrix **A** of dimension 332 by 332 with elements a_{ij} defined by:

$$a_{ij} = \begin{cases} 1 & \text{if area } i \text{ and } j \text{ have a common boundary for } i \neq j \\ 0 & \text{otherwise} \end{cases} \tag{6.1}$$

This definition of neighbours is based on sites sharing a common border or connected by a line. It is also an option to use distances between the sites to define neighbouring sites. Distances are used indirectly in the Delaunay triangulation and Dirichlet tessellation. White and Ghosh (2009) used the following notation for this: $a_{ij} = I(0 < d_{ij} < d_U)$, where $I()$ is the indicator function with the value 1 if the argument inside the indicator function is true and 0 if untrue. The d_{ij} is the distance between the centre of areas i and j, and d_U is the greatest distance between two sites that are considered adjacent (neighbouring). In principle, this definition allows us to avoid the Delaunay triangulation, and we can use the location of the sampling points and work with distances between sites, but we need to set a value for d_U based on a definition of neighbouring sites. To define the elements of **A,** White and Ghosh (2009) used:

$$a_{ij} = \begin{cases} 1 & \text{if } 0 < d_{ij} \leq d_L \\ f(d_{ij}, \psi) & \text{if } d_L < d_{ij} \leq d_U \\ 0 & \text{otherwise} \end{cases} \tag{6.2}$$

Two sites i and j in proximity have $a_{ij} = 1$. Sites that are separated by a distance between D_L and D_U have an a_{ij} value between 0 and 1 specified by a function f that depends on distance and a parameter ψ. Sites with a distance larger than d_U are labelled non-adjacent ($a_{ij} = 0$).

It is also possible to use covariates to define neighbours. For example, we could label all sites with the same depth as neighbours.

6.6.3 The CAR model*

Now that we can define adjacent sites, we introduce the conditional autoregressive model (CAR). This model is typically explained in the literature in the context of a Gaussian distribution and linear regression models, so we do the same. We will follow a working paper by Song and De Oliveiral (2012). The following text is technical, and you may skip this subsection on first reading. However, the subsection will provide better understanding of the options in the WinBUGS software routines for CAR models, and we advise reading it at some stage.

We start with a lattice D of the form $\{A_1, A_2, A_3, ..., A_n\}$. These A_is are areas, for example, those created by a Dirichlet tesselation. The areas do not overlap, and collectively they cover the lattice D. Suppose we measure a particular variable Y in each area A_i and denote the value by $Y(A_i)$. In the CAR we model $Y(A_i)$ conditional on the Y value at all other areas $A_1, A_2, A_{i-1}, A_{i+1}, A_{i+2}, \ldots, A_n$. A shorter notation for this is $Y(A_i) \mid Y(A_{-i})$, and this is modelled as:

$$Y(A_i) \mid Y(A_{-i}) = \mu_i + \sum\nolimits_{j=1}^{n} c_{ij} \times (Y(A_j) - \mu_j) + \varepsilon_i$$

The c_{ij} are covariance parameters with $c_{ii} = 0$. The Gaussian distribution is used for $Y(A_i)$; therefore the CAR model can also be written as

$$Y(A_i) \mid Y(A_{-i}) \sim N\left(\mu_i + \sum_{j=1}^{n} c_{ij} \times \left(Y(A_j) - \mu_j\right), \sigma_i^2\right)$$

where σ_i^2 is the conditional variance of $Y(A_i)$. This is an expression for the conditional distribution, but we would like to have an expression for the joint distribution of the Y values of all areas, denoted by $\mathbf{Y} = (Y(A_1), \ldots, Y(A_n))'$. Following a landmark paper by Besag (1974), the joint distribution is a Gaussian distribution with mean and variance given by:

$$\mathbf{Y} \sim N\left(\mathbf{u}, (\mathbf{I}_{nxn} - \mathbf{C})^{-1} \times \mathbf{M}\right)$$

The term $\mathbf{I}_{nxn}$ is an identity matrix of dimension $n \times n$. To simplify notation we will drop the subscript $n \times n$. In many papers on spatial correlation using CAR, the equation above is presented without an explanation. The mathematics explaining why the joint distribution takes this form is too complex to reproduce here, and the interested reader is referred to the paper by Besag. We will only show how it is used.

The mean $\boldsymbol{\mu}$ can be modelled as a function of covariates and regression parameters. The term $(\mathbf{I} - \mathbf{C})^{-1} \times \mathbf{M}$ is a covariance matrix, and it is crucial that its inverse exists to enable calculation of standard errors and confidence intervals. The matrix $\mathbf{M}$ is a diagonal with the elements σ_i^2. In order for $(\mathbf{I} - \mathbf{C})^{-1} \times \mathbf{M}$ to be a valid covariance matrix, $\mathbf{M}^{-1} \times \mathbf{C}$ must be symmetrical and $\mathbf{M}^{-1} \times (\mathbf{I} - \mathbf{C})$ must be positive definite. The first condition implies that $c_{ij} \times \sigma_i^2 = c_{ji} \times \sigma_j^2$ for all i and j.

Several classes of CAR models exist (Cressie and Kapat, 2008), but before we can introduce them we impose two additional conditions on $\mathbf{C}$ and $\mathbf{M}$, $\mathbf{M} = \sigma^2 \times \mathbf{G}$ and $\mathbf{C} = \phi \times \mathrm{W}$. The matrix $\mathbf{G}$ is diagonal with known positive values, and σ is unknown. ϕ is an unknown spatial parameter, and $\mathbf{W}$ is a known weight matrix (not necessarily symmetrical). The elements of $\mathbf{W}$ must be non-negative if two sites are neighbours, and its diagonal elements must be 0. If $\phi = 0$, $\mathbf{C} = \mathbf{0}$, and the model assumes independence. The larger the value of ϕ, the stronger the spatial correlation; ϕ is therefore of prime interest.

Three important CAR models are the homogeneous CAR model (HCAR), the weighted CAR model (WCAR), and the auto-correlated CAR model (ACAR). In the HCAR model we use $\mathbf{G} = \mathbf{I}_{nxn}$ and $\mathbf{W} = \mathbf{A}$, where $\mathbf{A}$ is defined as in Equation (6.1). We get:

$$\text{HCAR:} \qquad \mathbf{M} = \sigma^2 \times \mathbf{I}_{nxn} \qquad \text{and} \qquad \mathbf{C} = \phi \times \mathbf{A}$$

In the WCAR model we use the reciprocal of the number of neighbours of each site in $\mathbf{G}$, and therefore in $\mathbf{M}$, resulting in:

$$\mathbf{M} = \sigma^2 \times \mathbf{G} = \sigma^2 \times \begin{pmatrix} \frac{1}{N_1} & 0 & \cdots & 0 \\ 0 & \frac{1}{N_2} & & \vdots \\ \vdots & & \ddots & 0 \\ 0 & \cdots & 0 & \frac{1}{N_n} \end{pmatrix} \qquad \mathbf{C} = \phi \times \mathbf{G} \times \mathbf{A} = \phi \times \begin{pmatrix} \frac{1}{N_1} & 0 & \cdots & 0 \\ 0 & \frac{1}{N_2} & & \vdots \\ \vdots & & \ddots & 0 \\ 0 & \cdots & 0 & \frac{1}{N_n} \end{pmatrix} \times \mathbf{A}$$

where $N_i = \sum_{j=1}^{n} a_{ij}$

N_i is the number of neighbours of site i. Finally, in the ACAR model, we use the same $\mathbf{G}$ (and therefore $\mathbf{M}$) as in the WCAR model, but $\mathbf{C}$ is given by $\mathbf{C} = \phi \times \mathbf{G}^{½} \times \mathbf{A} \times \mathbf{G}^{-½}$.

In all three CAR models we comply with the condition that $\mathbf{M}^{-1} \times \mathbf{C}$ is symmetrical. The second condition imposes a restriction on the range of the values for the spatial parameter ϕ. To be precise, ϕ

must fall between the reciprocal of the smallest and largest eigenvalues of the matrix $\mathbf{G}^{-½} \times \mathbf{A} \times \mathbf{G}^{½}$. This sounds complicated, but the software can do it for us. In the next subsection we will present a small simulation study that should clarify the rationale of CAR models.

Cressie and Wikle (2011) give expressions for the partial correlations of $Y(s_i)$ and $Y(s_j)$ for the three models. In these expressions it becomes clear that the spatial parameter ϕ cannot be interpreted as a correlation parameter with values between 0 and 1. For example, for the WCAR model ϕ^2 lies between 0 and $N(\mathbf{s}_i) \times N(\mathbf{s}_j)$.

6.6.4 Applying CAR to simulated spatially correlated data

In this subsection we will create a small simulation study to illustrate how CAR works. We use 200 sites with the spatial coordinates taken from a uniform distribution (Figure 6.16A). The R code to generate this graph follows:

```
> N  <- 200
> x1 <- runif(N, 0, 1)
> y1 <- runif(N, 0, 1)
> plot(x = x1, y = y1, type = "p", xlab = "x", ylab ="y")
```

Distances between sites can be calculated using the `dist` function, and a histogram of these distances can be created to visualise the distribution of the distances. We use a simple linear regression model of the form:

$$Y_i = \alpha + \beta \times \text{Covariate}_i + \varepsilon_i$$

where the term ε_i is spatially correlated noise. We choose values for the intercept and slope and draw random values for the covariate and the noise term. We then calculate the Y_i values, apply a CAR model in WinBUGS to the simulated data, and compare the estimated parameters to the original values. For the intercept, slope, and covariate we use:

```
> alpha <- 2
> beta  <- 1
> Covariate <- runif(N, min = 0, max = 1)
> mu <- alpha + beta * Covariate
```

There are several options to generate spatially correlated noise. We could use one of the spatial correlation structures introduced in Chapter 7 in Zuur et al. (2009a), for example the exponential correlation structure. This involves specifying a range and nugget and calculating the sample variogram values. An alternative approach is to choose values for **C** and **M** and sample spatially correlated residuals from a normal distribution with mean 0 and covariance matrix $(\mathbf{I} - \mathbf{C})^{-1} \times \mathbf{M}$, and that is what we will do here. The following R code defines a matrix **A** with elements 0 and 1. The value 1 refers to two neighbouring sites, and we use:

$$a_{ij} = \begin{cases} 1 & \text{if } 0 < d_{ij} \leq d_U \\ 0 & \text{otherwise} \end{cases} \tag{6.3}$$

First we calculate Euclidean distances d_{ij} between all sites:

```
> Dist1 <- as.matrix(dist(cbind(x1, y1)))
```

For the threshold d_U in Equation (6.3) we use 0.15. The R code to calculate the matrix **A** is:

```
> d.U <- 0.15
> A <- matrix(nrow = N, ncol = N)
> A[,] <- 0
> for (i in 1:N){
    Neighb.i <- Dist1[i,] <= d.U
    A[i, Neighb.i] <- 1
    A[i,i] <- 0
  }
```

Most papers applying CAR models use the weighted CAR, and the software routines in WinBUGS require a normalised weighting matrix, which is the case for WCAR. We therefore use the WCAR.

```
> Nn <- rowSums(A)
> G <- diag(1/Nn)
> W <- G %*% A
> Z <- (G ^ 0.5) %*% A %*% (G ^ 0.5)
> lambdas  <- eigen(Z, only.values = TRUE)$values
> lambda.1 <- lambdas[1]
> lambda.n <- lambdas[N]
# phi must be between:
> c(1/lambda.n , 1/lambda.1)

-2.413567  1.000000
```

The underlying theory dictates that the value of the spatial parameter ϕ must fall between the reciprocal of the largest and smallest eigenvalues, which means that ϕ must be between -2.4 and 1. This information is required for the prior distribution of ϕ in the MCMC algorithm. We can now simulate spatially correlated residuals.

```
> sigma    <- 0.5
> phi      <- 0.9
> M        <- sigma^2 *G
> C        <- phi * W
> I.nn     <- diag(N)
> CovMat   <- solve(I.nn - C) %*% M
> MuEps    <- rep(0,N)
> library(MASS)
> SpatEps <- mvrnorm(1, MuEps, CovMat)
```

The `phi` is the spatial correlation parameter ϕ, and we set it arbitrarily to 0.9. The sigma is the σ in **M**. The function `mvrnorm` from the `MASS` package is used to draw N observations from a multivariate normal distribution with covariance matrix $(\mathbf{I} - \mathbf{C})^{-1} \times \mathbf{M}$. The data are then obtained by:

```
> Y <- mu + SpatEps
```

A histogram of `SpatEps` is presented in Figure 6.16C, and a scatterplot of the covariate versus Y in Figure 6.16D. We apply a linear regression model of the form $Y = \alpha + \beta \times X$, and a sample variogram of the residuals of this model shows a clear spatial correlation in the range of 0.3 (Figure 6.17). The R code is:

```
> tmp <- lm(Y ~ Covariate)
> summary(tmp)  #Not presented here
> library(gstat)
> E1      <- resid(tmp)
> mydata <- data.frame(E1, x1, y1)
```

```
> coordinates(mydata) <- c("x1", "y1")
> Vario1 <- variogram(E1 ~ 1, mydata, cutoff = 1, robust = TRUE)
> plot(Vario1, pch = 16, col =1)
```

At this point you may want to rerun the code to see how drawing another set of random numbers changes the results. For some runs there will be a much stronger spatial pattern in the residuals, and for other random samples there may be no spatial correlation.

We now apply a CAR model in WinBUGS to the simulated data. We pretend that we do not know the matrices **M** and **C**, and that we have only *Y* and *X* vectors. Based on the sample variogram of the residuals of the linear regression model we use a d_U of 0.3. Note that this is slightly larger than the value of 0.15 that we used to generate the spatially correlated residuals.

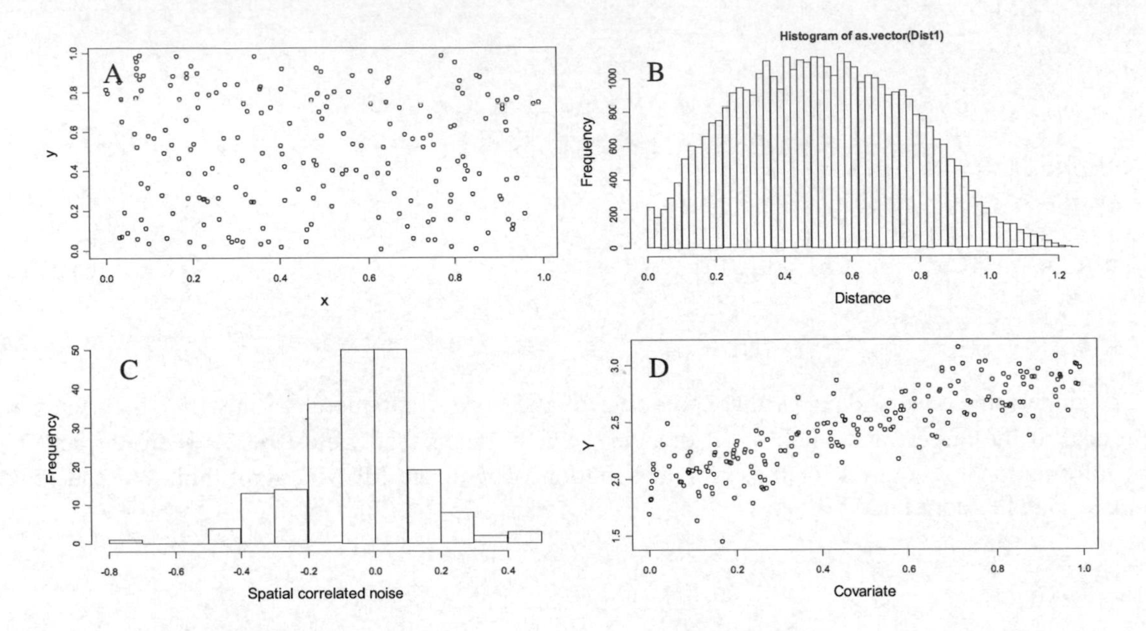

Figure 6.16. Simulated data. A: Simulated spatial positions of 200 sites. B: Histogram of distances between sites. C: Histogram of spatially correlated residuals. D: Scatterplot of simulated Y data versus covariate.

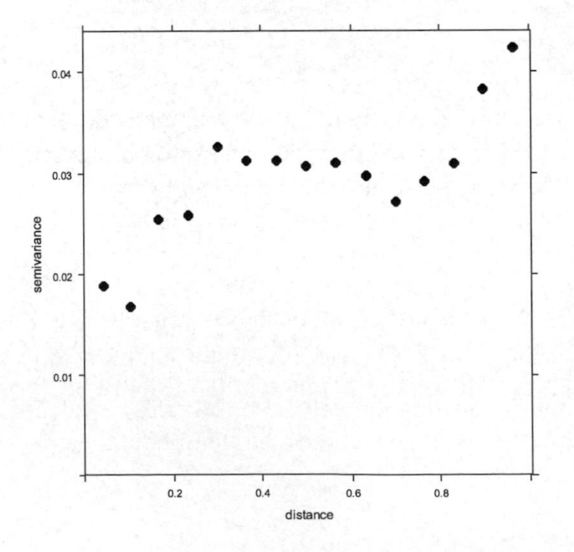

Figure 6.17. Sample variogram of the residuals from the linear regression model.

The next step is to prepare R code so that we can apply the CAR model. We will do this using the package `R2WinBUGS`, which means that we will run WinBUGS from R. The function we need for the weighted CAR is `car.proper`. The relevant text from the help file in the WinBUGS manual is:

Y[1:N] ~ car.proper(mu[], C[], adj[], num[], M[], tau, gamma)

where:

- mu[]: A vector giving the mean for each area. [...]
- C[]: A vector the same length as adj[] giving normalised weights associated with each pair of areas. [...]
- adj[]: A vector listing the ID numbers of the adjacent areas for each area. [...]
- num[]: A vector of length N (the total number of areas) giving the number of neighbours n_i for each area.
- M[]: A vector of length N giving the diagonal elements M_{ii} of the conditional variance matrix. [...]
- tau: A scalar parameter representing the overall precision (inverse variance) parameter.
- gamma: A scalar parameter representing the overall degree of spatial dependence. This parameter is constrained to lie between bounds given by the inverse of the minimum and maximum eigenvalues of the matrix M - 1/2 C M 1/2 (see appendix 1)

The CAR correlation is imposed on the response variable *Y*, so mu[] should be a vector of length N with values given by the fitted values: mu = $\alpha + \beta \times$ Covariate. Although **C** is a matrix of dimension *N*-by-*N* in our statistical equations, `car.proper` requires that it be a vector, and it should contain only the weights of adjacent sites, not the zeros of sites that are not adjacent. The vector `adj` gives the names of the neighbouring sites of each site. **C** must be converted to a vector of the same length as `adj`. We also need a vector `num` of length *N* that contains the number of neighbours of each site. The parameter tau is equal to $1/\sigma^2$, where σ^2 is the variance in the matrix **M**. The parameter gamma is our phi.

Let us start with the matrix **C**. The following code produces **C** in matrix format. We standardise the rows.

```
> C1 <- matrix(nrow = N, ncol = N)
> C1[,] <- 0
> Threshold <- 0.3  #This is d.U
> for (i in 1:N){
    Neighb.i <- Dist1[i,] < Threshold
    C1[i, Neighb.i] <- 1
    C1[i,i] <- 0 }
#Standardise the rows and call the resulting matrix C2
> C2 <- matrix(nrow = N, ncol = N)
> C2[,] <- 0
> RS <- rowSums(C1)
> for (i in 1:N){ if (RS[i] > 0) { C2[i,] <- C1[i,] / RS[i] } }
```

Now we determine the number of neighbours of each site and make a variable `ID` with the site names.

```
> NumNeighbours <- colSums(C1)
> ID <- 1:N
```

That was the easy part. Now it becomes more difficult. We create an object `Info` containing three variables: `num` is the number of neighbours, `adj` contains the names of all neighbouring sites, and `C` the weights of all neighbouring sites. We create them by first setting the values to NA and we calculate them with a loop. Inside the loop we will glue together old and new values, but to initiate the process we need to have these variables, thus the NA construction.

```
> Info <- list(num = NA, adj = NA, C   = NA)
```

We also add **M** to the object `Info`. It is in vector format as required by the `car.proper` function. Note that the definition of **M** follows the WCAR.

```
> Info$M    <- 1 / NumNeighbours
```

Next we calculate `num` (which is equal to `NumNeighbours`) and begin a loop to add the identity and weights of each site *i*. The construction with `c` concatenates old and new row information.

```
> Info$num <- rowSums(C2 > 0)
> for(i in 1:N) {
    cur.neighb  <- ID[C2[i,] > 0]
    Info$adj    <- c(Info$adj, cur.neighb)
    Info$C      <- c(Info$C, C2[i, cur.neighb]) }
```

Finally we need to remove the first elements of `adj` and `C`, as these were set to NA.

```
> Info$adj <- Info$adj[-1] # delete first NA
> Info$C   <- Info$C[-1]   # delete first NA
```

We now have all the information required and can prepare the code for WinBUGS. First we load the `R2WinBUGS` package and collect the variables.

```
> library(R2WinBUGS)
> win.data <- list(Y          = Y,
                   Covariate = Covariate,
                   N          = N,
                   num        = Info$num,
                   C          = Info$C,
                   M          = Info$M,
                   adj        = Info$adj)
```

The essential modelling code follows. We use diffuse priors for the intercept and slope. A uniform distribution is used for the prior of `phi`, which is the spatial correlation parameter, and it is bound by the reciprocal of the smallest and largest eigenvalues. The functions `min.bound` and `max.bound` calculate these values, but it is also possible to calculate them ourselves, as we did earlier, and to specify them directly in the `dunif` function. Note that such values will change for different random numbers.

```
#Model
sink("SimulateCAR.txt")
cat("
model{
  #Priors
  for (i in 1:2) { beta1[i]  ~ dnorm(0, 0.00001) }
  tau   <- 1 / (sigma * sigma)
  sigma   ~ dunif(0,100)

  lambda.low <- min.bound(C[], adj[], num[], M[])
  lambda.up  <- max.bound(C[], adj[], num[], M[])
  phi  ~ dunif(lambda.low, lambda.up)

  #Likelihood
  for (i in 1:N) {
     mu[i]         <- beta1[1]  + beta1[2] * Covariate[i]
  }
```

```
   Y[1:N] ~ car.proper(mu[], C[], adj[], num[], M[], tau, phi)
}
",fill = TRUE)
sink()
```

Initial values for all the variables are:

```
> inits <- function () {
    list(
      beta1    = rnorm(2, 0, 0.1),
      sigma    = rlnorm(1),
      phi      = 0)   }
```

The value of phi = 0 indicates no correlation. We want to obtain output from the following parameters:

```
> params <- c("beta1", "sigma", "phi")
```

We use 3 chains, each with 100,000 draws from the posterior; 5,000 iterations are used as burn-in, and the thinning rate is 50.

```
> nc <- 3          #Number of chains
> ni <- 100000     #Number of draws from posterior (for each chain)
> nb <-   5000     #Number of draws to discard as burn-in
> nt <-     50     #Thinning rate
```

Finally, we initiate the MCMC with the following code.

```
> SimOut <- bugs(data = win.data, inits = inits,
                 parameters = params, model = "SimulateCAR.txt",
                 n.thin = nt, n.chains = nc, n.burnin = nb,
                 n.iter = ni, debug = TRUE,
                 bugs.directory = MyWinBugsDir)
> print(SimOut, digits = 3) #Results not printed here
```

The mean of the posterior samples for the intercept and slope are close to the original values. However, for different random numbers the sigma and rho are not always close to the original values. When we change the definition of what constitutes a neighbouring site, i.e. change the value of the variable Threshold, values close to the original value of 0.15 produce mean values of the posterior for sigma and rho that are similar to the original values. This means that defining the distance between sites that constitutes a neighbour is crucial in this process. The sample variogram is a good tool to help decide what this value should be, but perhaps ecological knowledge is equally, if not more, important.

If we apply the methodology described above using neighbouring sites defined by Delaunay triangulation, the estimated spatial parameter is close to 0 for all simulated data sets. We therefore prefer to define neighbouring sites as a function of distance, based on a sample variogram or knowledge of the biology.

Now that we have explained the underlying principle of CAR models and provided a simple example, it is time to apply it to the real data set.

6.7 Analysis of species 3: *Sympterygia bonapartii*

6.7.1 Poisson or negative binomial distribution?

The GLM Poisson model for *S. bonapartii* data has an overdispersion of 34.64, and the NB GLM an overdispersion of 1.35. We could present the NB GLM. However, as before, we argue that the zeros in the counts consist of both true zeros and false zeros; therefore we also apply ZIP and ZINB GLM models. The ZIP GLM has an overdispersion of 4.99 and the ZINB GLM of 1.43. The overdispersion of 4.99 in the ZIP GLM may be due to spatial correlation in the residuals. The likelihood ratio test indicates that `fBottomType` is not significant at the 5% level in the ZIP GLM.

```
                               Log likelihood p-value
fYear in log link                      69.329   0.000
Temperature in log link               384.303   0.000
fBottomType in log link               372.204   0.000
fYear in logistic link                 21.579   0.000
Temperature in logistic link           10.961   0.001
fBottomType in logistic link            2.371   0.499
```

Instead of presenting an NB GLM or ZINB GLM, we will focus on the ZIP GLM and try to determine why it is overdispersed. The sample variogram of the Pearson residuals of the ZIP model shows clear spatial correlation (Figure 6.18). We continue the analysis by adding a spatial CAR correlation structure to the model.

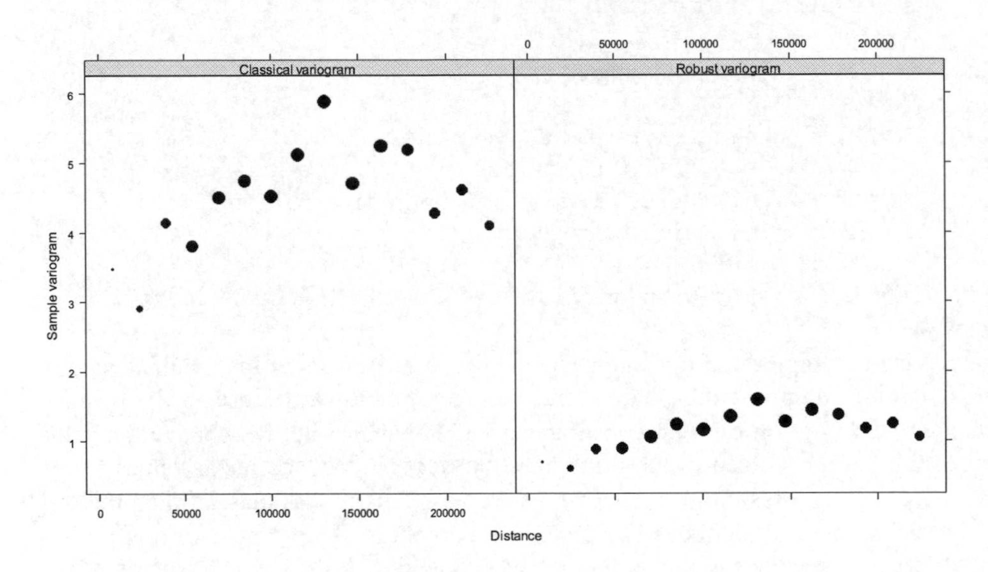

Figure 6.18. Sample variogram for the Pearson residuals of the ZIP model applied to *S. bonapartii* data. The horizontal axis corresponds to 250 km.

6.7.2 Adding spatial correlation to the zero inflated Poisson GLM

In the previous section we explained how to include a CAR correlation structure in a Gaussian model. Adding it to a Poisson GLM or a ZIP GLM is straightforward (see also Verhoef and Janssen, 2007). A potential starting model is:

$$SB_i \sim ZIP(\mu_i, \pi_i)$$
$$\log(\mu_i) = \alpha_1 + Temperature_i + fYear_i + fBottomType_i + offset(LogSA_i) + \varepsilon_i$$
$$\text{logit}(\mu_i) = \alpha_2 + Temperature_i + fYear_i + fBottomType_i + \gamma_i$$

In the previous section we applied the CAR correlation structure to the response variable, but in a zero inflated GLM model we apply the CAR to the residuals, so the residual correlation ε_i in the log link function is modelled with a CAR. This means that the residual ε_i at site i is modelled as a function of residuals at neighbouring sites.

$$E(\varepsilon_i) = \rho \times \sum_{j \neq i} c_{ij} \times \varepsilon_j$$

The c_{ij} is a weighting function and depends on the distance between sites. The parameter ρ measures the strength of the spatial correlation. If it is 0, there is no spatial correlation. The same correlation structure can be used for the residual term γ_i in the logistic link function. This means that we have:

$$\boldsymbol{\varepsilon} \sim N\left(\mathbf{0}, \sigma_\varepsilon^2 \times (\mathbf{I} - \rho_\varepsilon \times \mathbf{C})^{-1} \times \mathbf{M}\right)$$
$$\boldsymbol{\gamma} \sim N\left(\mathbf{0}, \sigma_\gamma^2 \times (\mathbf{I} - \rho_\gamma \times \mathbf{C})^{-1} \times \mathbf{M}\right)$$

The bold vector $\boldsymbol{\varepsilon}$ contains the individual values ε_1 to ε_n, where n is the sample size. The same holds for $\gamma = (\gamma_1, ..., \gamma_n)$. The ρ_ε and ρ_γ are the spatial correlation parameters for the log and logistic link functions, respectively. The techniques that we will apply involve extensive computing time, and the simpler the models the better. `fBottomType` was only weakly significant in the binary part of the ZIP model, and it may aid the estimation process if we omit it in the CAR ZIP model. Because the correlation parameters are difficult to calculate, it may be an option to impose a correlation structure only on the counts and not on the binary part. However, we will be courageous and use the two correlation terms and all three covariates in both link functions. If we encounter numerical problems we will simplify the model.

In the following subsection, we provide more detail on the implementation of the ZIP with CAR correlation structure in R.

6.7.3 Stetting up the required matrices for car.proper

Once again, it is best to close R and begin over. First we load the data and convert the spatial coordinates again:

```
> Skates <- read.table(file = "Skates.txt", header = TRUE)
> Skates$LogSA        <- log(Skates$SweptArea)
> Skates$fYear        <- factor(Skates$Year)
> Skates$fMonth       <- factor(Skates$Month)
> Skates$fBottomType <- factor(Skates$BottomType,
                levels = c(1, 2, 3, 4),
                labels = c("Mud", "Sand & mud", "Sand",
                     "Sand, shell, rest and/or tuff"))
> library(rgdal)
> UTM21S <- "+proj=utm +zone=21 +south"
> UTM21S <- paste(UTM21S, " +ellps=WGS84",sep = "")
> UTM21S <- paste(UTM21S, " +datum=WGS84",sep = "")
> UTM21S <- paste(UTM21S, " +units=m +no_defs", sep="")
> coord.UTM21S <- project(as.matrix(Skates[,
                c("Longitude", "Latitude")]), UTM21S)
> Skates$X <- coord.UTM21S[,1]
> Skates$Y <- coord.UTM21S[,2]
```

This code was explained in Section 6.2. We will use WinBUGS, and this program needs some adjustment of the categorical variables.

```
> BTf <- model.matrix(~ fBottomType - 1, data = Skates)
> Yef <- model.matrix(~ fYear - 1, data = Skates)
```

The function `model.matrix` creates the covariate matrix with dummy variables for the categorical variable `fBottomType`. This process is normally carried out automatically by regression functions, but to run the model in WinBUGS we need to do it ourselves. In this case, the function `model.matrix` converts the factor `fBottomType` with four levels into four dummy variables, and the -1 ensures that no intercept is added. To view the object `BTf`, use the `head(BTf)` command, which shows the first 6 lines.

The procedures that are carried out in WinBUGS benefit from continuous variables being centred around zero, or better, standardised so that the mean is 0 and the variance is 1. The function `scale` can be used for this. For our data this is relevant for the covariate `Temperature`.

```
> Skates$Tem.sc <- as.numeric(scale(Skates$Temperature))
```

Next we must select the format of the matrices **C** and **M**. As explained above we will use the weighted CAR model; thus **M** is a diagonal matrix with the reciprocal of the number of neighbours, and **C** is specified as

$$c_{ij} = \begin{cases} \dfrac{1}{n_i} & \text{if sites } i \text{ and } j \text{ are neighbours} \\ 0 & \text{otherwise} \end{cases}$$

where n_i is the number of neighbours of site i. Note that a site cannot be its own neighbour. Suppose site i has 4 neighbours, sites 2, 3, 4, and 5. This means that $n_i = 4$. The CAR equation for the residual at site 1 then becomes:

$$E(\varepsilon_1) = \rho \times \sum\nolimits_{j \neq i} c_{ij} \times \varepsilon_j = \rho \times \left(\frac{1}{4} \times \varepsilon_2 + \frac{1}{4} \times \varepsilon_3 + \frac{1}{4} \times \varepsilon_4 + \frac{1}{4} \times \varepsilon_5 \right)$$

Thus ε_1 is modelled as ρ times a weighted average of residuals of 4 neighbouring sites, and each neighbouring site has the same weight.

We first need to define what a neighbour is, and then use R code to calculate the c_{ij} values for all sites i and j. Based on the shape of the sample variogram in Figure 6.12, we could use a cut-off level between 30 and 60 km. Instead of using the sample variogram of residuals we can base our choices on biological knowledge. It is thought that skate movement is less than 20 km per day; therefore we can use a cut-off value of 20 km. In this case, any site j within a radius of 20 km of site i is considered a neighbour. Yet another approach is to set the cut-off level to 5% or 10% of the maximum distance or most frequent distances. In the latter case, setting it to 10 km may be an option. At this point it is useful to gain insight into the distances between the sites and how many neighbours each site has if we use a cut-off level of 10 km or 20 km. Figure 6.19 shows the number of neighbours for each site if we use a cut-off of 20 km. We do not want to choose a cut-off level such that no site has a neighbour, but neither do we want to define a cut-off so that every site has every other site as a neighbour.

We decide to use 20 km as the cut-off, so sites that are separated by less than 20 km are labelled neighbours. It may be useful to run CAR ZIP models with different values for this threshold and see how results (estimated parameters, DIC, spatial correlation in the residuals) change. Verhoef and Jansen (2007) defined a neighbour as any cell that is within 1 km, although their data was on seals, which may travel shorter distances.

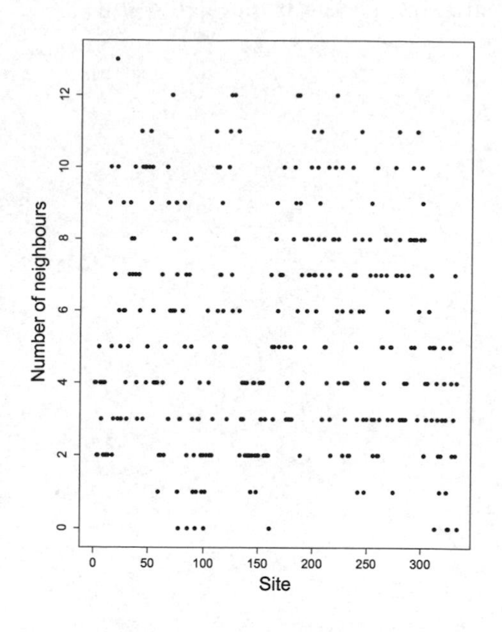

Figure 6.19. Number of neighbouring sites for each site are plotted along the *y*-axis, and the *x*-axis is the site identity. A distance of <20 km is used to define neighbour.

The following code was used to calculate distances between sites.

```
> Dist1    <- as.matrix(dist(Skates[,c("X", "Y")]))
> DistData <- data.frame(Dist = as.double(Dist1),
        From = rep(Skates$Station, each = nrow(Skates)),
        To   = rep(Skates$Station, nrow(Skates)))
```

The function `dist` calculates the distance in metres between the sites, and the `as.matrix` function ensures that the output is in the appropriate format. We create an object, `DistData` that contains the distances between each pair of sites, and the names of the sites in the pair. Use `head(DistData)` to see the first 6 rows:

```
> head(DistData)

       Dist From To
1      0.00    1  1
2  18350.01    1  2
3 293685.17    1  3
4 277583.59    1  4
5 258560.67    1  5
6 248862.39    1  6
```

The `From` and `To` are self-explanatory. The following code sets the threshold at 20 km and calculates a binary variable `Close` with values 0 (not close) and 1 (close). It is stored in the `DistData` object. The `tapply` function is used to calculate the number of neighbours (which is the sum of `Close` for each level of the `From` variable), and the `plot` function creates a visual display of the number of neighbours.

```
> Threshold        <- 20000
> DistData$Close <- DistData$Dist < Threshold
> DistData$Close <- as.numeric(DistData$Close)
> DistData[DistData$From == DistData$To , "Close"] <- 0
> NumNeighbours   <- tapply(DistData$Close, FUN = sum,
                             INDEX = DistData$From)
> NumNeighbours <- as.numeric(NumNeighbours)
> plot(NumNeighbours, xlab = "Site", ylab = "Number of neighbours")
```

R code to calculate C, M, adj and num was given in Subsection 6.6.4 and it is repeated below.

```
> N <- nrow(Skates)
> C1 <- matrix(nrow = N, ncol = N)
> C1[,] <- 0
> for (i in 1:N){
   Neighb.i <- Dist1[i,] < Threshold
   C1[i, Neighb.i] <- 1
   C1[i,i] <- 0 }
#Standardise the rows and call the resulting matrix C2
> C2 <- matrix(nrow = N, ncol = N)
> C2[,] <- 0
> RS <- rowSums(C1)
> for (i in 1:N){ if (RS[i] > 0) { C2[i,] <- C1[i,] / RS[i] } }
> NumNeighbours <- colSums(C1)
> ID <- 1:N
> Info <- list(num = NA, adj = NA, C   = NA)
> Info$M   <- 1 / NumNeighbours
> Info$num <- rowSums(C2 > 0)
> for(i in 1:N) {
   cur.neighb  <- ID[C2[i,] > 0]
   Info$adj    <- c(Info$adj, cur.neighb)
   Info$C      <- c(Info$C, C2[i, cur.neighb]) }
> Info$adj <- Info$adj[-1] # delete first NA
> Info$C   <- Info$C[-1]   # delete first NA
```

6.7.4 Preparing MCMC code for a ZIP with residual CAR correlation

We are now ready to prepare the code for WinBUGS. First we need the response variable, covariates, and other input data, and we store all required variables in the list win.data.

```
> win.data <- list(Y                  = Skates$SB,
                   Temperature        = Skates$Tem.sc,
                   N                  = nrow(Skates),
                   LogSA              = Skates$LogSA,
                   BT2                = BTf[,2],
                   BT3                = BTf[,3],
                   BT4                = BTf[,4],
                   Ye2                = Yef[,2],
                   Ye3                = Yef[,3],
                   Ye4                = Yef[,4],
                   num                = Info$num,
                   C                  = Info$C,
                   M                  = Info$M,
                   adj                = Info$adj)
```

Note that the variable Temperature inside the win.data list is the normalized temperature. We suggest you inspect the values in win.data for any irregularities. If a site does not have neighbours then the corresponding value in **M** will be Inf and this causes an error in WinBUGS. We suggest setting Inf values to 1 or define a neighbour for this particular site.

The MCMC code is:

```
> sink("modelSB.txt")
  cat("
```

```
model{
 #Priors
  for (i in 1:8) { beta1[i]    ~ dunif(-50, 50) }
  for (i in 1:8) { gamma1[i]   ~ dunif(-50, 50) }
 #for (i in 1:8) { beta1[i]    ~ dnorm(0, 0.001) }
 #for (i in 1:8) { gamma1[i]   ~ dnorm(0, 0.001) }
  tau.eps ~ dunif(0.0001, 5)
  tau.xsi ~ dunif(0.0001, 5)

 #lambda.low <- min.bound(C[], adj[], num[], M[])
 #lambda.up  <- max.bound(C[], adj[], num[], M[])
 #rho        ~ dunif(lambda.low,   lambda.up)
  rho        ~ dunif(0.0001, 0.9999)
  rho.xsi ~ dunif(0.0001, 0.9999)

 #Likelihood
 for (i in 1:N) {
    #Logit part
     W[i] ~ dbern(psim1[i])
     psim1[i] <- min(0.99999, max(0.00001,(1 - psi[i])))
     eta.psi[i] <- gamma1[1] +
                   gamma1[2] * Temperature[i] +
                   gamma1[3] * Ye2[i] +
                   gamma1[4] * Ye3[i] +
                   gamma1[5] * Ye4[i] +
                   gamma1[6] * BT2[i] +
                   gamma1[7] * BT3[i] +
                   gamma1[8] * BT4[i] + tau.xsi * XSI[i]
     logit(psi[i]) <- max(-20, min(20, eta.psi[i]))

    #Poisson part
     Y[i]         ~  dpois(eff.mu[i])
     eff.mu[i]   <-  W[i] * mu[i]
     log(mu[i]) <- max(-20, min(20, eta.mu[i]))
     eta.mu[i]   <- beta1[1]  +
                    beta1[2] * Temperature[i] +
                    beta1[3] * Ye2[i] +
                    beta1[4] * Ye3[i] +
                    beta1[5] * Ye4[i] +
                    beta1[6] * BT2[i] +
                    beta1[7] * BT3[i] +
                    beta1[8] * BT4[i] +
                    1 * LogSA[i] + tau.eps * EPS[i]
     EPS.mean[i] <- 0

    #Residuals
     EZip[i]    <- mu[i] * (1 - psi[i])
     VarZip[i] <- (1 - psi[i]) * (mu[i] + psi[i] * pow(mu[i], 2))
     PRES[i]    <- (Y[i] - EZip[i]) / sqrt(VarZip[i])

    #log-likelihood
    lfd0[i] <- log(psi[i] + (1 - psi[i]) * exp(-mu[i]))
    lfd1[i] <- log(1 - psi[i]) + Y[i] * log(mu[i])   -
               mu[i]   - loggam(Y[i]+1)
    L[i]     <- (1-W[i]) * lfd0[i] + W[i] * lfd1[i]
  }
```

```
  EPS[1:N] ~ car.proper(EPS.mean[], C[], adj[], num[], M[], 1, rho)
  XSI[1:N] ~ car.proper(EPS.mean[], C[], adj[], num[], M[], 1,
                        rho.xsi)
  dev <- -2 * sum(L[1:N])
 }
",fill = TRUE)
sink()
```

We include code to calculate the Pearson residuals and the deviance, which is needed for the DIC (see also Chapter 2). We experimented with different priors for the regression parameters and also with different precision parameters in the priors for these parameters. The results can be sensitive to these choices, and it is important that the prior distribution is not too flat, or the exponential function will result in excessively large values. The construction with the `max` and `min` functions ensures that the algorithm does not crash, but may result in non-convergence of the chains. The uniform distribution for the priors of the regression parameters produces more stable chains than the normal distribution, but the majority of books use the normal distribution as prior distribution for regression parameters in GLMs.

For the priors of the spatial correlation parameters ρ_ε and ρ_γ in the binary and count part of the model, we use a uniform distribution with values between 0 and 1. These values were taken from Verhoef and Janssen (2007). Choosing the initial values of these two parameters is crucial. The literature suggests that initial values must be set close to the maximum value. We began our attempts with initial values of 0 (implying no correlation), but this was not satisfactory. Only when we set the initial values to 0.5 does the MCMC produce sensible results. We also need to provide initial values of the regression parameters, and we use the estimated parameters from the ZIP obtained by the `zeroinfl` function. Therefore we need to run the `zeroinfl` function again. Note that the order of the covariates in the `zeroinfl` function and the MCMC code must match.

```
> library(pscl)
> M6 <- zeroinfl(SB ~ Tem.sc + fYear + fBottomType +
                      offset(LogSA) |
                      Tem.sc + fYear + fBottomType,
                 dist = "pois", link = "logit", data = Skates)
> Beta.Count    <- M6$coef$count
> Beta.Zero     <- M6$coef$zero
> SE.Beta.Count<- summary(M6)[1]$coefficients$count[,2]
> SE.Beta.Zero <- summary(M6)[1]$coefficients$zero[,2]
```

We have set the variance of the residual terms in the `car.proper` function to 1, but the pre-multiplication of `EPS[i]` by `tau.eps` models the variance. This construction tends to produce MCMC chains that are better mixed (Jay Verhoef, personal communication). The same was done for the correlation in the logistic link function.

The categorical variable `fBottomType` is modelled with the three dummy variables `BT2`, `BT3`, and `BT4`. The first level (`fBottomType1`) is used as baseline. We do the same for `fYear`. The mechanism for the MCMC implementation of a ZIP is discussed in Chapter 2 and also in Kéry (2010). The initial values are specified by:

```
> inits <- function () {
   list(
    beta1    = rnorm(8, Beta.Count[1:8], 0.1),
    gamma1   = rnorm(8, Beta.Zero[1:8],  0.1),
    W        = as.integer(win.data$Y>0),
    EPS      = rep(0, nrow(Skates)),
    rho      = 0.5,
    rho.xsi  = 0.5,
```

```
    tau.eps = 0.1,
    tau.xsi = 0.1) }
```

We specify from which parameters we want to store the MCMC samples:

```
> params <- c("beta1", "gamma1", "rho", "tau.eps", "rho.xsi",
              "tau.xsi", "dev", "EZip", "PRES")
```

Finally we need to specify the number of chains (nc), the number of draws from the posterior for each chain (ni), the number of draws to discard as burn-in (nb), and the thinning rate. We initially used 3 chains, a burn-in of 5,000 iterations, and 100,000 draws per chain. Computing time was a few minutes on a fast computer. However, the results showed correlated chains, and we therefore increased the number of chains to 1,000,000, the burn-in to 50,000 and set the thinning rate at 1,000. This means that we have 3 × (1,000,000 – 50,000) / 1,000 = 2,850 stored values for the posterior. Calculation took half a day. These numbers are model- and data set-specific, and you may need to change them for your own data. It may be advisable to use fewer iterations for initial test runs.

```
> nc <- 3         #Number of chains
> ni <- 1000000   #Number of draws from posterior (for each chain)
> nb <-   50000   #Number of draws to discard as burn-in
> nt <-    1000   #Thinning rate
```

It is now time to start the MCMC code. Be prepared to let the computer run for several hours, if not a day. Thc package R2WinBUGS (Sturtz et al. 2005) is used.

```
> library(R2WinBUGS)
> MyWinBugsDir <- "C:/WinBUGS14/"
> SB.out <- bugs(data = win.data, inits = inits,
            parameters = params, model = "modelSB.txt",
            n.thin = nt, n.chains = nc, n.burnin = nb, n.iter = ni,
            debug = TRUE, bugs.directory = MyWinBugsDir)
> print(SB.out, digits = 3)
```

The output is discussed in the next subsection

6.7.5 MCMC results for ZIP with residual CAR structure

Congratulations, you made it so far. Getting WinBUGS to run code properly is always a major challenge. Setting up the code may take a full day, and it will take the computer a full night to run it. It may be wise to set your computer to not update its operating system or any of its software packages and automatically reboot; if this happens you will need to run the code again. The print command gives the mean, median, and other summary statistics for the posterior distribution. The 95% credible interval is the Bayesian equivalent of a 95% confidence interval. The 2.5% and 97.5% percentiles can be used for a 95% credibility interval. If 0 is in this interval, the corresponding covariate can be considered unimportant. Results are:

```
Inference for Bugs model at "modelSB.txt", fit using WinBUGS,
 3 chains, each with 1e+06 iterations (first 50000 discarded), n.thin = 1000
 n.sims = 2850 iterations saved

          mean    sd   2.5%    25%    50%    75%  97.5%  Rhat n.eff
beta1[1]  5.391 0.451  4.492  5.104  5.400  5.690  6.250 1.002  2800
beta1[2]  0.024 0.239 -0.444 -0.138  0.033  0.186  0.496 1.001  2800
beta1[3] -1.161 0.391 -1.953 -1.423 -1.160 -0.896 -0.429 1.001  2300
beta1[4] -0.969 0.332 -1.631 -1.185 -0.962 -0.743 -0.330 1.001  2800
beta1[5] -0.553 0.409 -1.355 -0.824 -0.545 -0.277  0.230 1.001  2800
```

```
beta1[6]    -0.388   0.371  -1.117  -0.630  -0.391  -0.140   0.345 1.002  1300
beta1[7]    -0.392   0.517  -1.429  -0.743  -0.379  -0.035   0.598 1.002  1600
beta1[8]    -1.314   0.786  -2.899  -1.808  -1.316  -0.810   0.247 1.001  2800
gamma1[1]  -39.030   8.754 -49.648 -46.047 -41.165 -33.870 -18.014 1.003   940
gamma1[2]   15.445  11.711 -14.738   9.789  18.940  23.918  29.668 1.001  2800
gamma1[3]   -3.392  23.999 -47.245 -24.145   1.476  17.650  30.638 1.001  2800
gamma1[4]   -5.883  20.425 -47.395 -21.595   2.120  10.300  21.140 1.001  2800
gamma1[5]    9.973  21.624 -44.656   2.931  16.725  24.783  36.503 1.001  2800
gamma1[6]  -30.016  17.331 -49.230 -42.625 -34.815 -22.660  17.759 1.001  2800
gamma1[7]  -21.645  17.906 -48.718 -36.530 -22.440  -8.634  15.930 1.001  2800
gamma1[8]  -16.144  19.165 -48.310 -32.055 -16.395  -0.340  17.294 1.002  1400
rho          0.865   0.054   0.742   0.834   0.874   0.904   0.945 1.001  2800
tau.eps      3.596   0.276   3.090   3.400   3.581   3.784   4.180 1.001  2800
rho.xsi      0.503   0.287   0.026   0.253   0.508   0.750   0.977 1.001  2200
tau.xsi      2.487   1.452   0.115   1.212   2.520   3.743   4.868 1.006  2400
...
dev        853.045  27.659 808.245 835.200 851.500 868.775 905.677 1.012   430
deviance   845.515  23.134 803.622 829.200 844.850 860.175 893.855 1.001  2800

For each parameter, n.eff is a crude measure of effective sample size,
and Rhat is the potential scale reduction factor (at convergence, Rhat=1).

DIC info (using the rule, pD = var(deviance)/2)
pD = 267.7 and DIC = 1113.2
DIC is an estimate of expected predictive error (lower deviance is better).
```

Note that, among the regression parameters, only the two intercepts and two year parameters (β_2 and β_3, representing years 1999 and 2003, respectively) in the count model have 95% credible intervals that do not contain 0. These are printed in bold. These two years have a negative effect (lower abundance than in the baseline year 1998). For the binary model we obtained large negative estimated parameters. For example, the mean intercept is -39.03. The large negative values indicate that we could try a Poisson GLM with CAR residuals as well, and compare the DIC.

The samples of the posterior for the spatial correlation parameter ρ_ε have a mean of 0.86, and the 95% credible interval is 0.74 to 0.94. The spatial correlation parameter for the binary part is 0.50.

We can also plot the samples from the posterior distribution (Figure 6.20, Figure 6.21, and Figure 6.22). R code to produce these graphs was presented in Chapter 2.

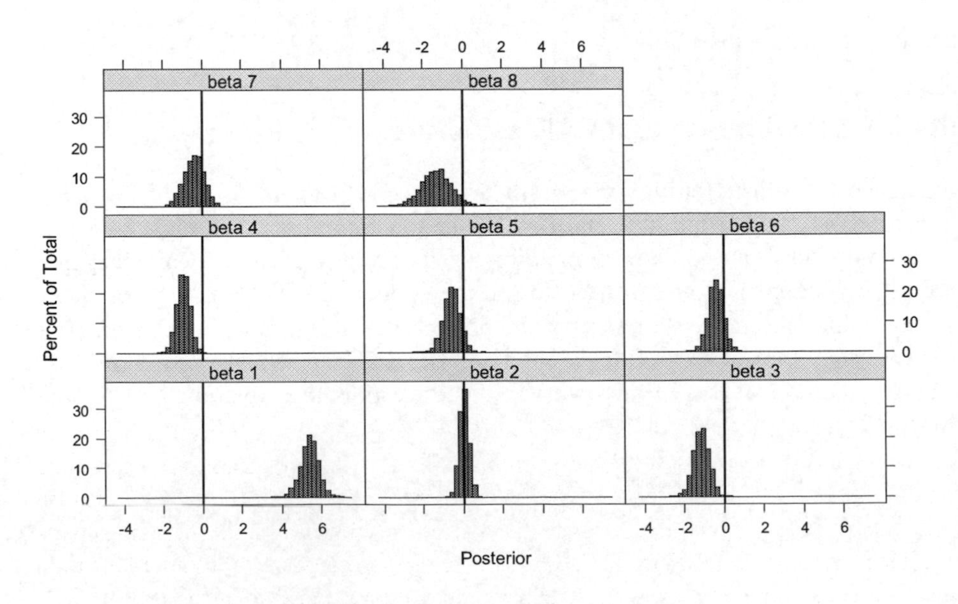

Figure 6.20. Histogram of samples from posterior distribution for the regression parameters in the count model.

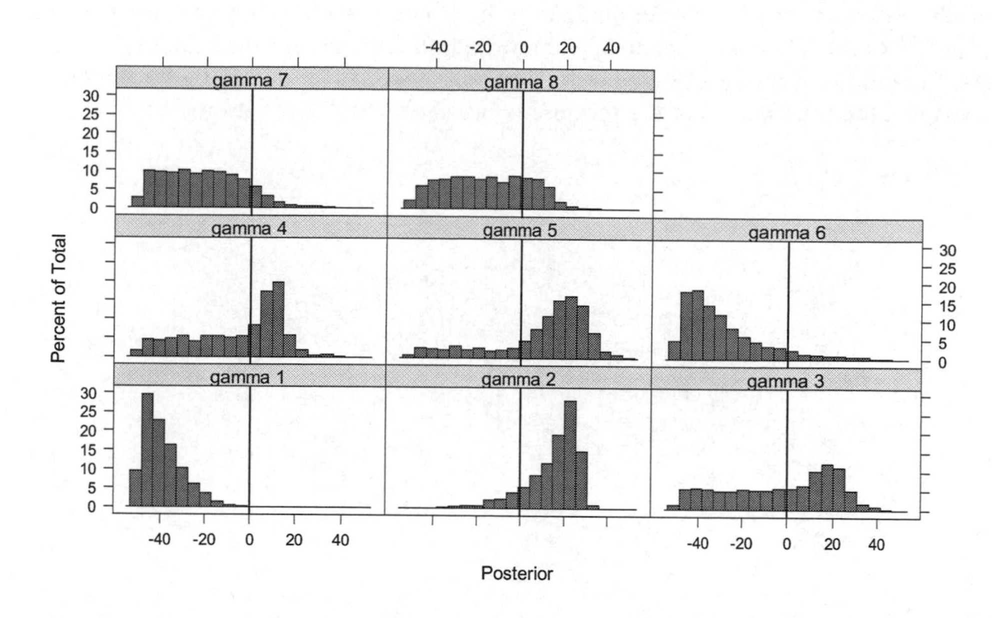

Figure 6.21. Histogram of samples from posterior distribution for the regression parameters in the binary model.

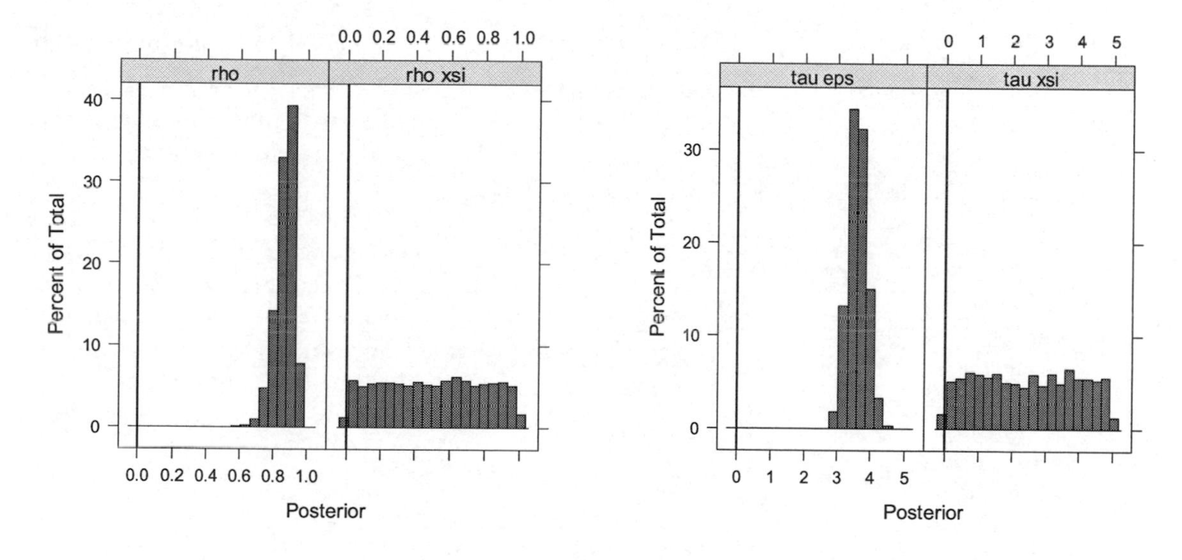

Figure 6.22. Histogram of samples from posterior distribution for the spatial correlation parameters and the precision parameter.

Mixing of the chains

When the MCMC algorithm is finished we must investigate whether the chains are properly mixed. We can plot the chains, examine auto-correlation functions, or use statistical tests to assess the mixing. The Gelman-Rubin statistic is a formal test to assess chain convergence (denoted by `Rhat` in the output) and compares the variation between chains to the variation within a chain. When convergence has been reached, it will have a value close to 1. Gelman (1996) suggests that a value less than 1.1 or 1.2 is acceptable. This test is applied to each of the model parameters. For our data, the Gelman-Rubin statistic is less than 1.2 for all parameters. Figure 6.23 provides a trace plot for the spatial correlation parameter ρ_ε, and Figure 6.24 contains the auto-correlation function for each chain. The graphs indicate proper mixing, and there is no strong no auto-correlation in the chains.

Initial MCMC runs with a lower thinning rate show auto-correlated chains, and therefore we use a thinning rate of 1,000. The mean, median, and most quartile values for the initial results obtained with the 100,000 iterations were similar to the final results obtained with 1,000,000 iterations. The only difference is that the final results of the former did not show auto-correlation.

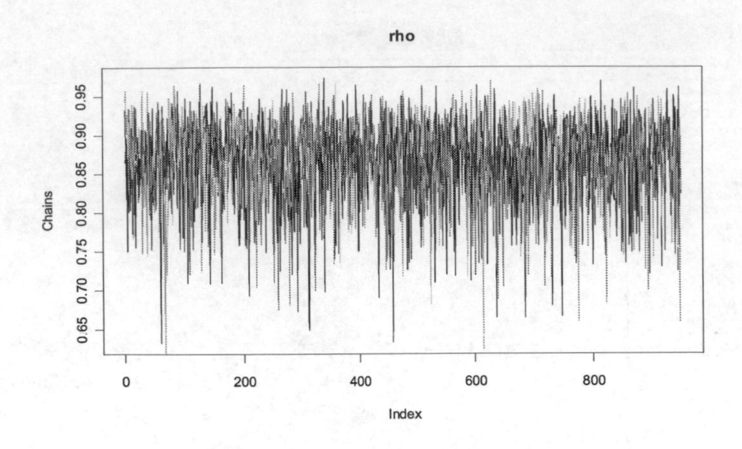

Figure 6.23. Generated draws from the parameter ρ_ε. The three chains are plotted.

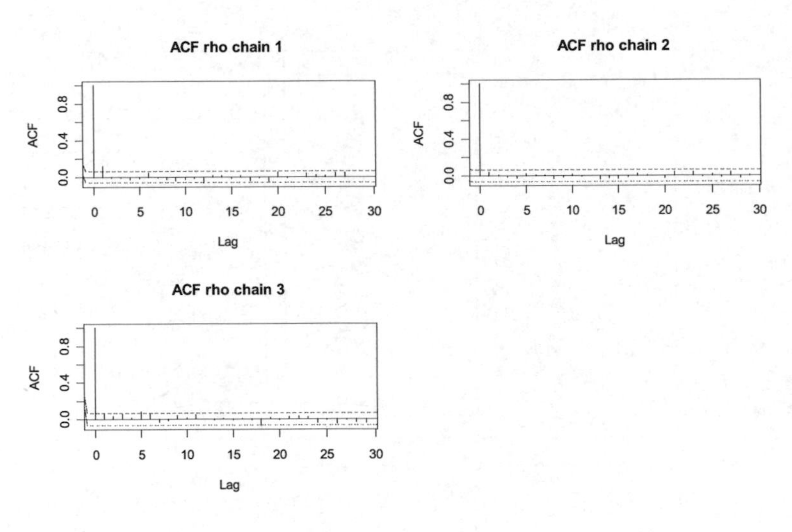

Figure 6.24. Auto-correlation functions for each chain of ρ_ε showing that the chains are not auto-correlated.

R code to produce Figure 6.23 and Figure 6.24 is given below. The major challenge is to determine which part of the object `out.SB` to access. Note that these graphs must be produced for all parameters.

```
> Chains <- out.SB$sims.array[,,paste("rho","", sep = "")]
> plot(Chains[,1], type = "l", col = 1,
       main = paste("rho", "", sep = ""))
> lines(Chains[,2], col = 2)
> lines(Chains[,3], col = 3)
> par(mfrow = c(2, 2))
> MyParam <- i
> Chains <- out.SB$sims.array[,,paste("rho","", sep = "")]
> acf( Chains[,1], main = paste("ACF ", "rho", "Chain 1", sep = ""))
> acf( Chains[,2], main = paste("ACF ", "rho", "Chain 2", sep = ""))
> acf( Chains[,3], main = paste("ACF ", "rho", "Chain 3", sep = ""))
```

Spatial correlation

As part of the model validation we must extract the Pearson residuals; for example, the mean of the posterior for each residual. These need to be plotted against the fitted values and each covariate, and we also need to inspect them for spatial and temporal correlation. As earlier, we make a sample variogram of the Pearson residuals (Figure 6.25). There is still some spatial correlation, but the range is reduced, and the values of the sample variogram are lower.

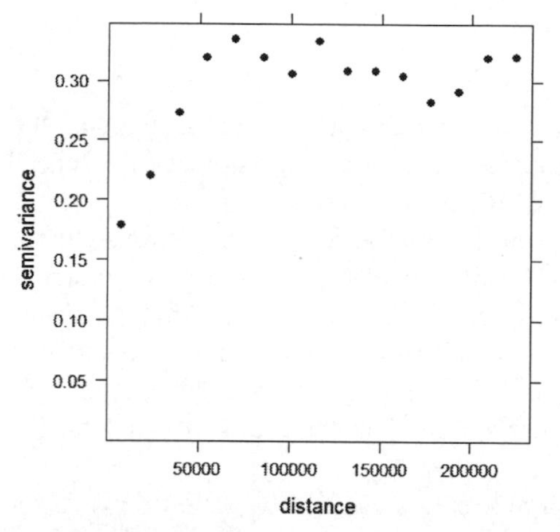

Figure 6.25. Sample variogram of Pearson residuals of the CAR ZIP.

6.8 Discussion

In this chapter we analysed count data for three skate species. For each species, we began the analysis by applying Poisson and negative binomial GLM models. All Poisson GLM models were overdispersed. Instead of presenting the results of NB GLMs we asked ourselves why we had over-dispersion, and we suspected zero inflation to be the cause. Therefore we also applied ZIP and ZINB GLM models. In all cases the zero inflated counterparts were better. Other factors that could potentially cause overdispersion are temporal and spatial correlation. None of the species data showed temporal residual correlation. Assessing whether there is spatial correlation was difficult. We used sample variograms with different settings (robust versus classical), tests such as the Moran *I* test with various settings and ZIP models with and without CAR correlation. Results were sometimes conflicting, but only for *S. bonapartii* did all tools indicated clear spatial correlation. For this species we applied a CAR residual correlation in the log and logistic link function. Our initial attempts resulted in poor mixing of the chains, probably due to the relatively small sample size, until we used 1,000,000 iterations and thinning rate of 1,000.

The definition of a neighbouring site requires further research. We semi-arbitrarily set the threshold at 20 km. Trying different values, and even different structures for the matrix **C** may be an interesting extension of this analysis.

The analysis of the *S. bonapartii* data is by no means finished. The CAR ZIP provides a baseline model, and if we were to write a paper on these data we would implement the following steps:

1. Extend the code for the CAR ZIP so that the DIC can be calculated (see Chapter 2).
2. Obtain the DIC for a CAR ZIP.
3. Modify the code so that a Poisson GLM with CAR residuals is fitted. Changing the code would be a simple operation, as we only have to delete the code related to the binary part of the model. Obtain a DIC for the Poisson CAR GLM.
4. Change the value that defines a neighbour; try lower values as well as higher values. Apply the CAR ZIP and Poisson CAR GLM and make a table of all the DICs. Also inspect the sample variogram for each model. Given that each run takes about a day, this process will consume 1 or 2 weeks computing time.

Our feeling is that the zero inflation and the spatial correlation are competing, something we will see in the zero inflated whale stranding time series (Chapter 9). In that example, sequential zeros are auto-correlated, but also contribute to zero inflation. An option is to set the spatial correlation parameter to a smaller range.

We would also try to vary the definitions of **C**. It may be an option to use weights proportional to distance rather than only close or not-close. Another aspect that could have caused trouble is correlation between temperature and one of the spatial coordinates (Figure 6.3).

6.9 What to present in a paper?

We used the paper from Verhoef and Janssen (2007) when writing this case study chapter, although our models are more easily implemented. Their paper serves as a good example of essential elements to present when using ZIP (and ZINB) GLMs with CAR residuals.

In the Introduction section of your paper you need to formulate the underlying questions. In the Methods section, mention the data exploration tools applied. An important aspect is how to deal with the collinearity. State that there is zero inflation and potentially spatial correlation. At this stage you need to explain a ZIP or ZINB GLM with CAR correlated residuals. Present the underlying equations. We strongly suggest presenting a table in which different types of models are compared via the DIC. Fit a CAR ZIP and a CAR Poisson GLM and try different values for the threshold defining a neighbour.

In the Results section present the MCMC results for each model and explain how you derived the optimal model. Comment on mixing of the chains.

Acknowledgement

We would like to thank Jay Verhoef for providing the MCMC code from VerHoef and Janssen (2007).

7 Zero inflated GLMs with spatial correlation–analysis of parrotfish abundance

Alain F Zuur, Mara Schmiing, Fernando Tempera, Pedro Afonso, and Anatoly A Saveliev

7.1 Introduction

Marine Reserves, areas set aside from fishing, have received increasing interest as a tool for ecosystem-based marine management. These reserves serve as a refuge for a large number of marine species, facilitating enhanced and undisturbed reproduction and growth. But how do we know which areas are essential for the survival and growth of fish and should be of priority for protection? Investigating the environmental variables driving fish abundance may provide the answer to this question. For this reason, underwater surveys were carried out in coastal areas of the Azores archipelago. The islands are the most recently formed and isolated islands of volcanic origin in the North Atlantic.

A large number of underwater fish sampling surveys were conducted around Faial Island and the adjacent passage that separates it from the west coast of neighbouring Pico island (Figure 7.1). From 1997 through 2004, fish were counted at different depths along 50 m transects. Parts of the study area are designated marine reserves with varying levels of protection status.

In this case study, we analyse the data of the parrotfish (*Sparisoma cretense*). The parrotfish is a typical herbivore fish inhabiting the shallow reef areas around the islands (e.g. Guénette and Morato 2001). This species is targeted by small scale gillnet fishing as well as a growing number of recreational fishers (e.g. Diogo 2007) and is known to be sensitive to overexploitation and fishing pressure (i.e. Tuya et al., 2006). Hence conservation management of this species is of major concern.

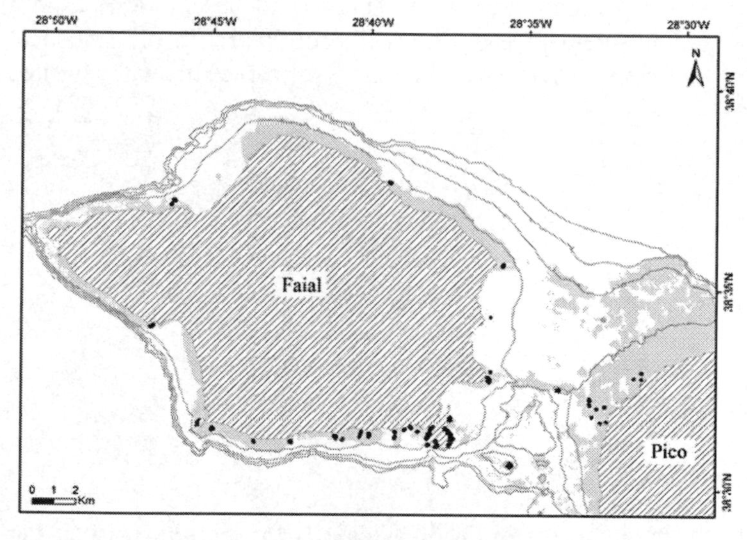

Figure 7.1. Distribution of 462 underwater transects (mid-points) surveyed from 1997 through 2004. The grey shaded areas indicate rocky seafloor. Depth is shown in 50 m contour increments up to 200 m.

We aim to discover which covariates influence parrotfish abundance. We also want to visualise the results of the statistical models as maps to better communicate the information to those involved in the creation of reserves. There are several issues that make the statistical modelling process difficult:

- The underwater surveys were not explicitly designed for application of spatial statistical models.
- Transects are not spread equally throughout the study area and are particularly concentrated in and around the marine reserve of Monte da Guia, on the southeast coast of Faial Island.
- Sampling effort differs per month due to weather conditions.

- Most environmental variables are mean values extracted from satellite imagery analysis and oceanographic models, and therefore multiple transects may have the same value of these covariates.
- The spatial correlation cannot be modelled with standard tools available in R, because we have a circular correlation, i.e. for biologically valid analysis, distances between transects cannot be measured across land.
- The counts of parrotfish are zero inflated.

These are sufficient complications to keep us busy for a while! Our data analysis approach follows the usual steps. First we will apply data exploration and try to remove any collinear covariates. The number of covariates in Table 7.1 is large, and, based on past experience with this type of data, we expect weak relationships between fish abundance and covariates, so we will use a threshold of 3 for variance inflation factors (VIF).

7.2 The data

7.2.1 Surveying

A number of sites were chosen, and a variable number of underwater transects were delineated at each site. A diver randomly selected the start and direction of the 50 m transects. The diver swam along this transect and counted the total number of fish per species in a 5 m wide path along the transect (Figure 7.2). Thus the experimental unit is 'transect,' which is represented in the data set as a row. A single dive was made at each site on a given day, during which one to seven transects at different depths and bottom types were surveyed. Some sites were revisited at a later time. This strategy is assumed to reduce the probability of recounting fish. Nevertheless, some transects are close together, so potentially the same fish may have been counted more than once within a dive. The majority of correlation in the data is likely to be between transects of the same dive (on the same day). Correlation among observations from multiple dives at the same site (but from different transects or sampling days) is thought to be of a lesser concern.

We define a numerical variable `Dive` that contains the same value for all observations from a dive. This lets us include `Dive` as a random intercept and allows for correlation in the statistical models. However, it is possible that the geographical area, and potentially the same site, was revisited later. This did not happen often.

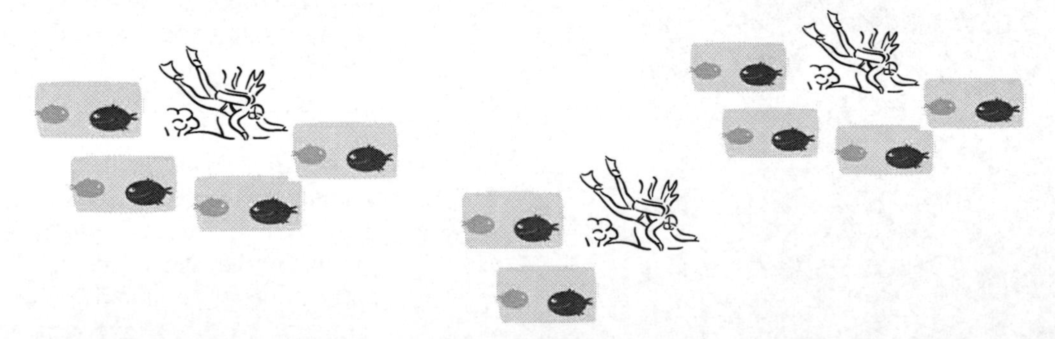

Figure 7.2. Setup of the experiment. A site is chosen and a diver counts the number of fish per species along 1 – 7 50 m transects at different depth strata/bottom type. In most cases, a single dive is made per day. Some of the sites are revisited the following year. The data are nested, in the sense that multiple observations are made during a dive. The data are too unbalanced to consider 2-way nested (multiple observations per dive and multiple dives per site).

The data are imported and checked for the correct coding with the usual set of commands:

```
> PF <- read.table(file = "ParrotFish.txt", header = TRUE)
> names(PF)

 [1] "ID"                     "Dive"              "x"
 [4] "y"                      "X.long"            "Y.lat"
 [7] "Date"                   "Year"              "Input_Month"
[10] "Depth"                  "Slope"             "Bottom_type"
[13] "DistCoa"                "DistRck"           "DistSed"
[16] "Current"                "Swell"             "Chla"
[19] "SST"                    "Ruggedness"        "Slope_slope"
[22] "Sparisoma_cretense"
```

The variables x and y are UTM coordinates for the transects, which are derived from the longitude and latitude variables X.long and Y.Lat, respectively. The UTM coordinates are required for the statistical models.

The variable Sparisoma_cretenese contains the counts of the adult parrotfish. Because the name is long to type in each time, we rename it ParrotFish using the code:

```
> PF$ParrotFish <- PF$Sparisoma_cretense
```

Before engaging in further data exploration, we visualise the sampling effort over time (Figure 7.3). The *x*-axis shows the years and the *y*-axis the months. A point is plotted in the figure to show when a sample was taken. A certain amount of jittering (adding random variation) was applied to avoid plotting multiple observations as a single dot. In the years 1998 – 2000 only, surveying took place January – April. Based on biological knowledge, we can expect seasonal differences in parrotfish abundance among sites. It therefore makes sense to use only the June – November data for the statistical analysis.

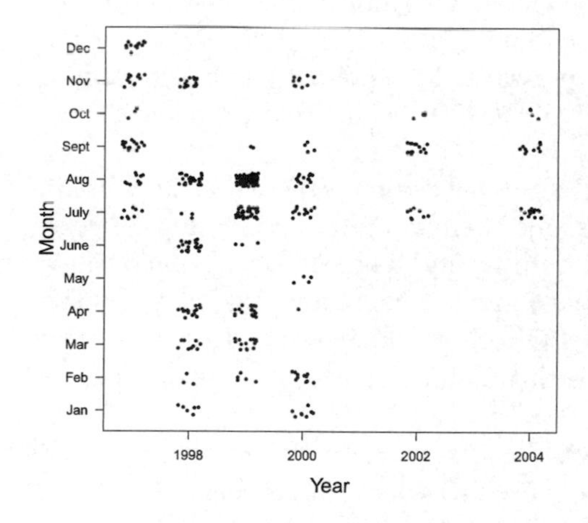

Figure 7.3. Survey month versus survey year. Jittering was applied to ensure that the difference between points sampled in the same month and year is apparent.

Figure 7.3 was created with the following code. The function jitter adds a small amount of random noise to avoid multiple observations being superimposed, the axes = FALSE option suppresses the axes, and the las = 1 option prints the labels horizontally.

```
> plot(x = jitter(PF$Year), y = jitter(PF$Input_Month), cex = 0.6,
       xlab = "Year", ylab = "Month", axes = FALSE, pch = 16)
> axis(1); box()
> axis(2, at = c(1, 2, 3, 4, 5, 6, 7, 8, 9, 10, 11,12),
       labels = c("Jan", "Feb", "Mar", "Apr", "May", "June", "July",
                  "Aug", "Sept", "Oct", "Nov", "Dec"), las = 1)
```

The June – November data are extracted with the code:

```
> PF2 <- PF[PF$Input_Month > 5 & PF$Input_Month < 12,]
> dim(PF2)

354   23
```

The dim function shows that we have 354 rows remaining. We lost 23% of the data (108 rows) by dropping the December – May data.

7.2.2 Response variable

We begin by creating a frequency plot to reveal possible zero inflation, using the plot and table functions. The results are presented in Figure 7.4; note that the data seem to be zero inflated.

```
> plot(table(PF2$ParrotFish), type = "h", xlab = "Frequencies",
       ylab = "Number of observations per frequency")
```

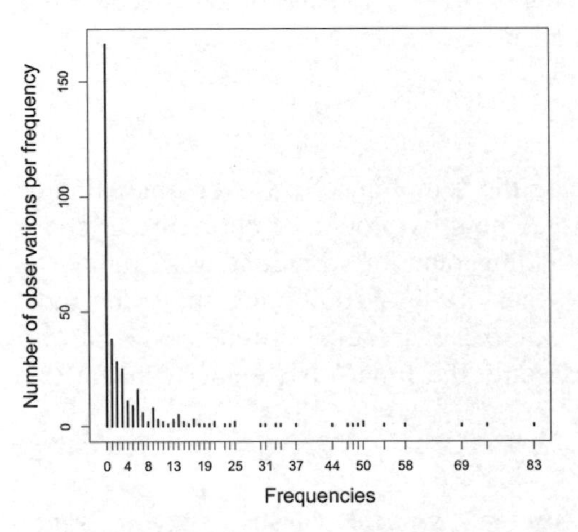

Figure 7.4. Frequency plot of the parrotfish observations. Note that there are considerably more zeros than other values, indicating potential zero inflation.

7.2.3 Covariates

The regional marine environment has been comprehensively characterised, particularly the shelves around Faial and Pico Islands to a depth of 150 metres (Tempera 2008, Tempera et al., 2012). These data are available for the Geographic Information System (GIS). A variety of toolboxes in ArcGIS (® ESRI) offer the possibility of estimating environmental covariates that have proven to help predict the distribution of marine animals (i.e. the Benthic Terrain Modeller (Wright et al., 2005)). Table 7.1 gives a short summary of the covariates.

We investigate the covariates for collinearity. However, before presenting VIFs we make a graph of each covariate versus the parrotfish count in order to assess the strength of the relationships. Strong relationships mean that we can tolerate a higher amount of collinearity, whereas weak relationships mean we should avoid even minimal collinearity. Figure 7.5 is a multi-panel scatterplot showing each covariate except Dive versus the number of parrotfish. A LOESS smoother was added. Note that none of the panels show a clear pattern, so we should ensure that collinear covariates are dropped as far as is possible.

The code to create Figure 7.5 is intimidating (see below). We first create a character string MyX with the names of the covariates that we intend to use. This is used to select the columns of the data frame PF2, which we convert first to a matrix, since the vector function cannot be applied to a data frame, and then to a vector. The variable ID contains the names of the variables (each repeated 354 times), and AllY contains the response variable repeated 14 times (this is the number of selected covariates). The xyplot function then draws the multiple panels and adds a LOESS smoother to each window. Note that warning messages (not shown here) appear, due to panels in which the covariate contains few unique values.

Table 7.1. Environmental covariates used in the analysis. All variables are continuous except for bottom type.

Covariate	Description	Dimension	Observations
Depth	Mean water depth	Metre	Mean depth of dive
Slope*	Steepness of the terrain	Degree	Raster derived from bathymetric grid with Spatial Analyst in ArcGIS; Raster resolution 5m
Swell*	Relative exposure to swell and waves	% of maximum	GIS-based fetch index weighted by swell statistics; Multiannual average 1989 – 2002; Raster resolution 200m
Chla*	Average surface chl-a concentration	mg/m^3	Multiannual weighted average (01/1999 - 10 / 2004). Raster resolution 0.412'×1.130' (Lat × Long)
SST*	Average sea surface temperature	ºC	Multiannual weighted average (04/2001 - 04/2006); Raster resolution 0.819'×0.684' (Lat×Long)
DistCoa	Distance of transect to shore	Metre	Shapefile derived from Spatial Join in ArcGIS
DistRck	Distance of transect to nearest hard bottom	Metre	Shapefile derived from Spatial Join in ArcGIS
DistSed	Distance of transect to nearest soft bottom	Metre	Shapefile derived from Spatial Join in ArcGIS
Current*	Relative exposure to currents	% of maximum	Maximum spring tide model; Raster resolution 500m
Ruggedness	Terrain ruggedness as the variation in 3D orientation of grid cells within a neighbourhood	Dimensionless, range from 0 (flat) to 1 (most rugged)	GIS raster derived by Vector Ruggedness Measure Toolbox (Sappington et al. 2007)
Bottom type*	Classification into hard and soft bottom	Dimensionless, with 0 = rock and 1= sediment	Backscatter data; Raster resolution 5m
Slope_slope	The slope of the slope	Degrees of degrees	Raster from Spatial Analyst in ArcGIS

* From Tempera (2008).

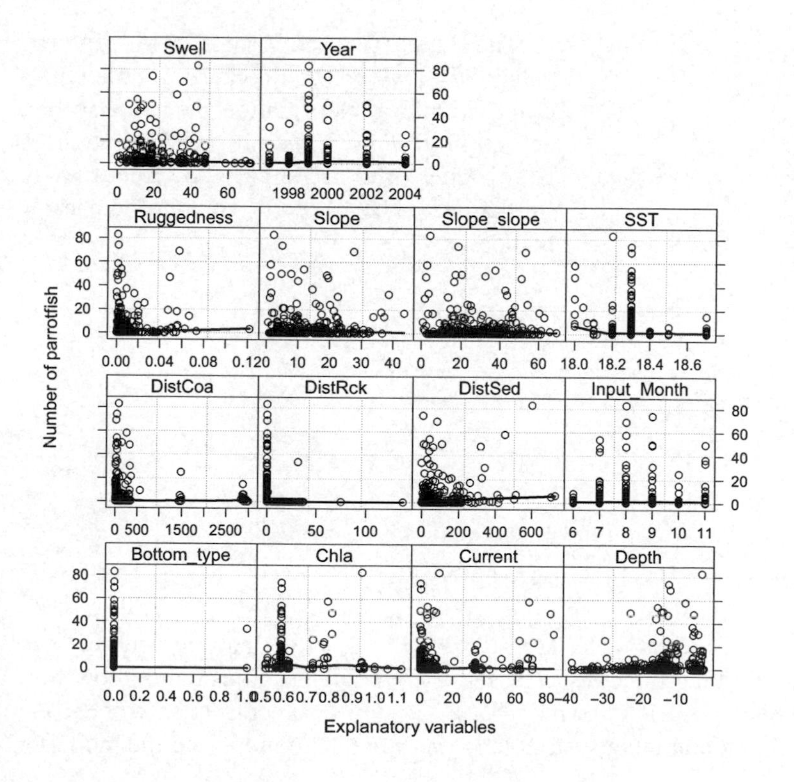

Figure 7.5. The number of parrotfish plotted versus each available covariate. A LOESS smoother was added to aid visual interpretation.

```
> MyX <- c("Year", "Input_Month", "Depth", "Slope", "Bottom_type",
           "DistCoa", "DistRck", "DistSed", "Current", "Swell",
           "Chla", "SST", "Ruggedness", "Slope_slope")
> AllX <- as.vector(as.matrix(PF2[,MyX]))
> ID   <- rep(MyX, each = nrow(PF2))
> AllY <- rep(PF2$ParrotFish, length(MyX))
> library(lattice)
> xyplot(AllY ~ AllX|ID, col = 1, xlab = "Explanatory variables",
        ylab = "Number of parrotfish",
        strip = function(bg = 'white', ...)
        strip.default(bg = 'white', ...),
        scales = list(alternating = TRUE,
                      x = list(relation = "free"),
                      y = list(relation = "same")),
        panel = function(x, y){
                  panel.grid(h = -1, v = 2)
                  panel.points(x, y, col = 1)
                  panel.loess(x, y, col = 1, lwd = 2)})
```

The final step before assessing collinearity is inspecting the covariates for outliers. A multipanel Cleveland dotplot is useful for this (Figure 7.6). The covariates DistCoa, DistRck, and Ruggedness each contain observations with considerably larger values than the majority of the observations. We log transform DistCoa and square root transform DistRck and Ruggedness. The choice to do this is subjective and against the principle of applying the same transformation to similar covariates (these have the same units), but the spread of DistCoa is much wider than of the other two, and a logarithmic transformation does not help for Ruggedness, as all its values are < 1.

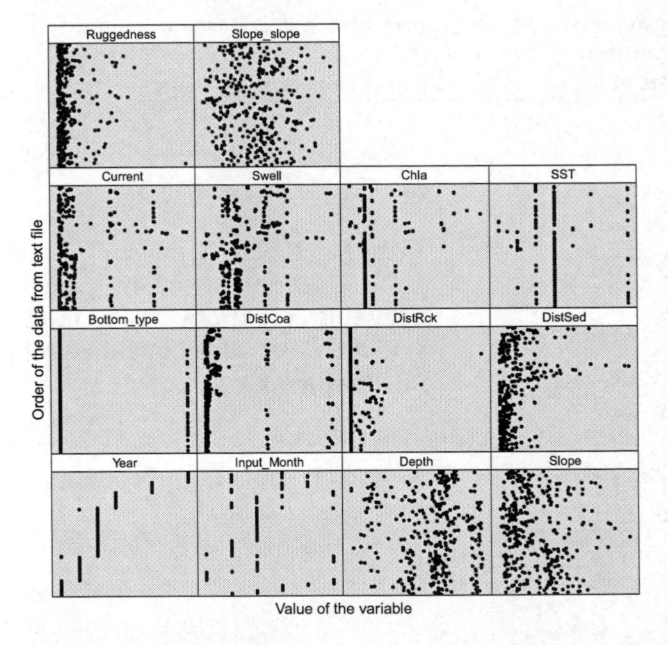

Figure 7.6. Multipanel dotplot of the covariates.

The latter two covariates also contain values of 0, which would cause trouble with a logarithmic transformation. It may also be an option to remove the observations having large values or to use slightly different transformations. The R code for the transformation is:

```
> PF2$LogDistCoa      <- log(PF2$DistCoa)
> PF2$SQgDistRck      <- sqrt(PF2$DistRck)
> PF2$SQRuggedness    <- sqrt(PF2$Ruggedness)
```

The VIFs are obtained by sourcing our library file HighStatLib.R. Note that the code below assumes that this file is stored in the working directory. The set of covariates is adjusted to allow for the transformations. We do not include a covariate for month, as it is sure to be correlated with SST. The variable Year was initially used as a continuous variable in the VIF calculations and the models, although we could consider it categorical, since it has only 6 unique values (1997 – 2004). The advantage of using it as a categorical variable is that we would allow for non-linear effects, but the disadvantage is that it consumes a large number of degrees of freedom.

```
> source("HighStatLib.R")
> MyX2 <- c("Year", "Depth", "Slope", "Bottom_type", "LogDistCoa",
            "SQDistRck", "DistSed", "Current", "Swell", "Chla",
            "SST", "SQRuggedness", "Slope_slope")
> corvif(PF2[, MyX2])

                  GVIF
Year          1.242852
Depth         1.924337
Slope         3.538202
Bottom_type   7.269756
LogDistCoa    5.781635
SQDistRck     7.021591
DistSed       1.685378
Current       3.802648
Swell         2.206684
Chla          2.071392
SST           1.701705
SQRuggedness  3.755381
Slope_slope   5.009839
```

Due to the weak patterns in Figure 7.5, we use a cut-off of 3 for the VIFs. Note that the VIF values indicate serious collinearity. After dropping a single variable at a time and recalculating VIF values, we remove the following covariates: `Bottom_type`, `LogDistCoa`, and `Slope_slope`. The VIF values for the remaining covariates are all less than 3.

Figure 7.7 shows a multipanel scatterplot for the remaining covariates. The lower diagonal panels contain Pearson correlation coefficients. The correlation between `Slope` and `SQRuggedness` is still high, at 0.7. We remove `SQRuggedness`. Correlation between `Current` and `Chla` is also high and we remove `Current`. The R code to make the pairplot in Figure 7.7 is:

```
> pairs(PF2[,MyX2], lower.panel = panel.cor)
```

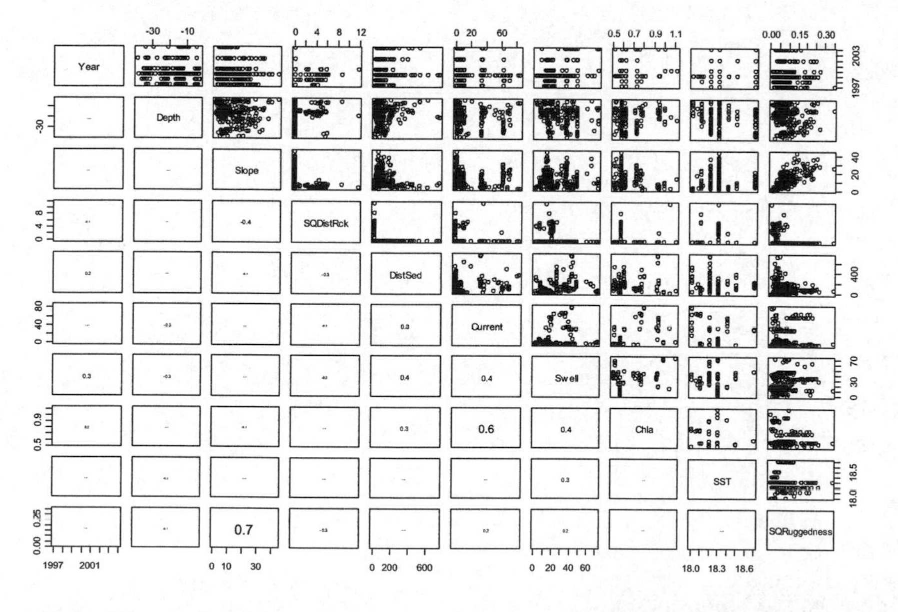

Figure 7.7. Multipanel scatterplot of the covariates.

The code requires that the `HighStatLib.R` function has been sourced. A final exploratory graph is created to investigate the number of transects per dive (Figure 7.8). For some dives only one transect was surveyed. This will lead to complications in the statistical analysis. There are 166 unique dives. The R code to produce Figure 7.8 is:

```
> plot(tapply(PF2$ParrotFish, FUN = length, INDEX = PF2$Dive),
       type = "h", xlab = "Dive", ylab = "Number of transects")
```

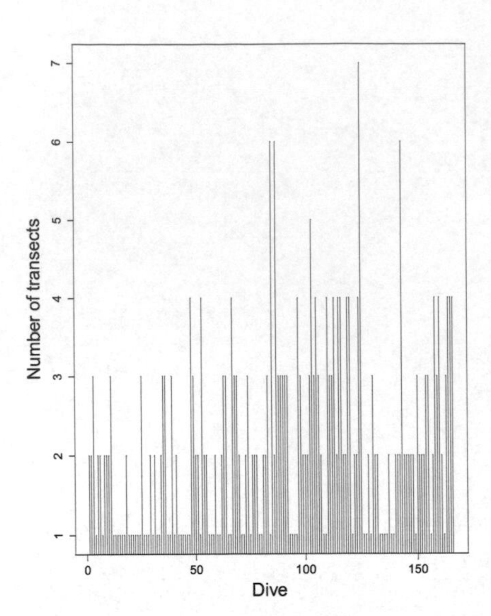

Figure 7.8. Number of sampled transects per dive.

7.3 Analysis of the zero inflated data ignoring the correlation

In this section we analyse the data without including a correlation structure in the model. We will use a generalised linear model (GLM) with a Poisson or negative binomial (NB) distribution, and then apply a zero inflated model.

7.3.1 Poisson and NB GLMs

We omit the covariate `Year`, as some of the other covariates may have changed over time, e.g. `SST`. The VIFs would have revealed any linear relationships, but not non-linear relationships. When the models have been applied, we will check whether the residuals contain non-linear year effects. The following Poisson GLM is applied:

$$\begin{aligned}
&ParrotFish_i \sim Poisson(\mu_i) \\
&E(ParrotFish_i) = \mu_i \quad \text{and} \quad \text{var}(ParrotFish_i) = \mu_i \\
&\log(\mu_i) = \alpha + \beta_1 \times Depth_i + \beta_2 \times Slope_i + \beta_3 \times SQDistRck_i + \beta_4 \times DistSed_i + \beta_5 \times Swell_i + \\
&\qquad \beta_6 \times Chla_i + \beta_7 \times SST_i
\end{aligned}$$

Fitting this model in R shows overdispersion of 24.88.

```
> M1 <- glm(ParrotFish ~ Depth + Slope + SQDistRck + DistSed +
            Swell + Chla + SST, family = poisson, data = PF2)
> PearsonRes <- resid(M1, type = "pearson")
> Dispersion <- sum(PearsonRes^2) / M1$df.resid
> Dispersion

24.87598
```

We apply the following negative binomial GLM:

$$ParrotFish_i \sim NB(\mu_i, k)$$

$$E(ParrotFish_i) = \mu_i \quad \text{and} \quad \text{var}(ParrotFish_i) = \mu_i + \frac{\mu_i^2}{k} = \mu_i + \alpha \times \mu_i^2$$

$$\log(\mu_i) = \alpha + \beta_1 \times Depth_i + \beta_2 \times Slope_i + \beta_3 \times SQDistRck_i + \beta_4 \times DistSed_i + \beta_5 \times Swell_i + \beta_6 \times Chla_i + \beta_7 \times SST_i$$

See Chapter 3 for a discussion on the notation $\alpha = 1/k$. The only difference between the Poisson and NB GLM is the variance structure. The NB GLM is also overdispersed:

```
> library(MASS)
> M2 <- glm.nb(ParrotFish ~ Depth + Slope + SQDistRck + DistSed +
                Swell + Chla + SST, data = PF2)
> PearsonRes <- resid(M2, type = "pearson")
> Dispersion <- sum(PearsonRes^2) / M2$df.resid
> Dispersion

2.433066
```

Overdispersion can be caused by a variety of factors: missing covariates; missing interactions; modelling an effect as linear when it is non-linear; ignoring temporal correlation, spatial correlation, or nested data structures; excessive number of zeros; or simply noisy data. Before proceeding to zero inflated models, one should first eliminate some of these potential causes, but, in our enthusiasm, we jump immediately to zero inflated models, partly motivated by the frequency plot in Figure 7.4. We will see later that there is an outlier as well as a missing covariate, and common sense predicts correlation in this data set. This means our decision to continue with zero inflated models at this stage is potentially too hasty.

Our rationale in presenting this analytic procedure in 'real-time' is to share our thinking processes. The price we pay is that later we will find it necessary to rerun some of the models without the outlier, etc., but this is how data analysis tends to work. At this stage we argue (partly wrong and partly right) that the overdispersion is due to the excessive number of zeros, and therefore zero inflated models may be more appropriate.

7.3.2 ZIP and ZINB GLMs

The parrotfish count contains a large number of zeros and, based on biological knowledge, we argue that these are a mixture of true and false zeros. See Chapters 2 – 4 for the definition and examples of true and false zeros. The parrotfish may be obscured from the diver's view, or it may be only temporarily not present. In both cases a zero is recorded, although the fish would normally be present. We argue that zero inflated Poisson (ZIP) or their negative binomial equivalent (ZINB) should be applied rather than the hurdle models.

The model formulation of the ZIP is given by:

$$ParrotFish_i \sim ZIP(\mu_i, \pi_i)$$

$$E(ParrotFish_i) = \mu_i \times (1 - \pi_i) \quad \text{and} \quad \text{var}(ParrotFish_i) = (1 - \pi_i) \times (\mu_i + \pi_i \times \mu_i^2)$$

$$\log(\mu_i) = \alpha + \beta_1 \times Depth_i + \beta_2 \times Slope_i + \beta_3 \times SQDistRck_i + \beta_4 \times DistSed_i + \beta_5 \times Swell_i + \beta_6 \times Chla_i + \beta_7 \times SST_i$$

$$\text{logit}(\pi_i) = \gamma_0 + \gamma_1 \times Depth_i + \gamma_2 \times Slope_i + \gamma_3 \times SQDistRck_i + \gamma_4 \times DistSed_i + \gamma_5 \times Swell_i + \gamma_6 \times Chla_i + \gamma_7 \times SST_i$$

The log-link is used to model the counts and the logistic link function models the probability of false zeros versus all other data. The model is implemented with the following code:

```
> library(pscl)
> M3 <- zeroinfl(ParrotFish ~ Depth + Slope + SQDistRck + DistSed +
                 Swell + Chla + SST,
                 dist = "poisson", link = "logit", data = PF2)
> summary(M3)
> Gammas.logistic <- coef(M3, model = "zero")
> X.logistic      <- model.matrix(M3, model = "zero")
> eta.logistic    <- X.logistic %*% Gammas.logistic
> p               <- exp(eta.logistic) / (1 + exp(eta.logistic))
> Betas.log       <- coef(M3, model = "count")
> X.log           <- model.matrix(M3, model = "count")
> eta.log         <- X.log %*% Betas.log
> mu              <- exp(eta.log)
> ExpY            <- mu * (1 - p)
> VarY            <- (1 - p) * (mu + p * mu^2)
> PearsonRes      <- (PF2$ParrotFish - ExpY)/ sqrt(VarY)
> N               <- nrow(PF2)     #MIND MISSING VALUES
> Dispersion      <- sum(PearsonRes^2) / (N - 16)
> Dispersion

 6.36154
```

The `zeroinfl` function from the `pscl` package (Zeileis et al., 2008) applies the ZIP. We then extract the estimated regression coefficients for both the binary process and the count process using the `coef` function, and manually calculate the expected value and variance of the ZIP. The equations for the expected value and variance come from Chapter 11 in Zuur et al. (2009a). The expected value and variance allow us to determine the Pearson residuals, which are required for assessment of the overdispersion. The 16 in the calculation represents the number of parameters in the model. The dispersion value is 6.361; thus this is not a good model. We can also obtain the residuals via existing functions. We double-check our results by extracting the residuals with the command:

```
> E3 <- resid(M3, type = "pearson")
> Dispersion3 <- sum(E3^2) / (N - 16)
> Dispersion3

6.36154
```

The vector `E3 - PearsonRes` represents the differences between the residuals produced by R and those we calculated. If our code is correct, the difference should be 0, and this is indeed the case. Our next step is to apply a ZINB GLM to allow for the extra overdispersion in the counts. Its mathematical formulation is:

$$ParrotFish_i = ZINB(\mu_i, \pi_i)$$

$$E(ParrotFish_i) = \mu_i \times (1 - \pi_i) \quad \text{and} \quad \text{var}(ParrotFish_i) = (1 - \pi_i) \times \mu_i \times (1 + \pi_i \times \mu_i + \frac{\mu_i}{k})$$

$$\log(\mu_i) = \alpha + \beta_1 \times Depth_i + \beta_2 \times Slope_i + \beta_3 \times SQDistRck_i + \beta_4 \times DistSed_i + \beta_5 \times Swell_i + \beta_6 \times Chla_i + \beta_7 \times SST_i$$

$$\text{logit}(\pi_i) = \gamma_0 + \gamma_1 \times Depth_i + \gamma_2 \times Slope_i + \gamma_3 \times SQDistRck_i + \gamma_4 \times DistSed_i + \gamma_5 \times Swell_i + \gamma_6 \times Chla_i + \gamma_7 \times SST_i$$

The mean and variance are again taken from Chapter 11 in Zuur et al. (2009a). In the equation for the variance for the ZINB in Zuur et al. (2009a) one of the terms has the wrong sign. The R code to fit this model and assess the dispersion is:

```
> M4 <- zeroinfl(ParrotFish ~ Depth + Slope + SQDistRck + DistSed +
                 Swell + Chla + SST, dist = "negbin",
                 link = "logit", data = PF2)
> E4 <- resid(M4, type = "pearson")
> Dispersion <- sum(E4^2) / (nrow(PF2) - 17)
> Dispersion

5.699788
```

The 17 indicates the number of parameters in the model. The ZINB is also overdispersed. To understand why we have overdispersion, we check for outliers. A graph of the Pearson residuals versus the fitted values (Figure 7.9) shows an outlier.

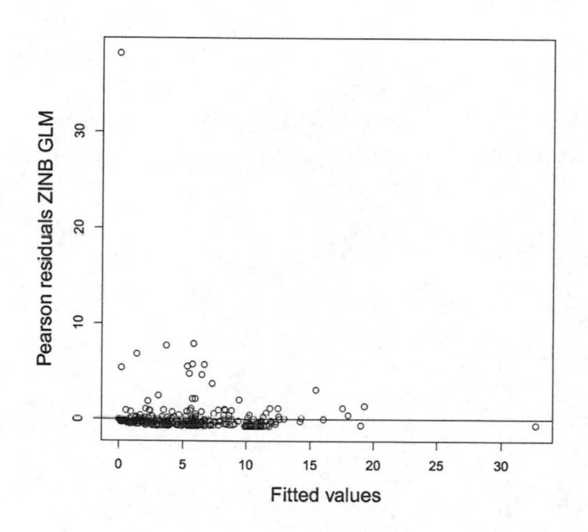

Figure 7.9. Pearson residuals of the ZINB model M4 plotted versus fitted values. Note the outlier.

The `identify` command was used to determine which observation is responsible for the large Pearson residual. It proved to be row 111, with 34 parrotfish. We drop this observation because the model is not able to fit it well, and it causes serious model misspecification (overdispersion in the ZINB). The R code to make Figure 7.9, identify the observation with the large residual, and delete row 111 is:

```
> plot(x = ExpY, y = E4, xlab = "Fitted values",
       ylab = "Pearson residuals ZINB GLM")
> abline(0, 0)
> identify(x = ExpY, y = E4) #Press escape
> PF3 <- PF2[c(-111), ]       #Remove row 111
```

Obviously, we need to refit the ZIP and ZINB with the new data set and check the new models for overdispersion.

```
> M3B <- zeroinfl(ParrotFish ~ Depth + Slope + SQDistRck + DistSed +
                  Swell + Chla + SST,
                  dist = "poisson", link = "logit", data = PF3)
> N3  <- nrow(PF3)
> E3B <- resid(M3B, type = "pearson")
> Dispersion3B <- sum(E3B^2) / (N3 - 16)
> Dispersion3B

5.446697

> M4B <- zeroinfl(ParrotFish ~ Depth + Slope + SQDistRck + DistSed +
                  Swell + Chla + SST,
                  dist = "negbin", link = "logit", data = PF3)
```

```
> E4B <- resid(M4B, type = "pearson")
> Dispersion4B <- sum(E4B^2) / (nrow(PF3) - 17)
> Dispersion4B

1.457008
```

The ZIP is still overdispersed, but the ZINB shows only marginal overdispersion. It may be reduced to 1 if we drop some of the non-significant terms in the model. If we apply a log-likelihood test to compare the ZIP with the ZINB (basically testing whether *k* is extremely large), we get:

```
> LikRatio       <- 2 * (logLik(M4B) - logLik(M3B))
> df             <- 1
> pval           <- 1 - pchisq(abs(LikRatio), df)
> Output         <- c(LikRatio, df, pval)
> names(Output) <- c("Lik.ratio", "df", "p-value")
> round(Output, digits = 3)

Lik.ratio         df   p-value
 1624.124      1.000     0.000
```

There is overwhelming evidence in favour of the ZINB. Results of the `summary(M4B)` command are presented below and indicate that various covariates are not significant at the 5% level.

```
Count model coefficients (negbin with log link):
              Estimate Std. Error z value Pr(>|z|)
(Intercept) 43.292341  15.863364   2.729  0.00635
Depth         0.054776   0.015222   3.598  0.00032
Slope         0.009860   0.012503   0.789  0.43036
SQDistRck    -0.579405   1.038126  -0.558  0.57676
DistSed       0.001182   0.001170   1.010  0.31229
Swell        -0.003256   0.010526  -0.309  0.75707
Chla         -0.922107   1.791903  -0.515  0.60683
SST          -2.203020   0.842757  -2.614  0.00895
Log(theta)   -0.707701   0.117427  -6.027 1.67e-09

Zero-inflation model coefficients (binomial with logit link):
              Estimate Std. Error z value Pr(>|z|)
(Intercept) 53.516699  43.019322   1.244  0.21349
Depth        -0.192230   0.067397  -2.852  0.00434
Slope         0.058073   0.069936   0.830  0.40633
SQDistRck     4.424162   2.650010   1.669  0.09502
DistSed      -0.005552   0.007747  -0.717  0.47359
Swell         0.165897   0.061954   2.678  0.00741
Chla         -4.851205   4.546358  -1.067  0.28595
SST          -3.385835   2.354814  -1.438  0.15048
Theta = 0.4928
Number of iterations in BFGS optimization: 176
Log-likelihood: -720.7 on 17 Df
```

`Depth` and `SST` are significant for the count model and `Depth` and `Swell` are significant for the binary model. The sign of the estimated regression coefficients indicate that more parrotfish are counted at shallower depths (`Depth` is expressed as a negative value in the data set) and lower SST. The probability of measuring false zeros increases at greater depths and higher values of exposure to sea swell.

7.3.3 Model selection for ZINB

The pros and cons of model selection were discussed earlier. Because we have dealt with collinearity using VIF values, we could present the model as it is and refrain from any model selection.

Alternatively, we could drop the covariate with the largest *p*-value, in this case `Chla` from the log-link function, refit the model, drop another non-significant covariate, and continue this process until all covariates are significant. This is a simple but controversial process. Because all covariates are continuous, we can drop them based on the *p*-values presented above. If categorical variables of more than two levels had been used, we would need the likelihood ratio test.

It is also an option to calculate the AIC for the full model and obtain new AICs while dropping each term in turn. The covariate with the highest AIC can be dropped, and a new selection round applied. This approach follows the `step` function as implemented for `lm` and `glm` models. Yet another option is to formulate 10 to 15 models *a priori*, and compare these. Our preferred option is the first: keep all covariates and refrain from model selection. This is only a good option if there is no collinearity. To demonstrate that, for these data, it doesn't matter which model selection approach is followed, we apply the second approach, dropping the least significant term and refitting the model, and continue until everything is significant.

Backwards selection by dropping the least significant terms in turn gives the following model:

```
> M5 <- zeroinfl(ParrotFish ~ Depth +  SST | Depth + SQDistRck +
                   Swell, dist = "negbin", link = "logit", data = PF3)
> summary(M5)

Count model coefficients (negbin with log link):
             Estimate Std. Error z value Pr(>|z|)
(Intercept) 38.30007   12.87715   2.974  0.00294
Depth        0.05736    0.01455   3.943 8.06e-05
SST         -1.95029    0.70434  -2.769  0.00562
Log(theta)  -0.72891    0.11966  -6.091 1.12e-09

Zero-inflation model coefficients (binomial with logit link):
             Estimate Std. Error z value Pr(>|z|)
(Intercept) -9.13054    2.71135  -3.368 0.000758
Depth       -0.19482    0.05395  -3.611 0.000305
SQDistRck    4.20112    1.91792   2.190 0.028491
Swell        0.10067    0.03993   2.521 0.011691
Theta = 0.4824
Number of iterations in BFGS optimization: 52
Log-likelihood: -724.9 on 8 Df
```

We obtain similar results to those of the full model, thanks to removing all collinear covariates. Using the code as for `M4B`, the dispersion is 1.29, which is acceptable.

```
> E5 <- resid(M5, type = "pearson")
> Dispersion5 <- sum(E5^2) / (nrow(PF3) - 8)
> Dispersion5

1.289365
```

7.3.4 Independence

The next step is to plot the residuals versus the spatial position (Figure 7.10) and create a sample variogram (Figure 7.11) to verify the assumption of independence. The graph with the residuals plotted versus their spatial coordinates shows a serious problem: a cluster of about 8 residuals with large positive values, as represented by large open circles. This may be an indication of a missing covariate. The sample variogram in Figure 7.11 is limited to 1000 m and is, in principle, faulty, as it calculates the shortest distance between points regardless of whether it crosses water or land. We will address this issue later. In the next subsection we will investigate any other patterns in the residuals.

The R code for Figure 7.10 uses the `xyplot` function from the `lattice` package.

```
> library(lattice)
> PF3$MyCex <- E5 / max(abs(E5))
> PF3$MyPch <- E5
> PF3$MyPch[E5 >= 0 ] <- 1
> PF3$MyPch[E5 < 0 ] <- 16
> xyplot(y ~ x, aspect = "iso", pch = MyPch, cex = 3 * MyCex,
         col = 1, data = PF3)
```

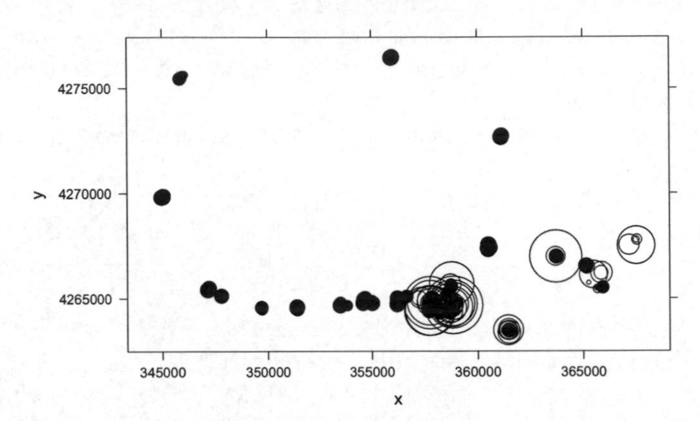

Figure 7.10. Pearson residuals plotted versus spatial position. The size of the symbol is proportional to the value of the residual. Open circles represent positive residuals and dots negative residuals.

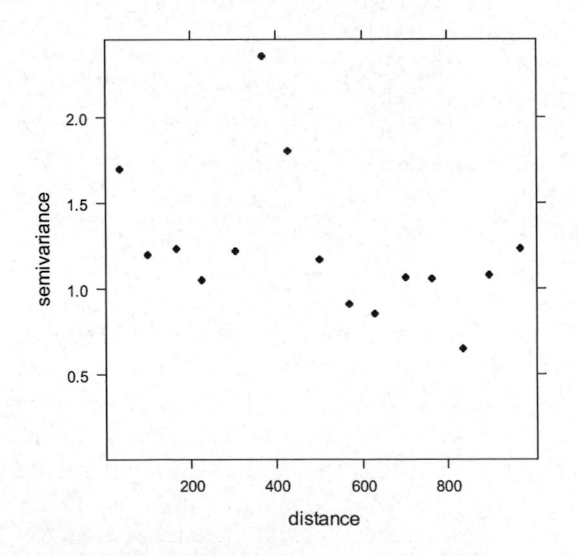

Figure 7.11. Sample variogram for the Pearson residuals. The sample variogram ignores land between some transects.

The variable `MyCex` represents the size of the dots. We arbitrarily multiply it by 3. For the sample variogram we use the `gstat` package, which is not part of the default R installation, so must be installed first.

```
> library(gstat)
> mydata <- data.frame(E5, PF3$x, PF3$y)
> coordinates(mydata) <- c("PF3.x", "PF3.y")
> Vario1 <- variogram(E5 ~ 1, mydata, cutoff = 1000)
> plot(Vario1, pch = 16, col =1)
```

7.3.5 Independence – model misspecification

Further model validation steps consist of plotting the Pearson residuals against each covariate in the model as well as each covariate not in the model. Any obvious patterns will indicate that we modelled a covariate effect as linear whereas it has a non-linear effect, or that we wrongly omitted a covariate. Besides examining these scatterplots, we can add a smoother, or base our assessment on the numerical output of a smoother. We therefore fit the following model for each covariate.

$$E_{5i} = \alpha + s(\text{Covariate}_i) + \gamma_i$$

where E_{5i} is the Pearson residual of model `M5` at observation *i*, and $s(\text{Covariate}_i)$ is a smoother. Note that this is an additive model with a Gaussian distribution for the error term γ_i. A significant smoother indicates residual patterns in the model, and further model improvement is required.

We use each covariate as a smoother in turn and obtain a strong residual pattern only with the model in which `Year` was used (Figure 7.12). This indicates that the ZINB model can be improved by adding a non-linear year effect. We can either start from scratch and add `Year` as a categorical variable to `M3B` or continue with `M5`.

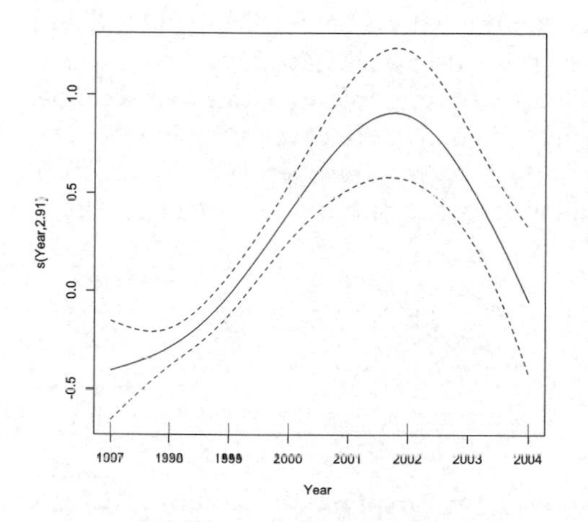

Figure 7.12. Smoother obtained with the Pearson residuals of model M5 as response variable and year as a smoother in an additive model using the Gaussian distribution. Note that there is a clear residual pattern related to year. No sampling took place in 2001 and 2003.

7.3.6 Adding year to the ZINB?

We refit the ZINB with `Year` as a categorical variable starting with `M3B` and `M4B`. When we apply a backwards selection, we encounter numerical convergence problems, which may be due to collinearity or the complexity of the model. To investigate whether any of the covariates is non-linearly related to `Year`, we fit an additive model on each covariate. `Year` is used as smoother in each model. The following additive model is used:

$$\text{Covariate}_i = \alpha + s(\text{Year}_i) + \gamma_i$$

A significant smoother indicates collinearity. Cross-validation is used to obtain the optimal degrees of freedom. Due to the limited number of years, we use an upper limit of 3 degrees of freedom. There is a significant non-linear year effect for all covariates except `Current`, `SQRuggedness`, and `SST`. This means that these covariates exhibit temporal patterns or, more likely, that a change over time in these covariates reflects a change in survey site position. Figure 7.13 shows the positions of the sites for each year. The positions of the sites are not consistent over time. This means that we cannot determine whether changes in the covariates and the response variable over time are real trends or are due to a change in survey location. A sensible approach would be to limit the analysis to those areas that were sampled in all years, but this would seriously reduce the size of the data set. We will not include year in the models due to collinearity and numerical estimation problems. This means that we must be cautious with the interpretation of the final results. The R code for Figure 7.13 is:

```
> PF3$fYear <- factor(PF3$Year)
> xyplot(y ~ x | fYear, aspect = "iso", col = 1, pch = 16,
         data = PF3)
```

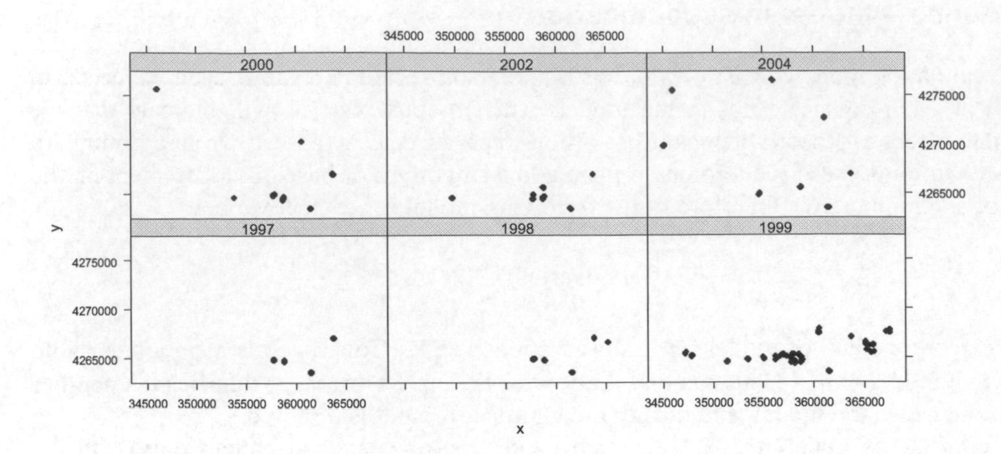

Figure 7.13. Positions of the sites by year.

7.3.7 Dive effect on the residuals

Figure 7.14 presents a boxplots of the Pearson residuals conditional on `Dive`. We would like to see each boxplot around the horizontal line crossing zero. For some dives, we have only 1 or 2 observations, so this graph is of limited use. However, for those dives with multiple transects, the boxplots of the residuals does not always cross the zero line, so there may be a dive effect in the residuals. This may be an indication to extend the ZINB model with a random intercept. However, since some dives include only one transect, we may encounter numerical challenges when fitting such a model. The R code to make Figure 7.14 is simple:

```
> boxplot(E5 ~ PF3$Dive, ylab = "Residuals", xlab = "Dive",
          varwidth = TRUE)
> abline(0, 0, lty = 3)
```

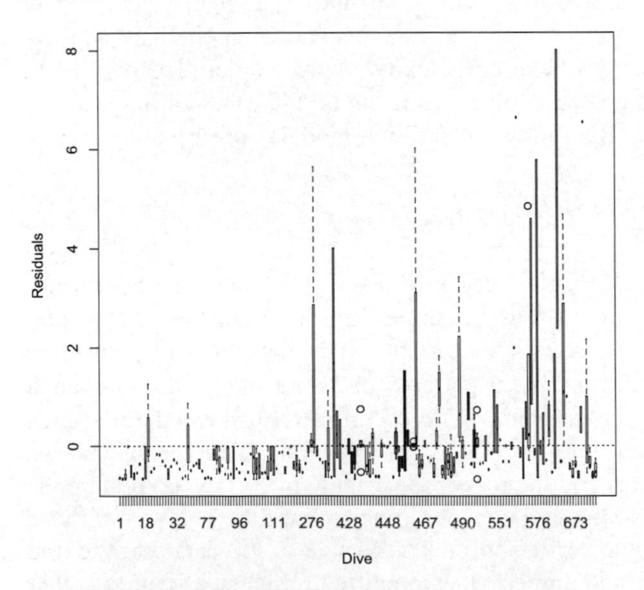

Figure 7.14. Boxplots of Pearson residuals of model `M5` conditional on dive.

7.3.8 Adjusting the sample variogram for land between transects

The sample variogram for the residuals calculated in Subsection 7.3.4 uses some distances between transects that are measured across land. A parrotfish cannot cross land (unless it is in a basket with a price tag on its belly), so we need to modify the calculations for the sample variogram.

We first need the distances between all transects. The variable `Distances` in the R code below contains the *x*- and *y*-coordinates of each transect. The data frame `DistID` contains these distances together with the information telling which transects we are using to calculate a distance. The best way to understand this is to inspect the values of `DistID` with the `head` function, which shows the first 10 rows. The `From` and `To` columns show which two transects were used to calculate distance. The remaining code plots the two histograms of the distances in Figure 7.15A. Note that in the study area transects are separated by up to 20 km. Panel B zooms in on the large number of distances of less than 1000 m. It is likely that these will be from transects sampled during the same dive, from plots in proximity, or from plots that have been revisited (sampled in other years or months).

```
> Distances <- as.matrix(dist(PF3[, c("x", "y")]))
> DistID <- data.frame(Dist = as.double(Distances),
                       From = rep(PF3$ID, each = nrow(PF3)),
                       To   = rep(PF3$ID, nrow(PF3)))
> head(DisID, 10)

         Dist  From     To
1     0.00000 10001  10001
2    63.70423 10001  10002
3    65.06473 10001  20001
4    95.15326 10001  20002
5    12.31141 10001  30001
6    42.81985 10001  30002
7    84.95904 10001  30003
8  1123.33261 10001  40001
9    78.05319 10001 110001
10  103.71956 10001 110002

> hist(DistID$Dist, main = "Long range distances", xlab="Distance")
> hist(DistID$Dist[DistID$Dist < 1000],
       main = "Short range distances", xlab = "Distance")
```

At this point it is interesting to visualise the position of some transects. We use GIS tools to chart the position of the 50 m transects with their 2.5 m bands in a particular area (Figure 7.16). There is considerable overlap. This graph ignores any time effect, plotting only the spatial positions, so the difference in time may be 3 or 4 years. However, spatial correlation can arise in various ways. The obvious correlation occurs if a diver counts a fish in transect A, and then goes to transect B and counts the same fish due to overlapping or close transects. Another cause of correlation between residuals is missing covariates, or lack of fine scale values of a covariate. For example, in Figure 7.16, sea surface temperature has the same value for all transects. Fine scale differences in fish abundance are therefore included in the residuals, and may cause residual correlation.

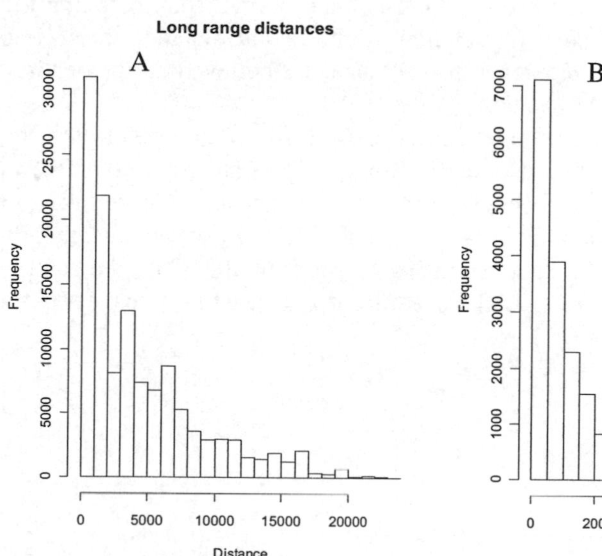

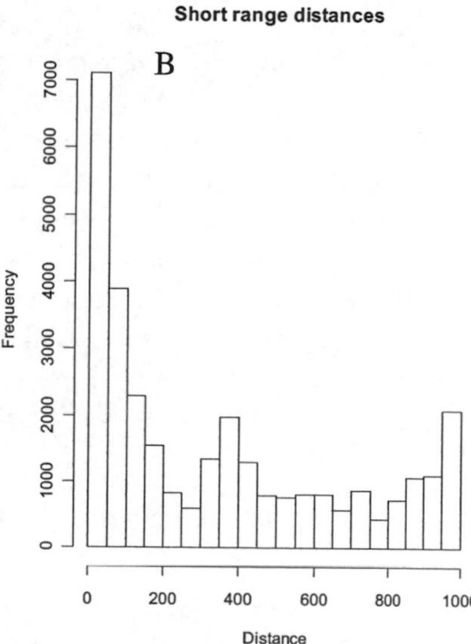

Figure 7.15 A: Histogram of distances between each combination of two transects. B: As panel A, but with distances of less than 1000 m.

We return to the calculation of the sample variogram. We now have code that calculates the distance between each combination of two transects, and we know which two transects belong to a given distance (see

DistID). Sample variograms can be obtained by various means, and we will use the robust variogram (Cressie 1993) and the Huberized variogram (Chiles and Delfiner 1999). The robust variogram reduces the effect of outliers in the residuals. The equations for the variograms are:

$$\hat{\gamma}_{Robust}(h) = \frac{1}{2}\left[\frac{1}{N(h)}\sum\nolimits_{N(h)} |Z(s_i) - Z(s_j)|^{1/2}\right]^4 / \left(0.457 + \frac{0.495}{|N(h)|}\right)$$

$$\hat{\gamma}_{Hub}(h) = \frac{1}{f(c)} \times \frac{1}{N(h)}\sum\nolimits_{H(h)} \min\left(\frac{1}{2}\left[Z(s_i) - Z(s_j)\right]^2, c^2\hat{\gamma}_{Hub}(h)\right)$$

where the summation is over all pairs of values $Z(s_i)$ and $Z(s_j)$ at sites s_i and s_j that have a distance lag h. $N(h)$ is the number of pairs that have a distance lag of h. For the Huberized variogram the values $c = 2.5$ and $f(c) = 0.978$ are selected (Chiled and Delfiner 1999).

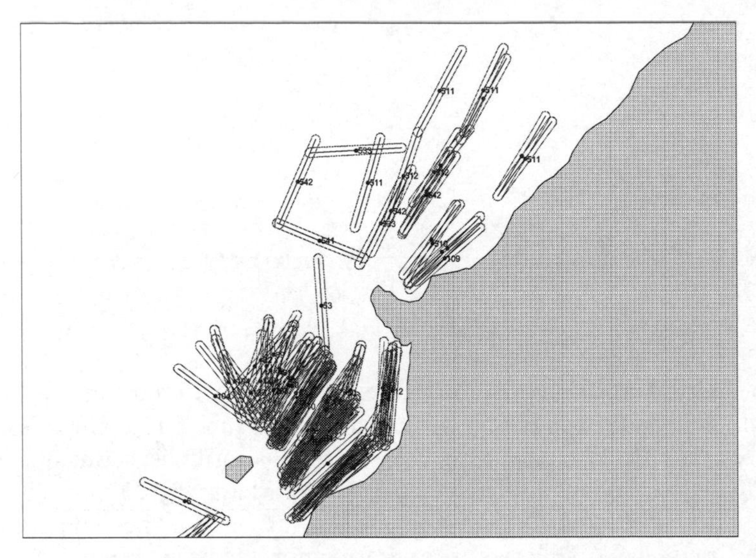

Figure 7.16. Position of some transects. A 50 m transect is represented as a bar (including 2.5 m buffer for both sides of the centreline).

When calculating these sample variograms we need to omit those distances measured across land. To do this we write R code that draws a straight line between two transects, A and B, and assesses whether any point on this straight line is on land. The code contains a loop that originates at transect A and takes small intermediate steps along the line towards transect B. At each step the code assesses whether the intermediate point is on land. To determine whether a point is on land, we use the `rgdal` package, which allows one to import GIS shapefiles with the contours of the islands. The functions `readOGR` and `point.in.polygon` are used for this. The code is complex and lengthy and is not presented here. Also, it has scope for improvement. The current code labels a path as crossing land even if there is only a small rock breaking the sea surface. It may be an option to improve the code by calculating the actual distance that a fish would have to swim in order to travel from A to B. However, for our data, we believe that the benefits would be minimal.

Figure 7.17 shows the resulting sample variograms. We limited the distances included to 500 m. There is no clear indication of spatial correlation. We also apply Moran's I test and again there is no evidence of spatial correlation. At this point we could conclude that there is no spatial correlation in the residuals and continue with the prediction of abundances using model M5. However, the sample variograms ignore the years. When we separate the data according to year and calculate the sample variogram for data of each year, there is evidence of spatial correlation for at least two years (Figure 7.18).

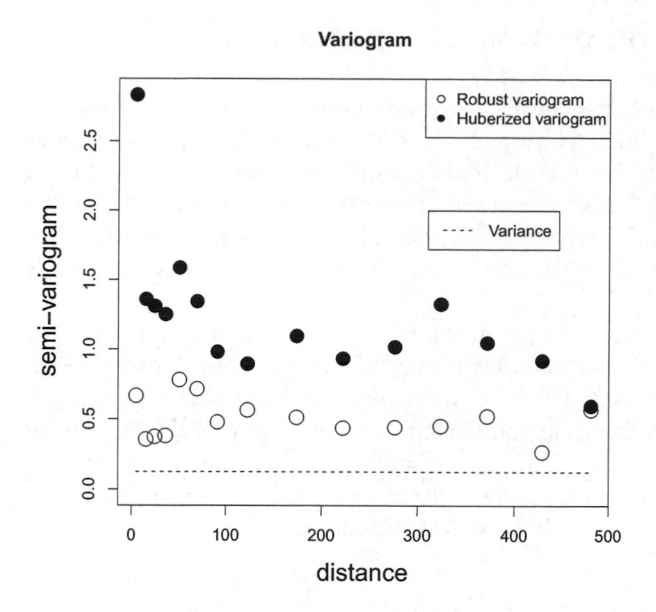

Figure 7.17. Robust and Huberized sample variogram for short distances. The dotted line is the sample variance. The robust estimator was based on quantiles. Data for all years were included.

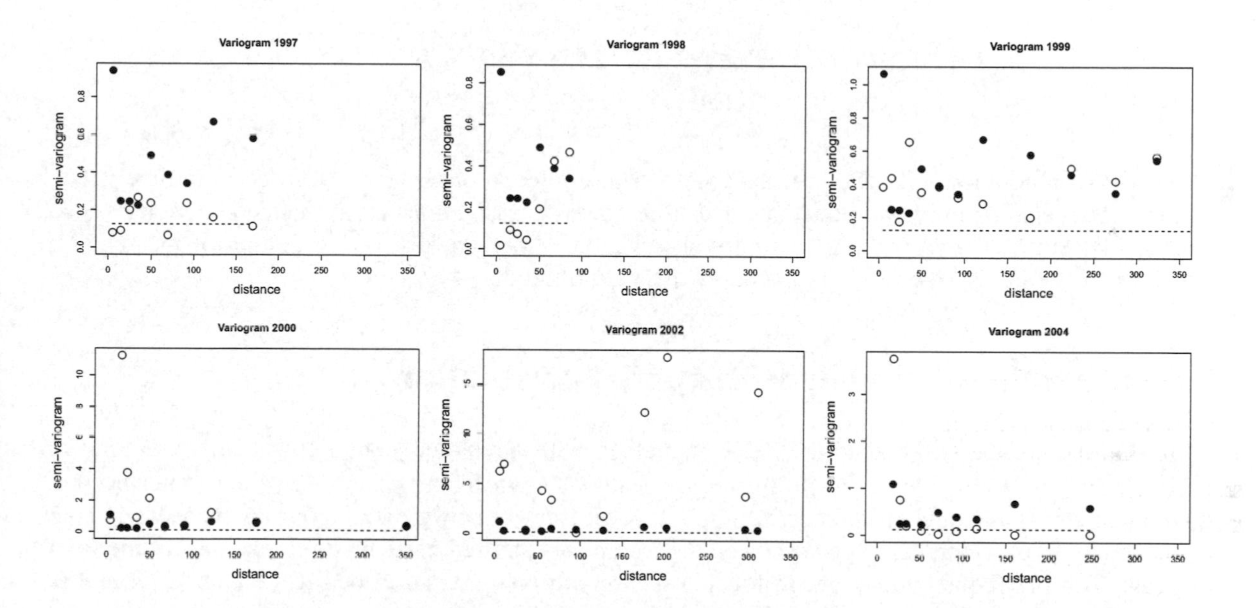

Figure 7.18. Robust (open rectangles) and Huberized (filled circles) sample variogram per year for short distances. The dotted line is the sample variance. The robust estimator was based on quantiles.

7.3.9 Which model to choose?

Many authors using zero inflated models present tables in which estimated parameters and standard errors obtained by Poisson, NB, ZIP, and ZINB GLMs are printed side by side. There is no point in doing this here, since both the Poisson and NB GLMs are overdispersed and should not be used. The choice is then between the ZIP and the ZINB. The likelihood ratio test shows that we should use the ZINB.

7.4 Zero inflated models with a random intercept for `dive`

A potential source of correlation is that we have transects located close together. It is therefore possible that a diver counted the same fish more than once. This introduces correlation between observations from the same dive. To deal with this pseudo-replication, total counts per dive can be taken, or the mean or median of counts per dive can be used. However, this suggestion was not acceptable to the involved biologists, because different strata (depths, bottom type) were sampled in a single dive. Instead we will assume that pseudo-replication occurred in only a limited number of cases. Note that this is a biological assumption.

There may still be correlation among observations from the same dive. Habitat in a particular bay may be favourable, resulting in high abundance at all transects surveyed during a dive. Or the covariate used may not have a fine scale difference. Sea surface temperature, for example, has the same value for all transects of a particular dive. Differences in abundance due to fine scale SST differences in transects of the same dive will be included in the residuals and may cause residual correlation. To model this type of correlation, zero inflated mixed effects models can be used. For example, model `M4` can be extended with random effects in the log- and logistic link functions:

$$\begin{aligned}\log(\mu_{ij}) &= \alpha + \beta_1 \times Depth_{ij} + \beta_2 \times Slope_{ij} + \beta_3 \times SQDistRck_{ij} + \beta_4 \times DistSed_{ij} + \\ &\quad \beta_5 \times Swell_{ij} + \beta_6 \times Chla_{ij} + \beta_7 \times SST_{ij} + a_j \\ \text{logit}(\pi_{ij}) &= \gamma_0 + \gamma_1 \times Depth_{ij} + \gamma_2 \times Slope_{ij} + \gamma_3 \times SQDistRck_{ij} + \gamma_4 \times DistSed_{ij} + \\ &\quad \gamma_5 \times Swell_{ij} + \gamma_6 \times Chla_{ij} + \gamma_7 \times SST_{ij} + b_j\end{aligned}$$

We adopt a notation with two indices, *i* and *j*. These refer to observation (transect) *i* in dive *j*. The random intercepts a_j and b_j introduce a correlation among observations from the same dive (See also Chapter 4). Some dives consist of a single observation, which makes the estimation of the random intercepts difficult. We therefore do not apply this type of model.

7.5 Zero inflated models with spatial correlation

In this section we will extend the ZINB model `M5` with a residual spatial correlation structure. In principle, we could decide not to do this, since results shown in the previous section indicated that there is no strong residual spatial correlation. However, for your own data there may be a strong spatial correlation, in which case you may need the techniques applied here. It is not material for the faint of heart! The pre-requisite knowledge for this section are methods described in Chapters 1, 2 and 6. The method was also presented in VerHoef and Jansen (2007), so you may want to consult that paper as well.

7.5.1 Adding a residual correlation structure to the ZINB

In this subsection, we will follow the approach presented in VerHoef and Jansen (2007). They analysed seal observations from a spatio-temporal point of view and applied a ZIP model with spatially correlated residuals. Temporal correlation was included by allowing some of the regression parameters to change over time. For the parrotfish data, the quality of the data does not lend itself to temporal correlation structures. Furthermore, we concluded with a ZINB instead of a ZIP model. We can still apply the VerHoef and Jansen approach to incorporate the spatial correlation. Their approach requires the addition of a spatially correlated error term in the log- and logistic link functions:

$$\log(\mu_i) = \beta_0 + \beta_1 \times Depth_i + \beta_2 \times Slope_i + \beta_3 \times SQDistRck_i + \beta_4 \times DistSed_i + \beta_5 \times Swell_i + \beta_6 \times Chla_i + \beta_7 \times SST_i + \varepsilon_i$$
$$\text{logit}(\pi_i) = \gamma_0 + \gamma_1 \times Depth_i + \gamma_2 \times Slope_i + \gamma_3 \times SQDistRck_i + \gamma_4 \times DistSed_i + \gamma_5 \times Swell_i + \gamma_6 \times Chla_i + \gamma_7 \times SST_i + \xi_i$$

Note that we re-converted to our original index notation. The only new components are the residuals ε_i and ζ_i. A conditional auto-regressive model (CAR) is used to allow for spatial correlation in these residual terms. CARs were discussed in Chapter 6, and we advise reading that section before continuing with this chapter. We will apply the CAR model to the residuals. For the residuals in the log-link function we have:

$$E(\varepsilon_i) = \rho_\varepsilon \times \sum_{j \neq i} c_{ij} \times \varepsilon_j$$

and for the residuals in the logistic link function we use:

$$E(\xi_i) = \rho_\xi \times \sum_{j \neq i} c_{ij} \times \xi_j$$

We can write this as (see Chapter 6):

$$\varepsilon \sim N\left(\mathbf{0}, \sigma_\varepsilon^2 \times (\mathbf{I} - \rho_\varepsilon \times \mathbf{C})^{-1} \times \mathbf{M}\right)$$
$$\xi \sim N\left(\mathbf{0}, \sigma_\xi^2 \times (\mathbf{I} - \rho_\xi \times \mathbf{C})^{-1} \times \mathbf{M}\right)$$

The definition of **M** and **C** is discussed below.

7.5.2 ZINB with a residual CAR correlation in R

Code to fit a ZINB with a residual CAR correlation structure is not available in packages such as `mgcv`, `nlme`, `lmer`, and `pscl`. We therefore formulate the ZINB in a Bayesian context and use the package `R2WinBUGS` to fit the model. Fitting ZIP and ZINB models within a Bayesian context was explained in detail in Chapter 4, and CAR GLM was discussed in Chapter 6.

The core of the code for the ZINB with spatial correlation follows. To reduce computing time we use the optimal ZINB model as starting point.

```
Model {
  #Likelihood
  for (i in 1 : N) {
    #Logit part
    W[i] ~ dbern(psim1[i])
    eta.psi[i] <- gamma[1] + gamma[2]*Depth[i] +
                  gamma[3]*SQDistRck[i] + gamma[4]*Swell[i] +
                  RE.bin[i]
    logit(psi[i]) <- max(-20, min(20, eta.psi[i]))
    psim1[i] <- min(0.99999, max(0.00001,(1 - psi[i])))

    #Negbin part
    p[i] <- size / (size + eff.mu[i])
    ParrotFish[i] ~ dnegbin(p[i], size)
    eta.mu[i]  <- beta[1] + beta[2]*Depth[i] + beta[3]*SST[i] +
                  RE.nb[i]
```

```
    log(mu[i]) <- max(-20, min(20, eta.mu[i]))
    eff.mu[i]  <- W[i] * mu[i]

    #CAR stuff
    EPS.mean[i] <- 0
  }

  #Priors
  for (i in 1:4) { gamma[i] ~ dnorm(0, 0.001) }
  for (i in 1:3) { beta[i]  ~ dnorm(0, 0.001) }
  size ~ dunif(0.2, 2)

  #Proper CAR prior distribution for spatial random effects:
  RE.bin[1:N] ~ car.proper(EPS.mean[],C[],adj[],num[],M[],tau.bin,
                           gamma.bin)
  RE.nb[1:N]  ~ car.proper(EPS.mean[],C[],adj[],num[],M[],tau.nb,
                           gamma.nb)

  #CAR priors:
  tau.bin    ~  dgamma(0.01, 0.01)
  tau.nb     ~  dgamma(0.01, 0.01)
  gamma.min <- min.bound(C[], adj[], num[], M[])
  gamma.max <- max.bound(C[], adj[], num[], M[])
  gamma.bin ~  dunif(gamma.min, gamma.max)
  gamma.nb  ~  dunif(gamma.min, gamma.max)
}
```

The code that is under the `Likelihood` heading is nearly identical to the ZIP code in Kéry (2010). The only difference is that we use a negative binomial distribution, and the terms `RE.bin` and `RE.nb` in the link functions refer to the residuals ζ_i and ε_i, respectively.

Diffuse (non-informative) priors are used for all regression parameters $\alpha, \beta_1, ..., \beta_7, \gamma_0, ..., \gamma_7$. This explains the

```
for (i in 1:n.gamma) { gamma[i]  ~ dnorm(0, 0.001)}
for (i in 1:n.beta)  { beta[i]   ~ dnorm(0, 0.001) }
```

in the code. The WinBUGS function `car.proper` carries out the residual CAR model; its input matrices `C[]` and `M[]` need to be specified before running the MCMC. These are the **C** and **M** matrices in vector format (see Chapter 6). We use a threshold of 10 metres to define neighbouring sites, and set all elements of **C** to 0 for non-neighbouring sites (distance ≥ 10 metres). For neighbouring sites (distance < 10 meters), the elements of **C** are set to the reciprocal of the number of neighbours (see also the GeoBUGS user manual). This means that the CAR correlation will capture small-scale correlation. The diagonal elements of **M** contain the reciprocal of the number of neighbours. This part of the code is difficult, and the relevant sections in the online GeoBUGS user manual are useful.

The variables `gamma.bin` and `gamma.nb` represent the spatial auto-correlation parameters ρ_ξ and ρ_ε, respectively. The variables `prec.bin` and `prec.nb` are the reciprocals of the square root of the variance terms for the residuals in the logistic and log-link function, respectively, σ_ζ and σ_ε. The `gamma.min` and `gamma.max` variables specify the range of the spatial correlation parameters (see also the help file of `car.proper`).

7.5.3 MCMC results

The chief point of interest in the numerical output is whether the spatial auto-regressive parameters ρ_ξ and ρ_ε are different from 0. If these parameters are 0, there is no spatial correlation (at least not the type of correlation that can be modelled with a CAR), in which case we can proceed with the ZINB model presented in Subsection 7.3.3.

For the MCMC settings, we use 3 chains with a burn-in of 5,000 iterations and a thinning rate of 1,000. The number of draws from the posterior distribution is 5,000,000, and the resulting 4,995 stored values are used to create 95% credibility intervals.

The MCMC chains mix well for some parameters, and all parameters pass the tests for stationarity. Posterior distributions for all the parameters follow. We are especially interested in the spatial auto-regressive parameters ρ_ξ and ρ_ε, which are the `gamma.bin` and `gamma.nb`.

```
Inference for Bugs model at "ZINBCAR.txt", fit using WinBUGS,
 3 chains, each with 5e+06 iterations (first 5000 discarded),
 n.thin = 1000, n.sims = 14985 iterations saved

            mean   sd   2.5%    25%    50%    75%  97.5% Rhat n.eff
gamma[1]   -15.6  9.2  -36.9  -21.0  -14.1   -8.7   -2.0    1   610
gamma[2]   -16.2  7.5  -33.0  -20.8  -15.3  -10.7   -4.3    1   550
gamma[3]    52.3 19.6   16.9   38.2   51.3   65.1   93.8    1   650
gamma[4]    13.6  6.6    3.1    8.7   12.8   17.6   28.8    1   500
beta[1]      1.6  0.1    1.4    1.6    1.6    1.7    1.9    1  2100
beta[2]      0.5  0.1    0.2    0.4    0.5    0.6    0.8    1 15000
beta[3]     -0.3  0.1   -0.5   -0.4   -0.3   -0.2   -0.1    1 15000
size         0.5  0.1    0.4    0.4    0.5    0.5    0.6    1  9600
gamma.bin    0.2  0.6   -0.9   -0.3    0.3    0.7    0.9    1  2000
gamma.nb     0.0  0.5   -0.9   -0.4    0.1    0.5    0.9    1 15000
tau.bin      0.6  9.1    0.0    0.0    0.0    0.0    0.5    1   700
tau.nb      38.9 51.8    2.5    8.6   20.0   47.3  185.3    1  4700
deviance  1405.5 23.1 1363.0 1389.0 1404.0 1421.0 1453.0    1  9500

For each parameter, n.eff is a crude measure of effective sample
size, and Rhat is the potential scale reduction factor (at conver-
gence, Rhat=1).

DIC info (using the rule, pD = var(deviance)/2)
pD = 266.2 and DIC = 1671.7
DIC is an estimate of expected predictive error (lower deviance is
better).
```

Results for the parameters ρ_ξ and ρ_ε indicate that 0 is within the 95% credibility interval, indicating that there is no strong spatial correlation. Note that the dispersion parameter k is similar to the one obtained by model `M5`. Using other values (e.g. 5 metres, 15 metres, 25 metres) for the distance threshold defining neighbouring sites gives similar results.

The bottom line is there is no strong spatial correlation; therefore we continue with the ZINB model presented in Subsection 7.3.3.

7.6 Predictions

Results of the ZINB model with CAR residual correlation indicate no spatial correlation. The same information was obtained by a sample variogram applied to the Pearson residuals from a ZINB in Subsections 7.3.4 and 7.3.8. We can conclude that there is no spatial correlation and predict fish abundance based on the ZINB model presented in Subsection 7.3.3. The code to predict values at the observed transects is simple:

```
> M5 <- zeroinfl(ParrotFish ~ Depth +  SST | Depth +  SQDistRck  +
                 Swell, dist = "negbin", link = "logit", data = PF3)
> Betas.logistic <- coef(M5, model = "zero")
> X.logistic     <- model.matrix(M5, model = "zero")
> eta.logistic   <- X.logistic %*% Betas.logistic
> p              <- exp(eta.logistic) / (1 + exp(eta.logistic))
> Betas.log      <- coef(M5, model = "count")
> X.log          <- model.matrix(M5, model = "count")
> eta.log        <- X.log %*% Betas.log
> mu             <- exp(eta.log)
> ExpY           <- mu * (1 - p)
> k              <- M5$theta
> VarY           <- (1 - p) * mu * (1 + mu * p + mu / k)
```

This is the same code used in Subsection 7.3.2 but now applied to `M5`. The code below prints the individual components: the probabilities of false zeros p, the mean of the count process μ, the fitted values *ExpY* of the ZINB, and the variance *VarY* of the fitted values. The dot size is proportional to the count value.

```
> par(mfrow = c(2, 2), mar = c(4, 2 ,2, 2))
> plot(x = PF3$x, y = PF3$y, xlab = "", ylab = "", type = "n",
       main = "Probability of false zeros")
> MyCex1 <- p / max(p)
> points(x = PF3$x, y = PF3$y, cex = 2 * MyCex1, pch = 16)
> plot(x = PF3$x, y = PF3$y, xlab = "", ylab = "", type = "n",
       main = "Count process")
> MyCex2 <- mu / max(mu)
> points(x = PF3$x, y = PF3$y, cex = 2 * MyCex2, pch = 16)
> plot(x = PF3$x, y = PF3$y, xlab = "", ylab = "", type = "n",
       main = "Fitted values ZINB")
> MyCex3 <- ExpY / max(ExpY)
> points(x = PF3$x, y = PF3$y, cex = 2 * MyCex3, pch = 16)
> plot(x = PF3$x, y = PF3$y, xlab = "", ylab = "", type = "n",
       main = "Variance ZINB")
> MyCex4 <- VarY / max(VarY)
> points(x = PF3$x, y = PF3$y, cex = 2 * MyCex4, pch = 16)
```

The resulting graph is shown in Figure 7.19. With a little extra work we could add the coastlines of the islands.

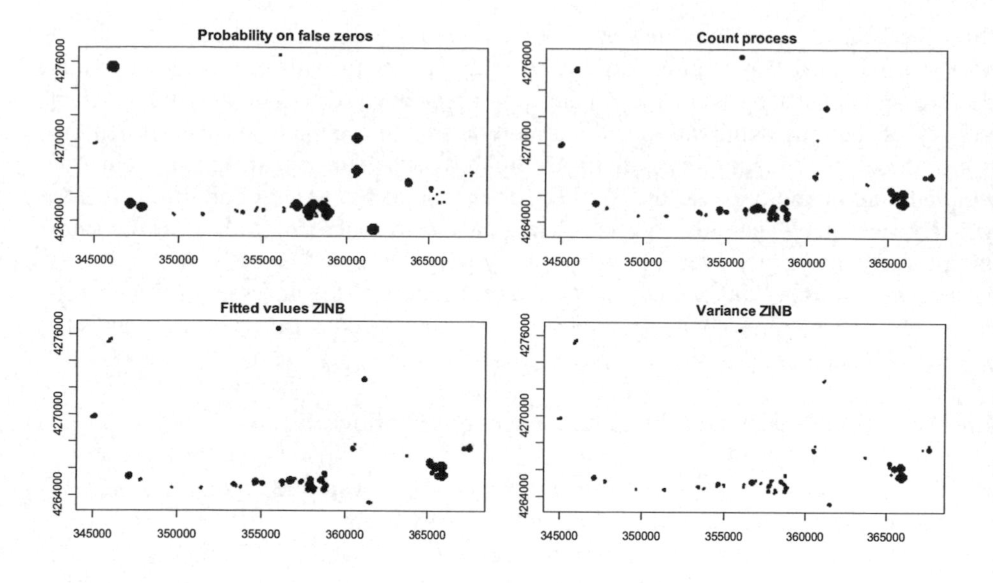

Figure 7.19. Probability of false zeros, expected values of the count process in the ZINB, fitted values of the ZINB, and the variance of the ZINB, plotted against position. Dot size is proportional to the count value.

7.7 Discussion

In this chapter we applied zero inflated models to parrotfish abundance. The main problem we encountered was residual patterns. To be more specific, we are under-predicting the counts for about 10 transects that are close together. Adding a spatial correlation structure to the ZINB model did not improve the model. In other words, the spatial correlation structure was not able to capture the residual patterns. Most likely there was a missing covariate; for example the size class of the parrotfish, the distribution of its food source (e.g. algae), or the protection level of a site. This means that we must adopt a cautious attitude when discussing the results; parameters may be biased and standard errors may be too small. However, some of the covariates are highly significant, and a missing covariate does not change the conclusion that such covariates are important.

One of the aims of this analysis was to provide an attractive visual representation of the predicted values. Figure 7.19 is not the best graph for this purpose. We could add colours, contour lines and the coastline.

There is an issue with changes in sampling locations over time, which is discussed in the next section.

7.8 What to write in a paper

When submitting a paper to a journal, it will first go to an editor who will decide if it is potentially appropriate for the journal, in which case it will be sent to referees. In this process there are crucial questions:

1. Is the subject of the paper of sufficient interest to the journal readership? Are the underlying biological questions of importance?
2. Has the data collection process been done correctly?
3. Has the statistical analysis been properly conducted and presented?
4. Do the results present novel and relevant information?

The statistical analysis of the parrotfish data was straightforward, although the statistical models that we applied may have seemed challenging. ZIP and ZINB models are now becoming standard tools in analysis of ecological data, and the ZINB with a residual CAR correlation structure has been used elsewhere. Provided it is properly presented, we do not foresee problems with the statistical components of a report on this data. The results make biological sense and should be of interest to journal readership. So, questions 3 and 4 are answered positively. As to question 1, it is matter of finding the appropriate journal. That leaves question 2. A primary problem with this data set is that the sampled locations were not consistent over time. Changes in abundance over time may then reflect differences in the ecology of the area being surveyed. To include this in a paper, you would need to argue that, based on biological knowledge, like is being compared with like. But if, in one year, the northern areas are sampled, and in another year the southern areas, and if the habitat conditions differ, this assumption may not be accepted. The only solution in that case is to focus on a subset of the sampling area that was monitored consistently over the entire survey period.

As to the statistics, we are confident that we can write a useful report of the analysis of the parrotfish data. There may be residual patterns, but provided we show and discuss this, and are cautious with the interpretation of the results (e.g. ignore the covariates with weak significance), it should not be grounds for rejection.

We would structure the statistical section of the Methods section as follows:

1. Data exploration. The keywords are zero inflation in the response variable, transformation of some of the covariates (outliers), and collinearity (VIF values were used).
2. A short discussion of the nature of the zeros before introducing the ZIP and ZINB models.
3. The basic ZINB model with formula explained in 2 or 3 lines. Many readers of the paper will be unfamiliar with it.
4. Discussion of the model selection approach.
5. Explanation of the model validation steps, including how we dealt with potential spatial correlation. Mention that a sample variogram was applied to the Pearson residuals, and that distances measured across land were excluded from the calculations.
6. Optionally, include the ZINB model with spatial correlation. The sample variogram shows that there is no strong spatial correlation, so we could omit mention of the ZINB with CAR residuals. Showing that we have created the model lets referees know that we have conducted a comprehensive analysis, and it would add only 1 or 2 lines to the paper. If for other species there is spatial correlation, the full ZINB with residual CAR correlation must be presented.

In the Results section, state which covariates were dropped due to collinearity. We could show a frequency plot to emphasise the zero inflation, although this information can be presented in the text by including the percentage of zeros in the abundance data. We would definitely include a graph showing the spatial position of the sites.

We would present the results of the likelihood ratio test comparing the ZIP with the ZINB. A table containing the results of the ZINB showing estimated parameters, standard errors, t-values, and p-values is needed. A colour version of Figure 7.19 may be worthwhile.

The discussion needs to mention that there were residual patterns, and that we are under-fitting the abundance in about 10 transects.

Finally, the problem with surveying different locations over time needs to be discussed. This may be the main challenge to the paper.

Acknowledgements

Part of this work was funded by the European project FREESUBNET, a Marie Curie Research Training Network (MRTN-CT-2006-036186) and the FCT scholarship SFRH/BD/66117/2009 of Mara Schmiing. Thanks are due to the SCUBA divers for their effort in sampling, with funding provided by the projects CLIPE (FCT – Praxis XXI/3/3.2/EMG/1957/95), MARÉ (LIFE B4-3200/98-509), MAREFISH (FCT-POCTI/BSE/41207/2001), OGAMP (INTERREG IIIb/MAC/4.2/A2 2001), and MARMAC (INTERREGIIIb-03/MAC/4.2/A1 2004). We thank Dr. Ricardo S. Santos, coordinator and Principal Investigator of the projects and co-promoter of Mara Schmiing's PhD for support. IMAR-DOP/UAz is Research and Development Unit no. 531 and LARSyS-Associated Laboratory no. 9 funded by the Portuguese Foundation for Science and Technology (FCT) through pluriannual and programmatic funding schemes (OE, FEDER, POCI2001, FSE) and by the Azores Directorate for Science and Technology (DRCT).

8 Analysis of zero inflated click-beetle data

Alain F Zuur, Rod P Blackshaw, Robert S Vernon, Florent Thiebaud, Elena N Ieno, and Anatoly A Saveliev

8.1 Introduction

Click beetles (*Elateridae*) belong to a group of species with economically important members. The larvae, or 'wireworms,' live in soil and attack a wide range of crops. Most notable of these is potato, but cereal grains, strawberries, carrots, and leeks are also vulnerable. Development time from egg to pupation is 3 to 5 years, depending on species and environmental factors.

Traditionally, crop damage has been associated with grass in the rotation cycle, since this provides a stable habitat in which the larval phase can be completed. More recently, in England, damage has occurred in all crop rotations, seemingly related to the use of minimal tillage techniques to replace traditional ploughing.

Three species have historically been assumed to be responsible for damage in the UK: *Agriotes lineatus*, *A. obscurus*, and *A. sputator*, but recently developed molecular methods of identification (Ellis et al., 2009) have shown that other species are commonly present in grasslands (Benefer et al., 2010). The species mentioned are known as 'European wireworm' in North America where they are invasive. Rather than the mixed populations generally found in the UK, they tend to exist as geographically isolated single species in Canada, reflecting introduction patterns. In British Columbia, *A. lineatus* and *A. obscurus* are found, with the latter dominant in the region where the data discussed here was collected.

Estimating numbers of larvae is a demanding exercise, and the population at which economic damage occurs is often below the detection threshold of any reasonable direct monitoring effort. Instead, there have been efforts to trap and count adults using sex pheromone baits. Underlying this approach are several assumptions about the temporal and spatial relationships of adult males to larvae. Adults are more mobile than previously believed (Hicks and Blackshaw, 2008; unpublished data), and sex pheromone monitoring methods do not deliver useful information. The research question has shifted from 'Can we use adult male catches to predict wireworm numbers?' to 'What are the characteristics of adult wireworm dispersal across agricultural landscapes?' The biological question we will focus on is whether we can identify differences in dispersal behaviours attributable to species or gender.

8.2 The setup of the experiment

In 2007, two similar areas (72 m x 72 m), located at the Pacific Agricultural Research Centre, Agassiz, British Columbia, were used as study sites. Field 9 was sown with spring wheat, while the second, Field 12, was left fallow, weeded regularly to keep free of plant growth, and kept level by periodic raking. In each field, pitfall traps were placed at the boundaries. Cross-traps and pitfall traps were also systematically positioned inside each field. We will not consider data from the boundary pitfall traps in this chapter.

Marked batches of beetles were released at the centre of each field on three occasions in Field 9 and four occasions in Field 12. At specified times all traps were inspected, and the number of beetles were counted and removed. The number of released beetles differed with respect to field, date, species, and sex (Table 8.1). The dates 8 and 10 May are close in time, and it is unlikely that environmental conditions of these two days differed appreciably, so we will consider them as a single 'release day.'

Table 8.1. Release dates and the number of released beetles per date (*N*), species, and sex, for each field.

Field 9 (Wheat)				Field 12 (Fallow)			
Date	Species	Sex	N	Date	Species	Sex	N
				8 May	*A. obscurus*	M	420
						F	420
10 May	*A. obscurus*	M	200	10 May	*A. obscurus*	M	200
25 May	*A. obscurus*	M	400	25 May	*A. obscurus*	M	400
	A. lineatus	M	336		*A. lineatus*	M	336
30 May	*A. obscurus*	M	257	30 May	*A. obscurus*	M	257
		F	513			F	513

8.3 Importing data and coding

The data are available in a tab-delimited ascii file and are imported with the `read.table` function. The `names` and `str` functions show the variable names and whether these are coded as numerical or as categorical variables.

```
> Beetles <- read.table("CentreMRR.txt", header = TRUE)
> names(Beetles)

[1]  "Designation" "Field"      "Group"      "Date"      "Trap"
[6]  "Time"        "Count"      "Xcoord"     "Ycoord"    "Distance"
[11] "Direction"   "Radians"    "Nrelease"

> str(Beetles)

'data.frame':   10573 obs. of  13 variables:
 $ Designation: Factor w/ 133 levels "a14","a2","a20",..: 2 6 ...
 $ Field      : int  1 1 1 1 1 1 1 1 1 1 ...
 $ Group      : int  1 1 1 1 1 1 1 1 1 1 ...
 $ Date       : int  1 1 1 1 1 1 1 1 1 1 ...
 $ Trap       : int  3 3 3 3 3 3 3 3 3 3 ...
 $ Time       : int  1 1 1 1 1 1 1 1 1 1 ...
 $ Count      : int  0 0 0 0 0 0 0 0 0 0 ...
 $ Xcoord     : int  1 1 1 1 1 1 6 6 18 18 ...
 $ Ycoord     : int  6 18 30 42 54 66 1 71 1 71 ...
 $ Distance   : num  46.1 39.4 35.5 35.5 39.4 ...
 $ Direction  : num  229 243 260 280 297 ...
 $ Radians    : num  4 4.24 4.54 4.88 5.19 ...
 $ Nrelease   : int  200 200 200 200 200 200 200 200 200 200 ...
```

The variable `Designation` is not used in the analysis, and all other variables are coded as numerical data. We therefore need to do some recoding to define the variables `Group`, `Field`, `Date`, and `Traptype` as categorical variables. An `f` preceding the categorical variable name reminds us that these are factors.

```
> Beetles$fGroup <- factor(Beetles$Group,
                           levels = c(1, 2, 3),
                           labels = c("A. obscurus m", "A. obscurus f",
                                      "A. lineatus"))
> Beetles$fField <- factor(Beetles$Field,
                           levels = c(1, 2),
                           labels = c("Wheat", "Fallow"))
> Beetles$fDate  <- factor(Beetles$Date,
                           levels = c(1, 2, 3),
                           labels = c("8-10", "25", "30"))
> Beetles$fTraptype <- factor(Beetles$Traptype,
                              levels = c(1, 2, 3),
                              labels = c("X", "Circular", "Linear"))
```

The categorical variable `fGroup` tells which rows in the object `Beetles` are male *A. obscurus*, female *A. obscurus*, or *A. lineatus*. All specimens of *A. lineatus* are male. To evaluate a sex or species effect we must decide what to compare. The options are:

- Drop the *A. lineatus* data and compare the male and female data from *A. obscurus* to assess whether there is a sex effect in the species *A. obscurus*.
- Drop the female data from *A. obscurus* and use the males from *A. obscurus* and *A.lineatus* to assess whether there is a species effect in the male data.
- Keep all three levels and assess whether they are different from one another.
- Make new variables, *Species* and *Sex*, and include the significance of their interaction. One sex/ species combination is not available, so we cannot do this.

Using all available data and investigating whether there is a species or sex effect is not acceptable, as an apparently 'significant' species effect may be a sex effect and vice versa. We initially intended the third approach, but other issues arose. Towards the end of the data exploration we will explain why we select options 1 and 2. The variable `fGroup` is of chief interest, as it allows us to answer the basic biological question of whether there is a species effect.

The variable `fDate` is the release date. Each trap was monitored at irregular time intervals. As an example, one of the traps was sampled at 1, 3, 21, 46, 50, 53, 70, 94, 118, 142, and 166 hours after release. This information is given in the variable `Time`.

The variable `Nrelease` contains the number of released beetles per date and species (Table 8.1). It will be used as an offset variable in the generalised linear models that will be applied later in this Chapter.

Finally, `fTrapType` refers to the trap type and this variable is further discussed in Subsection 8.4.1

8.4 Data exploration

8.4.1 Spatial position of the sites

Before continuing, we will examine the spatial position of the traps (Figure 8.1). The linear traps are placed at the boundary of the fields. If we want to state whether the species behave differently, whether species behave differently in the two fields (2-way interaction between species and field), or whether the trap type has a different effect on the species in each field (3-way interaction among species × field × trap type), we need to be able to compare like with like. With one set of traps at the boundary, and two types of traps spread throughout the field, we are not comparing like with like. When analyzing data obtained by different traps types, we could potentially conclude that there is a trap effect, but this may be due to the position of the traps. Hence we will remove the traps at the boundary data from the analysis. These are labelled as 'Linear' in `fTrapType`. R code to remove the traps at the boundary is as follows (the level `"Linear"` represent the boundary traps).

```
> Beetles2 <- Beetles[Beetles$fTraptype != "Linear",]
```

The remaining two levels of `fTrapType` are `X` and `Circular`, representing cross traps and pitfall traps, respectively. We assume that the data obtained from the X and that from the circular traps can be compared. Figure 8.1 was created with the R code:

```
> library(lattice)
> xyplot(Ycoord ~ Xcoord | fTraptype * fField, data = Beetles,
        aspect = "iso", pch = 16, col = 1, cex = 1)
```

The `aspect = "iso"` ensures that distances along the horizontal axis are the same as those along the vertical axis.

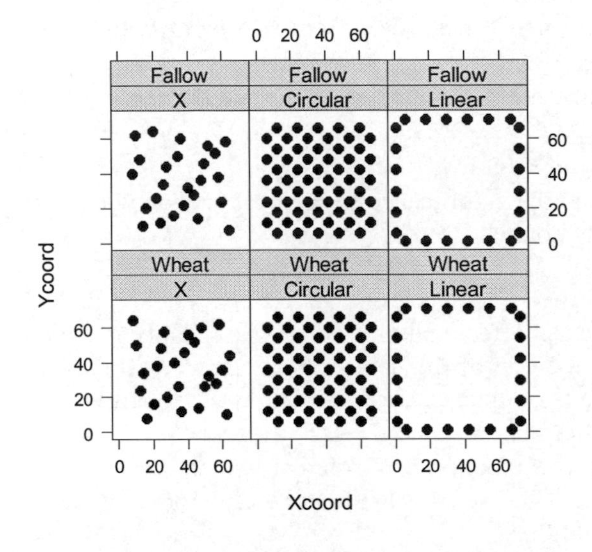

Figure 8.1. Position of the three types of traps (X, circular, and linear) in the two fields. The linear traps were placed at the boundary. The data obtained from these traps will not be used in the analysis.

8.4.2 Response and explanatory variables

The response variable is the number of beetles caught in each trap and is denoted by `Count`. Beetles are released at the centre point, and all traps are inspected at specified times. The variable `fTime` represents the time of sampling a trap, expressed as the number of hours since release. We have the spatial coordinates of the traps (`Xcoord`, `Ycoord`), trap type (`fTrapType`), species type and sex (`fGroup`), field type (`fField`), and the release date (`fDate`).

Release dates were 8-10 May, 25 May, and 30 May. On each of these dates beetles were released, and sampling of the traps started 1 hour later. The spatial coordinates of the traps are specific to a particular field. Beetles were released at a central point with coordinates 36, 36, and we know the coordinates of the trap in which they were caught, so we can calculate the distance travelled by applying elementary mathematics. This assumes that the beetle travelled in a straight line, which may not be the case. So the distance represents an average. In a similar way the direction to the point of capture from the release point can be derived.

It can be argued that distance and direction are response variables, but in this analysis we will model the counts. We are aiming for something of the form:

$$\text{Counts} = \text{Group} + \text{Field} + \text{Date} + \text{Trap type} + \text{Time} + \text{Direction} + \text{Interactions} + \text{Distance} + \text{noise} \quad (8.1)$$

Another type of model can be:

$$\text{Counts} = \text{Group} + \text{Field} + \text{Date} + \text{Trap type} + f(\text{Xcoord}, \text{Ycoord}) + \text{Interactions} + \text{noise} \quad (8.2)$$

The term f(Xcood, Ycoord) is a 2-dimensional spatial smoother giving the spatial pattern of the counts. This model can be extended by allowing an interaction between the spatial smoother and, for example, species, to see whether the two species exhibit different spatial patterns. Note that we dropped the time, distance, and direction variables in Equation (8.2). This is because we expect high collinearity between spatial position and time, for example. The interactions may involve field type, but we have limited biological knowledge of what to expect.

8.4.3 Viewing the data as time series

At each trap, on each of the three days, we basically have a time series, since counts took place at different points in time. The best way to comprehend this is to plot the data for one of the sampling dates, for instance 8-10 May, for one field. This produces Figure 8.2. Each panel represents a trap and shows the number of captured beetles over time. To keep the graph a manageable size, we use only the wheat field. It seems that once beetles are in a trap, more beetles are captured in the same trap. The same figure can be made for other days and for the fallow field. The figure implies that a potential way to analyse the data is to use the variable `TrapID` (see the R code below) as a random intercept in the models. This would introduce a correlation structure between sequential observations of a given trap on a given day in a given field. However, the large number of trap-day-field combinations with zero captured beetles may cause numerical problems in such a model.

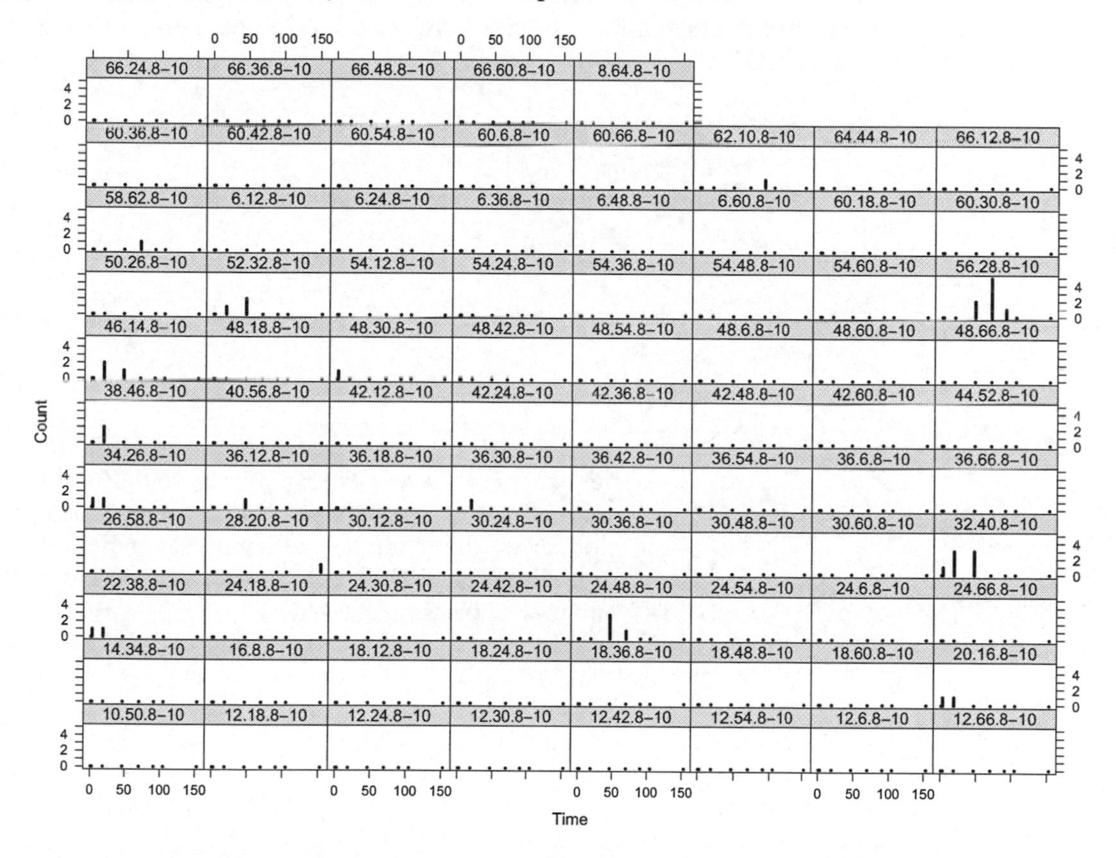

Figure 8.2. Presentation of the data from the wheat field as a time series for 8-10 May. We have multiple observations from each trap, made over time. The horizontal axis indicates time in hours, and the vertical axis is the number of beetles caught. Each panel represents a particular trap. Numbers of beetles are represented by vertical lines. To improve this graph, we can sort the panels according to their spatial position so that spatial-temporal patterns can be visualised. We leave this as an exercise for the reader. To do it, change the order of the levels in the variable `TrapID`.

The R code to create Figure 8.2 follows. The `paste` command concatenates the variables `Xcoord`, `Ycoord`, and `date`, and defines the unique trap on a specific day. We use the `subset` argument in the `xyplot` function to plot the wheat field data from 8-10 May.

```
> Beetles2$TrapID <- factor(paste(Beetles2$Xcoord, Beetles2$Ycoord,
                            Beetles2$fDate, sep = "."))
> xyplot(Count ~ Time | TrapID, type   = "h", col = 1, lwd = 3,
    data = Beetles2, subset = (fField == "Wheat" & fDate == "8-10")
```

Another way to visualise the time effect is by plotting the counts versus time (Figure 8.3). After 150 hours no trap has caught more than 2 beetles, and there is a clear downward trend in the numbers over time. The R code to generate this graph is:

```
> plot(x = jitter(Beetles2$Time), y = jitter(Beetles2$Count),
       xlab = "Time (h)", ylab = "Counts")
```

The `jitter` function adds random variation to the variables in the `plot` function so that plots of observations are not superimposed. It is interesting to observe the number of times that beetles were caught in traps as a function of time. This is similar to information in Figure 8.3, but expressed numerically. First we extract the data for which the counts are larger than 0; then the `table` and `cut` function are used to determine the number of times beetles were caught in 24 hour periods. In the first 24 hours, beetles were caught on 117 occasions.

```
> Beetles3 <- Beetles2[Beetles2$Count > 0, ]
> MyBreaks <- seq(0, 288, length = 13)
> table(cut(Beetles3$Time, breaks = MyBreaks))

    (0,24]    (24,48]    (48,72]    (72,96]   (96,120] (120,144](144,168]
       117         76         58         51         30         14        21
 (168,192] (192,216] (216,240] (240,264] (264,288]
        13          0          0          0          3
```

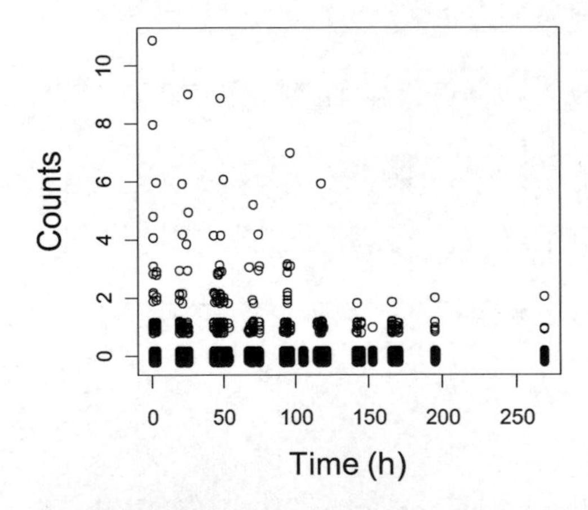

Figure 8.3. Counts per trap versus time. Both fields and all dates were used. Random variation was added to the counts and to avoid points being superimposed. The graph ignores differences in the number of beetles released.

8.4.4 Spatial and temporal patterns

Another way to visualise the temporal and spatial aspects of the data is shown in Figure 8.4. The left panel includes only the wheat field data and the right panel the fallow field data. Each panel corresponds to a 24 hour time period, and we plotted the spatial coordinates. The size of a dot is proportional to the number of beetles in a trap in that time period. The square in the centre is the release point. The spatial patterns of the two fields differ.

R code to produce Figure 8.4 follows. We first create a new variable, `Time2`, inside the `Beetles3` object to define the 24 hour periods. The `CexID` prints the dots proportional to the value of `Count`. The square root and multiplication by 2 is based on trial and error; without it we get a few large dots and many small dots. The `xyplot` function creates the graph. Change the `subset` argument for the fallow field graph.

```
> Beetles3$Time2 <- cut(Beetles3$Time, breaks = MyBreaks)
> Beetles3$CexID <- sqrt(Beetles3$Count)
> Beetles3$MyCex <- 2 * Beetles3$CexID / max(Beetles3$CexID)
> xyplot(Ycoord ~ Xcoord | Time2, data = Beetles3,
         panel = function(x, y, subscripts,...){
           panel.points(36,36, pch = 15, cex = 1, col = 1)
           panel.points(x, y, col = 1,
                        cex = Beetles3$MyCex[subscripts]) },
         subset = (fField == "Wheat"))
```

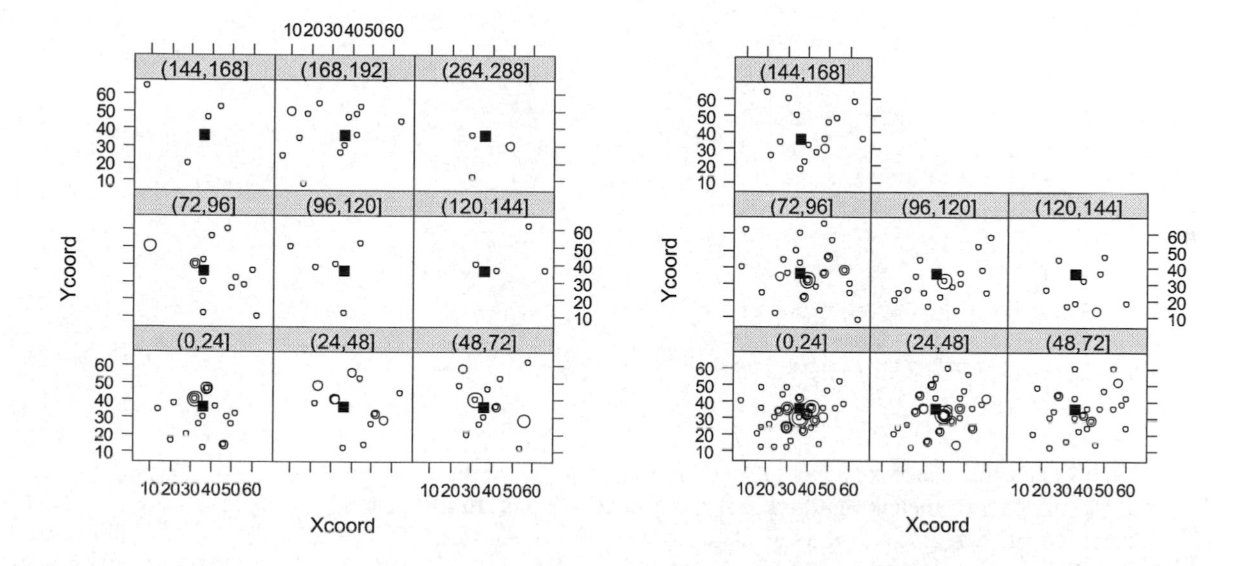

Figure 8.4. Left panel: Number of beetles per trap in the wheat field for different time periods. Right panel: Number of beetles per trap in the fallow field for different time periods. Each panel represents a 24 hour period. The size of a dot is proportional to the number of beetles captured. These graphs ignore the difference in released numbers. Solid squares mark the release point.

8.4.5 Big trouble

At this point in the data exploration we look at the effects of main terms and interactions. That means boxplots, design plots, and interaction plots. This is also the stage at which we encounter major obstacles, and decisions must be made. The difficulties are related to the experimental design. Unfortunately, not all groups were released on every release date, as can be seen from the results of the `table` function:

```
> table(Beetles2$fGroup, Beetles2$fDate)
```

```
               8-10  25   30
A. obscurus m 2210 1530 1020
A. obscurus f  935    0 1020
A. lineatus      0 1530    0
```

Note that *A. lineatus* specimens were not released on 8-10 May and 30 May, and female *A. obscurus* were not released on 25 May. This has some serious consequences for the statistical analyses. If we were to use all available data and apply a model with `fGroup` as covariate, a significant effect of `fGroup` may be due to the fact that *A. lineatus* was not released on 25 May. The same holds for a date effect. This comes back to comparing like with like. Testing for interactions is more difficult as well. Let us go one step further and present the results of the `table` command with 3 arguments:

```
> table(Beetles2$fGroup, Beetles2$fField, Beetles2$fDate)

, ,  = 8-10
                Wheat Fallow
  A. obscurus m   680   1530
  A. obscurus f     0    935
  A. lineatus       0      0

, ,  = 25
                Wheat Fallow
  A. obscurus m   935    595
  A. obscurus f     0      0
  A. lineatus     935    595

, ,  = 30
                Wheat Fallow
  A. obscurus m   510    510
  A. obscurus f   510    510
  A. lineatus       0      0
```

Each of the three tables shows how many samples (not counts) we have for a given day, field, and group level. A zero means that there were no observations. In other words, female *A. obscurus* were not released on 8-10 May in the wheat field, but were released in the fallow field. So, it is not only a problem to compare dates but also to compare fields. We will visualise the numbers with a Cleveland dotplot (Figure 8.5). Any empty combinations show that a given species or sex was not released on a given day or field.

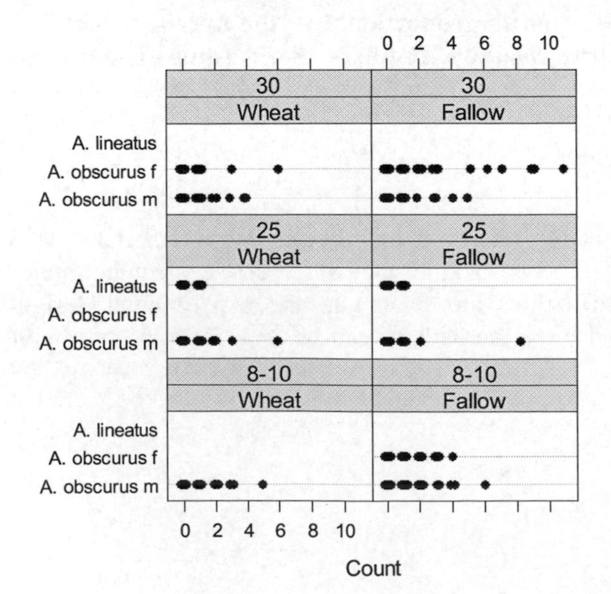

Figure 8.5. Cleveland dotplot of the `Beetle2` data. Empty combinations mean that no beetles were released. A small amount of random variation was added to the counts in order to avoid plots of multiple observations being superimposed.

There are 6 combinations with empty cells. This is a serious experimental design issue. As mentioned, a significant species effect in the models may be due to different release dates, or a significant field effect may be due to different species being released. We can expect some serious reservations to be expressed by referees if we ignore this and analyse the full data. We can do three things:

1. The first option is to conclude from Figure 8.5 that the experimental design is such that no further statistical analysis can be carried out. We can consider it a learning experience for future research. A harsh lesson.
2. We can ignore the problem and proceed with the analysis. Some of the studies dealt with in other chapters of this book have similar experimental design issues, and, during the refereeing process of spin-off papers, these were picked up and criticised.
3. We can analyse a subset of the data for which we can compare like with like.

We will select option 3. To analyse whether there is a species effect in the data, we will extract *A. lineatus* (all males) and the male *A. obscurus* data. Beetles of these species were not released on 8-10 May, and *A. lineatus* was not released in the fallow field on 30 May. Both species were released in both fields only on 25 May, and we will use these data. A compromise would be to include the 8-10 May release and argue that there cannot be a date effect. This is a questionable assumption, as dates may differ in such variables as temperature, wind speed, wind direction, light intensity, etc.

So to test for species effect we will use the 25 May data. These data are extracted and stored in an object called `Species`:

```
> I1 <- (Beetles2$fGroup == "A. obscurus m" |
       Beetles2$fGroup == "A. lineatus") & (Beetles2$fDate  == "25")
> Species        <- Beetles2[I1,]
> Species$fDate  <- factor(Species$fDate)
> Species$fGroup <- factor(Species$fGroup)
```

The symbols | and & refer to the Boolean OR and AND. The two `factor` commands ensure that the labels for the levels of the categorical variables are updated. So the data in the object `Species` can be used to test for a species effect in the males. To test for a male vs. female effect, we use only the male and female *A. obscurus* data from 30 May. These are extracted and stored in an object called `FM`:

```
> I2 <- (Beetles2$fGroup == "A. obscurus m" |
      Beetles2$fGroup == "A. obscurus f") & (Beetles2$fDate  == "30")
> FM        <- Beetles2[I2,]
> FM$fDate  <- factor(FM$fDate)
> FM$fGroup <- factor(FM$fGroup)
```

The data in the object `FM` can be used to test whether there is a sex effect in the *A. obscurus* data from 30 May. If we separate the data into two groups, the graphic output also duplicates. Because the basic question deals with the species effect, we will continue presenting graphs obtained using the object `Species` and summarise the results for `FM` at the end of the chapter.

8.4.6 Effects and interactions for the species data

A so-called design plot is presented in Figure 8.6. It shows mean count values for each level of categorical variable. We convert direction to a categorical variable to eliminate the requirement for smoothing techniques. Note that the mean count values for X-traps are higher than those for the circular traps. The R code to create the figure follows. The first 8 lines are used to convert the continuous variable `Radians` (direction) to a categorical variable `fDir` of levels north, south, east, and west. The plot is made with the `plot.design` function.

```
> Species$Degrees <- Species$Radians * 180 / pi
```

```
> MyBreaks <- c(45, 45 + 90, 45 + 2 * 90, 45 + 3 * 90)
> Dir <- vector(length = nrow(Species))
> Dir[Species$Degrees > 315 | Species$Degrees <= 45  ] <- "North"
> Dir[Species$Degrees >  45 & Species$Degrees <= 135 ] <- "East"
> Dir[Species$Degrees > 135 & Species$Degrees <= 225 ] <- "South"
> Dir[Species$Degrees >=225 & Species$Degrees <= 315 ] <- "West"
> Species$fDir <- factor(Dir)
> plot.design(Count ~ fGroup + fField + fTraptype + fDir,
              data = Species, cex.lab = 1.5)
```

The design plot also indicates that mean count values per species differ. This can be visualised in more detail with a Cleveland dotplot (Figure 8.7). Note that counts for *A. lineatus* are all either 0 or 1, whereas the *A. obscurus* data range from 0 to 8. Interaction plots (not shown here) to visualise potential interactions suggest an interaction of `fGroup` and `fField`.

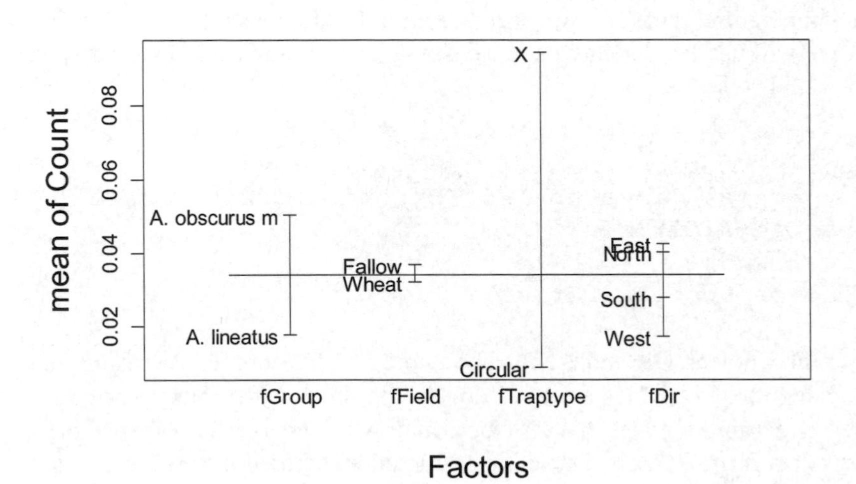

Figure 8.6. Design plot for the species data showing the mean count values per level for each categorical covariate. The graph ignores differences in numbers of beetles released.

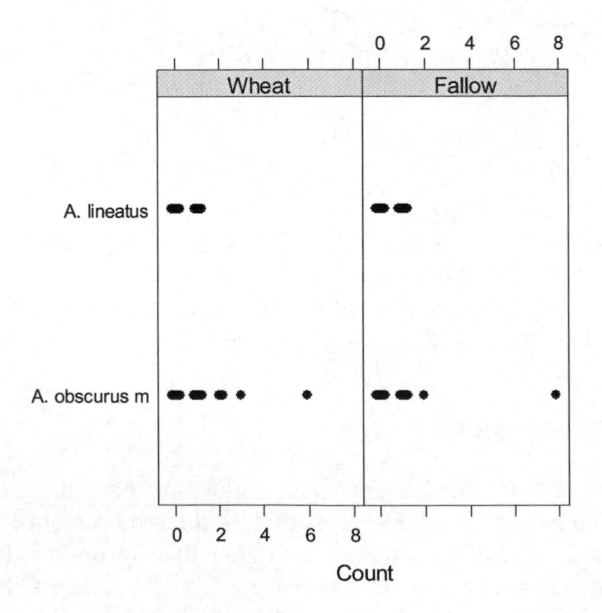

Figure 8.7. Cleveland dotplot of the species count data conditional on the variables `fGroup` and `fField`. The vertical axis shows the species and the horizontal axis the counts. Each panel represents a field type. A small amount of random noise was added to the count data, so that we can distinguish among multiple observations with the same value. The real values range from 0 to 8. The graph ignores differences in numbers of beetles released.

8.4.7 Zero inflation

The response variable `Count` in the `Species` object is clearly zero inflated, as can be seen from the `table` command:

```
> table(Species$Count)

   0    1    2    3    6    8
2976   75    6    1    1    1
```

This means that the majority of traps contain 0 captured beetles. We have only 84 non-zero observations. For the `FM` object we have:

```
> table(FM$Count)

   0    1    2    3    4    5    6    7    9   11
1941   72    7    9    3    2    2    1    2    1
```

We need to apply techniques that can cope with a large number of zeros, but we should try to keep the models simple, as we have only 99 non-zero observations.

8.4.8 Direction

A potential covariate is direction, expressed as `radians`. Because radians are measured on a circular scale, we created a so-called rose diagram (Figure 8.8). This is essentially a histogram. It appears that the beetles in the object `Species` dispersed evenly. To create the rose diagram we used the function `rose.diag` from the `CircStats` package. You will need to download and install this package, as it is not included in the main installation of R.

```
> library(CircStats)
> par(mfrow = c(1, 1), mar = c(3, 3, 2, 2))
> rose.diag(Species$Radians, bins = 12, prop = 2.5)
```

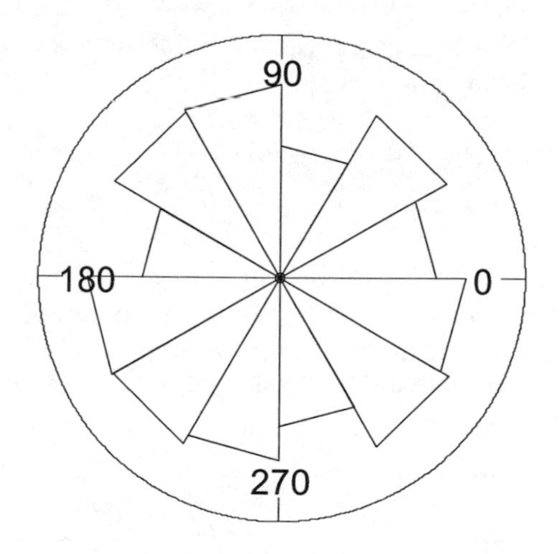

Figure 8.8. `Radians`. Histogram of the radian values. Lower left: Rose diagram. The data in the `Species` object was used.

8.4.9 Where to go from here; zero inflated models?

We enthusiastically started the data exploration with a data set containing 10,573 observations (i.e. sampling a trap at a particular hour) but quickly discovered that, in order to avoid confounding effects and compare like with like, we needed to separate the data into two considerably smaller data sets. The first data set, with 3,060 observations, is called `Species` and allows testing for a species effect. The second data set, `FM`, comprises 2,040 observations and allows us to evaluate a sex effect. The percent of zero counts in these two data sets is 97% and 95%, respectively. We have 84 non-zero observations for the `Species` data and 99 for the `FM` data.

So, what do we do next? To answer this question we need to determine why we have so many zeros. Further, what is our sampling unit? What are we trying to measure? What is it we want to know? The answer to the final question is that we want to know how beetles move through the field; therefore traps were placed in a large number of locations, and the number of beetles caught in traps was counted. A trap represents a small area, or site, and we hope that all beetles passing through the area are caught in the trap. So the sampling unit is a 'site,' and the capture rate is expected to correspond to the number of beetles passing through the area. The trap is our sampling device. We could have used a video camera, though this is impractical, or another sampling device. Whatever method of sampling we use, we intend it to adequately and precisely measure the number of beetles visiting a site. However, traps might not be 100% efficient in the sense that every beetle that passes by is caught.

So what about the zeros? These are traps that are being inspected at a specified times and are found to contain no beetles. Throughout this book we have talked about true zeros and false zeros. In the context of these data, a false zero is an observation of 0 beetles in a trap, whereas in reality there were beetles present. Reasons for this could be that the traps are too small, the beetles are too clever to be caught, they were affected by wind direction, etc. In all cases the beetles were present, but we did not record it. We argue that we have false zeros in the data, and this means that there is scope to apply zero inflated models, with the first candidate being is a zero inflated Poisson (ZIP) generalised linear model (GLM).

We have argued that discriminating between true and false zeros is a suitable biological rationale for ZIP GLM models, but in reality it is the lack of covariates that determines the probability of whether a zero is a false zero or not. If all 2976 zero observations in the `Species` data frame can be explained by the covariates in a Poisson GLM, such a model is sufficient. The ordinary Poisson GLM is a special case ZIP GLM; hence, if there is no need for a ZIP GLM, we will discover this during the modelling process and can revert to a Poisson GLM.

Due to the small number of non-zero observations, we will not use smoothers and will refrain from applying ZIP generalised additive models (GAM). For example, the covariate `Radians` could have been used as a smoother. Alternatively, it can be converted to a categorical variable with the levels north, south, east, and west, but this still requires 3 regression parameters.

8.5. ZIP GLM

A sketch of the process for the zero inflation is presented in Figure 8.9. In the zero inflated GLM we make a distinction between false and true zeros and the positive counts. We argued in the previous section that we may have false zeros, and these are in the upper branch. The lower branch contains the count process. Mathematical details of the ZIP GLM were given in Chapters 2 – 4 and are not repeated here.

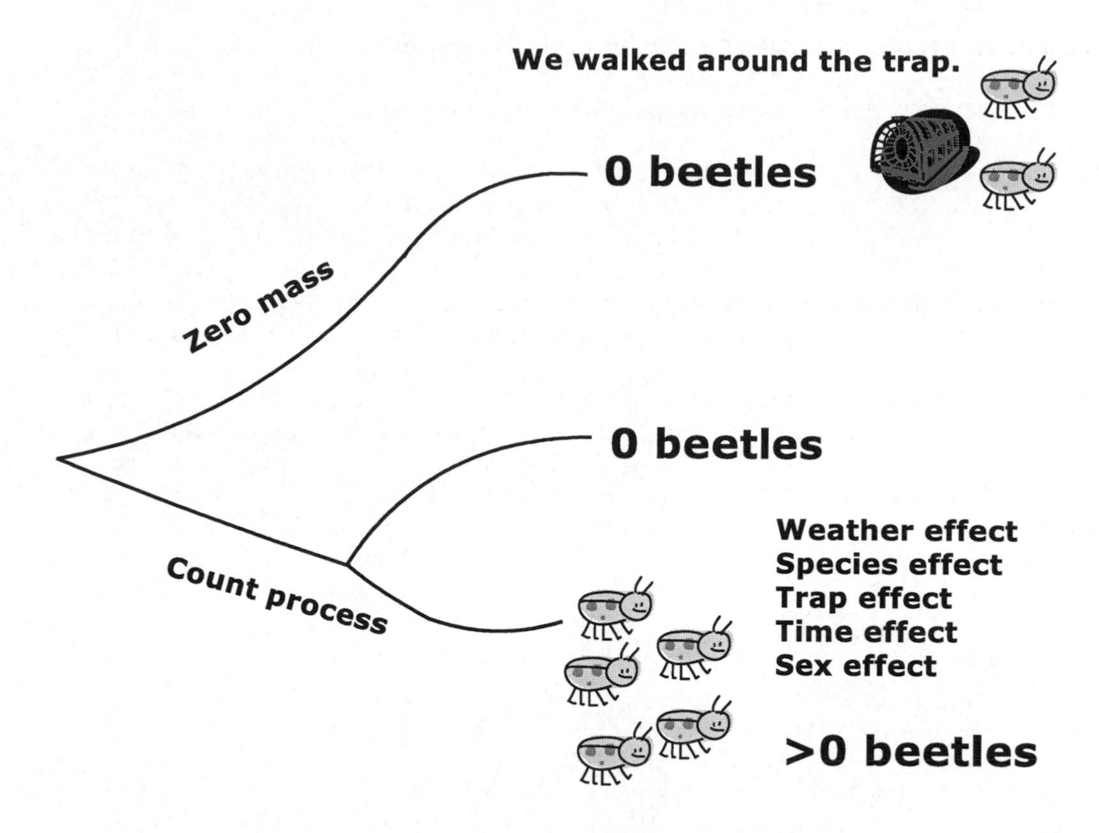

Figure 8.9. Sketch of the zero inflated data. The trap and beetles are for illustrative purpose only and do not match the real shape of the traps or beetles. Two processes are involved. The first models the false zeros versus all other data (true zeros and positive counts), and, in the second process, all data except the false zeros are modelled. The presence of false zeros is not a requirement for the application of ZIP models; see Chapter 10 for a discussion.

8.6 ZIP GLM without correlation for the species data

Before entering R code for a ZIP model, we need to decide which covariates to use. Obviously this should be driven by the underlying questions. For the count data we would like to know whether there is an effect of field, trap type, species, time, or direction. We are particularly interested in the interaction of field and species. This gives the following model for the count part of the ZIP:

$$\log(\mu) = fField \times fGroup + fTraptype + Time + fDir + offset(\log(\text{Nrelease})))$$

All the data are for male beetles, so `fGroup` has only two levels (species). The offset variable is needed because the number of released beetles differs per field and species.

Concerning the model for the false zeros, the problem is that we have little ecological knowledge to postulate what could be driving the results. We would not expect an interaction of field and species (coded by `fGroup`). Could there be a species effect? Would specimens of one of the species stand in front of a trap and decide not to go in? Unlikely. Similarly, we find it difficult to justify a field effect on the probability of false zeros. What about direction? The covariate `fDir` (orientation of the trap relative to the release point) may represent wind direction, sunlight, or other small-scale habitat effects. Perhaps travelling with the wind has a different effect on false zeros than travelling against it. We include time and trap type and expect that the latter, especially, will be important. This means that the binary part of the model is of the form:

$$\text{logit}(\pi) = fField + fTraptype + Time + fDir$$

The following R code calculates the offset variable and applies the ZIP:

```
> Species$LogRel <- log(Species$Nrelease)
> library(pscl)
> ZI1 <- zeroinfl(Count ~ fGroup * fField + fTraptype + Time +
                         fDir + offset(LogRel) |
                         fTraptype + Time + fDir, data = Species)
```

The first segment after the ~ symbol is for the count model and that after the | symbol gives the covariates for the binary model. The output is obtained with the `summary(ZI1)` command:

```
Count model coefficients (poisson with log link):
                           Estimate Std. Error z value Pr(>|z|)
(Intercept)                -6.14701    0.46531  -13.21  < 2e-16
fGroupA. lineatus          -1.57566    0.38284   -4.12  3.9e-05
fFieldFallow               -0.38751    0.30246   -1.28  0.20013
fTraptypeCircular           1.77608    0.49654    3.58  0.00035
Time                       -0.01324    0.00267   -4.95  7.3e-07
fDirNorth                   0.84937    0.43148    1.97  0.04901
fDirSouth                  -1.17263    0.53014   -2.21  0.02697
fDirWest                    0.05443    0.62701    0.09  0.93082
fGroupA. lineatus:fFieldFallow  1.39723    0.51213    2.73  0.00637

Zero-inflation model coefficients (binomial with logit link):
                 Estimate Std. Error z value Pr(>|z|)
(Intercept)       1.09042    0.67922    1.61    0.108
fTraptypeCircular 5.38470    0.67911    7.93  2.2e-15
Time             -0.01995    0.00481   -4.15  3.4e-05
fDirNorth         1.25405    0.66124    1.90    0.058
fDirSouth        -2.10754    0.96203   -2.19    0.028
fDirWest          0.80560    0.88343    0.91    0.362
```

For the count data, the interaction between `fGroup` and `fField` is significant, and `Time` is also significant. A likelihood ratio test indicates that direction is also significant. For the binary model all covariates are significant. To understand what the model represents, we sketch the fitted values. We take values of time from 1 to 275 hours and calculate predicted values for the counts for all groups, fields, directions, and trap types (Figure 8.10). The lines are the predicted counts represented by the lower branch in Figure 8.9. The circular traps have higher predicted counts, and within the first 50 hours there are considerable differences among the directions. In all panels of Figure 8.11, the lower line corresponds to south; thus there were fewer beetles captured in the trap south of the release point. Figure 8.11 shows the prediction probabilities of false zeros. Recall that these are zeros that are not true zeros, see the upper branch in Figure 8.9. The circular traps have a much higher probability of producing false zeros than do the X traps. The probability of false zeros in the southern-directed traps is lower. The differences between curves correspond to the numerical output. Note the negative estimate regression parameters for the level `fDirSouth`.

The R code to produce Figure 8.10 and Figure 8.11 is complex. We first create a data set for which we want to make predictions.

```
> NewData <- expand.grid(Time = seq(1, 275, length = 50),
                     fGroup = c("A. obscurus m", "A. lineatus"),
                     fField = c("Wheat", "Fallow"),
                     fTraptype = c("X", "Circular"),
                     fDir = c("East", "North", "South", "West"),
                     LogRel = mean(Species$LogRel))
```

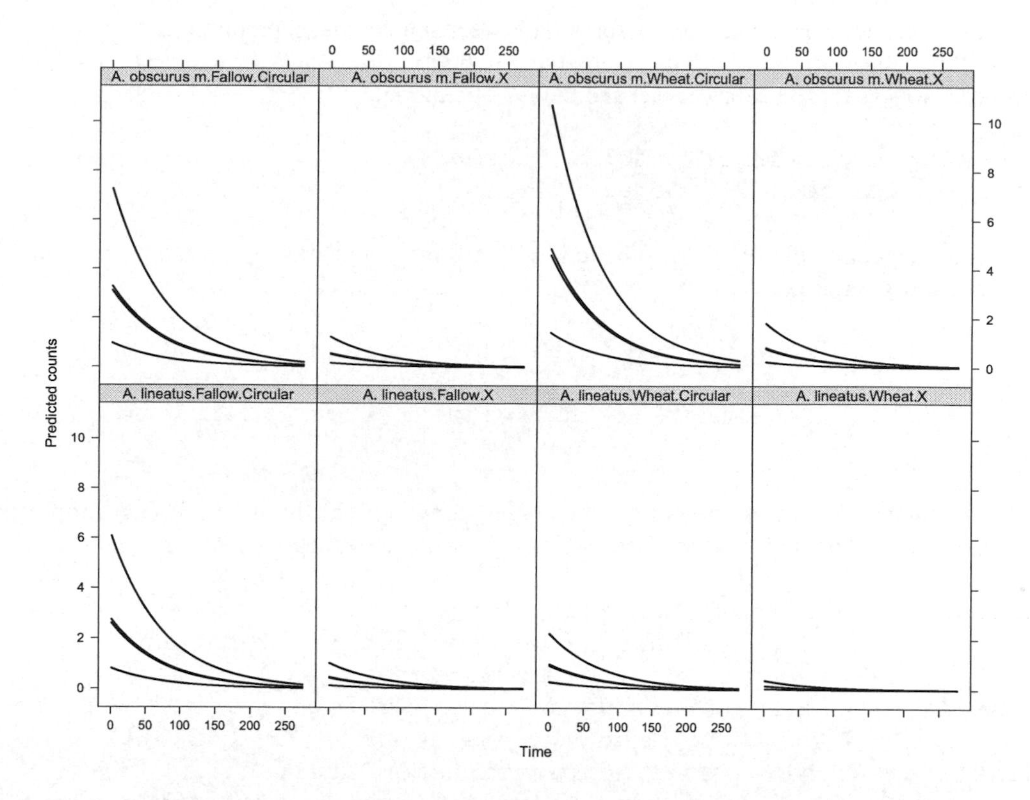

Figure 8.10. Predicted counts obtained by the ZIP model. The four lines correspond to the four directions. The lower line in each panel represents south. There is little difference between fitted values of east and west.

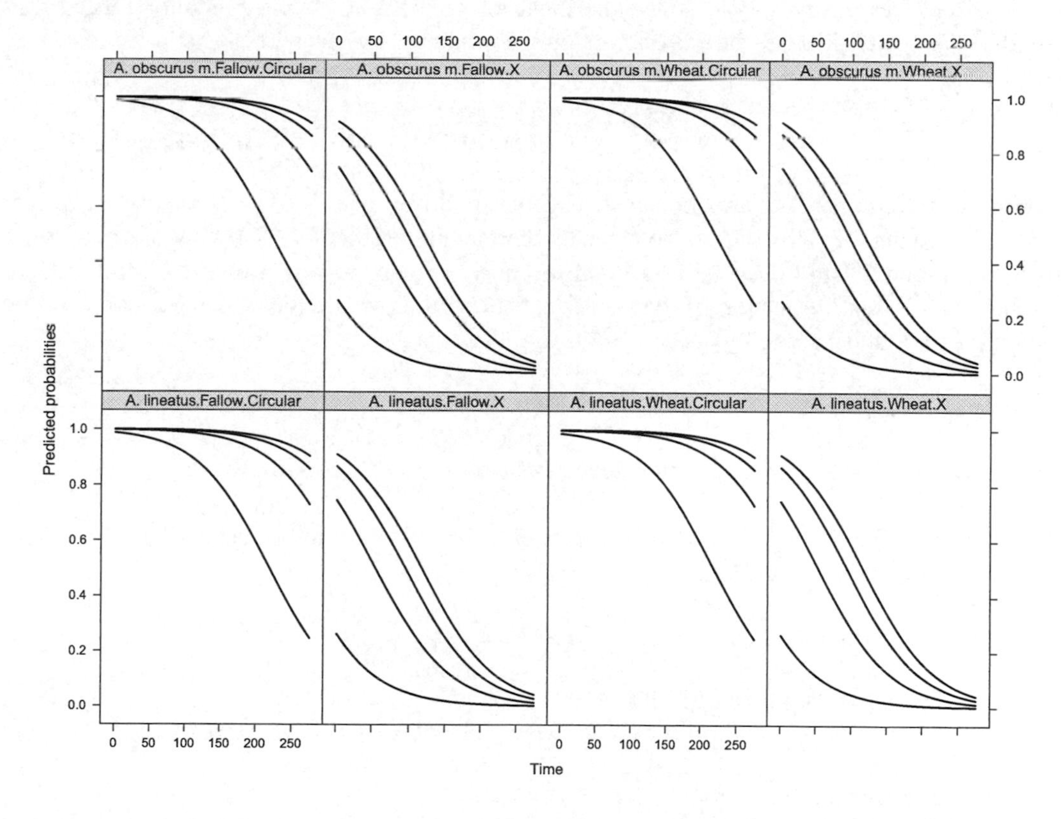

Figure 8.11. Predicted probabilities of false zeros obtained by the ZIP model. The four lines correspond to the four directions. The lowest curve in all panels represents south.

This object contains covariate values for which we want to make predictions. We will use the average of the offset. In order to calculate the predictions manually, we extract the estimated regression parameters for the count model and the logistic model.

```
> BetaCount <- coef(ZI1, model = "count")
> BetaZero  <- coef(ZI1, model = "zero")
```

We can now calculate the predictor functions for both the count model and the binary model. All we need is the design matrix:

```
> ModMatCount <- model.matrix(~ fGroup * fField + fTraptype +
                                Time + fDir, data = NewData)
> ModMatZero <- model.matrix(~ fTraptype + Time + fDir,
                              data = NewData)
```

These two matrices contain categorical covariates converted to dummies. Using simple matrix multiplication and the log and logistic link functions gives the predicted values:

```
> NewData1 <- data.frame(NewData,
                  etaCount  = ModMatCount %*% BetaCount,
                  etaZero   = ModMatZero  %*% BetaZero)
> NewData1 <- with(NewData1, data.frame(NewData1,
                  CountPred = exp(etaCount + 5.904288)))
> NewData1 <- with(NewData1, data.frame(NewData1,
                  ZeroPred  = exp(etaZero) / (1 + exp(etaZero))))
```

The variables `CountPred` and `ZeroPred` contain the predicted values for the count model and the binary model, respectively. We added the mean offset of 5.90 to the predictor function of the count model. We could skip this coding and obtain the predicted values with two commands:

```
> NewData1$PCount <- predict(ZI1, newdata = NewData, type = "count")
> NewData1$PZero  <- predict(ZI1, newdata = NewData, type = "zero" )
```

`PCount` and `CountPred` are the same, except for differences due to the offset. `PZero` and `ZeroPred` are identical. Why did we bother with the manual prediction? Why not use only these latter two R commands? Only because it is good to know we can do it ourselves! Plotting the multipanel scatterplot is now a matter of `lattice` magic. First we define a categorical variable for `fGroup`, `fField`, and `fTrapType`:

```
> NewData1$AllIDGroup <- paste(NewData1$fGroup, NewData1$fField,
                               NewData1$fTraptype, sep = ".")
> NewData1$AllIDGroup <- factor(NewData1$AllIDGroup)
```

A function to define lines inside a panel, used with the `groups` argument in the `xyplot`, produces the two multipanel scatterplots:

```
> MyLines <- function(xi, yi, subscripts,...){
   I <- order(xi)
   panel.lines(xi[I], yi[I], lwd = 2,
            col = as.numeric(NewData1$fDir[subscripts]))
  }
> xyplot(CountPred ~ Time | AllIDGroup,
        xlab = "Time", ylab = "Predicted counts",
        groups = fDir, layout = c(4,2),
        data = NewData1, type = "l", lwd = 2,
```

```
        panel = panel.superpose,
        panel.groups = MyLines)
> xyplot(ZeroPred ~ Time | AllIDGroup,
        xlab = "Time", ylab = "Predicted probabilities",
        groups = fDir, layout = c(4,2),
        data = NewData1, type = "l", lwd = 2,
        panel = panel.superpose,
        panel.groups = MyLines)
```

Thus far we have not applied model validation, so we do not know whether we can trust the results. There is no strong overdispersion:

```
> E1 <- resid(ZI1, type = "pearson")
> sum(E1^2) / (nrow(Species) - length(coef(ZI1)))

  1.3349
```

The residuals show a single atypically large value corresponding to the largest observation (Figure 8.12). We used the `identify` command. The outlier is observation number 2781. The code

```
> Species2 <- Species[-2781, ]
```

removes this observation, and rerunning the zero inflated model using the object `Species2` gives slightly modified regression parameters. However, the ecological conclusions are the same. The overdispersion drops from 1.33 to 0.83. We also make a sample variogram of the residuals. The tricky point is that the spatial coordinates of the traps are calculated with respect to each field. Hence we need to separate the residuals into two sets and make a sample variogram for each field. The results are presented in Figure 8.13, and do not show spatial correlation. When dropping row 2781, the shape of the sample variograms stays the same, but the absolute values (variance) of the sample variogram in the wheat field are reduced.

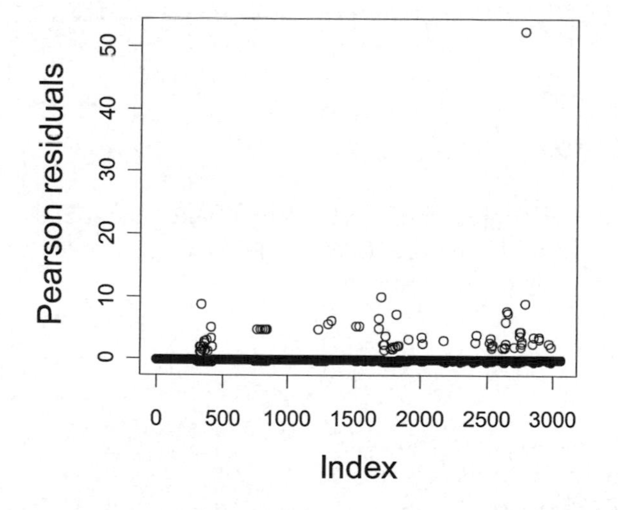

Figure 8.12. Index plot of the Pearson residuals. Note the large outlier. The graph was created with the `plot(E1)` command.

R code to produce the sample variogram follows. First the residuals and the spatial coordinates are stored in a data frame called `MyData`. The objects `ResFallow` and `ResWheat` are the Pearson residuals for the fallow and wheat fields, and a sample variogram is calculated for each of them. The remainder of the code extracts the relevant components and sets up the variables for a multi-panel scatterplot.

```
> library(gstat)
> MyData <- data.frame(E1 = E1, Xc = Species$Xcoord,
                       Yc = Species$Ycoord)
> coordinates(MyData) <- c("Xc","Yc")
> ResFallow <- MyData[Species$fField == "Fallow",]
> ResWheat  <- MyData[Species$fField == "Wheat",]
```

```
> V.Fallow  <- variogram(E1 ~ 1, data = ResFallow, robust = TRUE)
> V.Wheat   <- variogram(E1 ~ 1, data = ResWheat, robust = TRUE)
> Distance  <- c(V.Fallow$dist, V.Wheat$dist)
> Gamma     <- c(V.Fallow$gamma, V.Wheat$gamma)
> Field    <- rep(c("Fallow","Wheat"), c(nrow(V.Fallow),
               nrow(V.Wheat)))
> library(lattice)
> xyplot(Gamma ~ Distance | factor(Field), col = 1, pch = 16)
```

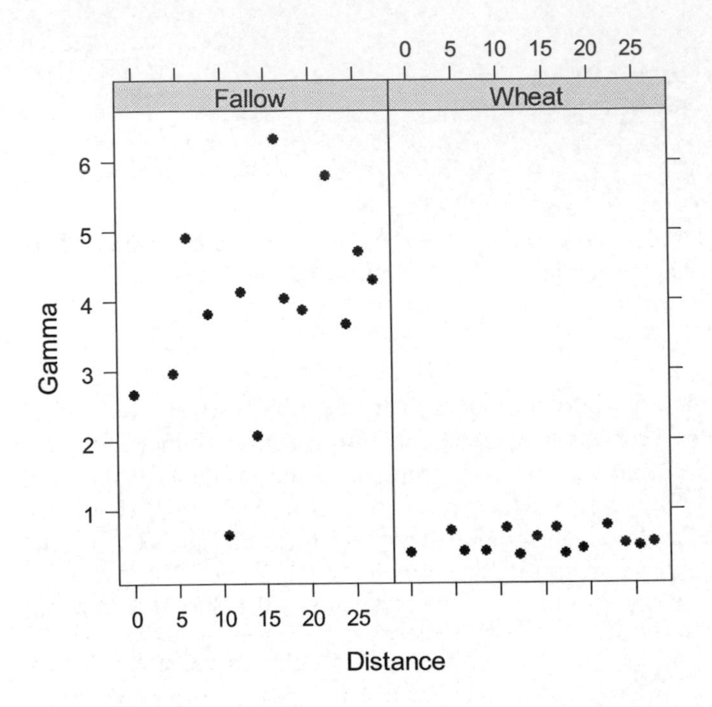

Figure 8.13. Sample variograms of the Pearson residuals of the ZIP GLM for the wheat and fallow fields.

We also plot the Pearson residuals versus trap identity, and noticed that there may be a small trap effect in the residuals. We therefore implement a ZIP GLM with random effects `Trap`. Such a model imposes a correlation structure on observations from the same trap. We run the code via the package `R2WinBUGS` in WinBUGS. We apply models with random effects in the predictor function of the count model and the binary model, and also in the count model only. The chains are highly correlated and show clear patterns. Most likely such a model is too complicated for the data structure; recall the small number of non-zero observations. We therefore select the ZIP GLM without further correlation structure or random effects.

8.7 Results for the female – male data

We fit the same ZIP model on the female and male data stored in the FM object. The R code to run this model and its numerical output are given below. First we define the variable direction (see the previous section) and apply the ZIP GLM to the female – male data from 30 May.

```
> FM$Degrees <- FM$Radians * 180/pi
> MyBreaks <- c(45, 45 + 90, 45 + 2*90, 45 + 3 * 90)
> Dir <- vector(length = nrow(FM))
> Dir[FM$Degrees > 315 | FM$Degrees <= 45  ] <- "North"
> Dir[FM$Degrees >  45 & FM$Degrees <= 135 ] <- "East"
> Dir[FM$Degrees > 135 & FM$Degrees <= 225 ] <- "South"
> Dir[FM$Degrees >=225 & FM$Degrees <= 315 ] <- "West"
> FM$fDir <- factor(Dir)
> FM$LogRel    <- log(FM$Nrelease)
> library(pscl)
```

```
> ZI1 <- zeroinfl(Count ~ fGroup * fField + fTraptype + Time +
                  fDir +offset(LogRel) |
                  fTraptype + Time + fDir, data = FM)
> summary(ZI1)

Count model coefficients (poisson with log link):
                            Estimate Std. Error z value Pr(>|z|)
(Intercept)               -5.4760507  0.3975770 -13.774   <2e-16
fGroupA. obscurus f       -0.9706745  0.3967419  -2.447   0.0144
fFieldFallow               0.0378486  0.3630478   0.104   0.9170
fTraptypeCircular         -0.2201789  0.2408303  -0.914   0.3606
Time                      -0.0007683  0.0028415  -0.270   0.7869
fDirNorth                 -0.4659411  0.2972196  -1.568   0.1170
fDirSouth                 -0.0935877  0.2755385  -0.340   0.7341
fDirWest                  -0.5922323  0.3487789  -1.698   0.0895
fGroupA.obscurusf:fFieldFal 1.1484355  0.4525863   2.537   0.0112

Zero-inflation model coefficients (binomial with logit link):
                   Estimate Std. Error z value Pr(>|z|)
(Intercept)        1.198612   0.277380   4.321 1.55e-05
fTraptypeCircular  1.399466   0.277323   5.046 4.50e-07
Time               0.003273   0.002822   1.160    0.246
fDirNorth          0.424842   0.334762   1.269    0.204
fDirSouth         -0.096822   0.379899  -0.255    0.799
fDirWest           0.266476   0.412286   0.646    0.518
```

A likelihood ratio test is applied to test whether direction is significant. It is not significant in the count model or in the binary model. There is also no significant time effect in the count model or binary model. The interaction of `fGroup` (representing the difference between males and females) with `fField` in the count model has a *p*-value of 0.01, which is not convincing, given that the *p*-values are approximate. The only term that is highly significant is `fTraptype` in the binary model. There is a higher probability of a false zero for the circular traps. A model validation shows that there are no outliers, but there is strong residual spatial correlation in the fallow field (Figure 8.14). This means that the model would need to be extended with a residual CAR correlation structure. However, given the small number of positive values and the fact that most terms are not significant, doing this would not be productive.

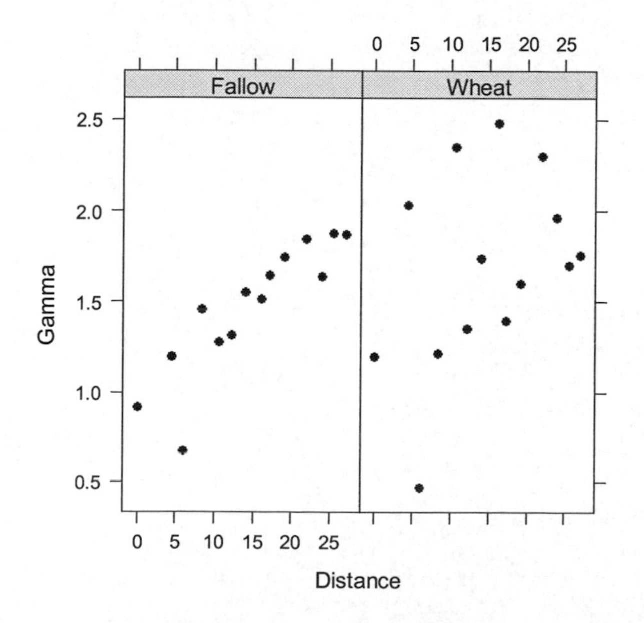

Figure 8.14. Sample variograms for the Pearson residuals of the ZIP model applied to the female-male *A. obscurus* data.

8.8 Discussion

We began the analysis with 10,573 observations, but, due to experimental design issues, concentrated on only a subset of the data. To deal with the large number of zeros, a zero inflated Poisson GLM was applied. When analyzing the male data of *A. obscurus* and *A. lineatus* we found a strong time effect in both the count and binary models. This means that absolute numbers decreased over time, along with the probability of measuring a false zero. For the count model there is a significant interaction of species and field. There is also a strong trap type effect in both parts of the model. Higher numbers are measured for the circular trap, and the probability of a false zero increases for this trap type. There is also a strong direction effect, with southerly showing lower predicted numbers and a lower probability of false zeros. The covariate representing direction may reflect small-scale habitat effects or variables such as wind direction, position of the sun, etc. There was no strong residual correlation in the ZIP model.

We also applied the same ZIP model on a second data set: males and females of *A. obscurus*. Only the trap type in the binary part was significant. However, this model showed strong spatial correlation in the wheat field but not in the fallow field, which possibly reflected a missing small-scale covariate. Surprisingly there was no time effect, and plotting the raw data versus time confirms this.

8.9 What to write in a paper?

In isolation, the results reported in this chapter may not be enough for a good publication, but perhaps it could be used in combination with other studies. The small-scale habitat effect inferred may play a role from a biological point of view. As to what to present, we face the daunting task of explaining why so many observations were dropped. The data is complicated in structure, and we advise presenting some data exploration graphs. Present the ZIP model and discuss the model validation and lack of spatial correlation. We strongly suggest presenting Figure 8.10 and Figure 8.11, as they clearly depict the results of the ZIP GLM.

9 Zero inflated GAM for temporal correlated sperm whale strandings time series

Alain F Zuur, Graham J Pierce, M Begoña Santos, Elena N Ieno, and Anatoly A Saveliev

9.1 Introduction

Information on migration patterns of sperm whales (*Physeter macrocephalus*) in the North Atlantic is preserved in historical records, notably for the North Sea (Smeenk 1997, 1999) and the Adriatic Sea (Bearzi et al., 2010) where sperm whale strandings have been documented since the 16th century. Most strandings on North Sea coasts occur during or after the southward migration from the feeding grounds, when some animals enter the North Sea instead of following their more usual route through deep waters to the west of the British Isles. There was much speculation about the high incidence of strandings on North Sea coasts in the 1990s (e.g. Robson and van Bree 1971; Klinowska 1985, Simmonds 1997, Sonntag and Lütkebohle 1998), with proposed causes including a range of anthropogenic factors (pollution, underwater noise) and the effects of solar activity or lunar cycles (Vanselow and Ricklefs 2005; Wright 2005).

Pierce et al. (2007) speculated that inter-annual variation in strandings of sperm whales in the North Sea could be related to the distribution and abundance of the squid that is their main prey, which in turn is expected to be strongly dependent on environmental conditions (Pierce et al., 2008). When a sperm whale enters the North Sea, which Smeenk (1997) characterized as a 'sperm whale trap,' it is argued that its limited ability to navigate the shallow waters over a gently sloping seabed with sandy substrate, coupled with the instinct to head south and west toward their breeding grounds, leads to stranding on North Sea coasts. Pierce et al. (2007) showed that long-term inter-annual variation in the incidence of sperm whale strandings on North Sea coasts is related to positive temperature anomalies; the incidence of strandings was shown to be higher in warmer periods.

In their statistical analyses, Pierce et al. (2007) converted annual numbers of stranded animals to incidence data, representing the presence or absence of strandings in a particular year. These data were modelled as a function of various covariates, e.g. sunspots, temperature, and the winter North Atlantic Oscillation (NAO) Index. In this chapter we will analyse the original count data. Due to the large number of years with no recorded strandings, potential auto-correlation, and non-linear patterns, we will be dealing with zero-inflated generalised additive models (GAM) with temporal correlation for count data.

Throughout this chapter we will use only the covariates 'annual temperature anomaly' and the winter North Atlantic Oscillation (NAO) index, and we will focus on the methodology and implementation of the models in R. Background information on the North Sea data can be found in Smeenk (1997, 1999) and Pierce et al. (2007). The aim of this chapter is to demonstrate fitting zero inflated Poisson (ZIP) GAMs with temporal correlation in R.

9.2 What makes this a difficult analysis?

We selected the data from the years 1563 (when the first recorded stranding occurred) to 2001, so we have a time series of 439 observations consisting of counts of sperm whale strandings per year. The data are zero inflated and temporally correlated, and the relationships among the covariates are weak and non-linear. We will apply a GAM with a zero inflated Poisson distribution and residual auto-regressive correlation. There is no existing software to fit such models in R; therefore, we will use WinBUGS. However, implementing a GAM in WinBUGS is a challenge even without adding zero inflation or residual temporal correlations, and we want to do all three! The implementation of GAMs using WinBUGS is described in detail in Crainiceanu et al. (2005), and the first part of this chapter is heavily based on that paper. In previous chapters we have seen how to implement a zero inflated distribution in generalized linear models; here we will do the same for GAMs. Finally, we will include a residual auto-regressive correlation structure within the context of a zero inflated GAM with a Poisson distribution.

9.3 Importing and data coding

The following code imports the data from a tab delimited ascii file and extracts the relevant variables. We use the data from 1563 to 2001, as there are no missing values in this time period. We use the annual temperature anomaly as covariate, because Pierce et al. (2007) found it to be the most important. To illustrate the coding for models with two covariates, we select a second, the winter NAO.

```
> SW <- read.table("Spermwhales.txt", header = TRUE)
> MyVar <- c("Year", "Nsea", "AnnualTTAnom", "WinterNAO")
> SW1 <- SW[SW$Year >= 1563 & SW$Year <= 2001, MyVar]
> str(SW1)

'data.frame':   439 obs. of  4 variables:
 $ Year        : int  1563 1564 1565 1566 1567 1568 ...
 $ Nsea        : int  1 0 0 1 0 0 0 0 0 3 ...
 $ AnnualTTAnom: num  -0.626 -0.216 -1.285 -0.07  ...
 $ WinterNAO   : num  -0.57 -0.1 -1.14 0.87 0.07 ...
```

The variable `Nsea` contains sperm whale strandings in the North Sea 1563 – 2001, and is based on a published data set (Smeenk, 1997, 1999; Pierce et al. 2007). These data refer to animals that became live-stranded or were washed ashore after dying. The variable `AnnualTTAnom` is the annual temperature anomaly, and `WinterNAO` is the North Atlantic Oscillation index for the winter months. The annual temperature anomalies are European temperature anomalies (Luterbacher et al., 2004; Xoplaki et al., 2005). The NAO series was reconstructed to the year 1500, using proxies such as ice and snow records and tree ring data (Luterbacher et al., 1999, 2001). A high NAO index corresponds to enhanced westerly winds and warmer temperatures during winter in the NE Atlantic area.

We standardise (subtract the mean and divide by the standard deviation) the two covariates, essential in Markov Chain Monte Carlo (MCMC) techniques. We shorten the covariate names as well.

```
> SW1$TempAn <- as.numeric(scale(SW1$AnnualTTAnom))
> SW1$NAO    <- as.numeric(scale(SW1$WinterNAO))
```

9.4 Data exploration

We first show a time series plot of the sperm whale strandings (Figure 9.1). Note the increase in strandings in the second half of the 20th century. Many years contain 0 strandings. A frequency plot indicates that the data are zero inflated (Figure 9.2). R code to create these two figures is simple:

```
> plot(x = SW1$Year, y = SW1$Nsea, type = "h", xlab = "Time (year)",
       ylab = "Number of strandings")
> Tab <- table(SW1$Nsea)
> plot(x = names(Tab), y = Tab, type = "h",
       xlab = "Observed values", ylab = "Frequency")
```

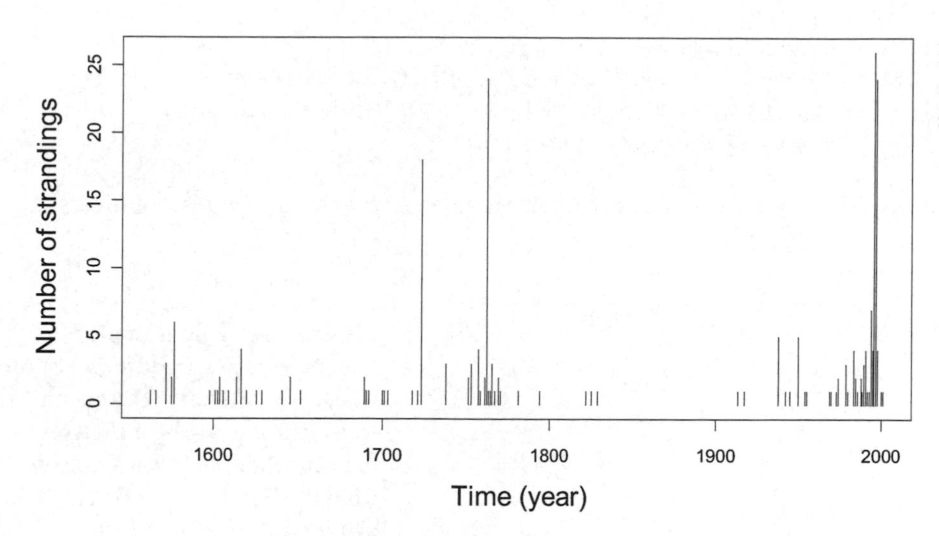

Figure 9.1. Time series of whale strandings.

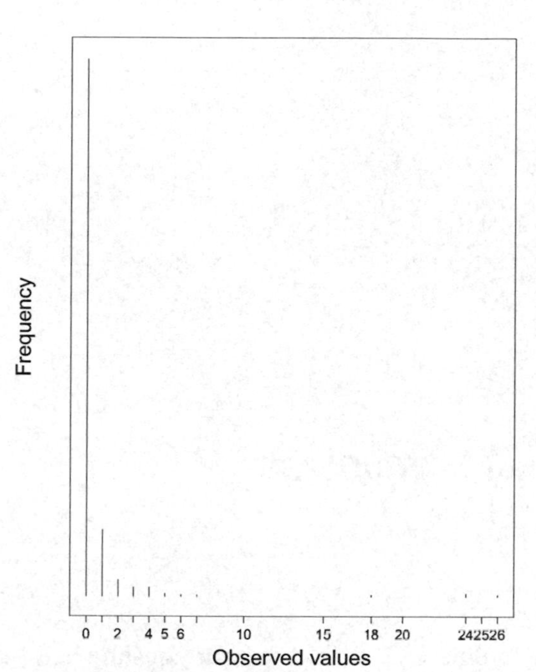

Figure 9.2. Frequency plot of the whale strandings. A value of zero was observed in 83% of the cases, making the data zero inflated.

A scatterplot of the whale strandings versus each of the two covariates is presented in Figure 9.3. A smoother obtained by a Poisson GAM was added to aid visual interpretation. Note that the patterns are weak. The R code to generate this figure follows. Although it looks daunting, we have seen similar code in previous chapters. We first concatenate the two time series with the covariates (X2), repeat the whale strandings twice (Y2), and create a vector of strings with the covariate names (ID2). These three vectors are used in the xyplot function of the lattice package. The only new element of the code is a Poisson GAM in the panel function to aid visual interpretation. The GAM ignores any zero inflation and overdispersion.

```
> MyVar <- c("TempAn", "NAO")
> X2 <- as.vector(as.matrix(SW1[ ,MyVar]))
> Y2 <- rep(SW1$Nsea, 2)
> ID2 <- factor(rep(MyVar, each = nrow(SW1)))
> library(lattice)
> library(mgcv)
> xyplot(Y2 ~ X2 | ID2, col = 1, cex.lab = 1.5,
    xlab = "Covariates", ylab = "North sea strandings",
    strip = function(bg = 'white', ...)
              strip.default(bg = 'white', ...),
    scales = list(alternating = T,
                  x = list(relation = "free"),
                  y = list(relation = "same")),
    panel = function(x, y){
```

```
    panel.grid(h = -1, v = 2)
    panel.points(x, y, col = 1)
    tmp <- gam(y ~ s(x), family = poisson)
    mydata <- data.frame(x = seq(min(x), max(x), length = 100))
    p1 <- predict(tmp, newdata = mydata, type = "response")
    llines(mydata$x, p1, lwd = 3, col = 1) })
```

There is some collinearity between the winter NAO and temperature anomaly series; the Pearson correlation is 0.51.

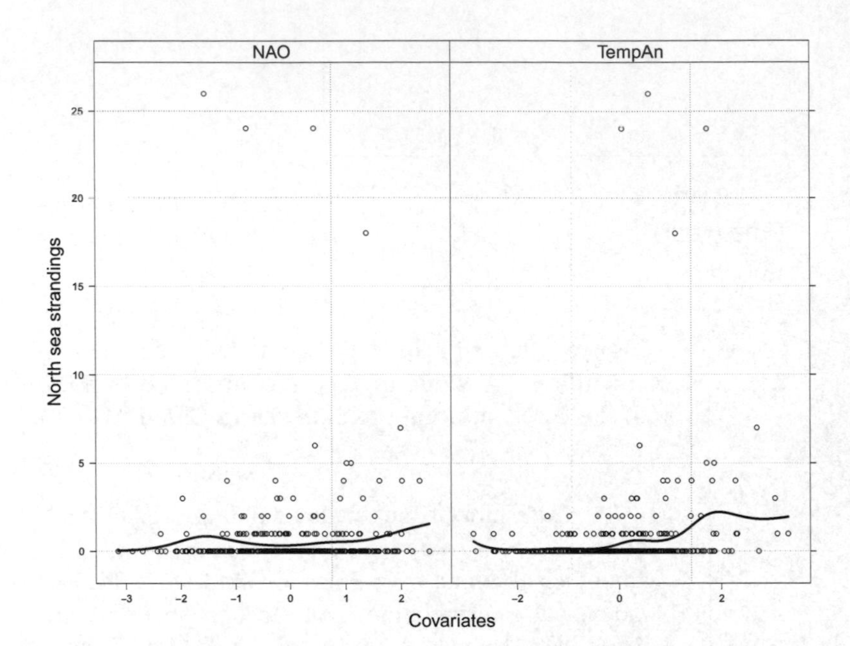

Figure 9.3. Scatterplot of sperm whale strandings versus the winter NAO and annual temperature anomaly. A smoother, using a Poisson distribution, was added to aid visual interpretation.

9.5 Poisson GAM with a single smoother in WinBUGS

9.5.1 Fitting a Poisson GAM using gam from mgcv

Before initiating WinBUGS, let us fit the model using 'ordinary' R code. The basic equation is:

$$\begin{aligned} &Y_s \sim \text{Poisson}(\mu_s) \\ &E(Y_s) = \text{var}(Y_s) = \mu_s \\ &\log(\mu_s) = \alpha + f(TempAn_s) \end{aligned} \tag{9.1}$$

where Y_s is the number of sperm whale strandings in year s, and $TempAn_s$ is the standardised temperature anomaly in year s. The Poisson distribution dictates that the mean and variance of the number of strandings in year s be equal to μ_s, and a log-link is used. The R code to fit this model is:

```
> library(mgcv)
> M1 <- gam(Nsea ~ s(TempAn), family = poisson, data = SW1)
> summary(M1)


Family: poisson
```

```
Link function: log
Formula: Nsea ~ s(TempAn)

Parametric coefficients:
            Estimate Std. Error z value Pr(>|z|)
(Intercept) -1.03126    0.09304  -11.08   <2e-16

Approximate significance of smooth terms:
            edf Ref.df Chi.sq p-value
s(TempAn) 8.043  8.704  135.3  <2e-16
R-sq.(adj) =  0.0247   Deviance explained = 14.5%

UBRE score = 1.1789  Scale est. = 1        n = 439
```

The estimated smoother is presented in Figure 9.4A, and shows a general increase in expected numbers of strandings for higher temperature anomalies. As a component of the model validation process we should extract Pearson residuals and create an auto-correlation plot (Figure 9.4B). The correlation is significantly different from 0 at time lag 1, though not strong. R code to produce Figure 9.4 follows. We must ensure that the residuals are arranged according to the time order and use `diff(SW1$Year)` to make certain that there are no missing years.

```
> plot(M1)
> E1 <- resid(M1, type = ”pearson”)
> acf(E1, main = "ACF residuals")
```

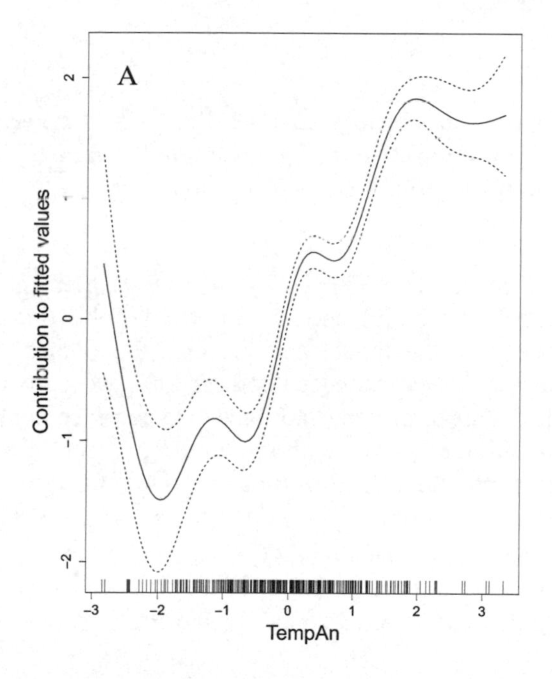

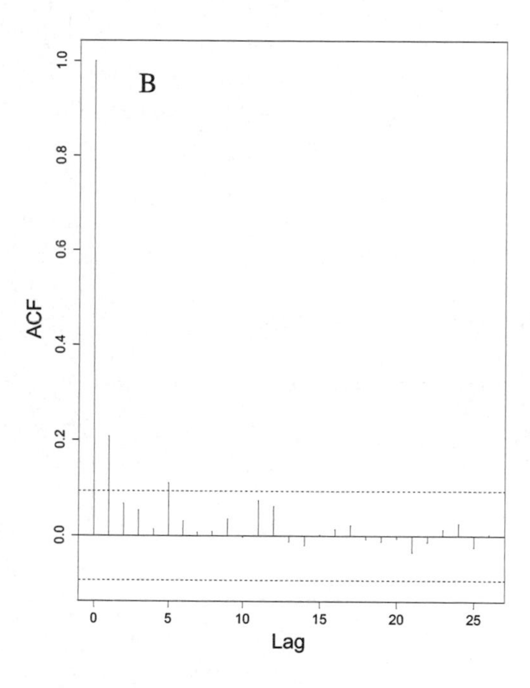

Figure 9.4. A: Estimated smoother for the model in Equation (9.1). B: Auto-correlation for the Pearson residuals from the model in Equation (9.1). The dotted lines represent 95% confidence intervals. Note that the correlation is significantly different from 0 at the 5% level at time lag 1.

To assess potential overdispersion, we calculate the sum of squared Pearson residuals and divide by the sample size minus the residual degrees of freedom. If this ratio is larger than 1, we have overdispersion. Here we have overdispersion of 7.90.

```
> sum(E1^2) / M1$df.res
```

```
7.904753
```

Overdispersion may be due to, among other factors, missing covariates, non-linear covariate effects that are modelled as a linear, correlation, zero inflation, or errors of distribution or link function. The residual correlation is not strong, and we are using a model that allows for non-linear relationships. Our first model improvement consists of allowing for zero inflation.

9.5.2 Fitting the binomial GAM from Pierce et al. (2007)

Pierce et al. (2007) used a similar time span for the data but converted the response variable to incidence data and applied a binomial GAM with temporal correlation. The underlying model is given in Equation (9.2), and the estimated smoother for the temperature anomaly is presented in Figure 9.5. The correlation between residuals separated by $s - t$ years is given by $\phi^{|t-s|}$.

$$
\begin{aligned}
& Y_s^{01} \sim Bin(\pi_s) \\
& E(Y_s^{01}) = \pi_s \quad \text{and} \quad \text{var}(Y_s^{01}) = \pi_s \times (1-\pi_s) \\
& \text{logit}(\pi_s) = \alpha + f(TempAn_s) + \varepsilon_s \\
& cor(\varepsilon_s, \varepsilon_t) = \phi^{|s-t|} \quad \text{and} \quad \varepsilon_s \sim N(0, \sigma_\varepsilon^2)
\end{aligned}
\tag{9.2}
$$

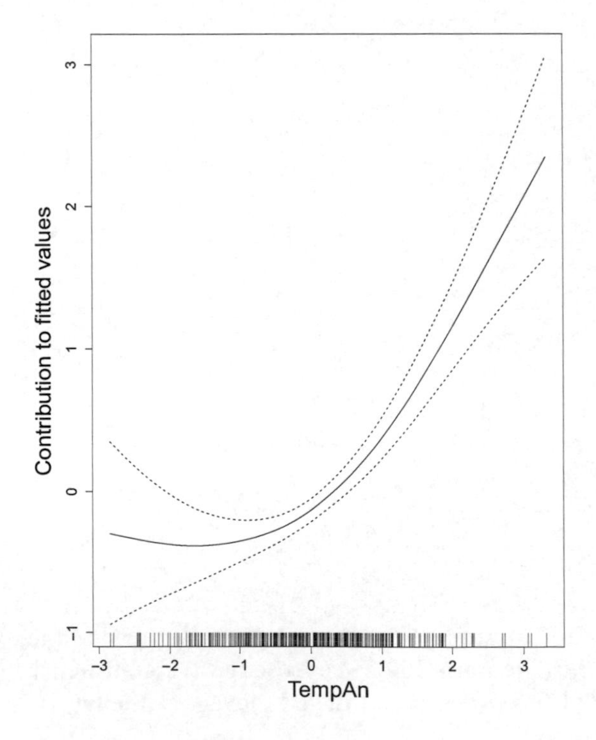

Figure 9.5. Estimated smoother for temperature anomaly for the binomial GAMM in Equation (9.2).

The variable $Y^{01}{}_s$ is 1 if, in year s, there is a recorded stranding and is 0 if not. The following R code runs the model and presents the numerical output and estimated smoother. First we convert the counts to presence/absence data and then apply the binomial GAM with an auto-regressive correlation structure. The `summary` command applied to the `$lme` object shows that the estimated value of ϕ is 0.14. The shape of the smoother indicates that the probability of stranding increases with higher temperature anomaly values.

```
> SW1$Nsea01 <- SW1$Nsea
> SW1$Nsea01[SW1$Nsea > 0] <- 1
```

```
> M2 <- gamm(Nsea01 ~ s(TempAn), correlation = corAR1(form =~ Year),
           family = binomial, data = SW1 )
> summary(M2$gam)  #Results not presented
> summary(M2$lme)  #Results not presented
> plot(M2$gam, cex.lab = 1.5)
```

9.5.3 Knots

Now that we have seen the results of an ordinary GAM with a Poisson distribution and the results of Pierce et al. (2007), we extend the Poisson model toward a ZIP GAM. Before doing this we will fit a Poisson GAM using WinBUGS, which means we need to introduce the concept of the knot. Two knots define a box (or window) along the graph horizontal axis (Figure 9.6). There are two issues to address: How many knots do we use and where do we place them? Ruppert (2003) suggested using an ample number of knots, typically between 5 and 20. We select 9, as it simplifies explanations, but that is not a rational justification. Crainiceanu et al. (2005) advised placing the knots at positions κ_1, κ_2, κ_3, ..., κ_K, where κ_i is the ith sample quantile of the covariate. Hence with 9 knots the sample quantiles are:

```
> num.knots <- 9
> Prob <- seq(0, 1, length = (num.knots + 2))
> Prob

0.0 0.1 0.2 0.3 0.4 0.5 0.6 0.7 0.8 0.9 1.0
```

Since we don't need the sample quantiles for probabilities 0 and 1, we drop the initial and final values:

```
> Prob1 <- Prob[-c(1, (num.knots+2))]
> Prob1

0.1 0.2 0.3 0.4 0.5 0.6 0.7 0.8 0.9
```

We now require the matching quantile values from the covariate. We could sort the covariate from small to large values and pick the 10th% value, the 20th% value, etc., but the function `quantile` does it for us:

```
> Knots <- quantile(unique(SW1$TempAn), probs = Prob1)
> Knots

    10%     20%     30%     40%     50%     60%
-1.3857 -0.8782 -0.5592 -0.2558 -0.0161  0.2528
    70%     80%     90%
 0.5117  0.8575  1.2381
```

So 10% of the unique covariate values are smaller than -1.3857, 10% range from -1.3857 to -0.8782, etc. Hence κ_1 = -1.3857, κ_2 = -0.8782, etc. Figure 9.6 was created with the following code:

```
> plot(y = SW1$Nsea, x = SW1$TempAn, xlab = "Temperature anomaly",
       ylab = "Sperm whale strandings", cex.lab = 1.5)
> abline(v = as.numeric(Knots), lty = 2)
```

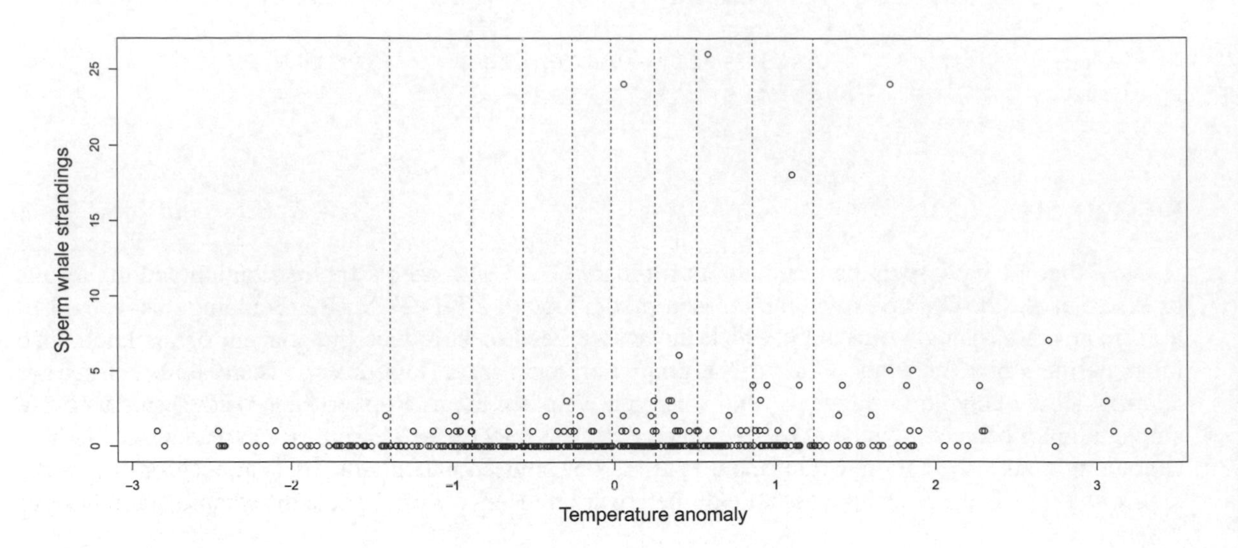

Figure 9.6. Scatterplot of temperature anomaly versus sperm whale strandings. A dotted line represents a knot.

A function can be written to automate this process (Ngo and Wand, 2004):

```
default.knots <- function(x, num.knots) {
   if (missing(num.knots)) num.knots <- max(5,
                 min(floor(length(unique(x)) / 4), 35))
   return(quantile(unique(x), seq(0, 1,
         length = (num.knots+2))[-c(1, (num.knots+2))]))   }
```

Although it looks intimidating, this function performs similar to our code. If the number of knots is not specified, N / 4 will be used, where N is the sample size. If this ratio is smaller than 5, 5 knots will be used. To call this function use:

```
> default.knots(SW1$TempAn, num.knots = 9)
```

It gives the same results.

9.5.4 Low rank thin plate splines

There are many options for modelling the smoother $f(\text{TempAn}_s)$ in Equation (9.1), a moving average smoother, LOESS smoother, regression spline, etc. See Wood (2006) for a detailed mathematical explanation or Chapter 3 in Zuur et al. (2009a) for a less math-oriented introduction. Crainiceanu et al. (2005) explain how to use a low rank thin plate regression spline for the smoothing function, and the remainder of this subsection is based on this publication. To minimize confusion for people reading the Crainiceanu paper along with this chapter, we will use their mathematical notation. We first derive the equations for a Gaussian model, and then show how to modify it for a Poisson, negative binomial, or binomial distribution.

The low rank thin plate regression spline for the smoothing function is given by:

$$f(TempAn_s,\theta) = \beta_0 + \beta_1 \times TempAn_s + \sum_{k=1}^{K} u_k \times |TempAn_s - x_k|^3 \qquad (9.3)$$

This equation can be written as:

$$
\begin{aligned}
f(TempAn_s, \theta) = \beta_0 + \beta_1 \times TempAn_s + \\
u_1 \times |TempAn_s + 1.385|^3 + u_2 \times |TempAn_s + 0.878|^3 + \\
u_3 \times |TempAn_s + 0.559|^3 + \cdots + u_9 \times |TempAn_s - 1.23|^3
\end{aligned}
\tag{9.4}
$$

We wrote out the summation in Equation (9.3) and filled in the values of the knots $\varkappa_i$. Thus the smoother is modelled as a straight line plus a set of cubic terms. Figure 9.7 provides a graphic illustration. Each line in the graph represents one of the cubic terms in Equation (9.4), but without the regression parameter u_i. So we calculate values of $|\text{TempAn}_s + 1.385|^3$ for a range of TempAn_s values, sketch it, and do it 8 times with different constant values.

The terms β_0, β_1, and u_1 to u_9 are unknown regression parameters that we need to estimate. To illustrate their effects, we will carry out a small simulation study. We use a random number generator to select values for β_0, β_1, and u_1 to u_9. This allows us to calculate the smoother $f(\text{TempAn}_s)$ for these simulated values by substituting them for the regression parameters in Equation 9.4. We do this process 16 times, and the 16 scenarios are shown in Figure 9.8. Note the wide range of shapes that the low rank thin plate spline can produce. By increasing the variation in the regression parameters u_1 to u_9, an even greater variation in shapes can be obtained.

Summarising, the low rank thin plate regression spline begins with a straight line given by $\beta_0 + \beta_1 \times \text{TempAn}_s$ and adds each of the cubic curves in Figure 9.7 multiplied by a regression parameter u_i.

To avoid overfitting, the following criteria, for a Gaussian model, is optimized:

$$
\sum_{i=1}^{N} \left(Y_i - f(TempAn_s, \theta)\right)^2 + \frac{1}{\lambda} \times \theta^T \times \mathbf{D} \times \theta \tag{9.5}
$$

where the λ is a smoothing parameter. See Chapter 3 in Zuur et al. (2007) for a detailed explanation. The first part of the equation measures the goodness of fit, and the second part is a penalty for model complexity. In the following subsection we follow Crainiceanu et al. (2005) and show how the expression in Equation (9.5) can be rewritten as a linear mixed effects model. If you are not interested in the mathematics, you can skip this and move to Subsection 9.5.6, as it contains a short summary of the mathematical derivations.

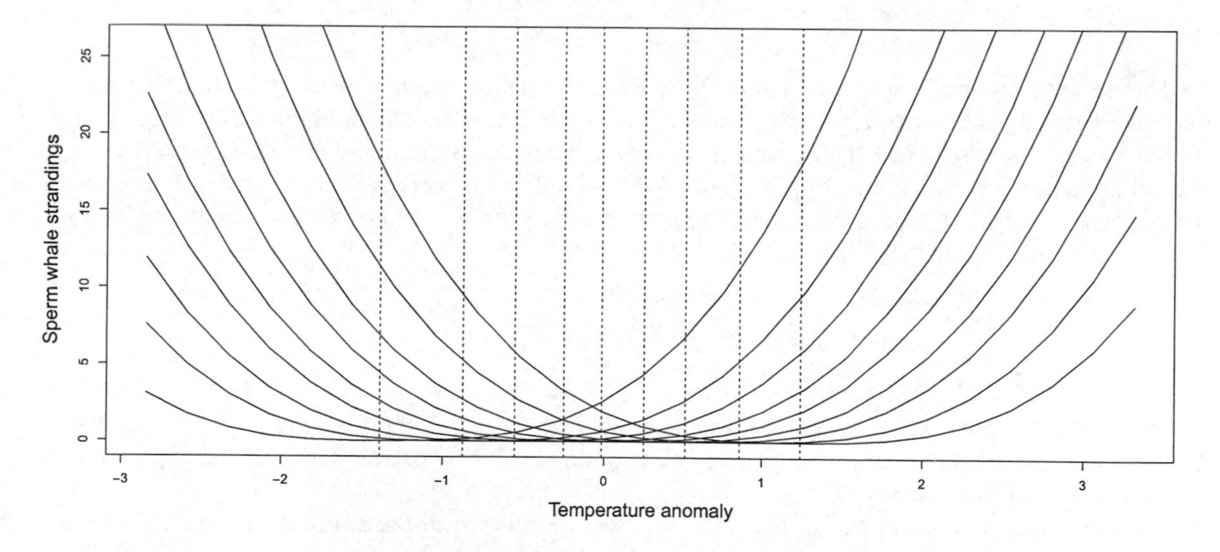

Figure 9.7. Plot of the effects of the cubic terms in the low rank thin plate regression spline. Each line represents a cubic term $|\text{TempAn}_s - \varkappa_i|^3$ for a value of $\varkappa_i$.

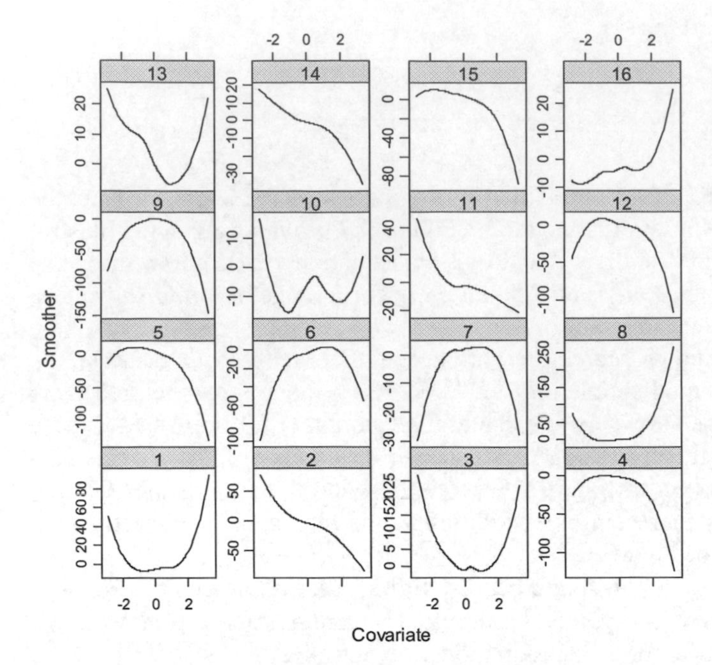

Figure 9.8. Each panel shows the estimated smoother for a set of randomly generated regression parameters β_0, β_1, and u_1 to u_9.

9.5.5 Mathematics for low rank thin plate splines

In this subsection we follow Crainiceanu et al. (2005) to write the expression in Equation (9.5) as a linear mixed effects model. The smoother in Equation (9.3) can be written as:

$$\begin{pmatrix} 1 & TA_1 \\ 1 & TA_2 \\ \vdots & \vdots \\ 1 & TA_{439} \end{pmatrix} \times \begin{pmatrix} \beta_0 \\ \beta_1 \end{pmatrix} + \begin{pmatrix} |TA_1 + 1.385|^3 & |TA_1 + 0.878|^3 & \cdots & |TA_1 - 1.23|^3 \\ |TA_2 + 1.385|^3 & |TA_2 + 0.878|^3 & \cdots & |TA_2 - 1.23|^3 \\ \vdots & \vdots & & \\ |TA_{439} + 1.385|^3 & |TA_{439} + 0.878|^3 & \cdots & |TA_{439} - 1.23|^3 \end{pmatrix} \times \begin{pmatrix} u_1 \\ u_2 \\ \vdots \\ u_9 \end{pmatrix}$$

This involves writing out the equation for each year *s* using standard matrix notation. To ensure that the equation fits on a single line we use *TA* instead of *TempAn*. The notation can be further simplified by using $\mathbf{X} \times \boldsymbol{\beta} + \mathbf{Z}_K \times \mathbf{u}$. **X** contains two columns; the first column contains only ones and the second is the temperature anomaly time series. The $\boldsymbol{\beta}$ contains β_0 and β_1. The columns of $\mathbf{Z}_K$ contain the components $|TA_i - x_k|^3$ and **u** the matching u_ks. With this matrix notation, we can rewrite the optimisation criteria in Equation (9.5) as:

$$\| \mathbf{Y} - \mathbf{X} \times \boldsymbol{\beta} - \mathbf{Z}_K \times \mathbf{u} \|^2 + \frac{1}{\lambda} \times \boldsymbol{\theta}^T \times \mathbf{D} \times \boldsymbol{\theta} \qquad (9.6)$$

The vector **Y** contains all the sperm whale strandings, and ||.|| is the Euclidean norm. So far, this has been a matter of changing the notation to matrix notation. We will now focus on the expression in the second part of Equation (9.6).

The parameter $\boldsymbol{\theta}$ contains β_0, β_1, and u_1 to u_9. The second part of the expression in Equation (9.6) is:

$$\frac{1}{\lambda}\times\begin{pmatrix} \beta_0 & \beta_1 & u_1 & \cdots & u_9 \end{pmatrix}\times\begin{pmatrix} \mathbf{0}_{2\times 2} & \mathbf{0}_{2\times 9} \\ \mathbf{0}_{9\times 2} & \mathbf{\Omega}_{9\times 9} \end{pmatrix}\times\begin{pmatrix} \beta_0 \\ \beta_1 \\ u_1 \\ \vdots \\ u_9 \end{pmatrix}$$

The notation $\mathbf{0}_{2\times2}$ means that we have a 2 by 2 matrix containing only zeros. If we write out the matrix and vector multiplications, all that remains is $1/\lambda$, the u_is, and $\mathbf{\Omega}$. The λ controls the penalty due to the second part of Equation (9.6). If λ is large, then $1/\lambda$ is small, and the penalty is small. Optimising the entire criterion means we focus on the sum of squares between the observed data and the smoother (between the || || symbols). On the other hand, if λ is small, $1/\lambda$ is large, and the penalty becomes more important in the optimisation criterion.

So, what about the matrix $\mathbf{\Omega}$? Its $(i, j)^{\text{th}}$ element contains $|\varkappa_i - \varkappa_j|^3$ and functions as a penalty for the $|\text{TempAn}_s - \varkappa_j|^3$ term.

The derivation in Crainiceanu et al. (2005) is based on a Gaussian model; the starting point is:

$$Y_s = f(TempAn_s) + \varepsilon_s$$

where ε_s is normally distributed with mean 0 and variance σ_ε^2. The next step is to divide the expression in Equation (9.6) on both sides by σ_ε^2, resulting in:

$$\frac{1}{\sigma_\varepsilon^2}\times\|\mathbf{Y}-\mathbf{X}\times\boldsymbol{\beta}-\mathbf{Z}_K\times\mathbf{u}\|^2+\frac{1}{\lambda\times\sigma_\varepsilon^2}\times\boldsymbol{\theta}^T\times\mathbf{D}\times\boldsymbol{\theta} \tag{9.7}$$

Defining $\sigma_u^2 = \lambda\times\sigma_\varepsilon^2$ gives:

$$\frac{1}{\sigma_\varepsilon^2}\times\|\mathbf{Y}-\mathbf{X}\times\boldsymbol{\beta}-\mathbf{Z}_K\times\mathbf{u}\|^2+\frac{1}{\sigma_u^2}\times\boldsymbol{\theta}^T\times\mathbf{D}\times\boldsymbol{\theta} \tag{9.8}$$

If we consider the parameter $\boldsymbol{\beta}$ as fixed, $\mathbf{u}$ as random with $E(\mathbf{u}) = 0$ and covariance $\text{cov}(\mathbf{u}) = \sigma_u^2 \times\mathbf{\Omega}^{-1}_K$, and assume that $\mathbf{u}$ and $\boldsymbol{\varepsilon}$ are independent, we obtain a penalized spline expressed in the notation of a linear mixed effect model. It turns out that our penalised low rank thin plate spline is equal to the best linear predictor:

$$\mathbf{Y}=\mathbf{X}\times\boldsymbol{\beta}+\mathbf{Z}_K\times\mathbf{u}+\boldsymbol{\varepsilon} \quad \text{with} \quad \text{cov}\begin{pmatrix}\mathbf{u}\\ \boldsymbol{\varepsilon}\end{pmatrix}=\begin{pmatrix}\sigma_u^2\times\mathbf{\Omega}^{-1}_{K\times K} & \mathbf{0}\\ \mathbf{0} & \sigma_\varepsilon^2\times\mathbf{I}_{N\times N}\end{pmatrix} \tag{9.9}$$

We can re-express this equation using $\mathbf{b} = \mathbf{\Omega}^{-1/2}_K \times \mathbf{u}$ as:

$$\mathbf{Y}=\mathbf{X}\times\boldsymbol{\beta}+\mathbf{Z}\times\mathbf{b}+\boldsymbol{\varepsilon} \quad \text{with} \quad \text{cov}\begin{pmatrix}\mathbf{b}\\ \boldsymbol{\varepsilon}\end{pmatrix}=\begin{pmatrix}\sigma_b^2\times\mathbf{I}_{K\times K} & \mathbf{0}\\ \mathbf{0} & \sigma_\varepsilon^2\times\mathbf{I}_{N\times N}\end{pmatrix} \tag{9.10}$$

This is our familiar expression for a linear mixed effects model. We can use the function `lme` from the `nlme` package to fit this. For an example see Ngo and Wand (2004).

9.5.6 Bypassing the mathematics

In the previous subsection we used not-too-demanding mathematics and showed that the expression in Equation (9.5) can be rewritten as:

$$\mathbf{Y} = \mathbf{X} \times \boldsymbol{\beta} + \mathbf{Z} \times \mathbf{b} + \boldsymbol{\varepsilon} \tag{9.11}$$

The vector **Y** contains all the Y_i values, and **X** is a design matrix. Its first column contains ones and the second column has the covariate. The regression parameter $\boldsymbol{\beta}$ contains β_0 and β_1. If 9 knots are used, **b** contains 9 regression parameters proportional to u_is, and **Z** is a design matrix containing the cubic terms $|\text{TempAn}_s - \varkappa_i|^3$ multiplied by constant terms based on the number of knots. The expression in Equation (9.11) is for a GAM with a Gaussian distribution. The conditions of $\boldsymbol{\beta}$ and **b** are:

$$\begin{aligned} &\beta \sim N(0, \sigma_\beta^2 \times \mathbf{I}_{2x2}) \\ &\mathbf{b} \sim N(0, \sigma_b^2 \times \mathbf{I}_{KxK}) \\ &\varepsilon \sim N(0, \sigma_\varepsilon^2 \times \mathbf{I}_{NxN}) \end{aligned} \tag{9.12}$$

where $\mathbf{I}_{K \times K}$ is a diagonal matrix of dimension K by K (K is the number of knots). We assume that the regression parameters and residuals are independent. If you have read Chapter 4 (or 2) in this book, you will recognize the expressions in Equations (9.11) and (9.12) as linear mixed effects models. The additive model can be fitted using the function `lme` or `lmer`.

The extension to a GAM with a Poisson distribution is simple:

$$\mathbf{Y} = e^{\mathbf{X} \times \boldsymbol{\beta} + \mathbf{Z} \times \mathbf{b}} \tag{9.13}$$

or, to allow for overdispersion, use:

$$\mathbf{Y} = e^{\mathbf{X} \times \boldsymbol{\beta} + \mathbf{Z} \times \mathbf{b} + \boldsymbol{\varepsilon}} \tag{9.14}$$

The term $\boldsymbol{\varepsilon}$ is the observation level random effect and allows for extra variation, which would otherwise be considered overdispersion (see Chapter 2). It can also be used in a negative binomial GLM or GAM. For a binomial GLM or GAM, $\mathbf{X} \times \boldsymbol{\beta} + \mathbf{Z} \times \mathbf{u}$ can be used in the logistic link function.

The following R code gives the **X** and the **Z** matrices. To fully understand the **Z**, we advise reading the previous subsection or Crainiceanu et al. (2005). The **X** is simple:

```
> N <- nrow(SW1)
> X <- cbind(rep(1, N), SW1$TempAn)
```

Type `head(X)` to convince yourself that the first column of **X** contains ones (for the intercept β_0) and the second column the covariate values (for β_1). We create a function to obtain **Z**; all steps are derived from the mathematics presented in the previous subsection.

```
GetZ <- function(x, Knots) {
  Z_K             <- (abs(outer(x, Knots,"-")))^3
  OMEGA_all       <- (abs(outer(Knots, Knots,"-")))^3
  svd.OMEGA_all  <- svd(OMEGA_all)
  sqrt.OMEGA_all <- t(svd.OMEGA_all$v %*% (t(svd.OMEGA_all$u) *
                         sqrt(svd.OMEGA_all$d)))
  Z <- t(solve(sqrt.OMEGA_all, t(Z_K)))
  return(Z)}
```

We would need to change the code if there were missing values in the covariate, but that is not the case here. To execute this function, we use:

```
> Z <- GetZ(SW1$TempAn, Knots)
```

We are now ready to initiate the code for WinBUGS.

9.5.7 WinBUGS code for GAM

There are many functions in R that are able to fit a GAM. For example, we could use the function `gam` from the `mgcv` package (Wood 2006) or `gam` from the `gam` package (Hastie and Tibshirani, 1990). In the previous two subsections we endeavoured to rewrite a simple smoother as a linear mixed effects model. We could use `lme` from the `nlme` package (Pinheiro and Bates, 2000) or `lmer` from the `lme4` package (Bates et al., 2011) to fit the models. However, we want to include zero inflation and temporal correlation in the model, so we will do it in WinBUGS using a package that allows us to run WinBUGS from within R: `R2WinBUGS`. This is not part of the base package, so you will need to download and install it. The process of running a model in `R2WinBUGS` is:

1. Prepare necessary files.
2. Run WinBUGS using short chains and spend half the day eliminating the mistakes.
3. Run WinBUGS using very long chains all day while you are at the beach.
4. Inspect and present results.

The files needed for WinBUGS follow. We assume that the matrices **X** and **Z** have been calculated as described in the previous subsection. The following code assembles the required variables:

```
> win.data <- list(Y          = SW1$Nsea,
                   NumKnots   = 9,
                   X          = X,
                   Z          = Z,
                   N          = nrow(SW1))
```

It contains the response variable, the number of knots, the **X** and **Z** matrices, and the sample size. The essential modelling code for WinBUGS is:

```
sink("modelgam.txt")
cat("
model{
 #Priors
 for (i in 1:2) { beta[i]  ~ dnorm(0, 0.00001) }
 for (i in 1:NumKnots) { b[i] ~ dnorm(0, taub) }
 taub ~ dgamma(0.01, 0.01)

 #Likelihood
 for (i in 1:N) {
     Y[i] ~ dpois(mu[i])
     log(mu[i])    <- max(-20, min(20, eta[i]))
     eta[i]        <- Smooth1[i]
     Smooth1[i] <- beta[1] * X[i,1] + beta[2]*X[i,2] +
                   b[1] * Z[i,1] + b[2] * Z[i, 2] +
                   b[3] * Z[i,3] + b[4] * Z[i, 4] +
                   b[5] * Z[i,5] + b[6] * Z[i, 6] +
                   b[7] * Z[i,7] + b[8] * Z[i, 8] +
                   b[9] * Z[i,9]
     PRes[i] <- (Y[i] - mu[i]) / pow(mu[i], 0.5)  }}
",fill = TRUE)
sink()
```

A Poisson distribution is used, and the predictor function follows Equation (9.13). The problem with an exponential model is that if one of the betas or *b*s selected in the simulation process is relatively large, exp(eta) becomes extremely large, and WinBUGS may close down with a vague error message. The construction `max(-20, min(20, eta[i]))` prevents large values in the predictor function. This, however, leads immediately to another problem: We used priors for β_0, β_1, and b_1 to b_9, as suggested by Crainiceanu et al. (2005), but their priors allowed for relatively large values of the parameters b_1 to b_9, resulting in high values of `eta[i]` and thus high values of the predictor function `eta`, so the `min` and `max` functions are activated, and, as a result, the draws in the chains do not change. We therefore used slightly different values in the gamma distribution. Note that `taub` equals σ_b^{-2}.

Crainiceanu et al. (2005) state that the notation for `b[1] * Z[i,1] + ... + b[9] * Z[i,9]` can be shortened using the inner product function `inner`, but that this increases computing time by a factor 5 to 10. The `PRes` term in the code denotes the Pearson residuals.

We must also specify initial values for the unknown parameters:

```
> inits <- function () {
   list(
    beta        = rnorm(2, 0, 0.2),
    b           = rnorm(9, 0, 0.2),
    taub        = runif(1, 0.1, 0.5))}
```

We have 2 betas and 9 *b*s. If a smoother with more or fewer knots is used, we need to change the 9. We also need to specify the information to store:

```
> params <- c("beta", "b", "Smooth1", "PRes")
```

Lastly, we need the number of chains, draws from the posterior, burn-in, and thinning rate:

```
> nc <- 3        #Number of chains
> ni <- 100000   #Number of draws from posterior
> nb <- 10000    #Number of draws to discard as burn-in
> nt <- 100      #Thinning rate
```

These settings take 3 × (100,000 – 10,000) / 100 = 2,700 draws from the posterior. The high thinning rate reduces the auto-correlation in the chains. It is wise to start with smaller values for `ni`, `nb`, and `nt` to check whether the code functions as intended and whether the chains converge, using smaller, less time consuming, values. We run the model with the following code:

```
> MyBugsDir <- "c:/WinBUGS14/"
> out1 <- bugs(data = win.data, inits = inits, debug = TRUE,
          parameters.to.save = params, model.file = "modelgam.txt",
          n.thin = nt, n.chains = nc, n.burnin = nb, n.iter = ni,
          codaPkg = FALSE, bugs.directory = MyBugsDir)
```

9.5.8 Results

When running models in WinBUGS we are destined to make mistakes, and determining what went wrong can be a nightmare. We need to ensure that every line of code is correct and that nothing can go wrong. The first problem we encountered was that the exponential of the predictor function produces extremely large values, causing WinBUGS to stop with a vague error message. The second problem was that, due to diffuse priors and the large values of the predictor function, only one chain changed. At one point, we made a mistake in the file name for the `model.file` argument. We made changes to the file modelgam.txt but did not rename it in the call to WinBUGS, and ran an older version.

The numerical output can be obtained using:

```
> print(out1, digits = 2)
```

It gives the mean, standard deviation, and a multitude of other numerical summaries of the samples from the posterior distributions for each parameter. We can inspect the chains for convergence using standard tools such as trace plots, auto-correlation plots, and the Gelman-Rubin convergence diagnostic. Code for this was provided in previous chapters and is not repeated here. If the chains are not well mixed, we can increase the thinning rate (or simplify the model). The chains for the smoothers are well mixed. For the regression parameters b_1 to b_9, we would prefer a higher thinning rate, but, when implemented, we encounter memory size problems. The relevant output from the `print` command is:

```
Inference for Bugs model at "modelgam.txt", fit using WinBUGS, 3 chains,
each with 1e+05 iterations (first 10000 discarded), n.thin = 100, n.sims =
2700 iterations saved

                mean   sd    2.5%     25%     50%     75%   97.5% Rhat n.eff
beta[1]        -1.24 0.37   -2.13   -1.42   -1.19   -1.03   -0.58 1.02    89
beta[2]         1.61 0.97    0.47    1.05    1.35    1.88    4.49 1.08    79
b[1]            0.20 0.14    0.00    0.11    0.18    0.27    0.55 1.02   210
b[2]            0.31 0.54   -0.20    0.02    0.13    0.42    1.69 1.12    74
b[3]           -0.12 0.45   -1.31   -0.22   -0.04    0.09    0.58 1.02   160
b[4]           -0.02 0.44   -1.08   -0.14    0.00    0.15    0.83 1.04   160
b[5]            0.41 0.64   -0.21    0.02    0.16    0.56    2.29 1.02   260
b[6]            0.13 0.44   -0.78   -0.04    0.09    0.29    1.11 1.03    94
b[7]            0.04 0.38   -0.81   -0.12    0.03    0.19    0.91 1.00   580
b[8]           -0.29 0.55   -1.88   -0.40   -0.10    0.03    0.25 1.08   120
b[9]           -0.08 0.12   -0.42   -0.13   -0.07   -0.02    0.15 1.05    63
Smooth1[1]     -1.98 0.25   -2.52   -2.14   -1.97   -1.82   -1.54 1.01   180
Smooth1[2]     -1.10 0.14   -1.36   -1.19   -1.10   -1.01   -0.80 1.01   400
Smooth1[3]     -2.16 0.68   -3.66   -2.59   -2.09   -1.67   -1.02 1.00   860
Smooth1[4]     -0.73 0.12   -0.96   -0.81   -0.74   -0.65   -0.46 1.00  1300
Smooth1[5]     -1.28 0.14   -1.58   -1.37   -1.20   -1.19   -1.03 1.00  2100
...
PRes[1]         2.34 0.39    1.69    2.08    2.31    2.57    3.24 1.01   190
PRes[2]        -0.58 0.04   -0.67   -0.60   -0.58   -0.55   -0.51 1.01   410
PRes[3]        -0.36 0.12   -0.60   -0.43   -0.35   -0.27   -0.16 1.00   800
PRes[4]         0.75 0.13    0.47    0.67    0.76    0.83    1.00 1.00  1200
PRes[5]        -0.53 0.04   -0.60   -0.55   -0.53   -0.50   -0.45 1.00  2100
...
deviance     1146.09 5.61 1134.00 1143.00 1147.00 1150.00 1156.00 1.01   460

For each parameter, n.eff is a crude measure of effective sample size, and
Rhat is the potential scale reduction factor (at convergence, Rhat=1).

DIC info (using the rule, pD = Dbar-Dhat)
pD = 6.3 and DIC = 1152.3
DIC is an estimate of expected predictive error (lower deviance is better).
```

We obtain the estimated values for β_0, β_1 and the random effects b_1 to b_9, and, inside the loop, we use these to calculate the smoother. Figure 9.9 contains the estimated smoother. The MCMC process has generated an entire distribution for each value of f_s, denoted by `Smooth1[]` in the code. We use the median of the posterior distribution as an estimator for the smoother and, as confidence bands, we use the $2.5^{th}\%$ and the $97.5^{th}\%$ quantile values for each value of f_s. Note that this is a point-wise confidence band. The required information can be extracted from the object `out1`:

```
> Sm1 <- matrix(nrow = N, ncol = 3)
```

```
> for (i in 1:N) {
    Sm1[i,] <- quantile(out1$sims.list$Smooth1[,i],
                   probs = c(0.025, 0.5, 0.975)) }
> colnames(Sm1) <- c("Lower", "Median", "Upper")
```

We now plot the smoother versus the covariate `TempAn`, but before we can do this we need to sort `TempAn` to avoid producing a spaghetti plot (see also Zuur et al., 2009b).

```
> I1      <- order(SW1$TempAn)
> SortX   <- sort(SW1$TempAn)
> Sm1Sort <- Sm1[I1,]
```

The object `SortX` contains the sorted covariate, and `Sm1Sort` contains sorted smoothers and confidence bands according to the order of `TempAn`. It would have been easier to sort the covariate before initiating the analysis, but this wouldn't help if there were multiple covariates. Plotting the smoother is now easy:

```
> plot(x = SortX, y = Sm1Sort[,2], type = "l", ylim = c(-4, 2),
      xlab = "Temperature anomaly", ylab = "Smoother")
> lines(SortX, Sm1Sort[,1], lty = 2)
> lines(SortX, Sm1Sort[,3], lty = 2)
```

That's it! We could also have used the mean and standard deviation of the posterior distribution to create the plot. These are extracted with:

```
> S1 <- out1$mean$Smooth1
> se <- out1$sd$Smooth1
```

Now multiply the standard errors by 2 and add or subtract them to or from the fitted values.

The shape of the smoother in Figure 9.9 is similar to that in Figure 9.5, perhaps more uniform due to the difference in the number of knots (see the help file of `gam`). The `gam` function in the `mgcv` package also performs cross-validation to estimate the optimal amount of smoothing.

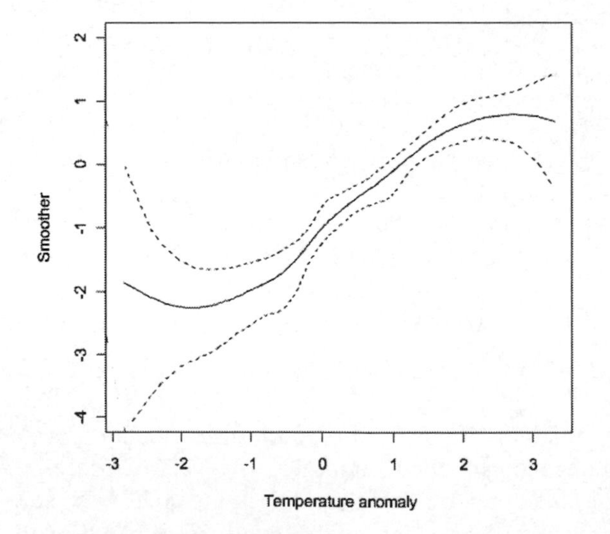

Figure 9.9. Estimated smoother (solid line) obtained by MCMC. The dotted lines are obtained by taking the 97.5th% and the 0.025th% quantiles from the posterior distribution of each `Smooth[i]` value.

The initial Poisson GAM model applied in Subsection (9.5.1) revealed overdispersion, but we need to verify it for the results obtained with MCMC as well. We are in the same situation as we encountered with the overdispersed owl data in Chapter 2. In that chapter we applied a Poisson GLM/GLMM and used several methods to assess the model for overdispersion.

The first option is to wait for the MCMC calculations, take the posterior mean value of `Smooth1[i]`, calculate the corresponding expected value `mu[i]`, and use this to calculate the Pearson residuals. With the Pearson residuals we can assess whether there is overdispersion.

A better approach is to calculate Pearson residuals inside WinBUGS, and that is what we did in the code above. When WinBUGS is complete, we can use the posterior mean for each `Pres[i]` to calculate the sum of squared Pearson residuals and divide this by $N - k$, where k is the number of parameters. This is done as follows:

```
> MPE1 <- out1$mean$PRes
> sum(MPE1^2) / (N - 11 - 1)

8.41064
```

We use 11 for the number of regression parameters, because we consider β_0, β_1, and $b_1 - b_9$ as parameters, but this is controversial. Some researchers consider the $b_1 - b_9$ to be residuals and count only the variance of the random intercepts as a parameter. We have done this in previous chapters. Regardless of whether we use 12 or 3 for k, the overdispersion is between 8 and 8.5. The extra '– 1' is for the variance of the b_1 to b_9.

A third approach is to fit the model with an extra residual term, see Equation (9.14). This only requires adding the extra residual term ε_s to the model and estimating the corresponding variance. A large value for this variance indicates overdispersion of the original model.

9.6 Poisson GAM with two smoothers in WinBUGS

The GAM with two additive smoothers is given by:

$$\begin{aligned} & Y_s \sim \text{Poisson}(\mu_s) \\ & E(Y_s) = \text{var}(Y_s) = \mu_s \\ & \log(\mu_s) = \alpha + f_1(TempAn_s) + f_2(NAO_s) \end{aligned} \tag{9.15}$$

This model can be fitted with the `gam` function from `mgcv` as follows:

```
> M3 <- gam(Nsea ~ s(TempAn) + s(NAO), family = poisson, data = SW1)
> summary(M3)
> par(mfrow = c(1, 2))
> plot(M3, cex.lab = 1.5)
```

The output from the `summary` command is given below, and the two smoothers are presented in Figure 9.10. We find the interpretation of smoother for the NAO index confusing, as we can't explain it in terms of biology, and its significance may be due to collinearity. The model still shows an overdispersion of 7.08.

```
             Estimate Std. Error z value Pr(>|z|)
(Intercept)   -1.1873     0.1013  -11.72   <2e-16

Approximate significance of smooth terms:
             edf Ref.df Chi.sq  p-value
s(TempAn) 8.383   8.864 163.67  < 2e-16
s(NAO)    6.174   7.319  69.06 3.37e-12

R-sq.(adj) = 0.0681   Deviance explained = 20.5%
UBRE score = 1.0572  Scale est. = 1   n = 439
```

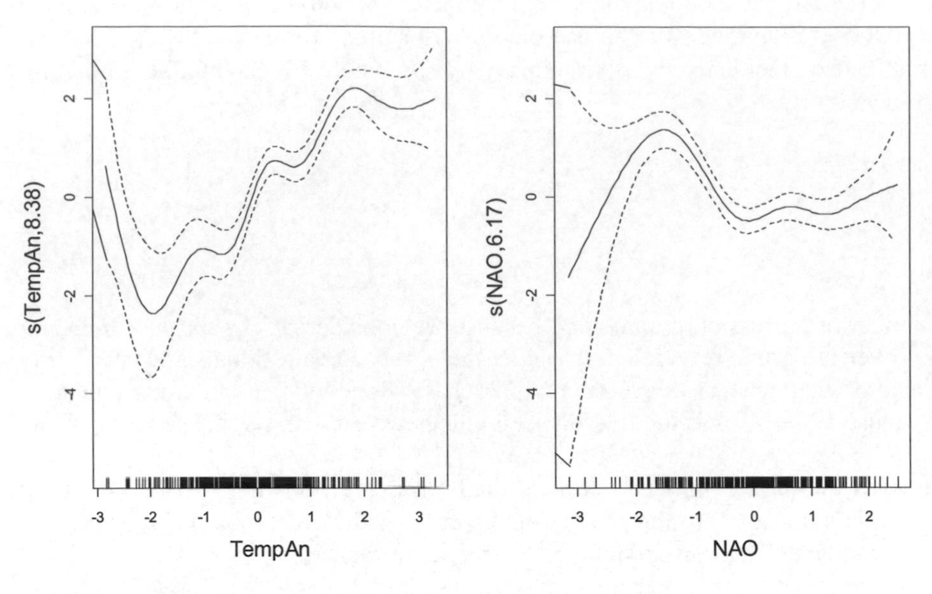

Figure 9.10. Estimated smoothers for the temperature anomaly and the NAO index.

To run the GAM in Equation (9.15) in WinBUGS we need to modify **X** and **Z** and include more betas and *b*s. Each smoother is now modelled by a low rank thin plate spline, as in Equation (9.3). The following code standardises the NAO index and calculates the matrix **X**:

```
> SW1$NAOstd <-  (SW1$NAO - mean(SW1$NAO))/sd(SW1$NAO)
> X <- cbind(rep(1, N), SW1$TempAn, SW1$NAOstd)
```

Hence **X** now comprises three columns. The first column contains ones, the second column contains the standardised TempAn_s time series, and the third column is the standardised NAO_s time series. Now we need the knots for the NAO series. To ensure consistent notation we rerun the code that gave us the knots and the Z for the temperature anomaly:

```
> Knots1 <- default.knots(SW1$TempAn, 9)
> Z1     <- GetZ(SW1$TempAn, Knots1)
> Knots2 <- default.knots(SW1$NAOstd, 9)
> Z2     <- GetZ(SW1$NAOstd, Knots2)
```

The `Z1` is the **Z** matrix for TempAn_s and is the same as that used in the previous subsection. `Z2` is the **Z** matrix for the standardised NAO series. We can either concatenate `Z1` and `Z2` using

```
> Z <- cbind(Z1, Z2)
```

and use `Z` in WinBUGS, or we can retain the `Z1` and `Z2` for ease of programming. We use the same number of knots for the NAO, but this is not necessary. The data for our code is given by:

```
win.data <- list(Y        = SW1$Nsea,
                 NumKnots = num.knots,
                 X        = X,
                 Z1       = Z1,
                 Z2       = Z2,
                 N        = nrow(SW1))
```

The model code is similar to that for the model with one smoother:

```
sink("modelgam2.txt")
cat("
model{
```

```
  #Priors
  for (i in 1:3) { beta[i]  ~ dnorm(0, 0.00001) }
  for (i in 1:NumKnots) {
    b1[i] ~ dnorm(0, tau1)
    b2[i] ~ dnorm(0, tau2)
  }
  tau1 ~ dgamma(0.01, 0.01)
  tau2 ~ dgamma(0.01, 0.01)

  #Likelihood
  for (i in 1:N) {
      Y[i] ~ dpois(mu[i])
      log(mu[i])    <- max(-20, min(20, eta[i]))
      eta[i]        <- beta[1] * X[i,1] + Smooth1[i] + Smooth2[i]
      Smooth1[i] <- beta[2] * X[i,2] +
                    b1[1] * Z1[i,1] + b1[2] * Z1[i, 2] +
                    b1[3] * Z1[i,3] + b1[4] * Z1[i, 4] +
                    b1[5] * Z1[i,5] + b1[6] * Z1[i, 6] +
                    b1[7] * Z1[i,7] + b1[8] * Z1[i, 8] +
                    b1[9] * Z1[i,9]
      Smooth2[i] <- beta[3] * X[i,3] +
                    b2[1] * Z2[i,1]  + b2[2] * Z2[i, 2] +
                    b2[3] * Z2[i,3]  + b2[4] * Z2[i, 4] +
                    b2[5] * Z2[i,5]  + b2[6] * Z2[i, 6] +
                    b2[7] * Z2[i,7]  + b2[8] * Z2[i, 8] +
                    b2[9] * Z2[i,9]
      PRes[i] <- (Y[i] - mu[i]) / pow(mu[i], 0.5)
       }
}
",fill = TRUE)
sink()
```

Finally, we specify initial values and tell WinBUGS which parameters to save.

```
inits <- function () {
   list(
    beta = rnorm(3, 0, 0.2),
    b1   = rnorm(9, 0, 0.2),
    b2   = rnorm(9, 0, 0.2),
    tau1 = runif(1, 0.1, 0.5),
    tau2 = runif(1, 0.1, 0.5))}
```

and

```
> params <- c("beta", "b1", "b2", "Smooth1", "Smooth2", "PRes")
```

We use the same chain length, thinning rate, and number of chains. To run the code we use:

```
> out2 <- bugs(data = win.data, inits = inits, debug = TRUE,
            parameters.to.save = params,
            model.file = "modelgam2.txt",
            n.thin = nt, n.chains = nc, n.burnin = nb, n.iter = ni,
            codaPkg = FALSE, bugs.directory = MyBugsDir)
```

The estimated smoothers are given in Figure 9.11 and show patterns similar to those of smoothers obtained by the gam function in mgcv. Any differences are likely to be due to differing numbers of knots. Mixing of the chains is not perfect, and it may be wise to run WinBUGS with a greater thinning rate.

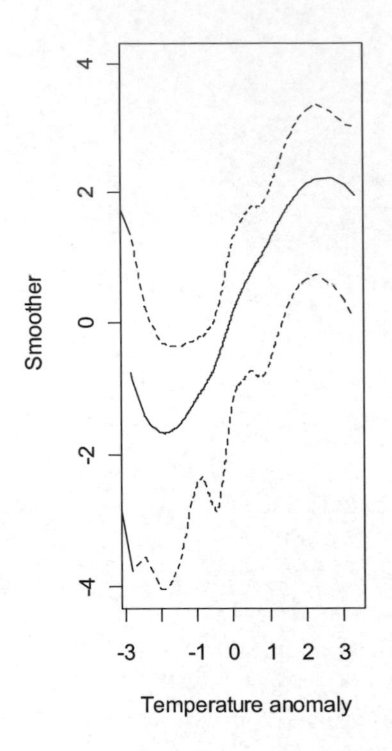

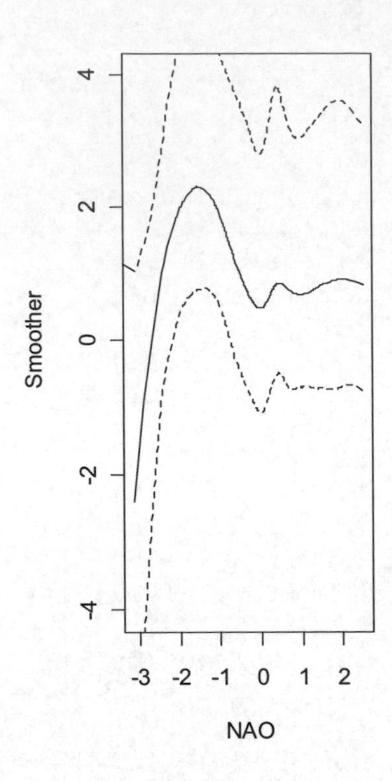

Figure 9.11. Estimated smoothers obtained by MCMC.

9.7 Zero inflated Poisson GAM in WinBUGS

9.7.1 Justification for zero inflated models

The frequency plot in Figure 9.2 showed that we have a large number of zeros, meaning that we have a large number of years with no recorded sperm whale stranding. The time series go back to the 16th century, and it may well be that in certain years there were strandings, but that these were not detected or not recorded. Coastal areas were less populated than in recent times.

In earlier chapters we explained that, in ZIP GLMs, we make a distinction between true zeros and false zeros. A false zero in this case may be a year in which no sperm whale strandings were recorded because they were not detected, the incident was not recorded, or perhaps the whale was miss-identified as another species (Figure 9.12). In these scenarios we have a 0 in the time series, but there may have been sperm whale strandings.

The ZIP is the perfect tool for separating the zeros into false and true zeros. Theoretically, a true zero is a year where there were no sperm whale strandings. In practise the true zeros are the zeros that we can relate to our covariates, and the false zeros are those for which no relationship is found.

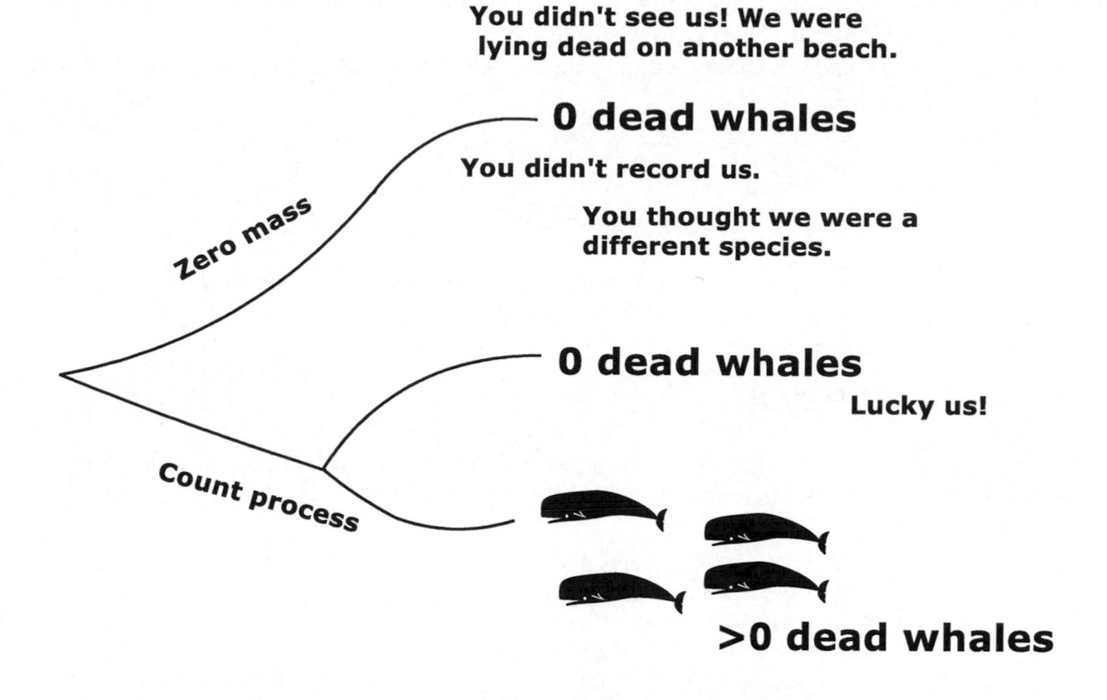

Figure 9.12. The sources of zeros for sperm whale strandings. An observation of 0 strandings in year *s* may be due to non-detected or non-recorded strandings. The presence of false zeros is not a requirement for the application of ZIP models; see Chapter 10 for a discussion.

9.7.2 Underlying equations for the ZIP GAM

The Poisson GAM in Equation (9.13) can be extended to a ZIP GAM (see also Chapter 2), which is given by:

$$Y_s \sim ZIP(\mu_s, \pi_s) \tag{9.16}$$

This expression states that Y_s follows a ZIP distribution, with mean μ_s for the count process and probability π_s that observation Y_s is a false 0. We will use the log-link function for the count process. Its predictor function contains the smoother of the annual temperature anomaly:

$$\log(\mu_s) = \alpha + f(TempAn_s) \tag{9.17}$$

For the probability that Y_s is a false zero we use a logistic link function, and its predictor function contains only an intercept:

$$\text{logit}(\pi_s) = \gamma \quad \Leftrightarrow \quad \pi_s = \frac{e^\gamma}{1+e^\gamma} \tag{9.18}$$

This means that we assume that the probability that Y_s is a false zero does not change over time. One may argue that coastal areas have become more populated over the past hundred years and that the probability of detecting stranded sperm whales has increased, but this would also increase the complexity of the model. Theoretically we could add a smoothing function of time, or use 'century' as a categorical variable in the logistic link function to model a change in the probability of false zeros.

The predictor function of the logistic link function contains an intercept. If it is highly negative, π_s converges to 0 and the ZIP GAM becomes a Poisson GAM.

Computing the mean and the variance of a variable following a ZIP distribution was demonstrated in Chapter 2 and is repeated below.

$$\begin{aligned} E(Y_s) &= \mu_s(1-\pi_s) \\ \text{var}(Y_s) &= (1-\pi_s)\times(\mu_s+\pi_s\times\mu_s^2) \end{aligned} \tag{9.19}$$

Inside our WinBUGS function we can easily calculate μ_s and $\pi_{s,}$ and therefore the mean and the variance, and using these we can calculate Pearson residuals.

9.7.3 R code for a ZIP GAM

In Section 9.5 we presented R code to fit Poisson GAMs in WinBUGS. The code can easily be adjusted to fit a ZIP GAM. We first need to define the input variables:

```
> X <- cbind(rep(1, N), SW1$TempAn)
> Knots1 <- default.knots(SW1$TempAn, 9)
> Z <- GetZ(SW1$TempAn, Knots1)
> win.data <- list(Y          = SW1$Nsea,
                   NumKnots = 9,
                   X          = X,
                   Z          = Z,
                   N          = nrow(SW1))
```

The matrix **X** contains two columns, the first for β_0 and the second for β_1 of the low rank thin plate spline. We use 9 knots, and Z contains the cubic terms as explained earlier. The initial values required for the WinBUGS code are produced by:

```
inits <- function () {
   list(
    beta   = rnorm(2, 0, 0.2),
    b      = rnorm(9, 0, 0.2),
    taub   = runif(1, 0.1, 0.5),
    gamma1 = rnorm(1, 0, 0.2),
    W      = as.integer(win.data$Y>0))}
```

The parameters we retain are:

```
> params <- c("beta",  "b", "Smooth1", "gamma1", "PRes")
```

The essential modelling code is:

```
sink("modelzipgam.txt")
cat("
model{
 #Priors
 for (i in 1:2) { beta[i]  ~ dnorm(0, 0.00001) }
 for (i in 1:NumKnots) { b[i] ~ dnorm(0, taub) }
 taub ~ dgamma(0.01, 0.01)
 gamma1 ~ dunif(-50,50) # dunif(-20,20) may be better

 #Likelihood
 for (i in 1:N) {
```

```
      W[i] ~ dbern(psim1)
      Y[i] ~ dpois(eff.mu[i])
      eff.mu[i]  <- W[i] * mu[i]
      log(mu[i]) <- max(-20, min(20, eta[i]))
      eta[i]     <- beta[1] * X[i,1] +  Smooth1[i]
      Smooth1[i] <- beta[2] * X[i,2] +
                         b[1] * Z[i,1] + b[2] * Z[i, 2] +
                         b[3] * Z[i,3] + b[4] * Z[i, 4] +
                         b[5] * Z[i,5] + b[6] * Z[i, 6] +
                         b[7] * Z[i,7] + b[8] * Z[i, 8] +
                         b[9] * Z[i,9]
      ExpY[i] <-  mu[i] * (1 - psi)
      VarY[i] <- (1 - psi) * (mu[i] + psi*pow(mu[i],2))
      PRes[i] <- (Y[i] - ExpY[i] ) / pow(VarY[i], 0.5)
       }
   psim1 <- min(0.99999, max(0.00001,(1 - psi)))
   logit(psi) <- max(-20, min(20, gamma1))
}
",fill = TRUE)
sink()
```

The algorithm for zero inflated models was introduced in earlier chapters. The only new addition is the inclusion of the smoother components. The following code runs the ZIP GAM:

```
> out3 <- bugs(data = win.data, inits = inits, debug = TRUE,
            parameters.to.save = params,
            model.file = "modelzipgam.txt",
            n.thin = nt, n.chains = nc, n.burnin = nb, n.iter = ni,
            codaPkg = FALSE, bugs.directory = MyBugsDir)
```

We used the same thinning rate, number of chains, and chain length as before. Rather than sampling values for gamma1, we could have sampled values for π directly (see Kéry, 2010), but our approach allows access to the significance of gamma1.

9.7.4 Results for the ZIP GAM

The first question we focus on is whether we need to include the zero inflation part; i.e. is π equal to 0? We assess this via the regression parameter gamma1. If gamma1 is highly negative, π converges to 0, and we should apply the Poisson GAM. The numerical output from WinBUGS is presented below and shows that the mean of the posterior distribution for gamma1 is 1.25, and the 2.5 and 97.5 percent quantiles are 0.96 and 1.35, respectively; hence the 95% credible interval does not contain 0. The value of 1.25 corresponds to $\pi = 0.77$. So the probability that a 0 is a false zero is 0.77.

```
> print(out3, digits = 2)

Inference for Bugs model at "modelzipgam.txt", fit using WinBUGS, 3 chains,
each with 1e+05 iterations (first 10000 discarded), n.thin = 100
 n.sims = 2700 iterations saved

           mean   sd  2.5%   25%   50%  75% 97.5% Rhat n.eff
beta[1]    0.58 0.29 -0.14  0.43  0.60 0.75  1.06 1.01   700
beta[2]    1.30 0.56  0.41  0.96  1.21 1.54  2.79 1.01  2700
b[1]       0.13 0.09 -0.04  0.07  0.12 0.19  0.33 1.00  2700
b[2]       0.05 0.29 -0.35 -0.10  0.00 0.12  0.81 1.04  1200
b[3]      -0.11 0.36 -1.09 -0.20 -0.06 0.05  0.46 1.01  1600
```

```
b[4]            -0.06  0.47  -1.12  -0.14  -0.02   0.11   0.58 1.04   340
b[5]             0.34  0.86  -0.26  -0.01   0.11   0.33   2.70 1.05   500
b[6]             0.00  0.52  -0.97  -0.08   0.04   0.17   0.70 1.09  1100
b[7]             0.06  0.33  -0.55  -0.09   0.03   0.17   0.84 1.01   720
b[8]            -0.13  0.29  -0.94  -0.23  -0.07   0.04   0.28 1.00  2600
b[9]            -0.15  0.09  -0.35  -0.19  -0.14  -0.09   0.02 1.02  2700
Smooth1[1]      -0.99  0.34  -1.71  -1.19  -0.97  -0.78  -0.36 1.00  2700
...
gamma1           1.25  0.15   0.96   1.15   1.25   1.35   1.53 1.00  2700
...
deviance       519.29 16.75 489.45 507.60 518.40 530.02 555.25 1.00  2700

For each parameter, n.eff is a crude measure of effective sample size, and
Rhat is the potential scale reduction factor (at convergence, Rhat=1).

DIC info (using the rule, pD = var(deviance)/2)
pD = 140.3 and DIC = 659.6
DIC is an estimate of expected predictive error (lower deviance is better).
```

The chains for gamma1 are well mixed, as can be seen in Figure 9.13, and an auto-correlation function of each chain (not presented here) does not show significant auto-correlation. Figure 9.14 shows a histogram of the samples from the posterior distribution for gamma1. The estimated smoother for the temperature anomaly is presented in Figure 9.15. Confidence bands are smaller compared to those of the smoothers obtained by the Poisson GAM in previous sections.

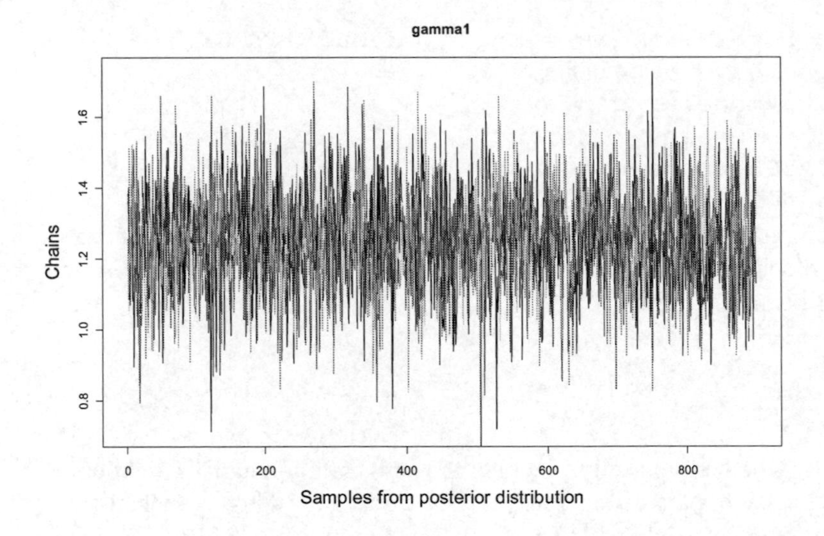

Figure 9.13. Plot of the three chains for the posterior distribution of gamma1.

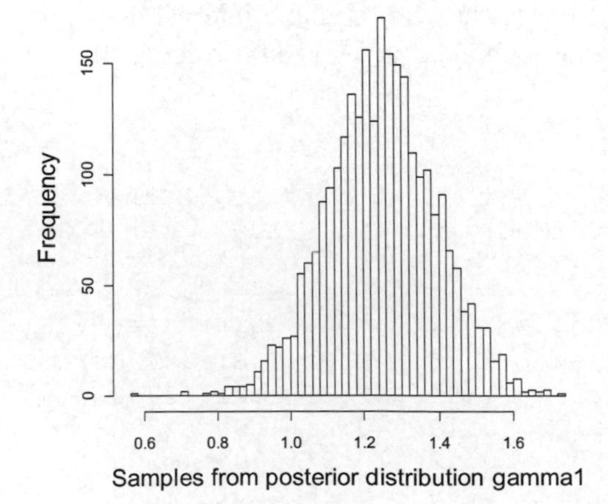

Figure 9.14. Histogram of the samples from the posterior distribution for gamma1.

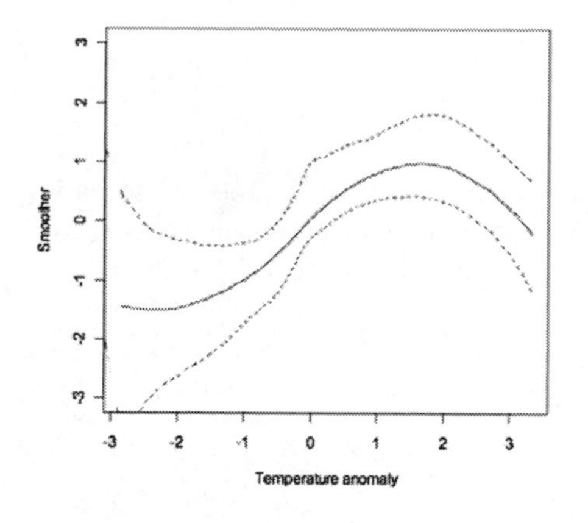

Figure 9.15. Estimated smoother for the temperature anomaly obtained by the ZIP GAM.

R code to generate Figure 9.13, Figure 9.14, and Figure 9.15 has been used and discussed elsewhere in this chapter and other chapters of this book, but to be comprehensive we present it again. The plot of the three chains is made as follows:

```
> Chains <- out3$sims.array[,,"gamma1"]
> plot(Chains[,1], type = "l", col = 1,
     main = paste("gamma1", "", sep = ""), cex.lab = 1.5,
     ylab = "Chains", xlab = "Samples from posterior distribution")
> lines(Chains[,2], col = 2)
> lines(Chains[,3], col = 3)
```

Each of the three chains can be used as input for the `acf` function. The histogram of the posterior distribution is made with the `hist` function:

```
> hist(out3$sims.list$gamma1, breaks = 50,
  main = "", cex.lab = 1.5,
  xlab = "Samples from posterior distribution gamma1")
```

Finally, the smoother is produced with the same code as in Section 9.5:

```
> Sm1 <- matrix(nrow = N, ncol = 3)
> for (i in 1:N) {
     Sm1[i,]<-quantile(out3$sims.list$Smooth1[,i],
                        probs=c(0.025, 0.5, 0.975)) }
> colnames(Sm1) <- c("Lower", "Median", "Upper")
> I1       <- order(SW1$TempAn)
> SortX    <- sort(SW1$TempAn)
> Sm1Sort <- Sm1[I1,]
> plot(x = SortX, y = Sm1Sort[,2], type = "l",
     xlab = "Temperature anomaly",
     ylab = "Smoother", ylim = c(-3, 3))
> lines(SortX, Sm1Sort[ ,1], lty = 2)
> lines(SortX, Sm1Sort[ ,3], lty = 2)
```

9.7.5 Model validation of the ZIP GAM

After applying any statistical model, whether it is a bivariate linear regression model or an advanced ZIP GAM, a model validation must be applied. Residuals are extracted and plotted versus each covariate in the model and each covariate not included in the model, and we inspect the residuals for patterns. For this data, we should also create an auto-correlation plot of the residuals to see whether there is significant temporal correlation. We will start with the mean value of each Pearson residual `Pres[i]` and use these for assessing overdispersion.

```
> MPE3 <- out3$mean$PRes
> sum(MPE3^2)/(N - 13)

2.647569
```

We subtracted 13 because we have 11 regression parameters and 1 variance for the spline and 1 regression parameter for the intercept in the logistic link function. Note that by adopting a zero inflation structure the overdispersion dropped from approximately 8.5 to 2.65. However, the residuals still exhibit significant temporal auto-correlation, as can be seen in Figure 9.16. A possible approach to resolving this is to include more covariates, but our initial analysis showed that this did not remove the residual auto-correlation. Hence we will extend the ZIP GAM in the next section with a residual correlation structure.

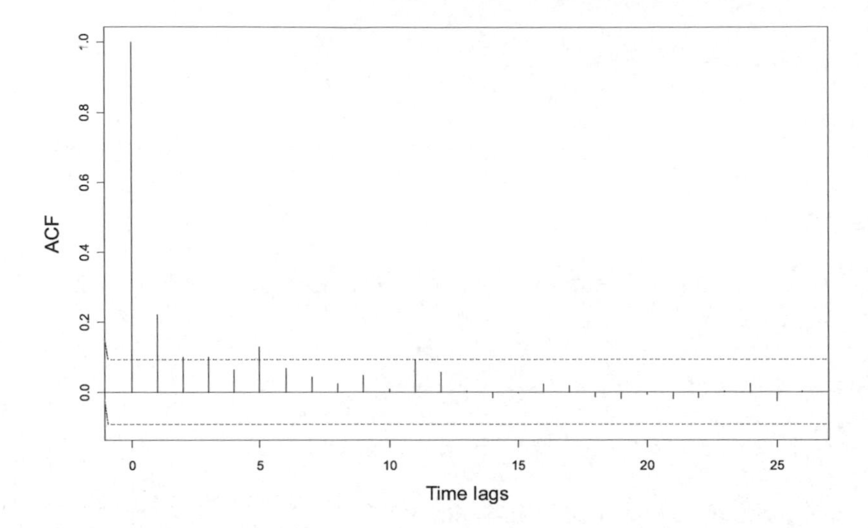

Figure 9.16. Auto-correlation of the Pearson residuals obtained by the ZIP GAM. The graph was produced with the `acf(MPE3)` command.

9.8 Zero inflated Poisson GAM with temporal correlation in WinBUGS

A statistician confronted with a time series of whale strandings will say, ‘If there are many sperm whale strandings in year *s*, it is likely that in the following year there will also be many strandings.’ This is positive auto-correlation. But why should this be the case?

If the population of sperm whales is low, and a large number of strandings occur in year *s*, the population is reduced, or perhaps they learn not to become stranded, and in the following year there may be fewer sperm whale strandings. This would mean negative correlation: high numbers followed by low numbers. Here it is important to use what biological knowledge we have available. From a biological point of view we would not expect negative correlation in the sperm whale strandings. Only a small proportion of whales enter the North Sea and their impact on population size will be negligible. Following the arguments in Pierce et al. (2007), the absolute number of stranded whales will be a function of (i) population size, which determines the number of whales that migrate to the Arctic feeding grounds) and (ii) local food abundance, which determines the number of these that decide to enter the North Sea as they return south. Depending on the relative importance of (i) and (ii), there will either be positive auto-correlation (high population one year means high the following year) or no auto-correlation (entirely driven by squid abundance which is unlikely to show auto-correlation, because they are short-lived and population size can vary widely from one year to the next).

If the population is large, a high number of strandings in year *s* may mean a high number of strandings in years *s* + 1. On the other hand, perhaps the whales show variation in annual migrations patterns, and strandings occur for a couple of years and then diminish. That is positive correlation.

Whatever scenario for sperm whale strandings we envision, we are analysing a time series, and there may potentially be correlation. Ignoring temporal correlation means that we may commit a type I error (identifying parameters or smoothers as significant, whereas in reality they are not significant), so we should allow for it in the model.

9.8.1 CAR residual correlation in the ZIP GAM

An option for including a temporal correlation structure in the ZIP GAM is to use a conditional auto-regressive correlation (CAR) structure. In Chapters 6 and 7 the CAR correlation was used for spatial data, but we can easily adapt it to time series data. A ZIP GAM with residual CAR correlation is formulated as:

$$\begin{aligned}\log(\mu_s) &= \alpha + f(TempAn_s) + \varepsilon_s \\ E(\varepsilon_i) &= \rho \times \sum_{j \neq i} c_{ij} \times \varepsilon_j\end{aligned} \qquad (9.20)$$

As explained in Chapters 6 and 7, the c_{ij} are weighting functions. The parameter ρ measures the strength of the correlation. A value of 0 indicates no correlation. This means that we have:

$$\boldsymbol{\varepsilon} \sim N\left(\mathbf{0}, \sigma_\varepsilon^2 (\boldsymbol{I} - \rho \times \mathbf{C})^{-1} \times \mathbf{M}\right)$$

The bold vector $\boldsymbol{\varepsilon}$ contains the individual values ε_1 to ε_N. We need to choose the structure of the matrices **C** and **M**. In Chapter 7 we discussed options for spatial data; we plotted sites on a lattice and defined neighbours as any two sites sharing a border. For a time series this is easier; two points are neighbouring if they are sequential in time. This means that every observation Y_s has two neighbours, Y_{s-1} and Y_{s+1}. The only observations having a single neighbour are the first and last in the series. We will use the same choices for **C** and **M** as verHoef and Janssen (2007). Each row in **C** contains zeros, except for the neighbouring elements. The value of those elements is the reciprocal of the number of their neighbours, which is 2 for all years except the first and last. Thus all elements of **C** equal 0 except for the $(i, i + 1)$ and $(i + 1, i)$ elements. These are above and below the diagonal and equal ½, so the expected value of ε_s is given by:

$$E(\varepsilon_s) = \rho \times \frac{\varepsilon_{s-1} + \varepsilon_{s+1}}{2}$$

M is a diagonal matrix containing the reciprocal of the number of neighbours. Other choices for **C** and **M** are possible (see Chapter 6). To implement this correlation structure in the ZIP GAM, we first need to calculate **C** and **M**. The matrix **C** is given by:

```
> C1 <- matrix(nrow = N, ncol = N)
> C1[,] <- 0
> N1 <- N - 1
> for (i in 1:N1) {
    C1[i, i+1] <- 1/2
    C1[i+1, i] <- 1/2 }
> C1[1,2] <- 1
> C1[439,438] <-1
```

The diagonal elements of this matrix will always be 0, since an observation cannot be the neighbour of itself. As explained in Chapter 7, WinBUGS does not accept C1 as direct input. It requires the number of neighbours for each year, a list of the identity of the neighbours, and the matching values of `C1`. This is accomplished as follows (see Chapter 7 for explanation). First we create an empty list:

```
> CAR.stuff <- list(adj = NA, C = NA)
```

The number of neighbours per year and **M** are easily calculated:

```
> CAR.stuff$num <- rowSums(C1> 0)
> CAR.stuff$M   <- 1/CAR.stuff$num
```

Although, in the covariance, **M** is a matrix, WinBUGS requires that it be a vector (the diagonal elements of **M**). The code below determines the names of the neighbours and converts C1 to the required format:

```
> ID <- 1:N
> for(i in 1:N) {
    neighb        <- ID[C1[i,] > 0]
    CAR.stuff$adj <- c(CAR.stuff$adj, neighb)
    CAR.stuff$C   <- c(CAR.stuff$C,   C1[i, neighb]) }
```

We remove the NAs that we used to create the variables inside the list:

```
> CAR.stuff$adj <- CAR.stuff$adj[-1]
> CAR.stuff$C   <- CAR.stuff$C[-1]
```

We can now start working with the WinBUGS code. First we need the data:

```
> win.data <- list(Y        = SW1$Nsea,
                   NumKnots = 9,
                   X        = X,
                   Z        = Z,
                   N        = nrow(SW1),
                   num      = CAR.stuff$num,
                   C        = CAR.stuff$C,
                   M        = CAR.stuff$M,
                   adj      = CAR.stuff$adj)
```

The essential model code is:

```
sink("modelzipgamCAR.txt")
cat("
model{

 #Priors
 for (i in 1:2) { beta[i]  ~ dnorm(0, 0.00001) }
 for (i in 1:NumKnots) { b[i] ~ dnorm(0, taub) }
 taub    ~ dgamma(0.01, 0.01)
 #psi    ~ dunif(0, 1)
 gamma1  ~ dunif(-50,50)
 rho     ~ dunif(0.0001, .9999)
 tau.eps ~ dunif(0.0001, 5)
 #Likelihood
 for (i in 1:N) {
     W[i] ~ dbern(psim1)
     Y[i] ~ dpois(eff.mu[i])
     eff.mu[i]   <- W[i] * mu[i]
     log(mu[i]) <- max(-10, min(10, eta[i]))
     eta[i]      <- beta[1] * X[i,1] +  Smooth1[i] + tau.eps * EPS[i]
```

```
     Smooth1[i] <- beta[2] * X[i,2] +
                       b[1] * Z[i,1] + b[2] * Z[i, 2] +
                       b[3] * Z[i,3] + b[4] * Z[i, 4] +
                       b[5] * Z[i,5] + b[6] * Z[i, 6] +
                       b[7] * Z[i,7] + b[8] * Z[i, 8] +
                       b[9] * Z[i,9]
     EPS.mean[i] <- 0
      }

  psim1 <- min(0.99999, max(0.00001,(1 - psi)))
  logit(psi) <- max(-20, min(20, gamma1))
  EPS[1:N] ~ car.proper(EPS.mean[], C[], adj[], num[], M[], 1, rho)
}
",fill = TRUE)
sink()
```

This is nearly the same code as for the ZIP GAM, except that CAR residuals have been added. Initial values for the parameters of the CAR part were taken from Chapter 6, and the same number of chains, chain length, and thinning rate are used as for the ZIP GAM. Mixing of the chains is acceptable. The relevant numerical output is:

```
Inference for Bugs model at "modelzipgamCAR.txt", fit using
WinBUGS,3 chains, each with 1e+05 iterations (first 10000
discarded), n.thin = 100 n.sims = 2700 iterations saved

            mean    sd   2.5%    25%    50%    75%  97.5% Rhat n.eff
beta[1]    -3.11  0.46  -4.11  -3.38  -3.07  -2.79  -2.29 1.00  2700
beta[2]     0.70  0.44  -0.06   0.42   0.68   0.97   1.67 1.01  1100
...
gamma1    -24.46 14.29 -48.59 -36.51 -24.82 -11.83  -1.73 1.00  2400
...
rho         0.86  0.10   0.59   0.82   0.89   0.93   0.97 1.03   140
tau.eps     2.01  0.33   1.45   1.78   1.97   2.21   2.73 1.02    96
deviance  388.07 20.11 348.35 374.20 387.55 401.50 428.30 1.00   590

DIC info (using the rule, pD = var(deviance)/2)
pD = 201.7 and DIC = 589.8
```

The histogram of the samples from the posterior distribution of ρ is presented in Figure 9.17. Its mean value is 0.86; thus there is a strong relationship between the residual at time s and the residuals at time $s - 1$ and $s + 1$. The estimated smoother for the temperature anomaly is presented in Figure 9.18. The shape of the smoother is similar to the pattern we have seen in the earlier analyses and in Pierce et al. (2007), so that is encouraging. There is, however, a problem with this analysis: the mean of the samples from the posterior distribution for $\gamma 1$ is a large negative, which means that the probability of false zeros converges to $\pi_s = 0$. This means that there is no need for a zero inflation component in the models. At first glance, this is puzzling, but it makes sense. The time series consists of a large number of *sequential* zeros. Because we have a high number of consecutive values that are equal, we have high temporal correlation. Because these values are the zeros, we have high correlation along with zero inflation, but these issues essentially arise from the same source. It is not surprising that the zero inflation disappeared. The correlation term modelled it. For this specific problem (i.e. a time series with many sequential zeros) we should not use zero inflation and temporal correlation in the same model. The choice is then between a Poisson GAM model with temporal CAR correlation and a ZIP GAM. The fact that the ZIP GAM still shows overdispersion is something we can easily deal with by including an extra residual term (see Equation (9.14)).

We refit the Poisson GAM with the residual CAR correlation, and the shape of the estimated smoother is nearly identical to that in Figure 9.18.

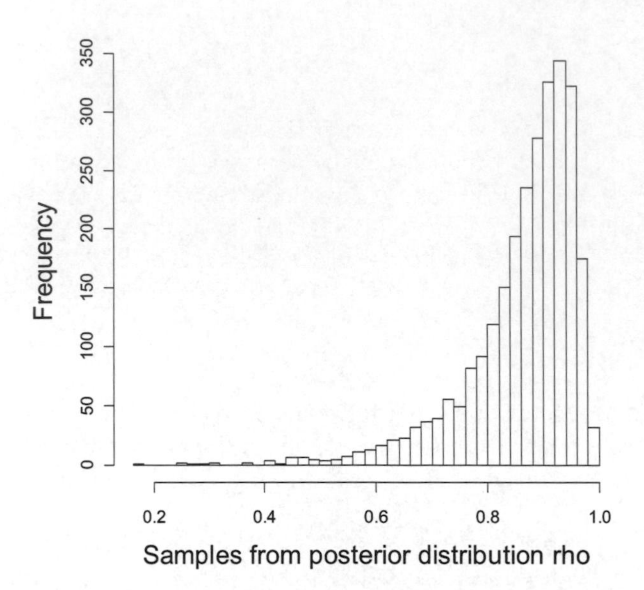

Figure 9.17. Posterior distribution of the temporal correlation parameter ρ.

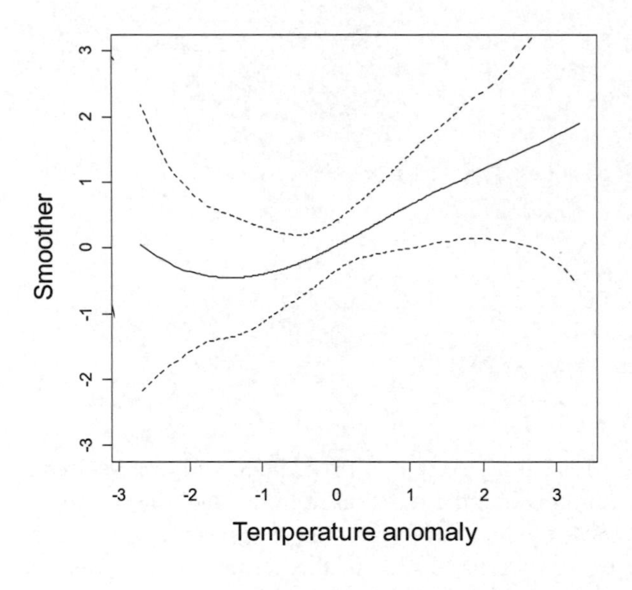

Figure 9.18. Estimated smoother for the temperature anomaly obtained by the ZIP GAM with CAR residual correlation.

9.8.2 Auto-regressive residual correlation in the ZIP GAM

Another way to include temporal auto-correlation in the ZIP GAM is to add an extra residual term in the log- or logistic link function, or both, and impose an auto-regressive temporal correlation structure on it. For simplicity we discuss this only for the count process.

We will use the same ZIP distribution for Y_s and the same log- and logistic link functions as in Equations (9.16), (9.17), and (9.18), except that the predictor function is extended with a residual correlation term:

$$\begin{aligned} \log(\mu_s) &= \eta_s + \varepsilon_s \\ \eta_s &= \alpha + f(TempAn_s) \\ \varepsilon_s &= \rho \times \varepsilon_{s-1} + \xi_s \\ \xi_s &\sim N(0, \sigma_\xi^2) \end{aligned} \tag{9.21}$$

This type of model (albeit in a Poisson GLM context) was also applied in Chan and Ledolter (1995). A detailed discussion of a Poisson GLM with the residual correlation structure in Equation (9.21) is presented in Kolbe (2004, unpublished dissertation), who also included equations for the co-variance and correlation of Y_s and Y_t. Useful references are Zeger (1988) and Chan and Ledolter (1995). Information on priors and other MCMC settings can be found in Czado and Kolbe (2004). Implementation in WinBUGS is straightforward, and the essential part of the code for the model in Equation (9.21) is given below.

```
sink("modelgamAR1.txt")
cat("
model{
 #Priors
 for (i in 1:2) { beta[i]  ~ dnorm(0, 0.001) }
 for (i in 1:NumKnots) { b[i] ~ dnorm(0, taub) }
 taub ~ dgamma(0.01, 0.01)
 taueps ~ dgamma(0.01, 0.01)
 rho ~ dunif(-0.99, 0.99)

 #For i = 1
 #Calcuate X* beta
 eta[1]      <- beta[1] * X[1,1] +  Smooth1[1]
 Smooth1[1] <- beta[2] * X[1,2] +
               b[1] * Z[1,1] + b[2] * Z[1, 2] +
               b[3] * Z[1,3] + b[4] * Z[1, 4] +
               b[5] * Z[1,5] + b[6] * Z[1, 6] +
               b[7] * Z[1,7] + b[8] * Z[1, 8] +
               b[9] * Z[1,9]
 #Calculate X * beta + eps
 eta1[1] ~ dnorm(eta[1], taueps)
 log(mu[1]) <- eta1[1]
 Y[1] ~ dpois(mu[1])

 for (i in 2:N) {
    #Calcuate X* beta
     eta[i]      <- beta[1] * X[i,1] +  Smooth1[i]
     Smooth1[i] <- beta[2] * X[i,2] +
                   b[1] * Z[i,1] + b[2] * Z[i, 2] +
                   b[3] * Z[i,3] + b[4] * Z[i, 4] +
                   b[5] * Z[i,5] + b[6] * Z[i, 6] +
                   b[7] * Z[i,7] + b[8] * Z[i, 8] +
                   b[9] * Z[i,9]
     #Calculate eta + rho * (log(u[i-1] - eta[i-1]) + EPS
     temp[i] <- eta[i] + rho * (log(mu[i-1]) - eta[i-1])
     eta1[i] ~ dnorm(temp[i], taueps)
     log(mu[i]) <- max(-10, min(10, eta1[i]))
     Y[i] ~ dpois(mu[i])
     PRes[i]      <- (Y[i] - mu[i] ) / pow(mu[i], 0.5)
}}
```

```
",fill = TRUE)
sink()
```

For the initial values we used:

```
> inits <- function () {
   list(
    beta        = rnorm(2, 0, 0.001),
    b           = rnorm(9, 0, 0.001),
    taub        = runif(1, 0.1, 0.5),
    taueps      = runif(1, 0.1, 5),
    rho         = runif(1,-0.99,0.99))}
```

9.9 Discussion

In this chapter we began with ordinary Poisson GAMs, developed code to fit these models in WinBUGS, and proceeded to ZIP GAMs and ZIP GAMs with temporal correlation. The ZIP GAM still indicated overdispersion, and the residuals showed minor auto-correlation; therefore we added a residual correlation structure. We did this with a CAR correlation, but we could have used an auto-regressive residual correlation structure, as in Equation (9.21).

Note that when an additional residual term is added to one of the link functions, we need to re-calculate expressions for the mean and variance of the response variable. Kolbe (2004) shows these expressions for a Poisson GLM, or a GAM, if a residual auto-regressive correlation term is added, but, for a ZIP, the expressions for the covariance and correlation of Y_s and Y_t are more complicated.

This leaves the question of whether we create a ZIP model or a model with temporal correlation. In this case, we can't do both. The temporal correlation is due to the sequential zeros and zero inflation, and the zero inflation may be due to the temporal correlation! Modelling one term produces the other. Perhaps a negative binomial zero inflated GAM can deal with the small amount of overdispersion remaining. On the other hand, the Poisson GAM with CAR correlation was satisfactory.

9.10 What to write in a paper?

This chapter is not so much an analysis of a data set as a demonstration of how to run WinBUGS from within R for a ZIP GAM with temporal correlation. As a biologist you may be more interested in knowing which covariates drive sperm whale stranding. Suggestions for writing a paper:

1. Justify the selected covariates.
2. Present data exploration graphs showing the time series plots.
3. Justify the application of smoothers rather than parametric terms.
4. Based on the nature of the data, we recommend applying either a GAM with temporal correlation or a ZIP GAM with temporal correlation. The choice between the Poisson and the ZIP depends on the number of zeros. You could compare the two and inspect the significance of the intercept term in the logistic link function.
5. Present the smoothers with 95% point-wise confidence bands and mention results of mixing of the chains. State the number of iterations used, thinning rate, and burn-in.

Acknowledgment

We would like to thank Chris Smeenk for providing the data from the sperm whale strandings and for his contributions to the discussion of the results and the formulation of the original hypothesis.

10 Epilogue

In this book we have taken the reader on a journey that started with a 'simple' Poisson generalised linear model (GLM) and ended with a zero inflated generalised additive model (GAM) containing a residual temporal correlation structure. The common problem throughout was zero inflation. Zero inflation is, by nature, not always straightforward, and we will attempt to address some of the questions and ambiguities that arise when dealing with this type of data.

10.1 An excessive number of zeros does not mean zero inflation

The response variable may contain a large number of zeros, but this does not necessarily mean that we have zero inflation. Figure 10.1A shows a scatterplot of simulated data. The horizontal band of points shows $Y_i = 0$ constituting 41.2% of the data. Figure 10.1B contains a frequency plot of the data and gives the same message: a lot of zeros. However, these data were simulated from a Poisson GLM:

$$Y_i \sim Poisson(\mu_i)$$
$$\log(\mu_i) = 2 + 2.5 \times X_{i1} - 10 \times X_{i2}$$

where the continuous covariate X_{i1} is taken from a uniform distribution with values ranging from 0 to 1, and X_{i2} is a categorical variable with values of 0 or 1. The horizontal band of zeros may give the impression that a zero inflated Poisson (ZIP) model should be applied, but this is not the case. This pattern of data may occur if X_{i1} is a covariate such as temperature (although the values on the horizontal axis are unrealistic for temperature) and X_{i2} is a treatment variable. The second level of the treatment variable is associated with most of the $Y_i = 0$ values.

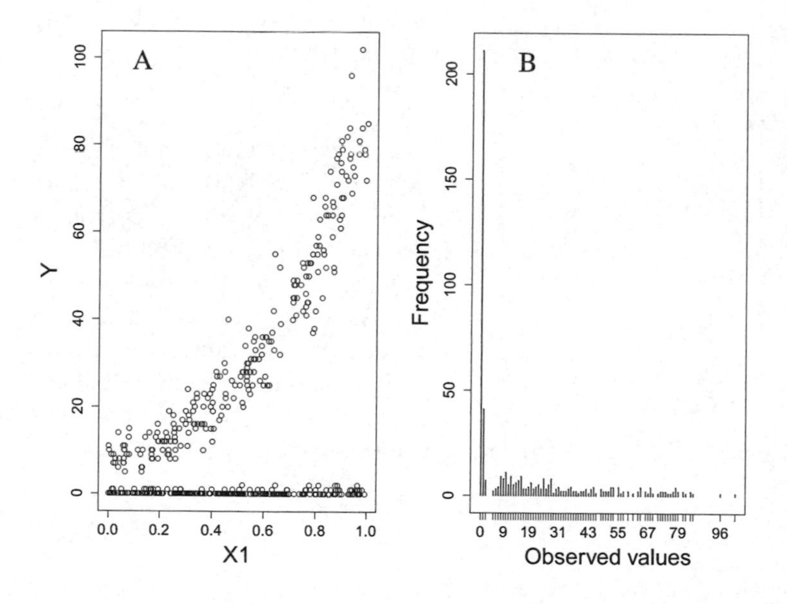

Figure 10.1. A: Scatterplot of Y versus X1. Note the horizontal band of points equal to 0. This raises the question whether we need a ZIP GLM. B: Frequency plot of the response variable.

The code below shows how the data is simulated. First we set the random seed to ensure that repeat runs of the code give consistent results. Regression parameters are chosen and N = 500 observations are drawn from a uniform distribution (`X1`) and binomial distribution (`X2`). The covariate `X2` takes the values 0 or 1. Using the log-link and Poisson distribution, `Y` values are created. A Poisson GLM is applied and the overdispersion is determined to be 1.19. This is sufficiently close to 1. Hence, the data in Figure 10.1A can be satisfactorily analysed with a Poisson GLM despite the zeros, which is not surprising, since we simulated it from a Poisson GLM.

```
> set.seed(123456)
> beta1 <- 2
> beta2 <- 2.5
> beta3 <- -5
> N     <- 500
> X1    <- sort(runif(N, min = 0, max = 1))
> X2    <- rbinom(N, size = 1, prob = 0.5)
> mu    <- exp(beta1 + beta2 * X1 + beta3 * X2)
> Y     <- rpois(N, lambda = mu)
> sum(Y==0) / N   #Gives: 0.412
> par(mfrow = c(1, 1), mar = c(5, 5, 2, 2))
> plot(X1, Y, cex.lab = 1.5)
> M1    <- glm(Y ~ X1 + X2, family = poisson)
> sum(resid(M1, type = "pearson")^2) / M1$df.res
```

```
1.19309
```

In this simulation example, excessive zeros resulted from the second level of the covariate `X2`, but we have seen that correlation may also cause excessive zeros. In Chapter 9 we analysed a time series of sperm whale strandings. If the process causing absence of strandings is auto-correlated, we are likely to measure a consecutive series of zeros. In this case the zeros are due to the inherent auto-correlation, which may be driven by unmeasured processes.

Yet another reason for an excessive number of zeros is the type of relationship (Figure 10.2A). The simulated data has 39.2% zeros, mainly shown at the lower left side of the `X1` gradient in the graph. A frequency plot is presented in Figure 10.2B.

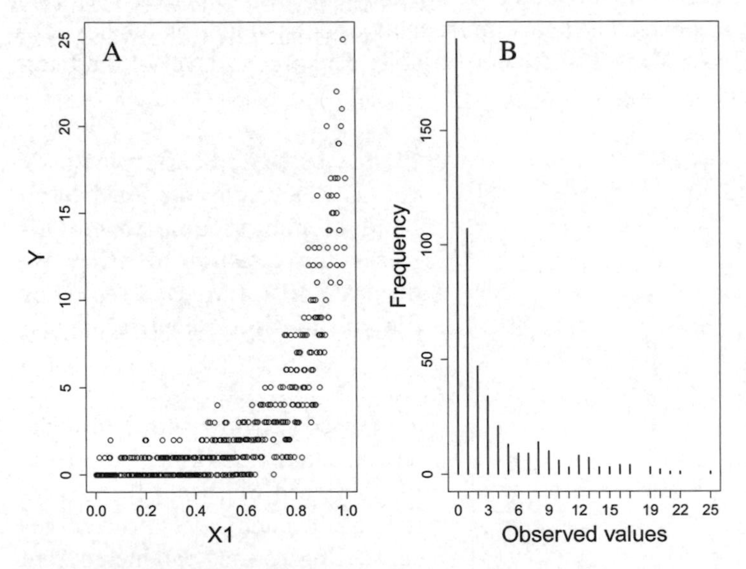

Figure 10.2. A: Scatterplot of Y versus X1. Note the large number of zeros at the left side of the X1-gradient. B: Frequency plot of the response variable.

Here we simulate data using a second order polynomial of `X1` in the predictor function. Applying a Poisson GLM gives a good fit.

The R code for this example follows. As before we set the random seed and choose values for the regression parameters and the covariate `X1`, which enters the predictor function as a quadratic term. The Poisson GLM has an overdispersion parameter of 1.07, which is sufficiently near to 1.

```
> set.seed(123456)
> beta1 <- -2
> beta2 <- 3
> beta3 <- 2
> N     <- 500
> X1    <- sort(runif(N, min = 0, max = 1))
```

```
> X2     <- X1 ^ 2
> mu     <- exp(beta1 + beta2 * X1 + beta3 * X2)
> Y      <- rpois(N, lambda = mu)
> sum(Y==0) / N   #Gives 0.392
> par(mfrow = c(1, 1), mar = c(5, 5 , 2, 2))
> plot(X1, Y, cex.lab = 1.5)
> M1 <- glm(Y ~ X1 + X2, family = poisson)
> sum(resid(M1, type = "pearson")^2) / M1$df.res

1.073593
```

Thus, under certain circumstances, the Poisson GLM is more than adequate to deal with an excessive number of zeros. It depends on what is driving the zeros, and where the zeros are located along the gradient. Thus a thorough understanding of the data and its possible implications is essential before carrying out statistical analyses.

Warton (2005) discussed the fact that many zeros does not imply zero inflation. He advocated using the negative binomial GLM when data contain many zeros, as this distribution can deal with overdispersion resulting from excessive zeros.

10.2 Do we need false zeros in order to apply mixture models?

Martin et al. (2005) concluded their paper by stating: 'Understanding how zeros arise and what type of zeros occur in ecological data are more than just semantics; failing to model zeros correctly can lead to impaired ecological understanding.' We have seen evidence of this, for example in Chapter 2 where we simulated zero inflated data and applied a Poisson GLM; estimated parameters were biased. But what if a data set has excessive numbers of zeros but the distinction between false and true zeros cannot be made?

The explanation of true and false zeros is useful from an ecological point of view, but, strictly speaking, is not necessary. ZIP and ZINB models, or any of their cousins such as the zero inflated Gamma model, can be applied regardless of the source of the excessive zeros. No papers published subsequent to the Lambert paper in 1992, which introduced zero inflated models, and before the 2005 paper from Martin et al., which introduced the concept of true and false zeros, address true and false zeros. Note that references to earlier papers using ZIP models are given in Lambert (1992), although these ZIP models were less complex.

If a distinction between false and true zeros cannot be made, we can follow Lambert (1992) and view the ZIP as a model that 'inflates the number of zeros by mixing a distribution with point mass at 0 with a Poisson distribution.' A distribution with point mass of 0 is defined as a distribution for which $P(X = 0) = 1$. The only possible value that can be observed is $X = 0$. Thus the ZIP is a combination of the 'point mass 0' and the Poisson distributions, similar to a weighted average where the weights are given by the probabilities π and $1 - \pi$. The estimated regression coefficients in the binary part then indicate which covariates drive the excess zeros (Hilbe, 2011). This interpretation implies that we can view the ZIP and ZINB as models that can deal with overdispersion due to excessive zeros without consideration of whether the zeros are true or false.

10.3 Were the false zeros in our examples really false?

In the previous section we discussed that in order to apply ZIP or ZINB models we do not, in principle, need to make a distinction between false and true zeros. In this section we briefly address each data set used in this book and discuss the strength of the case for false and true zeros.

In Chapter 2 we analysed sibling negotiation of barn owl chicks. Of the data sets used in this book, we think that these show the clearest distinction between false and true zeros. This is because sampling took place only during 15 seconds of a 15 minute period. The researchers now use technology that samples the behaviour continuously. In principle we can still apply zero inflated models.

The analysis applied on the otoliths in seal scat indicated that ordinary Poisson and negative binomial models can be applied. It is likely that there are false zeros; otoliths may have been dissolved, or can be in another (uncollected) scat of the same seal.

The offspring marmot data may contain false zeros as well, though it is harder to justify. Perhaps weather conditions made it difficult to detect the offspring.

The biologists collecting the Common Murre chapter argued that there were no false zeros. We could still have applied ZIP models using the Lambert interpretation, but because multiple smoothers were used it would have been an intensive programming job.

In Chapter 6 we analysed skate data that was measured using fishing trawls. In our opinion the zeros are a mixture of true zeros and false zeros, the latter due to fish being able to escape.

The parrotfish counts in Chapter 7 were obtained by underwater surveys. Parrotfish may swim away without being recorded when a diver approaches, constituting false zeros.

In Chapter 8 we analysed data of click beetles caught in traps. When presenting this analysis, we had reservations about defining the large number of zeros as false zeros. The sticking point may have been the definition of the experimental unit, which was the trap rather than the released beetle.

The sperm whale strandings data in Chapter 9 doubtlessly contain false zeros. Despite the huge efforts of the lead scientist (Chris Smeenk) to obtain all relevant records, it is common sense to suppose that, in some of the years with no recorded strandings, there must have been an unfortunate sperm whale stranded in some remote location along the coastline.

10.4 Does the algorithm know which zeros are false?

We will simulate data to respond to this question, one that is being asked in every course that we give on zero inflated models. Figure 10.3A shows a scatterplot of our simulated Y versus X variables. We will explain how the data were simulated; they are zero inflated. Figure 10.3B shows the fit of a Poisson GLM. The fit is poor with overdispersion of 27.35. The log likelihood of the Poisson GLM is -5098.17 using 2 degrees of freedom. We applied a ZIP model to these data:

$$
\begin{aligned}
&Y_i \sim ZIP(\mu_i, \pi_i)\\
&E(Y_i) = \mu_i \times (1-\pi_i)\\
&\log(\mu_i) = \beta_1 + \beta_2 \times X_i\\
&\text{logit}(\pi_i) = \gamma_1 + \gamma_2 \times X_i
\end{aligned}
$$

When fitting this model, the algorithm optimises the following log likelihood function:

$$
L = \sum\nolimits_i I_{Y_i=0} \times \log\left(\pi_i + (1-\pi_i) \times e^{-\mu_i}\right) + I_{Y_i>0} \times \log\left((1-\pi_i) \times \frac{e^{-\mu_i} \times \mu_i^{Y_i}}{Y_i!}\right)
$$

The indicator function I is 1 if its argument is true, and zero otherwise. L is the log likelihood of a zero inflated Poisson GLM. The individual components were given in Chapter 2. The function `zeroinfl` from the `pscl` package is used to obtain estimated regression parameters, and from these we can estimate the components π_i and μ_i. We have plotted these terms versus the covariate `X1` in Figure 10.3C and Figure 10.3D, respectively.

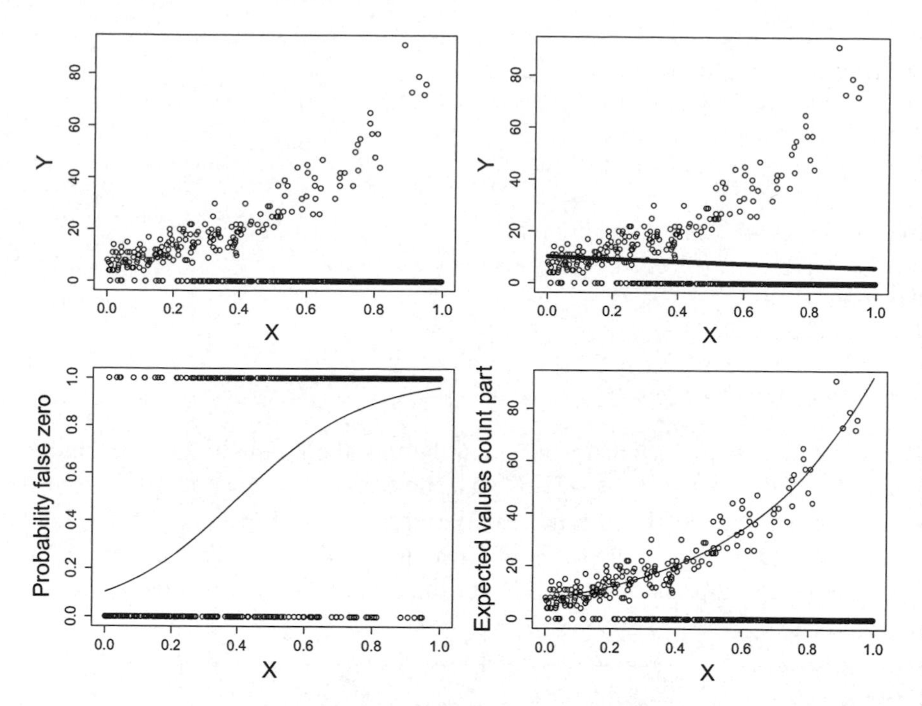

Figure 10.3. A: Scatterplot of Y versus X. B: Scatterplot of Y versus X with the fit of a Poisson GLM. C: Probability of false zeros. The higher the logistic curve, the higher the probability that a 0 is a false 0. D: Scatterplot of Y versus X with the expected values of the count data.

Figure 10.3C answers our question: We do not know which zeros are false zeros. We only know the probability of false zeros, given by the logistic curve in the figure. A high value for this curve means that the probability that the zero is a false zero is great. In this case, the zeros on the right side of the gradient have a high probability of being false zeros. These are also the zeros on the right side of the gradient in Figure 10.3D. The zeros with the higher probability of being false are on the right side of both graphs due to the manner of simulating the data.

To obtain the expected value of the ZIP we need to calculate the expected values $E(Y_i)$. These are plotted in Figure 10.4.

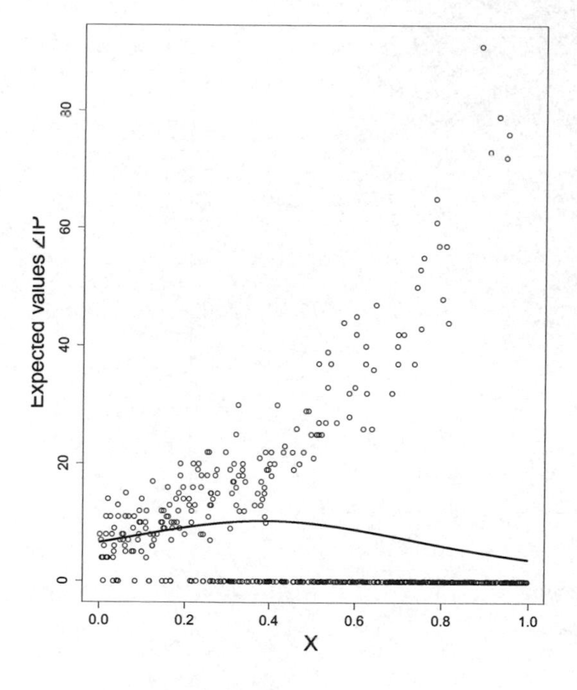

Figure 10.4. Scatterplot of Y versus X. The line represents the fitted values obtained by the ZIP. The overdispersion parameter for the ZIP was 1.03.

Essentially the fit of the Poisson GLM and ZIP GLM differ only minimally. The ZIP uses two components to fit the data, and these two components are multiplied to obtain the expected values. The end result is a model that is not overdispersed.

The data was created as follows: The random seed was set and regression parameters were selected. A covariate of 500 observations was drawn from a uniform distribution, and zero inflated data were simulated following the code presented in Chapter 2. The percent of zeros is 61%.

```
> set.seed(12345)
> beta   <- c(2, 2.5)
> gamma  <- c(-2, 5)
> N      <- 500
```

```
> X       <- runif(N, min = 0, max = 1)
> psi     <- exp(gamma[1] + gamma[2] * X) /
                (1 + exp(gamma[1] + gamma[2] * X))
> W       <- rbinom(N, size = 1, prob = 1-psi)
> mu      <- exp(beta[1] + beta[2] * X)
> mu.eff <- W * mu
> Y       <- rpois(N, lambda = mu.eff)
```

The ZIP model is fitted with:

```
> library(pscl)
> Z1 <- zeroinfl(Y~ X | X)
```

The estimated parameters are close to the original values and the log likelihood is -799.5, which is considerably better than that obtained with the Poisson GLM. To show results of the optimisation routine we create a contour plot for the log likelihood function. We vary the parameters β_2 and γ_2, and keep β_1 and γ_1 at their estimated values. The contour plot is presented in Figure 10.5. The algorithm for the ZIP attempts to find the maximum point on the surface of the log likelihood function.

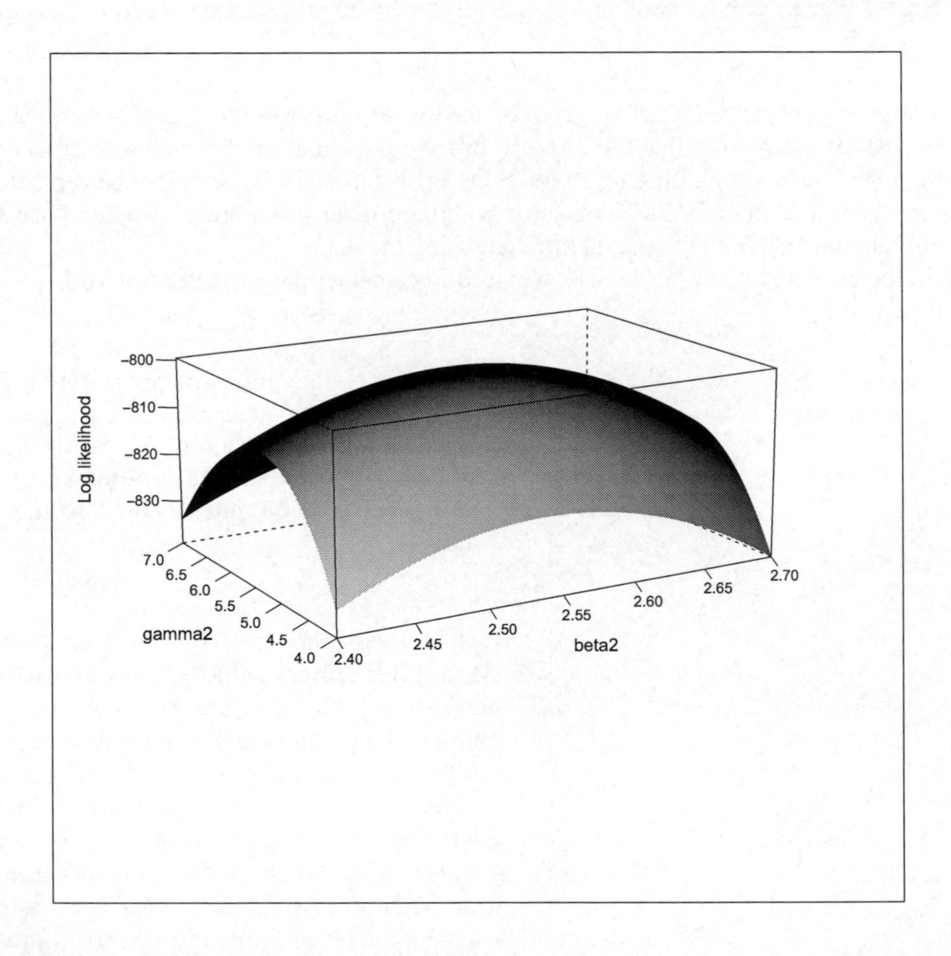

Figure 10.5. Contour plot of the log likelihood function for the ZIP as a function of β_2 and γ_2, while keeping β_1 and γ_1 fixed at their estimated values.

References

Akaike H (1973) Information theory as an extension of the maximum likelihood principle. Pages 267-281 in Petrov BN, Csaki F (editors). *Second International Symposium on Information Theory.* Akadémiai Kiadó, Budapest, Hungary.

Anderson DR, Burnham KP, Thompson WL (2000) Null hypothesis testing: Problems, prevalence, and an alternative. Journal of wildlife management 64: 912–923.

Armitage KB (1991) Social and Population Dynamics of Yellow-Bellied Marmots: Results from Long-Term Research. Annual Review of Ecology and Systematics 22: 379–407.

Armitage KB, Blumstein DT, Woods BC (2003) Energetics of hibernating yellow-bellied marmots (*Marmota flaviventris*). Comparative Biochemistry Physiology. Part A 134:101–114.

Austin MP, Meyers AJ (1996) Current approaches to modelling the environmental niche of Eucalypts: implications for management of forest biodiversity. Forest Ecology and Management 85: 95–106.

Bates D, Maechler M, Bolker B (2011) lme4: Linear mixed-effects models using S4 classes. R package version 0.999375-40. http://CRAN.R-project.org/package=lme4.

Bearzi G, Pierantonio N, Affronte M, Holcer D, Maio N, Notarbartolo di Sciara G (2011) Overview of sperm whale *Physeter macrocephalus* mortality events in the Adriatic Sea, 1555–2009. Mammal Review 41: 276–293.

Benefer C, Andrew P, Blackshaw RP, Knight M, Ellis J (2010) The spatial distribution of phytophagous insect larvae in grassland soils. Applied Soil Ecology 45: 369–274.

Besag J (1974) Spatial interaction and the statistical analysis of lattice systems (with discussions). Journal of the Royal Statistical Society, Series B 36: 192–236.

Bolker BM (2008) *Ecological Models and Data in R.* Princeton University Press.

Boucher JP, Guillen M (2009) A survey on models for panel count data with applications to insurance. *Review of the Royal Academy of Exact, Physical and Natural Sciences (RACSAM) Series A Matemáticas* 103: 277–294.

Browne WJ, Subramanian SV, Jones K (2005) Variance partitioning in multilevel logistic models that exhibit overdispersion. Journal Royal Statistical Society. Series A 168: 599–613.

Burnham KP, Anderson DR (2002) *Model selection and multimodel inference. 2nd ed.* Springer, Berlin.

Cameron AC, Trivedi PK (1998) *Regression Analysis of Count Data.* Cambridge University Press, Cambridge.

Chan KS, Ledolter J (1995) Monte-Carlo EM estimation for time series models involving counts. Journal of the American Statistical Association 90: 242–252.

Chiles JP, Delfiner P (1999) *Geostatististics: Modelling Spatial Uncertainty.* John Wiley-Interscience Publication. New York.

Clutton-Brock TH (1988) *Reproductive Success: Studies of Individual Variation in Contrasting Breeding System.* University of Chicago Press. Chicago.

Congdon P (2005) *Bayesian Models for Categorical Data.* John Wiley & Sons, Ltd. Chichester.

Crainiceanu CM, Ruppert D, Wand MP (2005) Bayesian analysis for penalized spline regression using WinBUGS. Journal of Statistical Software 14: 1–24.

Cressie N (1993) *Statistics for Spatial Data* (rev. ed.). Wiley. New York.

Cressie NAC, Kapat P (2008) Some Diagnostics for Markov Random Fields, Journal of Computational and Graphical Statistics 17: 726-749.

Cressie NAC, Wikle C (2011) *Statistics for Spatio-Temporal Data.* Wiley, Hoboken, NJ.

Czado C, Kolbe A (2004) Empirical study of intraday option price changes using extended count regression models. Discussion paper. Sonderforschungsbereich 386 der Ludwig-Maximilians-Universität München 403. http://hdl.handle.net/10419/31108.

Dalgaard P (2002) *Introductory Statistics with R*. Springer, New York.
Davison AC, Hinkley DV (1997) *Bootstrap Methods and their Applications*. Cambridge University Press. Cambridge.
Diogo H (2007) Contribution to the characterisation of recreational fishing activities on the islands of Faial and Pico, Azores. MSc thesis, Department of Oceanography and Fisheries, University of the Azores, Horta, Portugal.
Ellis JS, Blackshaw RP, Parker W, Hicks H, Knight M (2009) Genetic identification of morphologically cryptic agricultural pests. Agricultural and Forest Entomology 11: 115–121.
Elston DA, Moss R, Boulinier T, Arrowsmith C, Lambin X (2001) Analysis of aggregation, a worked example: numbers of ticks on red grouse chicks. Parasitology 122: 563–569.
Frase BA and Hoffmann RS (1980) *Marmota flaviventris*. Mammalian Species 135: 1–8.
Gelfand AE, Smith AFM (1990) Sampling-based approaches to calculating marginal densities. Journal of the American Statistical Association 85: 398–409.
Gelman A (1996) Inference and monitoring convergence, in: Wilks WR, Richardson S, Spiegelhalter DJ (eds). *Markov chain Monte Carlo in practice*. Chapman and Hall, London, pp 131–143.
Gelman A, Carlin JB, Stern HS, Rubin DB (2003) *Bayesian Data Analysis, Second Edition*. Chapman and Hall. London.
Geman S, Geman D (1984) Stochastic relaxation, Gibbs distributions and the Bayesian restoration of images. IEEE Transactions on Pattern Analysis and Machine Intelligence 6: 721–741.
Ghosh SK, Mukhopadhyay P, Lu JC (2006) Bayesian analysis of zero-inflated regression models. Journal of Statistical Planning and Inference 136: 1360 – 1375.
Gilks WR, Richardson S, Spiegelhalter DJ (1996) *Markov chain Monte Carlo in practice*. Chapman & Hall. London
Goldstein H, Browne W, Rasbash J (2002) Partitioning Variation in Multilevel Models. Understanding Statistics 1: 223–231.
Guénette S, Morato T (2001) The Azores Archipelago in 1997. In: Guénette S, Christensen V, Pauly D (Eds) Fisheries impacts on North Atlantic Ecosystems: Models and analyses (Fisheries Centre Research Report) Fisheries Centre, University of British Columbia, Vancouver, Canada, 9: 241–270.
Hardin JW, Hilbe JM (2007) *Generalized Linear Models and Extensions, Second Edition*. Stata Press, Texas.
Hastie T, Tibshirani R (1990) *Generalized additive models*. Chapman and Hall, London.
Hastings WK (1970) Monte Carlo sampling methods using Markov Chains and their Applications. Biometrika 57: 97–109.
Hicks H, Blackshaw RP (2008) Differential responses of three Agriotes click beetle species to pheromone traps. Agricultural and Forest Entomology 10: 443–448.
Hilbe JM (2007) *Negative Binomial Regression*. Cambridge University Press. Cambridge.
Hilbe JM (2011) *Negative Binomial Regression. Second Edition*. Cambridge University Press. Cambridge.
ICES (2009) Sandeel in Subarea IV excluding the Shetland area. In: ICES Advice 2009, Book 6. International Council for the Exploration of the Sea, Copenhagen, pp 15–27.
Johnson JB, Omland KS (2004) Model selection in ecology and evolution. Trends in Ecology and Evolution 19: 101–108.
Johnson NL, Kotz S, Balkrishan N (1994) *Continuous Univariate Distributions*. Wiley. New York.
Kéry M (2010) *Introduction to WinBUGS for Ecologists: Bayesian approach to regression, ANOVA, mixed models and related analyses*. Elsevier Inc.
Klinowska M (1985) Interpretation of the UK cetacean strandings records. Report of the International Whaling Commission 35: 459–467.
Kolbe A (2004) Statistical Analysis of Intraday Option Price Changes using extended Count Regression Models. Diplomarbeit thesis. Technische Universität München. Zentrum Mathematik.
Kuhnert PM, Martin TG, Mengersen K, Possingham HP (2005) Assessing the impacts of grazing levels on birds density in woodland habitats: a Bayesian approach using expert opinion. Environmetrics 16: 717–747.
Laird M, Ware JH (1982). Random-Effects Models for Longitudinal Data, Biometrics 38: 963–974.

Lunn DJ, Thomas A, Best N, Spiegelhalter D (2000) WinBUGS - a Bayesian modeling framework: concepts, structure, and extensibility. Statistics and Computing 10: 325–337.

Luterbacher J, Schmutz C, Gyalistras D, Xoplaki E, Wanner H (1999) Reconstruction of monthly NAO and EU indices back to AD 1675. Geophysical Research Letters 26: 2745–2748.

Luterbacher J, Xoplaki E, Dietrich D, Jones PD, Davies TD, Portis D, Gonzalez-Rouco JF, von Storch H, Gyalistras D, Casty C, Wanner H (2001) Extending North Atlantic Oscillation reconstructions back to 1500. Atmospheric Science Letters 2: 114–124.

Luterbacher J, Dietrich D, Xoplaki E, Grosjean M, Wanner H (2004) European seasonal and annual temperature variability, trends, and extremes since 1500. Science 303: 1499–1503.

Martin TG, Wintle BA, Rhodes JR, Kuhnert PM, Field SA, Low-Choy SJ, Tyre AJ, Possingham HP (2005) Zero tolerance ecology: improving ecological inference by modeling the source of zero observation. Ecology Letters 8: 1235–1246.

Massa AM, Lucifora LO, Hozbor MN (2004) Condrictios de la región costera bonaerense y uruguaya. In: Boschi EE (ed) El Mar Argentino y sus recursos pesqueros. Los peces marinos de interés pesquero. Caracterización biológica y evaluación del estado del estado de explotación. INIDEP Press, Argentina, pp. 85–99.

Menni RC, Stehmann MFW (2000) Distribution, environment and biology of batoid fishes off Argentina, Uruguay and Brazil. A review. Revista del Museo Argentino de Ciencias Naturales, nueva serie 2, 69–109.

Metropolis N, Rosenbluth AW, Rosenbluth MN, Teller AH, Teller E (1953) Equations of state calculations by fast computing machines. Journal of Chemical Physics 21: 1087–1091.

Mundry R, Nunn CL (2009) Stepwise model fitting and statistical inference: turning noise into signal pollution. American Naturalist 173: 119–123.

Murtaugh PA (2009) Performance of several variable-selection methods applied to real ecological data. Ecology Letters 12: 1061–1068.

Myers RA, Baum JK, Shepherd TD, Powers SP, Peterson CH (2007) Cascading effects of the loss of apex predatory sharks from a coastal ocean. Science 315: 1846–1850.

Ngo L, Wand MP (2004) Smoothing with Mixed Model Software. Journal of Statistical software 9: 1–54.

Ntzoufras I (2009) *Bayesian Modeling Using WinBUGS*. Wiley & Sons Inc. New Jersey.

Oli MK, Armitage KB (2008) Indirect fitness benefits do not compensate for the loss of direct fitness in yellow-bellied marmots. Journal of Mammalogy 89: 874–881.

Pebesma EJ (2004) Multivariable geostatistics in S: the gstat package. Computers & Geosciences 30: 683–691.

Pierce GJ, Santos MB, Smeenk C, Saveliev A, Zuur AF (2007) Historical trends in the incidence of strandings of sperm whales (*Physeter macrocephalus*) on North Sea coasts: An association with positive temperature anomalies. Fisheries Research 87: 219–228.

Pierce GJ, Valavanis VD, Guerra A, Jereb P, Orsi-Relini L, Bellido JM, Katara I, Piatkowski U, Pereira J, Balguerias E, Sobrino I, Lefkaditou E, Wang J, Santurtun M, Boyle PR, Hastie LC, MacLeod CD, Smith JM, Viana M, González AF, Zuur AF. (2008) A review of cephalopod–environment interactions in European Seas. Hydrobiologia 612: 49–70.

Pinheiro J, Bates D (2000) *Mixed effects models in S and S-Plus*. Springer-Verlag, New York.

Plummer M, Best N, Cowles K, Vines K (2006) CODA: Convergence Diagnosis and Output Analysis for MCMC. R News 6: 7–11. http://CRAN.R-project.org/doc/Rnews/.

Quinn GP, Keough MJ (2002) *Experimental design and data analysis for biologists*. Cambridge University Press. Cambridge.

R Development Core Team (2011) R: A language and environment for statistical computing. R Foundation for Statistical Computing, Vienna, Austria. ISBN 3-900051-07-0. URL http://www.R-project.org/.

Reed JM, Elphick CS, Ieno EN, Zuur AF (2011) Long-term population trends of endangered Hawaiian waterbirds. Population Ecology 53: 473–481.

Robson FD, van Bree PJH (1971) Some remarks on a mass stranding of sperm whales, Physeter macrocephalus Linnaeus, 1758, near Gisborne, New Zealand, on March 18, 1970. Z. Säugetierk. 36: 55–60.

Roulin A, Bersier LF (2007) Nestling barn owls beg more intensely in the presence of their mother than their father. Animal Behaviour 74: 1099–1106.

Ruppert D, Wand MP, Carroll RJ (2003) *Semiparametric Regression*. Cambridge University Press. Cambridge.

Ruxtona GD, Beauchamp G (2008) Time for some a priori thinking about post hoc testing. Behavioral Ecology 19: 690–693.

Sappington MJ, Longshore KM and Thomson DB (2007) Quantifiying Landscape Ruggedness for Animal Habitat Analysis: A case Study Using Bighorn Sheep in the Mojave Desert, Journal of Wildlife Management 7: 1419–1426.

Sarkar D (2008) *Lattice: Multivariate Data Visualization with R*. Springer, New York.

Simmonds MP (1997) The meaning of cetacean strandings. Bulletin de l'Institut Royal des Sciences Naturelles de Belgique Biologie 67: 29–34.

Smeenk C (1997) Strandings of sperm whales *Physeter macrocephalus* in the North Sea: history and patterns. Bull. Bulletin de l'Institut Royal des Sciences Naturelles de Belgique Biologie 67: 15–28.

Smeenk C (1999) A historical review. In: Tougaard S, Kinze CC (Eds), Proceedings from the workshop on sperm whale strandings in the North Sea: the event – the action – the aftermath. Rømø, Denmark 26–27 May 1998. Biological Papers 1, Fisheries and Maritime Museum, Esbjerg, pp. 6–9.

Sokal RR, Rohlf FJ (1995) *Biometry, 3rd edition*. Freeman, New York.

Song JJ, De Oliveira V (2012) Bayesian model selection in spatial lattice models. Statistical Methodology 9: 228–238.

Sonntag RP, Lütkebohle T (1998) Potential causes of increasing sperm whale strandings in the North Sea. Deutsche Hydrographisches Institut 8: 119–124.

Spiegelhalter DJ, Best NG, Carlin BP, Van der Linde A (2002) Bayesian measures of model complexity and fit (with discussion) Journal of the Royal Statistical Society, Series B 64: 583–639.

Spiegelhalter DJ, Thomas A, Best N, Lunn D (2003) WinBUGS User Manual, Version 1.4. MRC Biostatistics Unit, Institute of Public Health and Department of Epidemiology and Public Health, Imperial College School of Medicine, UK. http:/www.mrc-bsu.cam.ac.uk/bugs.

Sturtz S, Ligges U, Gelman A (2005) R2WinBUGS: A Package for Running WinBUGS from R. Journal of Statistical Software 12: 1-16.

Tempera F (2008) Benthic habitats of the extended Faial island shelf and their relationship to geologic, oceanographic and infralittoral biologic features, PhD dissertation, School of Geography and Geoscience, University of St. Andrews, St. Andrews, UK. 348 pages.

Tempera F, McKenzie M, Bashmachnikov I, Puotinen M, Santos RS, Bates R (2012) Predictive modeling of dominant macroalgae abundance on temperate island shelves (Azores, Northeast Atlantic). pp. 169–183. In: Harris PT, Baker EK (Eds.) *Seafloor Geomorphology as Benthic Habitat: GeoHAB Atlas of Seafloor Geomorphic Features and Benthic Habitats*. Elsevier Insights, London. DOI: 10.1016/B978-0-12-385140-6.00008-6.

Tuya F, Ortega-Borges L, Sanchez-Jerez P, Haroun RJ (2006) Effect of fishing pressure on the spatio-temporal variability of the parrotfish, *Sparisoma cretense* (Pisces: Scaridae), across the Canarian Archipelago (eastern Atlantic). Fisheries Research 77: 24–33.

Vanselow KH, Ricklefs K (2005) Are solar activity and sperm whale Physeter macrocephalus strandings around the North Sea related? Journal of Sea Research 53: 319–327.

Venables WN, Ripley BD (2002) *Modern Applied Statistics with S. Fourth Edition*. Springer, New York.

VerHoef JM, Jansen JK (2007) Space–time zero-inflated count models of Harbor seals. Environmetrics 18: 697–712.

Volker B, Su YS (2011) coefplot2: Coefficient plots. R Package version 0.1.1/r5. http:/R-Forge. R-project.org/projects/coefplot2/.

Warton DI (2005) Many zeros does not mean zero inflation: comparing the goodness-of-fit of parametric models to multivariate abundance data. Environmetrics: 16: 275–289.

White G, Ghosh SK (2009) A stochastic neighborhood conditional autoregressive model for spatial data. Computational Statistics and Data Analysis 53: 3033–3046.

Whittingham MJ, Stephens PA, Bradbury RB, Freckleton RP (2006) Why do we still use stepwise modelling in ecology and behaviour? Journal of Animal Ecology 75: 1182–1189.

Wood SN (2006) *Generalized Additive Models: An Introduction with R*. Chapman and Hall/CRC. London.

Wright AJ (2005) Lunar cycles and sperm whales (*Physeter macrocephalus*) strandings on the North Atlantic coastlines of the British Isles and Eastern Canada. Marine Mammal Science 21: 145–149.

Wright D, Lundblad JER, Larkin EM, Rinehart RW, Murphy J, Cary-Kothera L, Draganov K (2005) ArcGIS Benthic Terrain Modeler. Davey Jones Locker Seafloor Mapping/Marine GIS Laboratory and NOAA Coastal Services Center.

Xoplaki E, Luterbacher J, Paeth H, Dietrich D, Steiner N, Grosjean M, Wanner H (2005) European spring and autumn temperature variability and change of extremes over the last half millennium. Geophysical Research Letters 32: L15713.

Yee TW, Wild CJ (1996) Vector Generalized Additive Models. Journal of Royal Statistical Society, Series B 58: 481–493.

Zeger SL (1988) A regression model for time series of counts. Biometrika 75: 621–629.

Zeileis A, Kleiber C, Jackman S (2008) Regression Models for Count Data in R, Journal of Statistical Software 27: 1–25.

Zuur AF, Ieno EN, Smith GM (2007) *Analysing Ecological Data*. Springer. New York.

Zuur AF, Ieno EN, Walker N, Saveliev AA, Smith GM (2009a) *Mixed effects models and extensions in ecology with R*. Springer, New York.

Zuur AF, Ieno EN, Meesters EHWG (2009b) *A Beginner's Guide to R*. Springer, New York.

Zuur AF, Ieno EN, Elphick CS (2010) A protocol for data exploration to avoid common statistical problems. Methods Ecology and Evolution 1: 3–14.

Index

S

T

U

V

W

X

Z